Gene–Environment Interactions

Gene–Environment Interactions
Fundamentals of Ecogenetics

Edited by

LUCIO G. COSTA
DAVID L. EATON
Department of Environmental and Occupational Health Sciences
School of Public Health and Community Medicine
University of Washington
Seattle, WA

A JOHN WILEY & SONS, INC., PUBLICATION

Published by John Wiley & Sons, Inc., Hoboken, New Jersey.
Published simultaneously in Canada.

Library of Congress Cataloging-in-Publication Data is available.

ISBN-10 0-471-46781-2
ISBN-13 978-0-471-46781-6

Printed in the United States of America.

10 9 8 7 6 5 4 3 2 1

Contents

PART IV

Foreword

The interplay between heredity and environment has been of much interest in understanding the causes of human disease and disease susceptibility, but lack of methods to study the relative roles of nature and nurture made investigations difficult until relatively recent times. However, in the early 20th century a British physician, Sir Archibald Garrod, elucidated the role of inherited metabolic variation in some rare genetic diseases that he termed "inborn errors of metabolism." He further suggested that substances in certain foods and drugs may set off highly abnormal reactions in some people because of a specific hereditary susceptibility. In 1941, the one gene–one enzyme concept by U.S. biologists Beadle and Tatum provided the conceptual and methodologic framework necessary to study gene action. In the 1950s, rare adverse reactions to therapeutic drugs were shown to occur in individuals who had inherited an otherwise innocuous variant enzyme when exposed to the standard dose of a drug that required the enzyme for its metabolic breakdown. This insight set the stage for the emergence of pharmacogenetics as a field that studied the role of genetic variation in response to medicinal drugs.

Increasing recognition that variant genes may alter human responses to all kinds of environmental exposure led to the concept of ecogenetics—a field that studies the role of genetic variation in response to variable environments. Such environmental factors include chemical, physical, microbiological, and nutritional agents, as well as alcohol and tobacco. Pharmacogenetics deals with the organism's response to medicinal agents and therefore is part of the broader science of ecogenetics. Ecogenetics with its wide scope plays an increasingly important role in defining and elucidating the interaction of heredity and environment in predisposition and causation of various diseases. Early attention has been given to damage by foreign chemicals (xenobiotics) in some but not all exposed individuals. In this connection, the terms *toxicogenetics* and *toxicogenomics* have been used to refer to the role of genetic variation in toxicology so as to explain why certain toxicants produce biologic effects while others do not because of variable genetic makeup of the host.

Ultimate understanding of the mechanisms of ecogenetics requires knowledge about the underlying molecular and biochemical basis of gene action as well as an appreciation for genetic, epidemiologic, and statistical concepts. This book is the first attempt to cover the entire field of ecogenetics coherently in a single publication and includes approaches used in both laboratory and data analysis. Special emphasis is given to polymorphisms that may predispose to disease. Several common complex diseases where genetic and environmental factors interact are discussed separately. Genetic susceptibility in infectious disease and the role of genetics in nutrition are relatively novel areas of ecogenetics that are considered in separate chapters.

Ecogenetics raises a variety of problems that go beyond scientific methods and concepts. Ethical analysis and policy decisions affecting public health will often be required once genetic factors have been defined. Several chapters deal with this aspect of ecogenetics. The editors are to be congratulated on having enlisted a group of authors that are (or were) affiliated with the University of Washington. The book will not only be highly useful for college and graduate students as an introduction to this field, but can serve as a broad survey of an emerging discipline for molecular biologists, biochemists, environmental scientists, human and medical geneticists, public health experts and physicians.

ARNO G. MOTULSKY, M.D., D.SC.

Acknowledgments

We are grateful for the contributions of all of the authors to these chapters. We would like to especially acknowledge Drs. Arno Motulsky and Gil Omenn, and the late Sheldon D. Murphy, for their leadership and inspiration in bringing together at the University of Washington a strong interdisciplinary team of toxicologists, epidemiologists, geneticists, and other biomedical scientists in various aspects of ecogenetics research. Their foresight many years ago led to the acquisition of a novel grant from the Dana Foundation to foster the development of collaborations of University of Washington faculty in this area. We are grateful to the Dana Foundation for their courage to invest in this new area of research before it was even recognized as an important area of public health. The collaborations that developed from that initial Dana Foundation grant led to the successful development of an National Institute of Environmental Health Sciences Core Center of Excellence grant in Ecogenetics and Environmental Health (ES-07033), which is now entering its 11th year. The Center has served as the nexus for bringing together the chapter authors, all of whom are, or have been in the past, associated with the University of Washington. We thank the NIEHS for their continued support of this Center, and Dr. Ken Olden for his leadership in recognizing the importance of "gene–environment" interactions research and education in public health. This book would not have been possible without the excellent editorial assistance provided by Kris S. Freeman, who assisted in virtually all aspects of this book, including the preparation of chapter manuscripts and figures, and compilation of all contributions. We would also like to thank Azure Skye and Jennifer Gill for excellent administrative assistance throughout the project. Finally, we thank the editors at the Wiley press for their enthusiastic support and editorial assistance for this project.

Editors

L. G. C.
D. L. E.

Contributors

***Corresponding Authors**

***Jon P. Anderson,** LI-COR Inc., 4308 Progressive Avenue, PO Box 4000, Lincoln, NE 68504

***Melissa A. Austin,** Department of Epidemiology and Institute for Public Health Genetics, School of Public Health and Community Medicine, Box 357236, University of Washington, Seattle, WA 98195 (maustin@u.washington.edu)

***Theo K. Bammler,** Department of Environmental and Occupational Health Sciences, School of Public Health and Community Medicine, Box 354695, University of Washington, Seattle, WA 98195 (tbammler@u.washington.edu)

***Kate Battuello,** School of Law, Box 353020, University of Washington, Seattle, WA 98195 (kbatt@u.washington.edu)

Shirley Beresford, Department of Epidemiology, School of Public Health and Community Medicine, University of Washington, Fred Hutchinson Cancer Research Center, Seattle, WA

Richard P. Beyer, Department of Environmental and Occupational Health Sciences, School of Public Health and Community Medicine, University of Washington, Seattle, WA

Parveen Bhatti, Department of Environmental and Occupational Health Sciences, School of Public Health and Community Medicine, University of Washington, Seattle, WA

***Deborah Bowen,** Cancer Prevention Research Program, Fred Hutchinson Cancer Research Center, 1100 Fairview Ave. N, M3-B232, Seattle, WA 98109 (dbowen@fhcrc.org)

Wylie Burke, Department of Medical History and Ethics, School of Medicine, University of Washington, Seattle, WA

Christopher R. Carlsten, Occupational and Environmental Medicine Program Department of Medicine, School of Medicine, University of Washington, Seattle, WA

***Harvey Checkoway,** Department of Epidemiology, School of Public Health and Community Medicine, Box 357234, University of Washington, Seattle, WA 98195 (checko@u.washington.edu)

Toby B. Cole, Department of Environmental and Occupational Health Sciences, School of Public Health and Community Medicine, and Departments of Medicine and Genome Sciences, School of Medicine, University of Washington, Seattle, WA

*__Lucio G. Costa,__ Department of Environmental and Occupational Health Sciences, School of Public Health and Community Medicine, Box 354695, University of Washington, Seattle, WA 98195 (lgcosta@u.washington.edu)

Anneclaire De Roos, Departments of Epidemiology and Environmental and Occupational Health Sciences, School of Public Health and Community Medicine, University of Washington, and Fred Hutchinson Cancer Research Center, Seattle, WA

Brenda Diergaarde, Fred Hutchinson Cancer Research Center, Seattle, WA

*__David L. Eaton,__ Department of Environmental and Occupational Health Sciences, School of Public Health and Community Medicine, Box 354695, University of Washington, Seattle, WA 98195 (deaton@u.washington.edu)

*__Karen L. Edwards,__ Department of Epidemiology and Institute for Public Health Genetics, School of Public Health and Community Medicine, Box 357236, University of Washington, Seattle, WA 98195 (keddy@u.washington.edu)

Frederico M. Farin, Department of Environmental and Occupational Health Sciences, School of Public Health and Community Medicine, University of Washington, Seattle, WA

*__Elaine M. Faustman,__ Department of Environmental and Occupational Health Sciences, School of Public Health and Community Medicine, Box 354695, University of Washington, Seattle, WA 98195 (faustman@u.washington.edu)

*__Kelly Fryer-Edwards,__ Department of Medical History and Ethics, School of Medicine, Box 357120, University of Washington School of Medicine, Seattle, WA 98195 (edwards@u.washington.edu)

Clement E. Furlong, Departments of Medicine and Genome Sciences, School of Medicine, University of Washington, Seattle, WA

Gary K. Geiss, NanoString Technologies, Seattle, WA

Lindsay A. Hampson, National Institutes of Health, Bethesda, MD

*__Samir N. Kelada,__ Department of Environmental and Occupational Health Sciences, School of Public Health and Community Medicine, Box 354695, University of Washington, Seattle, WA 98195 (skelada@u.washington.edu)

*__Johanna W. Lampe,__ Fred Hutchinson Cancer Research Center, 1100 Fairview Ave N, M4-B402, Seattle, WA 98109 (jlampe@fhcrc.org)

Lawrence A. Loeb, Department of Pathology, School of Medicine, University of Washington, Seattle, WA

Anna Mastroianni, Institute for Public Health Genetics, School of Public Health Medicine, and School of Law, University of Washington, Seattle, WA

***Stephanie A. Monks,** Department of Statistics, 301 MSCS Building, Oklahoma State University, Stillwater, OK 74078-1056 (stephanie.monks@okstate.edu)

***Arno G. Motulsky,** Departments of Medicine and Genome Sciences, School of Medicine, Box 357730, University of Washington, Seattle, WA 98195 (agmot@u.washington.edu)

***Valle Nazar-Stewart,** Center for Research on Occupational and Environmental Toxicology, Oregon Health and Science University, 3181 SW Sam Jackson Park Road, L606, Portland, OR 97239-3098 (vns@bendbroadband.com)

***Gilbert S. Omenn,** Departments of Internal Medicine and Human Genetics, Schools of Medicine and Public Health, Box A510 MSRB I, University of Michigan, Ann Arbor, MI 48109-0656 (gomenn@umich.edu)

John D. Potter, Division of Public Health Sciences, Fred Hutchinson Cancer Research Center, Seattle, WA

Allan E. Rettie, Department of Medicinal Chemistry, School of Pharmacy, University of Washington, Seattle, WA

***Andrew J. Saxon,** Veterans Affairs Puget Sound Health Care System (S-116 ATC), 1660 S. Columbian Way, Seattle, WA 98108 (Andrew.Saxon@med.va.gov)

Steven M. Schwartz, Department of Pathology, School of Medicine, University of Washington, Seattle, WA

***David R. Sherman,** Department of Pathobiology, School of Public Health and Community Medicine, Box 357238, University of Washington, Seattle, WA 98195 (dsherman@u.washington.edu)

***Helen E. Smith,** Department of Pharmaceutics, School of Pharmacy, Box 357610, University of Washington, Seattle, WA 98195 (ehsmith@u.washington.edu)

***Kenneth E. Thummel,** Department of Pharmaceutics, School of Pharmacy, Box 357610, University of Washington, Seattle, WA (thummel@u.washington.edu)

***Thomas L. Vaughan,** Department of Epidemiology, School of Public Health and Community Medicine, Seattle, WA; Epidemiology Program, Fred Hutchinson Cancer Research Center, 1100 Fairview Ave. N, M4-B874, Seattle, WA 98109 (tvaughan@u.washington.edu)

Catherine K. Yeung, Department of Medicinal Chemistry, School of Pharmacy, University of Washington, Seattle, WA

Part I

1

Introduction

Lucio G. Costa and David L. Eaton
University of Washington, Seattle, WA

Although risks from exposure to exogenous chemicals depend on the intrinsic properties and dosage of the chemical, there is growing recognition that these risks may be significantly influenced by variations in target sites, biotransformation enzymes, and repair responses in the host. The host responses and target molecules are often specified genetically, and may have significant genetically determined variations. Significant discoveries relating genetically determined enzyme variations to adverse drug reactions were made in the 1950s, when the term "pharmacogenetics" was coined. The extrapolation that genetic variations would be expected to affect responses to any kind of environmental and xenobiotic agent, not just drugs, lead to "ecogenetics," which can be defined as the study of critical genetic determinants that dictate susceptibility to environmentally influenced adverse health effects.

With more recent advances in molecular sequencing technology, mutations [e.g., single-nucleotide polymorphisms (SNPs)] can be identified. Biochemical, clinical, and epidemiologic studies can then follow to assess whether such genetic polymorphisms have phenotypic consequences, and whether associations exist with specific disease outcomes and environmental exposures.

Gene-Environment Interactions: Fundamentals of Ecogenetics, edited by Lucio G. Costa and David L. Eaton

1.1. GENE–ENVIRONMENT INTERACTIONS: FROM PHARMACOGENETICS TO ECOGENETICS

In the 1950s, in a seminal paper, Arno G. Motulsky stated that "genetically conditioned drug reactions not only are of practical significance, but may be considered pertinent models for demonstrating the interaction of heredity and environment in the pathogenesis of disease" (Motulsky 1957). The first examples provided (e.g., heritable variations of plasma cholinesterase or primaquine-induced hemolysis, now known as *glucose-6-phosphate dehydrogenase deficiency*) paved the way for the rapid development of the field of "pharmacogenetics" (Vogel 1959; Kalow 1962; 1990), which studies individual differences in response to pharmacologic treatment as exhibited by drug toxicity or lack of therapeutic effect. [For a more recent review of the history and current state of the field of pharmacogenetics, see Meyer (2004).]

Today, the field of pharmacogenetics is transforming the way that the pharmaceutical industry assesses the efficacy and safety of potential new drugs. Alan Roses, a pharmaceutical industry leader in genetics and genomics, gained public notoriety for stating that "The vast majority of drugs—more than 90 percent—only work in 30 or 50 percent of the people" (Connor 2003). Roses (2004) notes that pharmacogenetics "is becoming the first drug-discovery pipeline technology to affect the structure and economics of the pharmaceutical industry." He adds that laboratory testing for common genetic differences in people (genotyping) is already being used to identify human disease-associated drug targets, and that "A new age for the treatment of diseases with safer and more targeted medicines is now beginning" (Roses 2004).

The concept that genetic variations would be expected to affect response to any kind of environmental and xenobiotic agent, not just drugs, was first made by Brewer (1971), who introduced the term "ecogenetics" (Motulsky 1991; Costa et al. 1993; Costa 2000; Omenn 2001). As discussed more recently (Eaton et al. 1998; Omenn 2001), there are clinical, policy, and scientific reasons for our increased attention to individual susceptibility. Susceptibility genes, unlike major genes associated with specific diseases (e.g., phenylketonuria or cystic fibrosis), are neither necessary nor sufficient to cause disease, but modify the risk when there is appropriate exposure (Eaton et al. 1998). Indeed, in most cases, the genetic difference does not result in a qualitatively different response, but rather induces a shift in the dose–response relationship (Kelada et al. 2004).

1.2. GENETIC POLYMORPHISMS

The term "genetic polymorphism" defines monogenic traits that exist in the normal population in at least two phenotypes, neither of which is rare. When the frequency of a specific genetic variation reaches 1% or more in the population, it is referred to as a *polymorphism*. A polymorphism may have no effect (i.e., be "silent") or may be considered functional if it results in altered function, stability, and/or level of expression of the resulting protein (Kelada et al. 2004). Functional polymorphisms include point mutations in the coding region of the gene, resulting in amino acid substitutions that in turn affect protein function or stability; duplicated genes, which result in higher protein levels; completely or partially deleted genes, resulting in no gene product; or splice-site variants that result in truncated or alternatively spliced protein products (Kelada et al. 2004).

Mutations in the regulatory regions of genes may affect the level of protein expression, and mutations in other noncoding regions may affect mRNA stability or splicing. Furthermore, even nonfunctional or silent mutations may be important, because they may be either in linkage disequilibrium with other functional polymorphisms or associated with still unknown functions (Kelada et al. 2004).

Most DNA polymorphisms occur as single-nucleotide polymorphisms (SNPs). As of June 2004 there were nearly 20 million SNPs identified in the NIH dbSNP database, of which about 800,000 are considered frequent (http://www.ncbi.nlm.nih.gov/SNP/snp_summary.cgi).

Determining which of these SNPs are functional is a huge challenge. Functional SNPs located in exons [complementary SNPs (cSNPs)] are the easiest to locate and study. Of the 800,000 frequently occurring SNPs, over 120,000 are cSNPs, 40% of which (about 50,000) are expected to be functional. However, the number of SNPs affecting protein function, including level of expression, is believed to be much greater and, as mentioned earlier, will include many polymorphisms outside exons (Kelada et al. 2004). Many of the functional SNPs located outside exons are logically assumed to occur at exon–intron boundaries (splice-site variants), or in regulatory protein binding sites (*cis* elements). However, more recent studies have demonstrated that SNPs located in introns that seem far removed from any functional aspect of a gene (called "deep intronic variants") may also have functional significance, perhaps through alterations in splicing, or the activation of "cryptic exons," exons that are transcribed infrequently or selectively (Pagani and Baralle 2004). Even the redundant third base in many codons may have functional significance, even if it does not result in a change in amino acid (e.g., is a "synonymous" variant), by altering mRNA stability and/or splice efficiency. Thus, a major challenge in the field of ecogenetics is the correct identification of genetic variation that is biologically meaningful.

1.3. THE ENVIRONMENTAL GENOME PROJECT

In 1997, as part of the Human Genome Project, the National Institute of Environmental Health Sciences (NIEHS) started a comprehensive effort to identify genetic polymorphisms in genes involved in environmentally induced disease, known as the "Environmental Genome Project" (EGP). Ken Olden, then director of NIEHS, liked to describe the relationship between genes and the environment this way: "Genes load the gun, the environment pulls the trigger." A loaded gun in itself causes no harm; it is only when the trigger is pulled that the potential for harm is released (Olden and Wilson 2000). Genetic susceptibility creates an analogous situation, where the loaded gun is equivalent to one or a combination of susceptibility genes, and the force pulling the trigger is an environmental exposure. The key objective of the EGP is to identify alleles that confer susceptibility to the adverse effects of environmental agents.

The EGP was proposed for development in three phases (Table 1.1). In the first phase, 554 candidate "environmental response" genes were identified, and are being resequenced to identify common genetic polymorphisms (SNPs). As of September 2004, 347 of these genes had been resequenced from 100 ethnically diverse human DNA samples, and hundreds of common SNPs had been identified. Such genes include those involved in xenobiotic

Table 1.1 Phases of the Environmental Genome Project

Phase 1	Develop a sample repository
	Resequence 200 candidate genes
	Develop new technologies for variant identification
	Develop databases of polymorphic variations
	Consider ethical, legal, and social implications
Phase 2	Functional studies of allelic variants
	Initiate population-based studies
	Refine databases
Phase 3	Implement genetic epidemiology studies
	Develop animal models
	Develop cell models
	Understand dose–response relationship
	Risk assessment

Source: Adapted from Olden and Wilson (2000).

metabolism, DNA repair, cell cycle regulation, cell death, oxidative metabolism, signal transduction, and immune and inflammatory response. (At the time of this writing, much of this resequencing is taking place at the University of Washington; see http://egp.gs.washington.edu for updated information.) In the second phase, the functional implications of polymorphisms in both the coding and the regulatory regions of these genes are being studied. A third phase involves the development of animal and cell models and the implementation of population-based genetic epidemiologic studies to investigate how a risk for a specific disease is altered by a particular genotype and environmental exposure. Considerations of the ethical, legal, and social implications (ELSI) of such research are an integral part of the EGP (see Chapters 23–26). Overall, the EGP will provide the ability to understand the combination of environmental and genetic components of important human diseases (cardiovascular disease, cancer, neurological disorders, asthma, etc.) and to identify and protect "at risk" subgroups.

1.4. A TEXTBOOK ON GENE–ENVIRONMENT INTERACTIONS

The growing realization that the vast majority of human diseases arise through a complex interplay of genetics and environment (gene–environment interactions) has advanced the field to the point that it has become a separate discipline in its own right. This volume is conceived as a textbook to be used primarily by students in the fields of toxicology, genetics, epidemiology, pharmacology, and other public health, medical, and general life science disciplines, as well as for practicing clinicians, public health officials, and regulators.

This book is divided into four sections. The first section covers fundamental aspects of ecogenetics, such as history of the discipline, a discussion of the molecular laboratory tools currently available to assess genotypes, the use of such measurements in molecular epidemiology studies, and the statistical issues involved in their analysis.

The second section focuses on a number of key genetic polymorphisms that are relevant for ecogenetics, including enzymes of phase I and phase II metabolism, enzymes involved in DNA repair, as well as receptors and ion channels. This section is by no means an exhaustive coverage of all possible polymorphisms relevant to gene–environment interactions, but is intended to highlight the characteristics of selected, widely studied genotypic–phenotypic differences, and discuss how these variations can influence responses to exogenous chemicals.

In the third section, gene–environment interactions are discussed through a disease-based approach to address the question of how genetic polymorphisms can influence susceptibility to various diseases. The chapters in that section cover important disease conditions such as various types of cancer, neurodegenerative diseases, cardiovascular disease, chronic pulmonary diseases, infectious diseases, diabetes, and obesity. Diseases were selected that pose major health issues, causing high morbidity and mortality in the population, and for which information on the role of gene–environment interactions in the etiology of the disease is continuously emerging. Two chapters in this section also address the issue of diet and nutrition and their interaction with genetic backgrounds, as well as genetic determinants of addiction to alcohol, tobacco, and drugs of abuse.

The final section of this book discusses the ethical, legal, and social issues that must be considered when investigating and evaluating genetic polymorphisms in human populations, as well as the impact of ecogenetics on risk assessment, regulatory policies, and medicine and public health.

2

Ecogenetics: Historical Perspectives

Gilbert S. Omenn
University of Michigan, Ann Arbor, MI

Arno G. Motulsky
University of Washington, Seattle, WA

2.1. INTRODUCTION

The origins of the field of ecogenetics lie in pioneering studies of the biochemical genetics of metabolic diseases beginning approximately 100 years ago. Sir Archibald Garrod described familial "inborn errors of metabolism," including alkaptonuria and albinism (Garrod 1902, 1909). His colleague Bateson, who had coined the term "genetics," noted that the familial occurrence of these metabolic disorders often reflected first-cousin marriages of unaffected parents, which suggested a recessive Mendelian pattern of inheritance. Garrod clearly articulated the concept of chemical individuality and foresaw the role of genetic factors in diseases, although he did not actually use the term "gene." He suggested that minor metabolic differences leading to variable amounts of intermediate or final products of metabolism could be responsible for more common diseases. Later, (Garrod 1931) he broadened the concept to explain adverse reactions in only some persons and incorporate environmental exposures:

> substances contained in particular foods, certain drugs, and exhalations of animals or plants produce in some people effects wholly out of proportion to any which they bring about in average individuals. Some effects vary from a

Gene-Environment Interactions: Fundamentals of Ecogenetics, edited by Lucio G. Costa and David L. Eaton

slight and temporary discomfort [to] morbid syndromes which amount to severe or fatal illness.

Despite his prominence in holding a prestigious chair at Oxford, Garrod's ideas and publications attracted little attention in medical circles. The Oxford University Press reissued the 1909 and 1931 books with extensive commentaries by the editors [Harris (1963) and Scriver and Childs (1989), respectively].

Genetically determined individuality represents a fundamental property of living organisms. An important desirable effect of genetic mutations during evolution is to provide variability of response for the members of a species or population, so that, as new environmental challenges are encountered, at least some individuals are more likely to survive. In human evolution, technological developments have exposed animals and humans to wholly new agents. Individuals' reactions depend on the ways in which these agents are handled in the body and how they act on specific tissues. Any variation in reactions to the agents will depend on the preexisting variation in the relevant enzymes or receptors. Some of the preexisting variability may prove to be highly adaptive and useful in the presence of new environmental agents, whereas other genetic variants may predispose to undue vulnerability as new agents are encountered. Detailed biochemical understanding is required to predict the effects of any particular agent. Even a brief compilation of common polymorphisms of blood types, red blood cell enzymes, blood plasma proteins, and leukocyte histocompatibility types demonstrated (Omenn and Motulsky 1972) that any two persons, except for identical twins, will have only a one-in-a-billion chance of the same profile for even such a short list of biochemical markers. Thus, the uniqueness of individuals about which poets write has a firm genetic basis!

In the 1950s the British geneticist J. B. S. Haldane published *The Biochemistry of Genetics* (Haldane 1954), covering microorganisms, plants, animals, and humans. He predicted that "the future of biochemical genetics applied to medicine is largely in the study of diatheses and idiosyncrasies, differences of innate makeup which do not necessarily lead to disease, but may do so." Dramatic examples of genetically determined variation in response to and metabolism of drugs soon appeared, signaling the emergence of the field of pharmacogenetics.

2.2. PHARMACOGENETICS

A major problem in Southeast Asia during World War II was the appearance of hemolytic anemia in some African-American soldiers after taking the antimalarial drug primaquine (Beutler 1957). This adverse response was shown to be inherited as an X-chromosome-recessive trait caused by deficiency of the enzyme glucose-6-phosphate dehydrogenase (Carson et al. 1956).

A second example was the observation that administration of the muscle relaxant succinylcholine prior to surgery occasionally led to prolonged apnea and even death, if the drug's effect on the patients went unrecognized in the operating room. The cause of the adverse reactions was found to be low activity of the circulating enzyme pseudocholinesterase, which is required to metabolize and inactivate succinylcholine (Lehmann and Ryan 1956). Kalow and Staron (1957) showed that the variant enzyme could be differentiated from the normal enzyme with certain inhibitors, permitting recognition of heterozygotes as well as homozygotes. Morrow and Motulsky (1968) developed a simple bedside test with a dye indicator for the variant enzyme.

These findings reached the general medical community through a seminal paper in JAMA outlining the confluence of genetics, biochemistry, and pharmacology (Motulsky 1957). The paper defines drugs as specific environmental agents that trigger disease (the adverse reaction to the drug) only in genetically susceptible individuals. Concurrently, an American biochemist, R. J. Williams, reinforced the Garrodian concept of biochemical individuality with notable emphasis on differences in nutrition and metabolism (Williams 1956).

The term "pharmacogenetics" was applied to this phenomenon by Vogel (1959) and expanded with many examples of human pharmacology by Kalow (1962) and mouse strain pharmacology by Meier (1963). Vesell and others showed the influence of genes through studies comparing monozygotic (identical) and dizygotic (fraternal) twin pairs; identical twins had much more similar plasma levels and rates of elimination of common drugs than did fraternal twins (Vesell and Page 1968). The heritability coefficients for elimination rate for phenylbutazone, antipyrine, bishydroxycoumarin, and ethanol were 0.99, 0.98, 0.97, and 0.99, on a scale where 1.0 indicates total genetic control of the observed variation in half-life of each drug (Vesell 1973). Those drugs had a nearly continuous distribution of variation among unrelated individuals.

In contrast, other drugs show a bimodal (or sometimes trimodal) distribution of response, side effects, elimination rates, or plasma levels, suggesting a simple Mendelian pattern (Omenn and Motulsky 1978). These drugs include antimalarials and sulfa drugs, which produce hemolytic anemia in individuals deficient in glucose-6-phosphate dehydrogenase; isoniazid, phenelzine, hydralazine, dapsone, and sulfadimidine. which produce side effects in individuals with slow acetylation (inactivation) of the drug; and dapsone and chloroquine, which cause cyanosis (purplish coloration of the skin) in people with methemoglobin reductase deficiency. These are all rather common traits. Other clinical situations reflect rare pharmacogenetic traits, such as the succinylcholine sensitivity mentioned above (1 in 2500 individuals), malignant hyperthermia from general anesthetics (1 in 20,000), and warfarin resistance, in which 25 times the conventional dose is required to "thin the blood" (very rare).

Despite compelling examples of single-gene pharmacogenetic relationships of clear clinical significance, and the certainty of interindividual variation in drug metabolism, pharmacogenetics research has had only a limited impact on drug development, clinical pharmacology, and medical practice so far. In fact, Burrill (2003, pp. 135–142) quoted George Poste on the application of pharmacogenomics to the diagnostic and therapeutic discovery and development process as being "far more complex than putting a person on the moon . . . what we are doing today is like trying to go to the moon without understanding gravitational fields and planetary orbits." Poste is credited with launching the corporate pharmacogenomics age in 1993 when he approved a $125 million collaboration between SmithKline Beecham and Human Genome Sciences for the first genomics database subscription. A decade later, Burrill concluded that pharmacogenomics remains a key pathway to "predict-and-prevent medicine," from genotypes to molecular and cellular functional assays, to pharmacokinetics, to pharmacodynamics and drug responses, to clinical outcomes. He listed 10 corporate pharmacogenomic alliances announced in 2001–2002, and 23 companies considered "pharmacogenomic suppliers." The NIH National Institute of General Medical Sciences and other institutes have established a Pharmacogenetics Research Network and database (PharmGKB) with an industry liaison group; and the Food and Drug Administration has a Pharmacogenomics Working Group in the Office of Clinical Pharmacology and Biopharmaceuticals.

Ironically, many pharmaceutical companies seem to view the potential to identify subsets of patients most likely to respond to mechanistically targeted new drugs as a "pharmacogenomic nightmare." While new drugs targeted to well-characterized subpopulations of patients should increase the therapeutic margin (ratio of dose for toxicity to dose for efficacy) for these patients and might even permit demonstration of clinical benefit in trials involving smaller numbers of patients, the "market" for the drug might be only, say, 15% of lung cancer patients, instead of all of them. We can imagine—even predict—that a constructive resolution of this situation may emerge from pharmacogenomic and proteomic analyses. If a particular receptor or set of receptors is involved in, say, a certain subset of adenocarcinomas of the lung and of other organ sites (like colon, pancreas, prostate, and estrogen–receptor–negative breast), the appropriately targeted patient markets could be the sum of those subsets, but nearly all patients, instead of just a minority, might benefit while with current approaches the others experience only toxicity (Omenn and Motulsky 2003).

Cautionary views about pharmacogenetics and pharmacogenomics have been presented also by others, including the Nuffield Council on Bioethics (2003). Optimism about development of "personalized medicine" raises concerns about burdens on clinicians' time with patients, premature application of incompletely understood findings and tests, disclosure of information affecting relatives, generalizations about racial predispositions (Omenn and Motulsky 2003), and compromise of anonymity of samples (see Section 2.7). The Council recommended that the UK Genetic Testing Network take responsibility for advising on the sale of pharmacogenetic tests directly to patients, using a case-by-case approach and opposed advertising to patients about tests not directly sold to them. The Council also opposed the use of pharmacogenetic information in setting insurance premiums.

A recent review discussed issues affecting priority and standards in pharmacogenetic research (Need, Motulsky, and Goldstein 2005). In most pharmacogenetic responses, complex genetic veriation rather than single gene control is expected. Emphasis on clinically useful diagnostic tests with appropriate definition of phenotypes is stressed. Special attention should be given to severe and fatal drug reactions. A detailed tabulation of 34 drugs that were withdrawn since 1990 provides hints towards possible genetic associations. The linkage disequilibrium approach using HapMaps to identify *common* pharmacogenetic variants is promising while alternate methods such as selective re-sequencing are required to find rare variants causing adverse drug reactions. FDA guidelines (FDA 2005) for submission of pharmacogenetic and pharmacogenomic data on new drugs are fully covered. Ultimate applications to medical practice will require involvement by academics, clinicians and industry researchers.

2.3. FROM PHARMACOGENETICS TO ECOGENETICS

In 1938 Haldane noted that "potter's bronchitis" might be caused by work-related environmental exposures in those potters with a "constitutional" predisposition to injury to the respiratory passages (Haldane 1938). Thirty years later Motulsky (1968) summarized examples of variation in susceptibility to chemicals, radiation, infections, and such environmental variables. Examples include elevated ambient temperature triggering hemolysis in deer mice who have the red blood cell trait of hereditary spherocytosis (Anderson and Motulsky 1966). In his Allan Award Lecture before the American Society of Human Genetics, Motulsky (1971) mentioned pharmacogenetics as one of the least developed areas of human genetics.

Soon after, Brewer coined the term "ecogenetics" to extend the concept of genetic variation beyond drugs to xenobiotics (chemicals) and other

Table 2.1 Classes of Environmental Agents with Known Ecogenetic Variation

Class of Agent	Nature of Effect	Ecogenetic Factors
Pharmaceutical agents	Biotransformation	Acetylation; CYP variants
	Target site susceptibility	G6PD deficiency
	Toxicity from chemotherapy	Thiopurine methyltransferase
Inhaled pollutants	Emphysema	α-1-Antitrypsin deficiency
	Lung cancer	Arylhydrocarbon hydroxylase induction and CYP variants
	Bladder cancer	Nicotine metabolism, acetylation differences
Pesticides	Neurotoxicity	Paraoxonase (PON1) variation
Foodstuffs	Lactose intolerance	Intestinal lactase turned off at weaning in most humans
	Celiac disease	Sensitivity to wheat gluten
	Atherosclerosis	Hyperlipidemias, hyperhomocysteinemia
	Hemolysis from fava bean ingestion	G6PD deficiency
	Thyroid goiter	Phenylthiocarbamide nontasters
Food supplements	Iron deposition disease	Iron absorption increases hemochromatosis or thalassemia gene
Stimulants	Wakefulness from caffeine	Uncertain
Alcohol	Flushing syndrome	Aldehyde dehydrogenase deficiency
Physical agents	Tolerance for heat, cold, humidity, motion, sunlight	Mechanisms unspecified, UV DNA damage repair
Metal poisoning	Minamata neurologic disease	Organic mercury ingestion (?)
Infectious diseases	Infections, autoimmune disorders, malaria, ankylosing spondylitis	Defects in cellular or humoral immunity, HbS, G6PD, thalassemia, HLA-B27

environmental agents and exposures (Brewer 1971). Clearly, on a population basis, ecogenetics is a major component of public health genetics (Omenn and Motulsky 1978; Calabrese 1984; Grandjean 1991; Khoury et al. 2000; Omenn 2000a, 2000b, 2002).

Examples abound of genetic variation in responses to various types of environmental agents, as outlined in Table 2.1, based on Omenn and Motulsky (1978) and Motulsky (2002).

2.4. NUTRITIONAL ECOGENETICS

One of the most frequently overlooked categories of environmental agents is food. Genetic variability of metabolic pathways affects biochemical and cellular processes involved in nutrition, thereby modifying nutritional requirements and susceptibility to diet-mediated diseases. Common examples are cholesterol and other dietary lipids in relation to coronary heart disease, and sensitivity to salt (NaCl) in kidney mechanisms that lead to high blood pressure. Reflecting Garrod's original observations of inborn errors of metabolism, proof of concept for the role of genes in nutritional disorders can be noted in phenylketonuria, ornithine transcarbamylase deficiency, hypophosphatemic rickets, and other striking, but rare, single-gene disorders. It is possible that individuals who are heterozygotes for these recessive disorders, as well as other individuals with other intermediate variants in the same genes, may be at increased risk to develop metabolic problems under conditions of stress, infection, or malnutrition.

Some quite common disorders have been studied extensively, including lactose intolerance, favism, hyperhomocyst(e)inemia, dyslipidemias, and alcohol sensitivity. The first four of these topics are discussed in more detail in Chapter 19, on genetic variation, diet, and disease susceptibility; while the latter is covered in Chapter 20, on genetic determinants of addiction to alcohol, tobacco, and drugs of abuse.

2.5. FROM ECOGENETICS TO RISK ASSESSMENT

Genetic and other host variation in susceptibility is an important component of the characterization of risk from exposures to environmental agents, together with the dose–response potency of the potentially hazardous agent and detailed evaluation of exposure pathways and levels. The roles of these components in a comprehensive framework for risk assessment and risk management are discussed in Chapter 25.

Workers coming to occupational medicine clinics are prone to ask, "Why me, doc?" when told that their medical problem is, or may be, related to exposures on the job. Furthermore, they sometimes add, "I'm no less careful than the next person." It is incumbent on the caring physician to investigate whether this patient has some detectable predisposition that can be managed by work practices, including protective devices, or medications. Regulators have a mandate to address differences in susceptibility, as in the charge from the Occupational Safety and Health Act of 1970 that health standards must be set so that "no employee will suffer material impairment of health or functional capacity even if such employee has regular exposure to the hazard . . . for the period of his working life." [OSHA 1970, Section

6(b)(5)]. The Clean Air Act of 1970 and its Senate report language require the Environmental Protection Agency to set standards for criteria air pollutants (such as ozone and lead) so that there will be an adequate margin of safety even for the most susceptible subgroup in the population (Lave and Omenn 1981, pp. 14–15).

2.6. THE MODERN MOLECULAR DEVELOPMENTS IN GENETICS

Whole genome studies in humans have expanded the concepts of pharmacogenetics and ecogenetics to pharmacogenomics and toxicogenomics. The Human Genome Project, initiated in the late 1980s, progressed with remarkable speed to the essentially complete sequencing of the 3 billion basepairs of the human DNA by 2003. Fittingly, the project was completed in time for the celebration of the 50th anniversary of the Watson and Crick paper in *Nature* proposing the double-helix structure for DNA.

The capability of technologies, such as microarrays, to analyze very large numbers of genes simultaneously makes feasible the search for multiple genes that influence a particular phenotype or trait, including responses to drugs and other environmental agents. Conversely, it is very useful to apply global gene expression and protein expression methods to analyze what effects chemicals have on the host cells and organs. These methods generate "biomarkers" of susceptibility, exposure, and early effects [National Research Council (NRC) 1989].

The recognition that genetic variations influence responses to potentially harmful environmental exposures led to the establishment of the Environmental Genome Project by the National Institute for Environmental Health Sciences of the NIH (www.niehs.nih.gov/envgenom/). One strategy is the selection of candidate genes likely to play major roles in the biologic action of specific environmental agents, hence "environmental response genes" or "molecular fingerprints." Polymorphisms with functional consequences might underlie and predict variation in such responses. Another strategy is the development of transgenic and knockout mice engineered to carry polymorphisms or deletions in specific environmental response genes. This work is highly complementary to the Human Genome Project's original aim to sequence a standard or composite human; here the emphasis is on interindividual variation, with the aim of protecting individuals at higher-than-average risk.

The National Center for Toxicogenomics at NIEHS, for example, created a ToxChip with 2090 unique human genes from 65,000 nonredundant clusters in UniGene, in turn selected from 750,000 human sequences in

GenBank. These genes included 72 related to apoptosis, 90 oxidative stress, 22 peroxisome proliferators—responsive, 12 Ah receptor battery, 63 estrogen—responsive, 84 housekeeping, 76 oncogenes and tumor suppressor genes, 51 cell cycle control, 131 transcription factors, 276 kinases, 88 phosphatases, 23 heat-shock proteins, 30 cytochrome P450s, and 349 receptors. During 1999–2002, in work conducted at the University of Washington and University of Utah Genome Centers, the Project resequenced 123 of its list of 554 environmentally responsive genes and identified more than 1700 SNPs in these genes (www.niehs.nih.gov/envgenom/). Some of the methods used in this project may become common in the near future in reference laboratories for preclinical and clinical pharmacology and toxicology.

Of course, genes code for proteins, so the natural follow-through of the Human Genome Project is the elucidation of variation in structure and expression levels of the proteins that carry out the myriad functions of the cells and organs of the body. Large-scale analyses of proteins are called "proteomics," a rapidly emerging field with substantial potential to discover molecular biomarker patterns for predicting and monitoring responses to environmental agents (Omenn and Motulsky 2003).

2.7. PUBLIC HEALTH AND CLINICAL IMPLICATIONS OF ECOGENETICS

There are already many known specific inherited traits and disorders that make affected individuals particularly susceptible, or relatively resistant, to certain environmental agents. Even so, the extent of risk in various circumstances and the numbers of people at elevated risk are not well established at present, leading to a major research agenda for the future. It can be assumed that only the tip of an iceberg has been revealed, representing a relatively few of the very many traits providing differential susceptibility or resistance to specific agents. Thus, policies that discourage or bar individuals from employment or recreation because of one trait, while others with undescribed susceptibility are allowed those personal opportunities, raise serious issues of justice and ethics (see Chapter 22, on ethical issues in ecogenetics; Chapter 23, on social and psychological aspects of ecogenetics; and Chapter 24, on legal issues).

No longer should we debate "nature versus nurture" or "genetics versus environment," nor should we presume, either actively or by omission, that genes by themselves "determine" common clinical disorders and outcomes of care. Understanding disease risks, predicting therapeutic and adverse effects of treatments, and creating comprehensive clinical and public health interventions require attention to gene–environment interactions. Nongenetic variables, as noted above, include diet, metabolism, behaviors, medical diagnoses, and various chemical, infectious, and physical agent exposures.

Some of these variables require data from individuals that may be difficult to access due to concerns about patient privacy and compliance with the Health Insurance Portability and Accountability Act of 1996 (HIPAA 1996). If privacy concerns can be addressed, and if the data has been systematically recorded, it may be available in patient charts or administrative datasets related mostly to billing for clinical care. Even so, the available information may not be linked to the results from genetic tests. Other variables may have to be estimated from population samples and environmental monitoring and modeling. Computer links to health department epidemiologic surveillance (possibly using the current EPI-INFO program from the CDC (CDC 2005)), EPA pollution monitoring, census tract marketing, healthcare and pharmacy utilization, and lifestyle databases will have to be devised, all with great care to protect confidentiality and privacy and to respect cultural sensitivities.

There are many scientific and technical hurdles to establishing these linkages, conducting appropriate studies, and applying the knowledge gained in educating healthcare professionals, patients, and communities. However, the biggest hurdle is not scientific; it is the policy position that genetic information is uniquely different from other kinds of personal medical information, mostly because of well-justified fears about discrimination in employment and insurance and implications for family members. Murray (1997) has called this position "genetic exceptionalism" (see Chapters 21–24).

Obstacles to research using existing data include special requirements by institutional review boards that require repeated consent for ancillary analyses of genetic markers in blood specimens collected in large clinical prevention trials (Thornquist et al. 2002). In addition, legislative proposals to regulate or prohibit access to personal genetic information (Annas et al. 1995), let alone links among essential datasets with personal identifiers, would leave many important public health and clinical questions unanswerable. After all, despite the special features of family history, other kinds of medical information are highly sensitive and often involve family members—including information about sexually transmitted infections, tuberculosis, exposures to environmental chemicals, interpersonal violence/abuse, smoking, and mental illnesses. The Social Issues Committee of the American Society of Human Genetics (SISASHG 1998) has provided good guidance on handling genetic and other sensitive medical information. The framework now exists for building linkages to other datasets that could energize and inform both educators and practitioners about the interactions of genes, environment, and behaviors.

3

Tools of Ecogenetics

Theo K. Bammler, Federico M. Farin, and Richard P. Beyer
University of Washington, Seattle, WA

3.1. INTRODUCTION

Human populations exhibit a considerable degree of genetic heterogeneity. The relatively recent unraveling of the human genome (Lander et al. 2001; Venter et al. 2001) has been a catalyst for researching this variability in the context of human disease. It has also generated a demand for the development of novel methods that allow both qualitative and quantitative measurement of genetic variation in a high-throughput fashion. In the following we provide an overview of various technologies designed to measure variations of nucleic acids.

3.2. DNA VARIANCE AND METHODS OF DETECTION

3.2.1. DNA Sequencing

The human genome contains different types of DNA variants including single-nucleotide polymorphisms (SNPs), tandemly repeated sequences, and insertion/deletion polymorphisms. SNPs are single-base variations located at specific locations in a DNA sequence that occur at a frequency of >1% in a particular population. Tandemly repeated sequences are multiple copies of a particular base sequence arranged in tandem on a chromosome. These tandem sequences range in size from 2 to over 80 nucleotides (nt).

Gene-Environment Interactions: Fundamentals of Ecogenetics, edited by Lucio G. Costa and David L. Eaton

The technique that has been utilized for the discovery of genomic variants is DNA sequencing. One of the first techniques that influenced the field of genome sciences was the Sanger method (Sanger et al. 1977) of sequencing DNA. This method was described in 1977 and led to the development of an automated fluorescence-based DNA sequencer in 1986 (Smith et al. 1986). This high-throughput method for DNA sequencing led to the initiation of the Human Genome Project in 1990. This scientific effort led to the dramatic improvement of DNA sequencing-related instrumentation and software, increasing sequencing speed, read length, and base-calling precision. The most recent DNA sequencers are able to identify 1.5 million bases in 24 h.

The Sanger method (Sanger et al. 1977) of DNA sequencing is based on the interruption of enzymatically synthesized DNA using a dideoxynucleotide (ddNTP). Dideoxynucleotides are incapable of forming the next bond in the DNA chain, thereby terminating the synthesis of that DNA chain when a ddNTP is incorporated. A brief summary of this method is as follows—a DNA polymerase copies a strand of DNA by inserting deoxynucleotides (dNTPs) until the random insertion of a terminator base, a ddNTP occurs that halts the DNA synthesis process. This results in a group of DNA fragments of different lengths, depending on the location at which the terminal base was inserted. These DNA fragments are subsequently separated by size using gel electrophoresis with the shortest fragments separated first at the bottom of the gel and the largest fragments located at the top of the gel. Different fluorescent dyes are conjugated to each ddNTP so, each DNA fragment will fluoresce a particular color depending on the particular terminating base (A, C, G, or T).

Cycle sequencing is a technique used to carry out DNA sequencing of defined DNA fragments and is a modification of the Sanger method (Wen 2001). Cycle sequencing uses a thermostable DNA polymerase that can be repeatedly heated to 95°C and still retain DNA polymerase activity. Consecutive cycles of heating and cooling the cycle sequencing reaction in a thermal cycler facilitate the annealing of the primers and the synthesis of new DNA strands. Therefore, less starting template DNA is required compared to conventional sequencing reactions.

In addition to the significant developments of molecule-based techniques that improve DNA sequencing throughput, tremendous progress has also been made in DNA instrument technology. The acrylamide slab-gel-based platform has been the dominant technology for DNA sequencing. Initial slab-gel-based platform was used in conjunction with radiolabeled DNA fragments and autoradiograms. The introduction of automated DNA sequencing was a significant technological advance. A scanning fluorescence detection system detects fragment bands as they drift past a region at the bottom of the gel. Deconvolution of these fluorescent data is performed by base-calling software.

Further improvements in efficiency, sensitivity, speed, and accuracy have been made since the mid-1990s; for example, the ABI PRISM 377 Automated DNA Sequencer [Applied Biosystems, Inc. (ABI), Foster City, CA] uses the four-color fluorescent dye chemistry together with CCD (charge-coupled device) imaging resulting in read lengths of 650–750 bases. This instrument can process 96 samples in less than 12 hours. However, this platform suffers from the common drawbacks of slab gel instrumentation; gel casting, gel loading, and lane tracking.

The need for faster sequencing runs, longer reads, and higher throughput, led to development of capillary-array-based systems. These capillary-array-based platforms deliver a single sample to each capillary, thereby eliminating the time-consuming tracking process. In addition, the high surface-to-volume ratio of the capillary readily dissipates heat and therefore permits higher operating voltages and subsequent faster runtimes.

The MegaBACE 1000 (Molecular Dynamics) is a 96-capillary array-based platform that utilizes a scanning confocal imaging system for multicolor fluorescence detection. In this instrument, a confocal optical system scans back and forth over the 96-capillary array, measuring fluorescence from each capillary by focusing a laser beam into the center of the capillary and collecting the resulting fluorescence. The fluorescence signal is color-separated through mirrors and filters, and collected with photomultipliers. These filters and mirrors can be quickly exchanged, facilitating the detection of fluorescence from various commercially available fluorescent dye conjugated primers and terminator chemistries.

The ABI 3700 is another capillary-array-based DNA sequencing system developed to analyze 96 samples per run. This platform incorporates a single, replaceable array of 104 capillaries, 8 of which are spares that can be used should some capillaries become functionally compromised during operation. In addition, the ABI 3700 is equipped with a two-channel robotic pipettor that increases the sample loading capacity, accommodating up to four 96-well or 384-well plates in an unattended 24-h period. In this instrument, a fixed laser excites the fluorescent dyes by illuminating the entire row of capillary ends, and the resultant two-dimensional fluorescence data (wavelength in one dimension and capillary number in the other) is measured by a CCD camera. The beneficial characteristics of this detection system includes very high sensitivity and the software-derived "virtual" filter wavelengths that permit the application of any dye sets without having to manually exchange elements in the instrument.

These two DNA sequencing platforms or the more recently developed smaller-scale versions, represent the majority of affordable commercially available instruments in high-throughput sequencing facilities.

3.2.2. The Polymerase Chain Reaction (PCR)

The polymerase chain reaction (PCR) is a key method in the fields of molecular biology and molecular diagnostics. The basic PCR method was invented by Kary Mullis (Mullis et al. 1986) and initially described as a technique to diagnose sickle cell anemia (Saiki et al. 1985). As an indicator of the importance of this technique to biology, Mullis received the Nobel Prize in Chemistry in 1993 for this achievement.

PCR is a procedure used for the amplification of a defined segment of a DNA strand. Beginning with only a very small amount of DNA template, a specific DNA region can be multiplied by over a millionfold in a single enzymatic reaction within a matter of hours. In view of the specificity and sensitivity of PCR, this technique has been utilized for a wide range of applications in virtually every area of biology, including ecogenetic studies.

PCR is basically DNA synthesis in a test tube. Therefore, to perform PCR amplification, all the components required for DNA replication are necessary. The reaction mixture contains double-helix DNA, two target-specific oligonucleotide primers, deoxynucleotide triphosphates (dNTPs), a DNA polymerase, and a magnesium-containing buffer. The double-stranded DNA serves as the template or target that you wish to amplify. The primers are short, single-stranded pieces of DNA (17–30 bases containing oligonucleotides) that attach (hybridize) to their complementary sequence on the template. A pair of primers will bind to either side of the target DNA segment furnishing initiation sites for DNA synthesis. The two primers used in a PCR are frequently described as the *forward* and *reverse* or *sense* and *antisense primers*, respectively. DNA polymerase is the enzyme used to produce new strands of DNA. DNA polymerase adjoins nucleotides onto the end of an annealed primer. The added bases are complementary to the DNA strand. DNA polymerase requires magnesium for activity. Therefore, magnesium is usually supplied to a PCR amplification in the form of magnesium chloride. In addition, since DNA polymerase requires a relatively stable pH in the reaction mixture for maximum activity, a buffer is incorportated that resists change in pH. dNTPs are the four nucleotides used by DNA polymerase to synthesize DNA from an annealed primer.

PCR is carried out in three distinct steps; denaturation, annealing, and extension. Each round of denaturation, annealing, and extension is known as a *cycle*. Each cycle begins with a denaturation step in which the reaction mixture is heated to 92–96°C to denature the double-stranded DNA. The denaturation step is followed by an annealing step in which the temperature is dropped (typically to between 55 and 65°C) for the primers to bind to their complementary DNA sequences on the template DNA. In the final step, the temperature is raised to ~72°C, the optimal temperature for the thermostable DNA polymerase to synthesize a complementary strand in the 5′ → 3′ direction (extension). Successive cycles are performed until an adequate quantity of target DNA is synthesized. Each PCR product made represents a template for amplification in the next PCR cycle (Figure 3.1). Therefore, there is an exponential increase in the accumulation of PCR product as a function of the cycle number.

The use of thermostable DNA polymerases in PCR has allowed automation of these PCR steps using programmable machines called *thermocyclers*. Thermostable DNA polymerases are enzymes isolated from thermophilic (heat-loving) Archaea bacteria that grow, for example, in geysers at a temperature of >110°C. When these enzymes are utilized in PCR, they are resistant to heat-related degradation during the denaturation step. Therefore, there is no need to add new DNA polymerase after each cycle. One of the first thermostable DNA polymerases was isolated from *Thermus aquaticus* and is referred to as *Taq*. Taq polymerase is commonly used for PCR. Other DNA polymerases, including *Pwo* and *Pfu*, obtained from other thermophilic Archaea bacteria, possess proofreading characteristics (mechanisms that check for errors).

The efficiency of the PCR process is dependent on the temperature parameters of the denaturation, annealing, and extension cycling steps, as well as the components of the PCR reaction. The efficiency of PCR is measured in terms of its specificity, yield, and fidelity. If a low annealing temperature is used in the PCR, the primers may hybridize to nonspecific targets as a result of looping out of primers at low temperature. The incorporation of thermostable DNA polymerases allows repeated annealing and extension at an elevated temperature, significantly decreasing primer mismatch resulting in specific amplification of the selected target DNA region. In addition, these polymerases permit amplification of much bigger DNA fragments as compared to 400 bp with Klenow's fragments.

PCR specificity is dependent primarily on the uniquely designed oligonucleotide primers of sufficient length so that their sequence is virtually distinct in the genome. The specificity of a PCR primer should be checked by comparing the primer sequence with a genomic DNA database using a service such as BLAST sequence analysis available through the National Center for Biotechnology Information (http://www.ncbi.nlm.nih.gov). In addition, various additives can be included in the PCR reaction to increase PCR specificity (Cheng et al. 1994; Vardaraj and Skinner 1994). It has been suggested that these reagents augment specificity by lowering strand separation temperature of DNA. At lower temperature, looping out of primers allows for the formation of mismatched primer template complexes and permit the synthesis of nonspecific mismatched targets. This can be avoided by "hot start" PCR (Erlich et al. 1991). Hot-start PCR is a method that

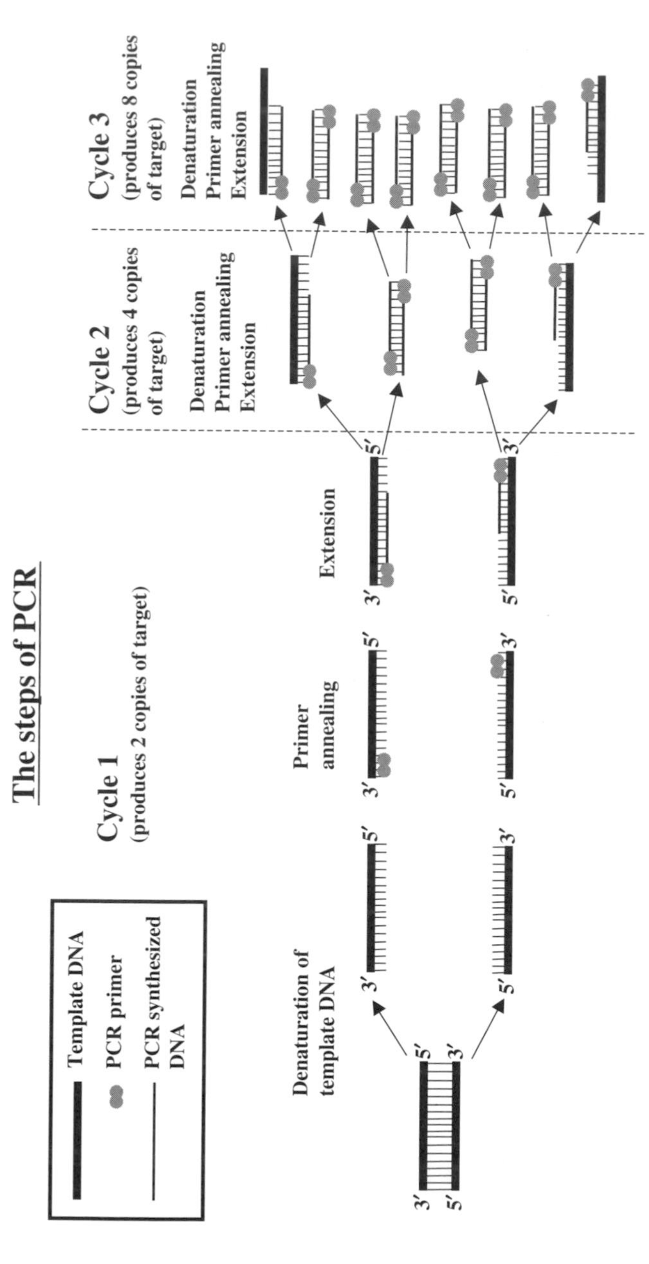

Figure 3.1. Schematic overview of the polymerase chain reaction (PCR).

generally results in more specific PCR products. Template DNA and primers are combined at a temperature above the threshold of nonspecific binding of primer to template. All the PCR reaction components are added for the extension reaction except one critical reagent (usually the thermostable polymerase). Before the PCR cycles are initiated, the absent component is added to permit the reaction to occur at higher temperature. Because of a lack of nonspecific hybridization of primers to template, the amplified DNA products tend to be more distinct since the primers don't have a chance to anneal nonspecifically.

The yield of PCR varies significantly depending on the nature of the target sequence, primer sequence, and reaction conditions. To maximize PCR yield, molar excess of primers and dNTPs should always be contained in the reaction. However, if the quantity of primers and dNTPs is excessively large, PCR is prone to low specificity and fidelity, respectively.

The rate of nucleotide misincorporation during PCR is termed *PCR infidelity*. The use of Taq polymerase introduces an infidelity factor (error rate) in PCR because it lacks 3′–5′ exonuclease proofreading activity. Errors in synthesizing DNA are attributed to this enzyme and result in mutations in the subsequent DNA sequence of the PCR products. The 3′–5′ proofreading activities of other thermostable DNA polymerases such as *Pfu* or *T4* polymerase are reported to improve PCR fidelity (Ling et al. 1991) and can significantly decrease the frequency of errors that occur in the newly synthesized DNA fragments. DNA polymerase combinations consisting of both *Taq* and *Pfu* are now commercially available and furnish high fidelity and accurate PCR-based amplification of DNA. In general, PCR infidelity can be attenuated by decreasing the annealing or extention time, minimizing dNTP and Mg ion concentration and maximizing annealing temperature. However, the PCR reaction parameters that diminish misincorporation will adversely effect PCR yield.

Analysis of a PCR product or amplicon is typically addressed using agarose gel (typically 1–1.5% w/v) electrophoresis followed by staining with ethidium bromide. The PCR product(s) will be represented as band(s) on the gel and will correspond to specific-size DNA fragments that are visible under UV transillumination. A PCR product is of a defined length according to the location of the primers on the DNA template. The size of a PCR product can be estimated by comparing its gel migration characteristics with appropriate molecular weight markers, known as a DNA "ladder." DNA ladders are mixtures of DNA fragments of known sizes that are loaded next to the PCR products in the gel and simultaneously run.

The specificity of the PCR and identification of the gel bands containing the target sequence can be checked using Southern blotting and followed by hybridization of a conjugated probe. Briefly, after electrophoresis, the

PCR amplicon is transferred (blotted) onto a nitrocellulose or nylon membrane. The blotted DNA is denatured to single strands and fixed onto the membrane by heating at 80°C or by UV crosslinking. The membrane is exposed to a solution of labeled (radioactive or fluorescent) complementary single-stranded oligonucleotide known as a *probe*. Under the appropriate hybridization conditions, annealing occurs between the labeled probe and complementary single-stranded sequence of the PCR amplicon present on the membrane. Optimal hybridization parameters will maximize specific binding of the labeled probe and minimize background caused by nonspecific binding. The position of hybridized probe can be visualized by autoradiography or chemiluminescense.

The fundamental technique of PCR has been altered in several ways that have extended its applications.

3.2.2.1. Multiplex PCR. Multiplex PCR is a technique that is designed to amplify two or more DNA templates simultaneously in the same reaction. An optimized multiplex PCR method can be more cost- and time-effective compared to carrying out individual PCRs and subsequent analyses of the PCR products. This method has been used by many investigators to identify gene deletions (Chamberlain et al. 1988) and genomic polymorphisms (Shuber et al. 1993). Organized approaches for the development of new multiplex PCR-based assays have been proposed (Henegariu et al. 1997; Markoulatos et al. 2002).

The development of a multiplex PCR can present numerous impediments, including poor sensitivity and specificity, and/or preferential amplification of certain specific targets over others (Mutter et al. 1995). All PCR components require optimization for a successful multiplex PCR assay: the relative amounts of the primers; the balanced concentrations of the PCR buffer, magnesium chloride, dNTPs, template DNA, Taq DNA polymerase; and the cycling temperatures. A multiplex PCR assay should provide specific and distinct PCR products or amplicons.

A specific example for a multiplex PCR assay in the field of ecogenetics is the procedure for simultaneously characterizing the homozygous glutathione *S*-transferase M1 null (GSTM1*0/GSTM1*0) and the glutathione *S*-transferase T1 null (GSTT1*0/GSTT1*0) genotypes (Arand et al. 1996; Abdel-Rahman et al. 1996). Glutathione *S*-transferases (GSTs) belong to a superfamily of detoxication enzymes that contribute substantial protection against some environmental toxicants. Individuals with the homozygous deletion polymorphism for the GST genes lack the respective enzyme function, may be at increased risk for many types of cancer, and may show an increased susceptibility to adverse health effects from exposure to some environmental pollutants. The multiplex PCR assay utilized in these studies contains specific PCR primers for the GSTM1 and GSTT1 genes. In addition, these assays contain an internal positive control consisting of PCR primers for the albumin or CYP1A1 gene. In this particular multiplex PCR assay, the absence of a specific PCR product indicates the presence of the null genotype. Therefore, an internal positive control is required for validation of this PCR.

3.2.2.2. Allele-Specific PCR. Allele-specific PCR (AS-PCR) is an assay used to detect SNPs in a genomic DNA sample. AS-PCR is a multiplex PCR method and therefore incorporates 3 (or 4) primers: one allele-specific primer for each allele that is complementary to the DNA template with the variant nucleotide at the 3′ end, and a common primer. The samples are amplified twice, once using each individual allele-specific primer (and the common primer). The results from each allele-specific amplification are used to establish a genotype for each sample. The specificity of AS-PCR is based mainly on the differential efficiency of the DNA polymerase in extending 3′ matched and 3′ mismatched primers but also depends on the amplification kinetics of each assay. This method of genotyping requires substantial optimization but uses inexpensive unmodified oligonucleotide primers and instrumentation.

An example of an AS-PCR assay in an ecogenetic-based study is the method used to genotype a 5′ flanking polymorphism in the cytochrome P450 2E1 (CYP2E1) gene (Sohda 1999). CYP2E1 has been established as an ethanol-metabolizing enzyme but is also significant in the metabolic activation of numerous low-molecular-weight carcinogens, including *N*-nitrosamines. Many studies have investigated the association between CYP2E1 genotypes and the development of cancer in various organs. In this method, the C1-R primer recognizes "allele C1" (wildtype allele), while the other primer, C2-F, is specific for "allele C2" (mutant allele). A "common primer" is not used in this assay, but both non-allele-specific primers are included in the amplification resulting in another PCR product that serves as a positive control.

3.2.2.3. PCR-Based Analyses of Tandemly Repeated Sequences. Tandemly repeated sequences are unique regions in the human genome that serve as effective markers for genotyping and linkage analysis. The DNA sequences of the repeated units represent a relatively wide diversity of size (2–80 bp), sequence, and genome distribution. In view of their highly polymorphic nature and abundance in the genomes of higher organisms, tandemly repeated sequences are extremely valuable as genetic markers. Based primarily on the size range of the repeated core sequence, tandemly repeated sequences have been classified into two groups—variable-number tandem repeats (VNTRs) and short tandem repeats (STRs).

VNTRs are tandemly arranged repeated sequences with lengths that typically range from 10 to 80 bp. VNTRs are highly polymorphic as to the number of repeats at a given locus and are present fairly frequently in the genome. Longer VNTRs are characterized by greater variability and, therefore, are more useful for genotyping. VNTRs occur every few kilobases on average, but they are not evenly distributed throughout the genome. VNTRs are clustered together more toward the telomeric ends of chromosomes compared to other regions in the genome.

Short tandem repeat (STR) sequences, or microsatellites, involve shorter tandemly repeated core sequences (2–10 bp). These allelic variants are repeated up to a usual maximum of 60 times but can be repeated as many as hundreds of times at different genetic loci. It is common to detect loci with more than 10 alleles and heterozygote frequencies above 0.60 when examining a relatively small number of samples. In view of the high variability of STRs and the relative ease of scoring microsatellites, these polymorphisms are considered to be important genetic markers. In addition to being highly variable, microsatellites are also densely distributed throughout eukaryotic genomes, constituting a good marker for very-high-resolution genetic mapping. Common repeats used for typing and linkage analysis are "CA" or "ACTT" sequences.

One of the most widely used methods for evaluating tandemly repeated polymorphisms consists of the PCR amplification of the tandemly repeated genomic region using a single fluorescently labeled primer followed by high-resolution electrophoresis-based analyses of the resultant PCR fragments. The most widely used instrument platforms for this type of genetic analysis are the automated DNA sequencers manufactured by companies such as Applied Biosystems, Inc. The Applied Biosystems' automated DNA sequencer platform has established software and a unique four-dye labeling system that allows the use of an in-lane size standard plus the analyses of at least three tandemly repeated polymorphisms per lane (or capillary). Inclusion of an in-lane size standard permits accurate and precise size determinations for DNA fragments even if alleles differ by as little as one basepair.

3.2.2.4. PCR-RFLP. Restriction fragment length polymorphism (RFLP) is a method in which a specific region of DNA may be differentiated by analysis of patterns derived from cleavage of that segment of DNA using a particular restriction endonuclease. Restriction endonucleases are enzymes that cleave DNA only within specific nucleotide sequences depending on the particular enzyme used. These particular nucleotide

Table 3.1 Examples of Restriction Enzymes

Enzyme	Organism from Which Derived	Target Sequence (Cut at *) 5′ > 3′
Ava I	*Anabaena variabilis*	C* C/T C G A/G G
Bam HI	*Bacillus amyloliquefaciens*	G* G A T C C
Bgl II	*Bacillus globigii*	A* G A T C T
Eco RI	*Escherichia coli* RY 13	G* A A T T C
Hind III	*Haemophilus influenzae* Rd	A* A G C T T
Kpn I	*Klebsiella pneumoniae*	G G T A C * C
Pst I	*Providencia stuartii*	C T G C A *G
Sma I	*Serratia marcescens*	C C C * G G G
Taq I	*Thermophilus aquaticus*	T * C G A
Xma I	Xanthamonas malvacearum	C * C C G G G

sequences or enzyme recognition sites are usually 4–6 bp in length. Restriction enzymes are derived from a wide variety of bacterial genera and are thought to be part of the cell's defenses against invading bacterial viruses. These enzymes are designated by using the first letter of the genus, the first two letters of the species, and the order of discovery (for examples, see Table 3.1).

The PCR-RFLP technique involves PCR amplification of a specific region of DNA sequence surrounding the polymorphic site, followed by digestion of the PCR products with an appropriate restriction enzyme, and then separation and visualization of the resultant DNA fragments by gel electrophoresis. If the genetic polymorphism results in a gain or loss of the restriction site, a different restriction digestion profile can be observed. For example, investigators have determined the 5′ flanking region polymorphism in the CYP2E1 gene using the endonuclease, *Pst*I, in a PCR-RFLP assay (Salama et al. 1999). In this assay, the PCR product was 410 bp in length. If the mutant allele is present, digestion of this allele will occur with *Pst*I and will result in two fragments of 290 and 120 bp in length. The wild-type allele is characterized by the absence of a *Pst*I restriction site. This study also examined the Tyr113His polymorhism in the microsomal epoxide hydrolase gene. For this assay, the specific PCR product was 231 bp. The Try113 allele remains undigested when exposed to the restriction enzyme *Tth111*I while the His113 allele is cleaved by *Tth111*I and results in two fragments of 209 and 22 bp.

Genetic variants were traditionally described by the specific restriction enzyme used to detect the polymorphism. For example, the CYP1A1*2A allele was originally described as the CYP1A1 *Msp*I polymorphism because these investigators utilized the *Msp*I restriction enzyme to characterize this genetic variant (Spurr et al. 1987). The particular restriction enzyme used for a specific PCR-RFLP genotyping assay depends on the sequence surrounding or adjacent to the DNA variant site. Table 3.1 shows examples of restriction enzymes.

Therefore, a constraint of the PCR-RFLP genotyping method is the requisite that the polymorphism alters a restriction enzyme cutting site. Many point mutations do not create or abolish a restriction site and, therefore, cannot be detected by this straightforward approach. However, using a modfied PCR primer can introduce a restriction digestion site, thereby expanding the use of this technique (Haliassos et al. 1989).

Gel-based genotyping assays are relatively simple and are helpful when analyzing a small number of samples. But these procedures are labor-intensive, are time-consuming, and require skilled laboratory staff for the development and execution of these assays as well as for the final manual genotype identification. Furthermore, enzymatic digestion in the process of RFLP often gives rise to ambiguous results, possibly due to partial digestion or the presence of heteroduplexes within the PCR products that are refractory to digestion. Many laboratories still use this type of gel-based genotyping assay. However, these are difficult to apply in a high-throughput format for large-scale molecular epidemiologic studies. For example, to characterize the CYP1A1*2A allele using a PCR-RFLP approach would most likely be a 2-day process. During the first day, PCR amplification of the genomic region surrounding the variant site would be addressed followed by an overnight digestion of the PCR products with the restriction endonuclease, *Msp*I. The following day, the subsequent PCR fragments would be separated and visualized using gel electrophoresis and the genotypes would be manually identified. This 2-day process can be shortened to 4h using other genotyping methods such as TaqMan allelic discrimination assays.

3.2.2.5. TaqMan-Based Genotyping. The TaqMan allelic discrimination assay uses the 5′ nuclease activity of Taq polymerase to detect a fluorescent reporter signal generated after PCR amplification (Livak et al. 1995). For single-nucleotide polymorphism (SNP) genotyping, two TaqMan probes and one pair of PCR primers (for specific PCR amplification of the region surrounding the polymorphic site) are used in the PCR reaction. The two TaqMan probes have a different nucleotide at the polymorphic site, with one probe complementary to the wildtype allele and the other to the variant allele. TaqMan probes are oligonucleotides that are usually 12–20 nucleotides in length. Different 5′ reporter dyes such as the fluorescent molecule 6-carboxyfluorescein (6FAM) and tetrachloro-6-carboxyfluorescein (TET) are covalently linked to the wildtype and variant allele probes, and a common quencher dye [6-carboxytetramethylrhodamine (TAMRA)] is attached to the 3′ end of both probes. When the probes are intact, fluorescence is quenched because of the physical proximity of the reporter and quencher dyes. During the PCR annealing step, the TaqMan probes hybridize to the targeted polymorphic site. During the PCR extension phase, the 5′ reporter dye is cleaved by the 5′ nuclease activity of the Taq polymerase, leading to an increase in the characteristic fluorescence of the reporter dye. Specific genotyping is determined by measuring the signal intensity of the two different reporter dyes after the PCR reaction (Figure 3.2). In addition to detecting SNPs, small gene deletions and insertions can also be identified by this method.

3.2.3. SNP Chips

The availability of the sequence of the human genome (Lander et al. 2001; Venter et al. 2001) allows for investigation of the role of genetic sequence variation as the basis of human disease. Single-nucleotide polymorphisms (SNPs) are the most common form of variation in the human genome where one of the four nucleotides is substituted for another nucleotide; SNPs have been estimated to occur approximately every 1200 bp in the human genome (Sachidanandam et al. 2001; Reich et al. 2003), and more than 2 million common SNPs have been identified by public efforts (www.ncbi.nlm.nih.gov/SNP). Because of their genetic stability, high rate of occurrence, and even distribution across the genome, SNPs are well suited for the detection of genetic variation underlying human diseases. This potential is being realized by the commercial development of high-density oligonucleotide arrays by Affymterix, designed to detect more than 10,000 common SNPs. Zhao and colleagues demonstrated that such high-density SNP arrays (~10,000 SNPs) are able to detect copy number changes including amplifications and homozygous deletions with a resolution approaching 300 kbp (Zhao et al. 2004). The next generation of Affymterix SNP arrays detect more than 100,000 common SNPs allowing for a genomewide scan of cytogenetic changes with high resolution and specificity.

3.2.4. RT-PCR Method for Measuring Gene Dosage

Variations in gene dosage due to deletions and amplifications have been associated with many diseases including cancer and developmental abnormalities (e.g., Down syndrome, Prader Willi syndrome, Angelman syndrome). Changes in gene copy number may be responsible for the under- or overexpression of genes related to a specific disease phenotype. Gene amplification significantly contributes to the pathogenesis of many solid tumors, including breast cancer, presumably due to overexpression of

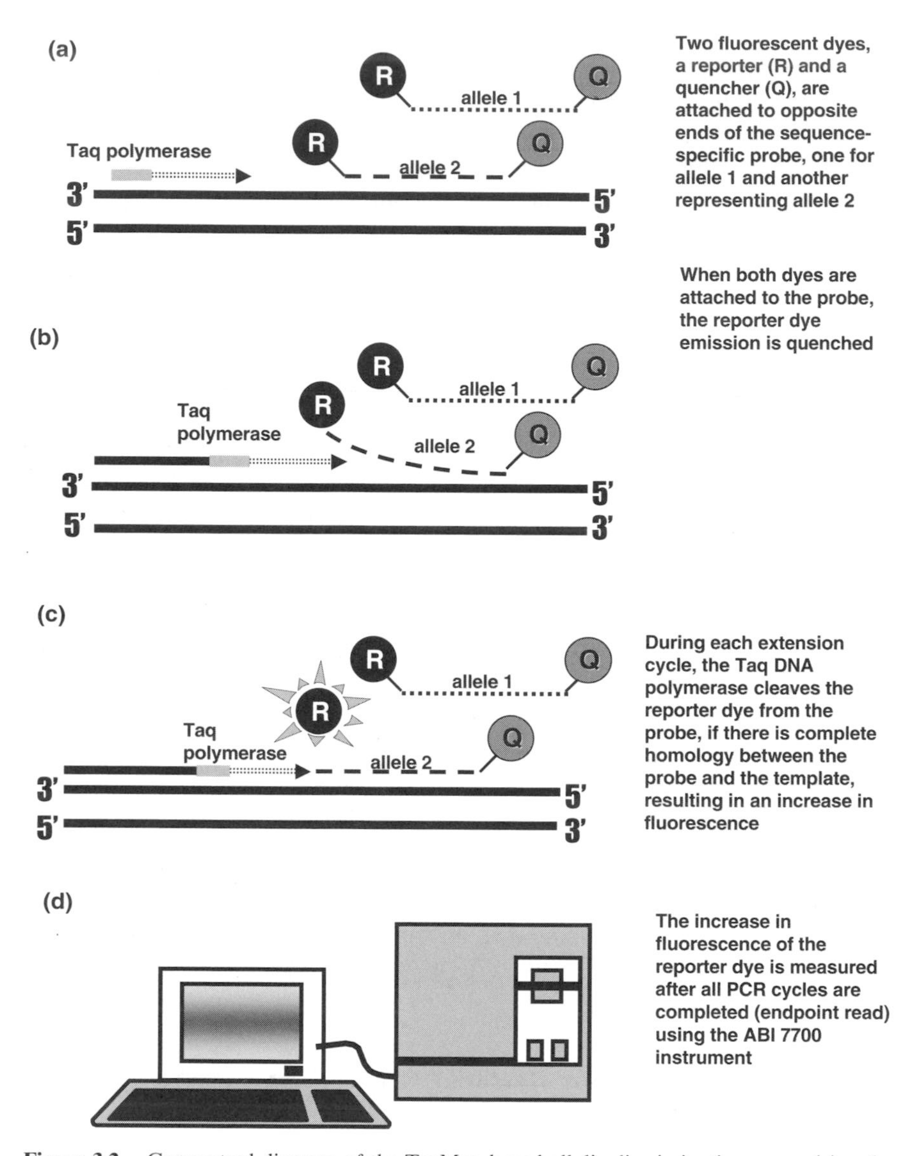

Figure 3.2. Conceptual diagram of the TaqMan-based allelic discrimination assay: (a) *polymerization*—two fluorescent dyes, a reporter (R) and a quencher (Q), are attached to opposite ends of the sequence-specific probe, one for allele 1 and another representing allele 2; (b) *strand displacement*—When both dyes are attached to the probe, the reporter dye emission is quenched; (c) *cleavage*—during each extension cycle, the Taq DNA polymerase cleaves the reporter dye from the probe, if there is complete homology between the probe and the template, resulting in an increase in fluorescence; (d) *endpoint measurement*—the increase in fluorescence of the reporter dye is measured after all PCR cycles are completed (endpoint read) using the ABI 7700 instrument.

the amplified target genes that grants a selective advantage. Gene amplification is a prevalent occurrence in the progression of human cancers, and amplified oncogenes have been demonstrated to have diagnostic, prognostic and therapeutic significance. In addition, telomeric chromosome rearrangements leading to subtle deletions and duplications have been associated with mental retardation, congenital anomalies, miscarriages, and hematologic malignancies. Therefore, gene dosage assays are important for the molecule-based evaluation of diseases characterized by either deletion or amplification of a specific DNA region containing particular genes.

Quantitative real-time PCR methods for the assessment of gene dosage have been developed (Boehm et al. 2004; Wilke et al. 2000). In these assays, a "normalizing," internal standard consisting of a disomic copy of a particular gene that is distinctly separate from the genomic region of interest is coamplified with the target DNA. This PCR-based amplification is carried out in a real-time manner, and quantitative measurements rely on the fact that the fractional cycle number (Ct), at which the amount of an amplified target (amplicon) reaches an established threshold, is directly proportional to the amount of starting target. A higher or lower initial copy number of the genomic DNA target will correspond to a consequential earlier (lower cycle number) or later (higher cycle number) increase in fluorescence, respectively. The ratio of the amount of both PCR products (derived from the internal standard or the actual sample) indicates whether there is duplication, deletion, or no change in the target DNA. Although very good results are obtained with this strategy, the selection of an appropriate internal standard gene is a significant decision.

3.2.5. Comparative Genomic Hybridization (CGH) for Measuring Gene Dosage

In 1992, Kallioniemi and colleagues reported on a methodology called *comparative genomic hybridization* (CGH) that they used for molecular cytogenetic analysis of tumors (Kallioniemi et al. 1992). CGH utilizes test (e.g., tumor derived DNA) and reference DNA (e.g., DNA derived from healthy matching control tissue) *targets*, which are fluorescently labeled and simultaneously hybridized to normal metaphase chromosomes on a glass slide that serve as *probes*. Since the test and reference DNA samples are labeled with two different fluorescent dyes, the ratio of the fluorescent intensities of the dyes measured is indicative of the relative DNA copy number in test relative to the reference DNA. The main limitation of metaphase CGH is its relatively low resolution at 5–10 Mbp. The resolution and specificity of this method were improved by substituting whole metaphase chromosome targets on the glass slide with much smaller mapped genomic clones in the

100 kbp size range, and this technique was called *matrix-based CGH* or *array-based CGH* (Solinas-Toldo et al. 1997; Pinkel et al. 1998). Pollack and coworkers reported the first genomewide array-based CGH of DNA deletions and amplifications in breast cancer samples using immobilized cDNA clones as hybridization probes (Pollack et al. 1999). The resolution and specificity of this technology were further improved by using oligonuclotide probes (Carvalho et al. 2004).

Array-based CGH is a powerful tool to detect and map genetic aberrations including deletions and amplifications at a genomewide scale. These aberrations are among the major pathogenetic mechanisms underlying genetic disorders. Array-based CGH can be used both as a research tool to identify genetic aberrations and correlate them with specific disease and a diagnostic tool for disorders with established chromosomal imbalance. Array-based CGH and expression arrays are complementary technologies. For example, Pollack and coworkers carried out array-based CGH hybridizations and gene expression profiling in parallel using the same set of cDNA probes on human breast cancer samples (Pollack et al. 2002). This study reported a correlation between gene copy number and relative expression; more than 60% of amplified genes also showed increased expression levels.

For more detailed information on CGH, the reader is referred to literature reviews by Albertson and Pinkel (2003), Mantripragada et al. (2004), and Snijders et al. (2003) covering this topic.

3.3. METHODS TO DETERMINE METHYLATION OF SPECIFIC GENOMIC REGIONS

Numerous studies have demonstrated that gene regulation also relies on epigenetic information that is not encoded in the DNA sequence of A, C, G, and T nucleotides. At least two different categories of epigenetic information can be inherited with chromosomes. One group is DNA methylation, in which a nucleic acid base is modified by an enzyme, DNA methyltransferase. In eukaryotes, DNA methylation usually transpires at the C5 position of cytosine, and these resultant methylcytosines are commonly located in cytosine–guanine (CpG) dinucleotides. About 56% of human genes and 47% of mouse genes are associated with CpG islands (Antequera and Bird 1993). The other group of epigenetic information is related to alterations in chromatin proteins, characteristically occurring around the histone tails.

Aberrant DNA methylation of the CpG site is one of the initial and most common alterations in cancer. Several investigators proposed that aberrant methylation on the CpG sites of the tumor suppressor gene is closely linked with carcinogenesis. However, recent studies have demonstrated that differential methylation of specific genes can be detected in phenotypically normal upper aerodigestive tract tissues of smokers (Zochbauer-Muller et al. 2003), suggesting that aberrant DNA methylation may also represent an important mechanism for certain xenobiotic-induced diseases. In this regard, a study has shown that hypermethylation of an important xenobiotic metabolizing enzyme, GSTP1, is present in some preneoplastic foci of prostate cancer (Nakayama et al. 2003).

Many techniques have been developed for the rapid and sensitive analysis of DNA methylation. One of these methods is methylation-specific PCR (Smith et al. 2003). A bisulfite modification step that creates sequence differences in the DNA is a universal component of the PCR-based methods to detect methylation. On exposure to bisufite, unmethylated cytosines of a CpG dinucleotide are converted to uracils, while methylated cytosines of a CpG dinucleotide are protected from this particular nucleotide conversion. Methylation-specific PCR assays utilize primers that distinguish between the methylated and unmethylated bisulfite-modified DNA and unmodified DNA, thereby identifying sequence differences by amplification.

The design of PCR primers specific for methylated DNA consists of cytosines that are conserved because of their methylations positioned in the 3′ end. For the design of PCR primers that will amplify unmethylated DNA only after bisulfite modification, thymidines, derived from transformed cytosines, are situated in the 3′ end. This approach represents a type of allele-specific PCR. A high-throughput obstacle of methylation-specific PCR is that only the few CpG sites that are situated within the template sequence to which the primers bind can be investigated for a given primer pair. Therefore, when characterizing aberrant DNA methylation of multiple CpG sites in numerous samples, this type of technique is laborious and time-consuming.

Large-scale analysis of candidate genes has so far been hampered by the lack of a high-throughput approach for analyzing methylation patterns. However, there are now commercially available (University Health Network, Princess Margaret Hospital, Ontario Cancer Institute) cDNA-based microarrays that can discern aberrant DNA methylation of specific human genomic CpG sites simultaneously in a high-throughput fashion. These CpG island microarrays contain 12192 CpG island clones from the Sanger Institute, and information on these clones can be found at www.sanger.ac.uk/HGP/cgi.shtml. In addition, investigators have described oligonucleotide-based microarray techniques that are both versatile and sensitive in revealing hypermethylation in defined regions of the genome (Gitan et al. 2002; Shi et al. 2003).

3.4. METHODS TO DETERMINE GENE EXPRESSION AT THE mRNA LEVEL

The publication of the human genome in 2001 (Lander et al. 2001; Venter et al. 2001) marked a milestone in the field of biology, and genomes of many other organisms have been completed. Never before in history has such a vast amount of genetic information been available. The genome is an organism's blueprint and forms the fundamental basis for understanding its biological processes such as growth, development, differentiation, homeostasis, aging, and disease. For information contained in a gene to impact a biological process, it needs to be transcribed into messenger RNA (mRNA), which in turn needs to be translated into the corresponding protein. Proteins are the "business end" of biological processes (Figure 3.3).

The genome itself is relatively static, and its sequence alone does not readily provide information as to where and when in an organism individual genes are expressed or what their function is. The precise temporal and

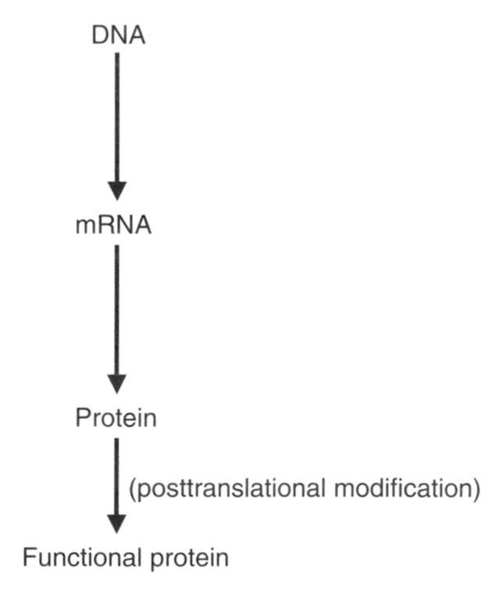

Figure 3.3. The central dogma of biology—genetic information flows from DNA via mRNA to protein.

spatial control of gene expression governs every biological process. For example, an organism reacts to toxic chemical exposure by switching on genes and producing the encoded proteins that combat toxicity. Once the toxic chemical has been eliminated, expression of these genes is switched off. Several methods have been developed to determine qualitatively as well as quantitatively the primary gene product, mRNA, in order to investigate function and regulation of genes.

For example, Northern and slot blot analysis (Thomas 1980) assesses the level of a specific mRNA species. In more recent years, a reverse transcription coupled polymerase chain reaction (RT-PCR) assay has been developed that also assesses levels of specific mRNA species.

3.4.1. Northern Blot Analysis

Northern blot analysis consists of the electrophoretic separation of total RNA on an agarose gel followed by transfer of the RNA to a nitrocellulose membrane. This membrane is subsequently hybridized with a labeled probe specific for a certain gene. The amount of expression of a particular gene is quantified with radio imaging or other densitometric detection systems. Aspects about the quality of the RNA samples, mRNA size, and possible alternative splicing can also be ascertained by Northern blot analysis.

3.4.2. Slot Blot Analysis

The slot blot method is similar to the Northern-based technique, except that there is direct application of total RNA onto a nitrocellulose membrane with no electrophoretic separation of RNA. The throughput of slot blot analysis is higher than that of the Northern method. However, when examining many samples together with the expression of numerous genes, both methods are labor-intensive and time-consuming compared to more recent techniques such as real-time reverse transcriptase polymerase chain reaction (RT-PCR) assays, and DNA-based microarrays.

3.4.3. TaqMan-Based Quantitation of mRNA Levels

A fluorescence-based real-time automated sequence detection system (such as the ABI Prism 7700) allows quantitation of specific mRNA species. These instruments monitor PCR at every cycle using a closed-tube fluorogenic 5′ nuclease-based assay, and utilize a fluorogenic probe consisting of an oligonucleotide with both a reporter and a quencher dye attached. During PCR, the probe anneals to the target of interest specifically between the forward and reverse primers. If hybridization occurs, the probe is cleaved by the 5′ nuclease activity of the Taq DNA polymerase. This separates the reporter from the quencher dye, and a sequence-specific fluorescence signal is generated. The signal is proportional to the amount of PCR product formed and is measured continuously throughout the entire PCR assay (real time). This technique further increases the specificity of the PCR reaction beyond the level achieved with gene-specific PCR primers. A mismatch between probe and target substantially reduces the efficiency of hybridization and cleavage; therefore, a mismatched probe does not contribute appreciably to the measured fluorescent signal. In addition, since fluorescence signal is generated only if the target sequence for the probe is amplified during PCR, no signal is generated by nonspecific amplification. The ABI 7700 instrument has a dynamic linear range of at least five orders of magnitude, reducing the need for serial dilutions. There is no electrophoresis or post-PCR processing. Sequence-specific signal is generated and detected—in solution—during the PCR. This direct detection method, together with the 96-well microtiter plate format, enables extremely high throughput. The fluorescent reporter signal is normalized to an internal reference dye and then PCR amplification plots for five serial dilutions of an established reference sample spanning a 10,000-fold dilution range is used to create standard curves. Standard curve lines are generated from *xy* (scatter) plots of the log of the input amount (dilution value of the reference sample) as x and C_T value (the PCR cycle at which an increase in reporter fluorescence exceeds an established threshold) as y. To normalize for differences in the amount of total RNA added to each RT reaction, a normalizing control gene (GAPDH, β-actin, 18S RNA, etc.) is selected as an endogenous RNA control. Each measured relative expression level for the target gene of interest is divided by the endogenous control gene level to obtain the normalized mRNA expression level. The relative quantity of target cDNA in each sample is calculated from the experimentally determined C_T value.

3.4.4. Global Gene Expression Profiling

While Northern and slot blot analyses and TaqMan methods assess expression of one or a few genes at a time, they are not suitable for measuring mRNA levels of thousands of genes in one experiment. In the mid-1990s, Dr. Brown's research team was instrumental in developing a technology that allowed the measurement of a large number of human mRNAs simultaneously (Schena et al. 1995, 1996). Dr. Brown's team successfully printed more than 1000 human cDNAs with a high-speed robotic device on glass microscope slides and named this device a "microarray" or "DNA chip."

Since then, microarray technology has evolved and we will provide an overview in the following.

3.4.4.1. Basic Principles Underlying Gene Expression Profiling. While proteins are the cellular macromolecules that ultimately impact biological processes, it is technically still challenging to measure protein levels on a large scale (i.e., all the proteins expressed in a cell, or its proteome). However, DNA microarray technology has been developed that allows assessment of mRNA levels of thousands of genes simultaneously. This process is commonly referred to as *DNA microarray analysis* or *gene expression profiling* and attempts to quantify the transcriptome, that is, all mRNA species expressed in a cell.

Two complementary nucleic acid strands forming a helix can be separated, or denatured, by heating to break the hydrogen bonds between bases. Lowering the temperature allows repeated annealing of the two strands of nucleic acids following the basepairing rules (A is always paired with T, and C is always paired with G), and this process is referred to as *hybridization*. DNA microarray technology is based on the hybridization properties of nucleic acids to monitor mRNA levels on a global scale.

In contrast to DNA, mRNA is a single-stranded molecule composed of nucleotides with ribose as the sugar component instead of deoxyribose and uracil (U) instead of thymine as one of the four bases. Messenger RNA molecules also anneal to their complementary DNA strand following similar basepair ruling as DNA with the exception that A pairs with U.

3.4.4.2. What Is a cDNA Microarray? There are essentially two types of DNA microarrays used for gene expression profiling: cDNA and oligonucleotide-based microarrays. cDNA (complementary DNA) is synthesised from a messenger RNA template by the enzyme reverse transcriptase. A cDNA is so called because its sequence is the complement of the original mRNA sequence. However, when double-stranded cDNA is synthesized, it contains both the original sequence and its complement. A typical cDNA microarray is a glass slide with thousands of different cDNA clones attached to it. The first step in the production of an array is the selection of *probes* to be printed on the array. The probes can be full-length cDNAs or partially sequenced cDNAs, also referred to as *expressed sequence tags* (ESTs), derived from cDNA libraries of interest. An array is generated with a high-speed robotic device capable of depositing a few nanoliters of a solution of cDNA clones that have been amplified by the polymerase chain reaction. This procedure results in a glass slide with thousands of "spots" organized in an orderly pattern, each "spot" containing millions of identical copies of one specific cDNA clone. cDNA arrays allow for assessment

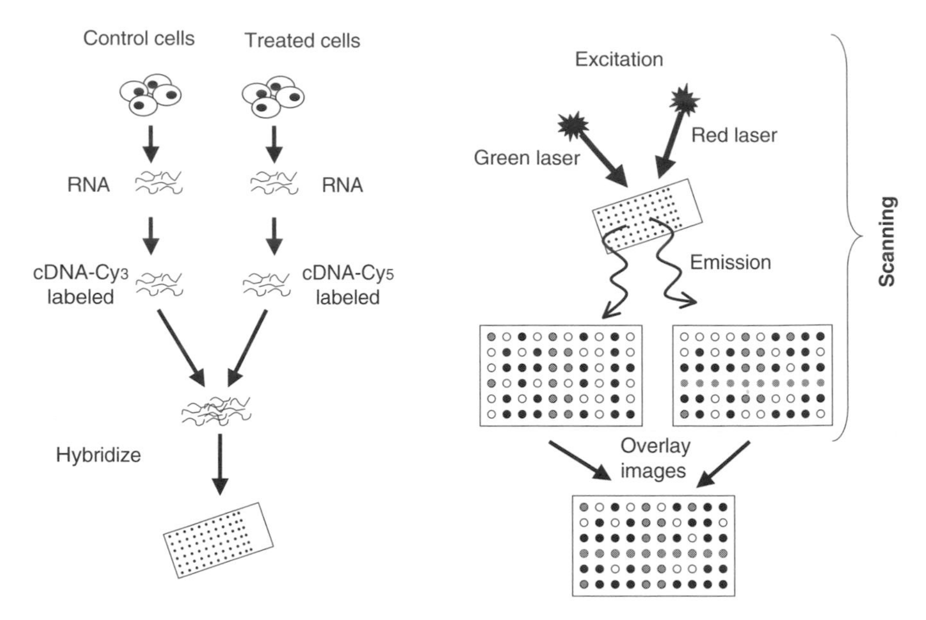

Figure 3.4. Overview of a cDNA microarray experiment.

of mRNA levels of thousands of genes simultaneously. The goal of a typical microarray experiment is to compare gene expression profiles of two or more samples (for an overview, see Figure 3.4).

3.4.4.2.1. RNA Processing and Hybridization. A simple microarray experiment could, for example, address the issue of how exposure of cultured cells to one dose of a certain chemical affects global gene expression relative to the unexcposed control cells at a given timepoint. The first step is to isolate RNA from both the exposed (E) and the control cells (C) and assess its quality. High-quality RNA samples without signs of degradation are critical for successful microarray experiments.

The second step is aimed at generating fluorescently labeled *targets* that are representations of the cellular mRNA samples. Several different protocols can be used to achieve this. One commonly used method is direct labeling of cDNAs. An oligo-dT primer is used to reverse-transcribe mRNAs to their complementary cDNAs in the presence of the four nucleotides: dATG, dCTG, dGTG, and dTTG. A portion of one of the four nucleotides is labeled with one type of fluorescent molecule, for example, the cyanine dyes Cy3 or Cy5 (e.g., Cy3-dCTP or Cy5-dCTP). This results in fluorescently labeled Cy3 or Cy5 cDNAs for the exposed and the control samples, respectively. One disadvantage of the direct labeling method is the relatively large amount of RNA required [1–2 μg of poly(A) mRNA or 30–100 μg of total RNA] to achieve an adequate fluorescent signal. In many cases, these large amounts of RNA are simply not available. Several alternative methodologies that address this issue have been developed, some of which involve linear RNA amplification (Phillips and Eberwine 1996; Beier et al. 2004). Until this point, the exposed and the control samples have been processed separately and cDNAs corresponding to one sample (e.g., exposed sample: ES) have been, for example, labeled with Cy3, whereas cDNA corresponding to the other sample (control sample: CS) have been labeled with Cy5. Equal amounts of both samples are now mixed and hybridized to a single microarray, usually overnight. Following hybridization, the microarrays are washed, dried, and scanned.

The fluorescent signals are detected by scanning the microarray separately for Cy3 and Cy5 with a confocal microscope with appropriate lasers for optimal excitation of the two fluorescent dyes. (Figure 3.4).

The more mRNA of a specific gene was present in a sample, the more fluorescently labeled cDNA (or cRNA) was generated, which then could hybridize to its corresponding cDNA probe immobilized on the microarray. This, in turn, translates into a proportionally high fluorescent signal. Therefore, the fluorescent signal measured is approximately proportional to the amount of mRNA present in the original RNA sample.

Scanning of each cDNA array that has been hybridized with two fluorescently labeled samples (e.g. the exposed sample labeled with Cy3 and the control sample labeled with Cy5) results in two fluorescent image files (usually in a TIFF file format): one derived from the Cy3 fluorescent signal and the other from the Cy5 fluorescent signal. Specialized software is required to use the TIFF image files and extract fluorescent intensities of each spot. This procedure is discussed in more detail below.

The two Cy3- and Cy5-derived fluorescent images are overlaid by specialized software, and the resulting pseudoimage is what many publications in the microarray field show. These pseudoimages show spots with different shades of essentially three colors: green, indicating genes that were expressed at higher levels in the Cy3 relative to the Cy5-labeled samples; red, indicating genes that were expressed at higher levels in the Cy5 relative to the Cy3-labeled sample; and yellow, indicating genes that were expressed at approximately equal levels in both the Cy3- and Cy5-labeled samples.

3.4.4.2.2. Technical Aspects of cDNA Arrays Affecting Measurement of mRNA Levels. Two samples labeled with two different fluorescent dyes (e.g., Cy3 and Cy5) are hybridized to a single array to minimize experimental variation inherent in the comparison of independent hybridizations. For example, it is technically difficult to reproducibly deposit uniform spots containing the same amounts of cDNAs onto a glass slide. In reality, many cDNA spots on an array are not unifom but are shaped like doughnuts, crescent moons, and other configurations, and different amounts of cDNAs are deposited. Let us consider a hypothetical case in which the chemical treatment increased mRNA levels of "gene A" tenfold relative to the control. Let us further assume that the same two fluorescently labeled samples were cohybridized to two different cDNA arrays, array 1 and array 2, and that the cDNA spot corresponding to gene A on array 1 was a perfect disk, whereas that on array 2 had the shape of a crescent moon and contained only 50% of cDNA probe relative to the spot on array 1. Let us also assume that all other factors were equal when arrays 1 and 2 were hybridized, washed, and scanned. The total fluorescently labeled cDNA that hybridized to the spot corresponding to "gene A" in both the chemically exposed sample and the control sample would have been approximately 50% measured with array 2 compared to array 1. However, the *ratio* of Cy3 to Cy5 fluorescent signals would have been approximately the same for array 1 and array 2 because both fluorescently labeled samples would have been affected proportionally. However, if only one fluorescently labeled sample would have been hybridized each to array 1 and array 2, the total fluorescent signal measured with array 2 would have had a 50% error attached to it. Because two fluorescently labeled samples are hybridized simultaneously to cDNA arrays, they are referred to as a *two-color* or *two-channel* microarray platform.

For the sake of simplicity, so far we have assumed that the two fluorescent dyes Cy3 and Cy5 themselves introduce little variation in cDNA microarray data. However, due to differences in bulkiness of the molecules, the Cy3 and Cy5 fluorescent dyes differ in their efficiency by which the reverse transcriptase incorporates them into cDNA. In addition, Cy5 is more susceptible to photobleaching than Cy3. To control for these confounding factors, a so called *dye flip experiment* is performed that requires two arrays per sample pair. For example, the exposed sample is labeled with Cy3 and the control sample is labeled with Cy5, and both samples are cohybridized to array 1. On array 2, the two samples are labeled the opposite; the exposed sample is labeled with Cy5 and the control sample is labeled with Cy3, and they are cohybridized. The results from both arrays are combined and analyzed so that the confounding dye effects cancel themselves out. Some of these technical issues do not apply to commercially produced *one-color* or *one-channel* oligonucleotide arrays described below.

3.4.4.3. What Is an Oligonucleotide Array? An oligonucleotide array is a glass slide with thousands of oligonucleotide probes, rather than cDNA, attached to its surface. The typical length of oligonucleotides used for microarrays ranges between 25 and 60 nucleotides. In contrast to cDNA arrays, these relatively short oligonucleotide probes can discriminate between genes that share a high degree of sequence homology. The two major strategies being employed in manufacturing oligonucleotide arrays are synthesizing the oligonucleotides (1) directly on the glass slide or (2) off the slide, purifying the synthesized product and then spotting it on the glass slide. For example, Affymetrix and Agilent synthesize oligonucleotides directly on the glass slide with a photolithographic or noncontact inkjet technology, respectively, whereas Amersham synthezises, purifies, and spots the oligonuclotide on the glass slide. Commercial array manufacturers have stringent quality control measures in place that control precisely the amount of each oligonucleotide present on the array. The capability to accurately and reproducibly deposit oligonucleotide probes greatly reduces variation inherent in the comparison of data obtained by independent hybridizations using different arrays. Oligonucleotide arrays meeting these manufacturing standards can be used as *one-color* or *one-channel* platforms, where a single fluorescently labeled sample can be hybridized per array. This approach also eliminates fluorescent dye bias effects seen in two-color platforms. Oligonucleotide arrays manufactured by Affymetrix or Amersham are examples of one-channel array platforms. Another distinguishing feature is the number of oligonucleotide probes per gene present on the array. Agilent and Amersham array platforms use a single oligonucleotide probe per gene. In contrast, Affymetrix uses several perfectly matched (PM) oligonucleotide probes for each gene as well as mismatched probes (MM; same sequence as perfect match except one mismatch nucleotide in the middle; see Figure 3.5). The idea behind the mismatched probes is to measure background signal and specially developed algorithms are employed to take this information into consideration when analyzing Affymetrix data.

3.4.4.4. Experimental Design Considerations in Microarray Studies. The design of any experiment should be driven by the specific biological question posed. The goal of a good design is to allow analysis of the data to be as straightforward and powerful as possible given the specific research objective and the constraints of biological and economical factors. When designing a microarray experiment, one should consider several principles, including robustness, simplicity, extendibility, randomization, and replication.

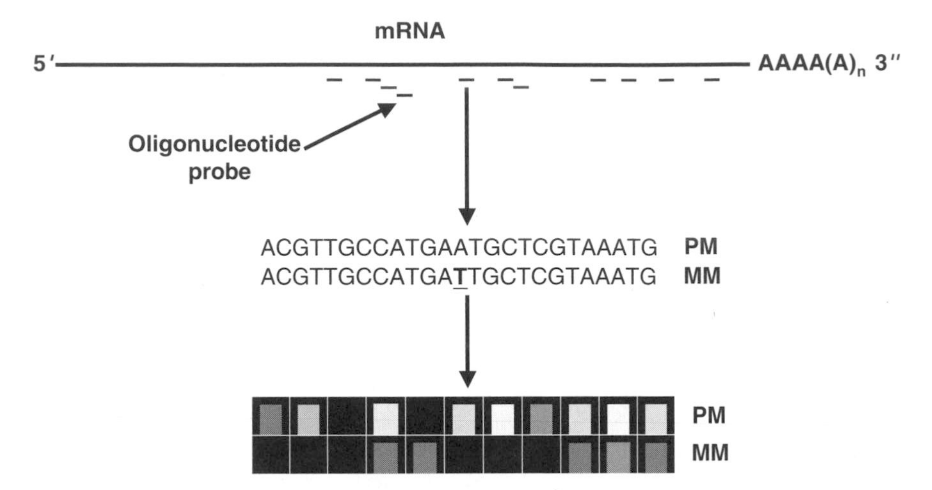

Figure 3.5. Design of an Affymetrix microarray gene chip. (PM, perfectly matched oligonucleotide probe; MM, mismatched oligonucleotide probe).

For two-channel microarrays, it is important to decide which samples are being paired for hybridization on the same array. This decision has important implications for the relative variance that can be achieved when comparing gene expression profiles of two different samples. The best variance is obtained when the two samples are being hybridized to the same array, allowing a direct comparison. Depending on the number of different types of samples, this cannot always be achieved. Another possibility is to compare samples through a common reference (Figure 3.6). Advantages of a reference design are its simplicity and ease with which the experiment can be extended at a later timepoint as long as the same reference sample is still available. In addition, the reference design is robust; if one array fails, only the failed array is affected. The major disadvantage is that the reference occupies 50% of all arrays and the gene expression profile of the reference may be of little interest to the study. So-called loop designs offer an alternative; they do not require a reference. Loop designs hold the potential to achieve relative variances that are the same as or better than those of the reference design using a comparable number of arrays. Loop designs are not as easily extendible as reference designs and may be less robust, depending on the specific construction of the loop. Pairing of samples is not an issue that needs to be considered for one-channel platforms, since only one sample is hybridized to one array.

Randomization is a fundamental principle in statistical analysis and should be applied to microarray experiments as much as is practical. As

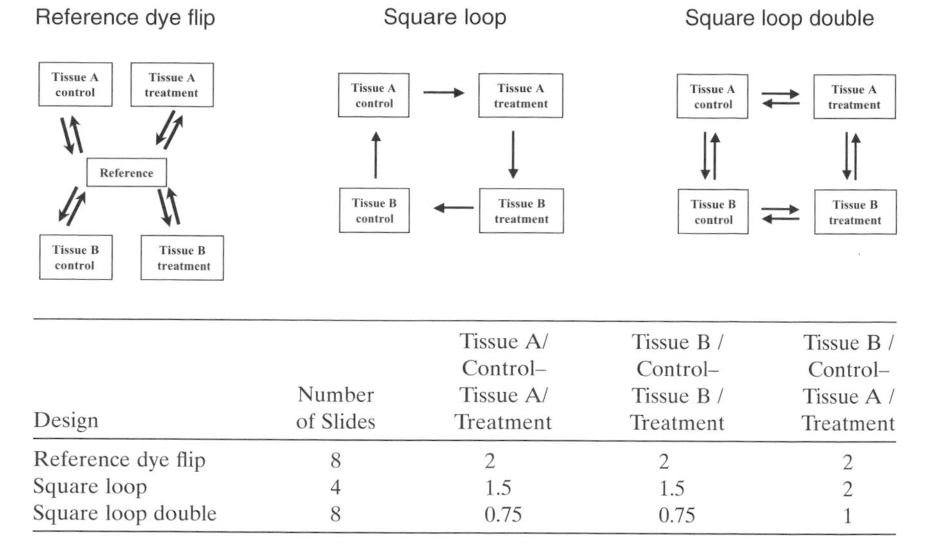

Design	Number of Slides	Tissue A/ Control– Tissue A/ Treatment	Tissue B / Control– Tissue B / Treatment	Tissue B / Control– Tissue A / Treatment
Reference dye flip	8	2	2	2
Square loop	4	1.5	1.5	2
Square loop double	8	0.75	0.75	1

Figure 3.6. Experimental design options for cDNA microarray experiments—relative variance comparisons for differential gene expressions.

pointed out by Kerr (2003), microarray experiments often involve several stages, and randomization should be applied to each stage. The goal of a typical experiment could be to assess the effect of a certain drug treatment on global gene expression in a certain strain of mice. Individual animals should be assigned randomly to the different treatment and control groups. Microarray experiments are sensitive to variation introduced by a number of technical artifacts, for example, the manufacturing process of the arrays themselves. To guard against such a bias, arrays should be chosen randomly, if possible.

Another important consideration is replication. There are two major types: technical replication and biological replication. Technical replication is designed to estimate the error introduced by the various steps in the microarray assay. For example, if the same RNA is used and its gene expression profile is determined with several different microarrays, the error associated with the various technical aspects, including fluorescent labeling, hybridization, washing, the quality of the arrays, and scanning, can be determined. However, such an experiment will not provide any information on the biological variability of the sample. Gene expression profiles of individuals randomly sampled from a population need to be determined before inferences about that population can be made. Microarray experiments are relatively expensive, and financial resources are often the limiting factor in how many replicates can be afforded. In most research contexts, biological replication will provide more relevant information than will technical replication. Excellent reviews on experimental design issues have been published by Churchill (2002), Kerr (2003), and Yang and Speed (2002).

3.4.4.5. Processing and Analysis of Microarray Data

3.4.4.5.1. Preprocessing: Diagnostics and Normalization. Once a set of microarrays has been hybridized and scanned, the scanner images need to be preprocessed before any statistical analysis can be carried out. This preprocessing phase involves visual inspection of the scanner images, spot quantification, slide diagnostics and quality control, and, finally, normalization.

The type of microarray determines which spot quantification software to use. Commercial oligonucleotide arrays manufacturers usually offer their own quantification software. Amersham CodeLink arrays require the CodeLink Expression Analysis program, Agilent arrays use Agilent's Feature Extractor program, and Affymetrix GeneChips require the use of GCOS. For two-color cDNA arrays there are a variety of open-source programs as well as commercial programs. The TIGR Website, http://www.tigr.org, is an excellent source for freely available software offerings. Commercial programs include Axon's GenePix, BioDiscovery's Imagene, or CISRO's Spot, to name only a few examples. Most of these programs allow visual inspection of the slide images to check for defects or damage.

After quantification of the scanner images, the next step involves some type of diagnostics and/or quality control (QC). Nearly all of the spot quantification programs mentioned previously also perform QC tasks, such as calculating mean signal strength, background thresholds, and control spot statistics. Summaries of these calculations provide a way to identify and exclude poor-quality arrays.

Normalization is the process of removing the uninteresting variability within the quantified images, that is, the variability that is not due to the biological differences of the samples. This process requires specialized software. There are many programs available that will take a spot quantification file as input, and perform normalization, as well as other statistical tasks discussed further in the next section. The specialized programs mentioned above have these statistical capabilities in addition to their image analysis features. However, there are also open-source programs available that allow for the custom analysis of data. For example, Bioconductor (http://www.bioconductor.org) allows the input of the raw quantification data. There are many normalization routines available for any type of microarray platform, from Affymetrix to two-channel cDNA arrays. Bioconductor can also perform more sophisticated multichip normalizations, such as Robust Multichip Average (RMA) and GC-RMA.

Further discussions of the preprocessing of microarrays can be found in Yang and Speed (2002). For Affymetrix array analysis, RMA has been demonstrated to be more robust than other available methods as assessed by measuring spike-in RNA samples of known concentrations. Furthermore, GC-RMA, which uses information about the GC content of the oligonucleotide sequences improves on the RMA algorithm to maintain precision while also improving accuracy (Wu and Irizarry 2004).

3.4.4.5.2. Statistical Methods for Identifying Differential Gene Expression. There are four main steps involved in the identification of differentially expressed genes. The first step is the choice of the appropriate statistical model, which is used to calculate the average intensities of gene expression for each gene across replicates and the sample variance for each gene. Appropriate statistical models will depend on the experimental design and can span the range from fixed-effect models to mixed-effect models if higher-order structure is present in the data such as groupings of covariates (Churchill 2002).

Calculation of the test statistic is the second step. If just two groups are being compared, then (Student's) t tests are useful. For more than two groups, some type of analysis of variance approach is more appropriate (Churchill 2002; Yang and Speed 2002, Kerr et al. 2000, Kerr and Churchill 2001; Lee et al. 2000). For data that contains more structure, such as balanced and unbalanced data, or when the within-group (intragroup) correlation is important in grouped data, then mixed-effect models are useful. Many investigators prefer modified versions of the standard t statistic that use information borrowed from all the genes on the array to estimate the individual gene variance (Storey and Tibshirani 2003; Efron and Tibshirani 2002). As demonstrated by Qin and Kerr (2004) in the analysis of spike-in experiments with two-color cDNA arrays, t tests performed the worst among test statistics for correctly identifying the rankings of the spike-in genes, while modified versions of the t statistic were shown to be much more robust. Furthermore, it was found that for some datasets the simple median of the log ratio across arrays had the best performance at correctly ranking the spike-in genes.

Unadjusted p-value calculation is the third step. This involves the calculation of both the null distribution for the test statistics and the selection of rejection regions (one-sided or symmetric).

The fourth and last step is to apply a method that controls the number of false-positive genes identified. When testing a hypothesis, one can make either a type I error (calling the change in gene expression significant when it is not, a false positive), or a type II error (calling the change in gene expression not significant when it is, a false negative). However, in microarray analysis, multiple hypotheses are being tested, so it is not clear how best to specify the overall error rate. As pointed out by Storey and Tibshirani (2003), there are a spectrum of choices. At one end are unadjusted p values, which result in far too many false positives (applying a p-value cutoff of 0.01 to an array with 50,000 genes implies that there are possibly 500 genes that are false positives). At the other end is the standard Bonferroni correction to control the familywise error rate, which is far too conservative and results in large numbers of false negatives. What has been found to be most helpful in the microarray context is the false discovery rate (FDR) or the positive false discovery rate (pFDR) or q value. With this approach, the investigator can control the number of false positives in the number of genes called significant, rather than controlling the number of false positives out of all the genes present, by examining the distribution of p values. A detailed discussion of these issues has been provided by Storey and Tibshirani (2003).

Commercial analysis programs that do most or all of the steps described above are now available. One popular program is SAM [statistical analysis of microarrays; see Storey and Tibshirani (2003)]. However, identification of differential gene expression is a very active research field. To stay most current with rapid changes in the field, we also recommend using the microarray statistical research tool Bioconductor (http://www.bioconductor.org/), where many new algorithms first appear.

3.4.4.5.3. Biological Interpretation. As a generic example, assume that the investigator chose a suitable experimental design, the RNA samples were processed and hybridized, the arrays were scanned, and the raw data were normalized and statistically analyzed. This procedure resulted in a list of genes that are differentially expressed at a chosen level of statistical significance. Now it's time to interpret and explore the data in a biologically meaningful way. A typical experiment can result in hundreds or even thousands of differentially expressed genes, and it can be challenging to make sense of the data. One approach is to categorize the differentially expressed genes into functional groups on the basis of gene annotations and then identify putative biological pathways. Several software tools are available to perform these tasks in an automated fashion including GenMAPP [Doniger et al. (2003); http://www.genmapp.org/] and GoMiner [Zeeberg et al. (2003); http://discover.nci.nih.gov/gominer/]. Another useful approach is to investigate promoter regions of differentially expressed genes to identify common or hierarchical transcription regulatory networks. PAINT is a software tool designed to facilitate identification of common transcriptional regulatory networks [Vadigepalli et al. (2003); http://www.dbi.tju.edu/dbi/tools/paint/]. GenMAPP, GoMiner, and PAINT are all freely available to the academic research community via the World Wide Web.

Another popular approach is to use cluster analysis to group genes displaying similar behavior into clusters (Eisen et al. 1998). Several clustering algorithms are being used in gene expression profiling, including *k*-means clustering, self-organizing maps, and model-based approaches. It is important to recognize that cluster analysis will identify clusters in a given list of differentially expressed genes regardless of whether the genes assigned to a cluster share a meaningful biological connection. Therefore, a clustering method needs to be chosen carefully to ensure that it is consistent with the goal of the analysis, and the results need to be interpreted critically in the appropriate biological context. More in-depth discussions of cluster analysis have been provided by Shannon et al. (2003) and Slonim (2002). In addition to commercial packages, numerous open-source software programs, including Bioconductor (http://www.bioconductor.org/), TIGR MultiExperiment Viewer (http://www.tigr.org/software/), and Eisen Lab's Cluster (http://rana.lbl.gov/EisenSoftware.htm), to name just a few, provide a wide range of clustering algorithms.

3.4.4.6. Applications of Gene Expression Profiling. Every nucleated cell of an organism contains two copies of the entire genome (all genes) of that individual (except germ cells). However, only a small percentage of all genes are expressed in a specific cell type or tissue and the gene expression profile is characteristic for the cell type or tissue. Endogenous (e.g., hormonal fluctuations) or exogenous perturbations (e.g., environmental chemicals, lifestyle factors such as drug, alcohol, smoking, and diet) cause precisely controlled transient changes in the gene expression profile. Similarly, a healthy cell or tissue can be distinguished from a diseased or chemically challenged cell or tissue by its transcriptional fingerprint. Although traditional methods for assessing the expression of a single or a few genes are not suitable for measurement and recognition of global gene expression patterns, microarray technology has been successfully applied for this purpose in many scientific disciplines, including biomedical and toxicological research.

Gene expression profiling has had a major impact on the field of cancer research. Histological and immunohistochemical methods provide information on the origin, type, and stage or grade of tumors, which, until relatively recently, still provided the most relevant prognostic information. However, these methods offer very limited insight into the underlying genetic and biochemical events that affect clinical outcome. Great progress has been made in discovering fundamental molecular mechanisms involved in the development and progression of many different cancer types, although the general genetic fingerprint leading to tumorigenesis is still elusive. It is widely recognized that patients with similar tumors, as judged by histopathology and other clinical measures, may respond differently to the same treatment and may have different clinical outcomes. As outlined above, the simultaneous measurement of thousands of genes in a specific tissue or population of cells greatly facilitates the identification of specific gene expression patterns that are associated with disease. This information has been used to improve tumor classification, diagnosis, prognosis, and clinical outcome. For example, Golub and coworkers were among the first to correlate gene expression profiles with tumor classification (Golub et al. 1999). These scientists were able to classify human acute leukemia samples into acute myeloid leukemia and acute lymphoblastic leukemia samples based on their gene expression profiles alone. In a more recent study, Dhanasekaran and colleagues used microarray analysis to identify a signature expression profile of healthy prostate, benign prostatic neoplasia, localized prostate cancer, and metastatic prostate cancer (Dhanasekaran et al. 2001). These and other studies have demonstrated that the transcriptional fingerprint is potentially more reliable than traditional pathological methods for classification of cancers. Gene expression profiling combined with traditional methods (morphology, staging and grading of cancer) has led to improved diagnosis and clinical outcome prediction.

Microarray technology also had a profound effect on the field of toxicology and the pharmaceutical industry in predicting toxicological and pharmacological effects of unknown compounds. The underlying hypothesis is that compounds sharing a similar mechanism of action or toxicological response also share a similar gene expression profile. Several in vitro and in vivo studies have demonstrated that this hypothesis has merit (Waring et al. 2001a, 2001b; Hamadeh et al. 2002), although the predictive value of expression patterns alone remains problematic because changes in gene expression are highly time-, dose-, and tissue-dependent (Waters et al. 2003). Public and private efforts have focused on establishing databases with gene expression profiles of compounds with well-characterized toxicological and pathological endpoints. These databases can be interrogated using gene expression profiles of compounds with unknown toxicological or pharmacological effects. A high degree of similarity between the transcriptional fingerprint of the tested compound with an archived compound is suggestive of similar biological effects.

3.4.4.7. Considerations in Gene Expression Profiling Technology. While microarray technology is a powerful tool facilitating the understanding of biological systems, one has to be aware of its limitations. Gene expression profiling measures mRNA, an intermediate step in gene expression, and not the functional products, which are proteins. No protein can be synthesized unless its corresponding mRNA has been generated first. Furthermore, changes in mRNA levels do not necessarily correlate with changes in corresponding protein levels. In addition, many proteins require posttranslational modifications (e.g., phosphorylation) in order to be functional. Therefore, one has to be cautious in interpreting gene expression data. Measuring cellular protein levels on a global scale (proteomics) is still challenging and currently not yet available to the majority of the research community. Until proteomic technologies become more easily accessible, microarrays remain the best established tool to investigate gene expression on a global scale. In contrast to oligonucleotide-based arrays, cDNA arrays with partial or full-length cDNA probes do not differentiate between highly homologues members of a gene family because of cross-hybridization. Microarray technlogy generates semiquantitative data and expression levels of interesting, differentially expressed genes need to be confirmed by an independent quantitative method such as RT-PCR.

4

Epidemiologic Approaches

Harvey Checkoway and Parveen Bhatti
University of Washington, Seattle, WA

Anneclaire De Roos
University of Washington and Fred Hutchinson Cancer Research Center, Seattle, WA

4.1. INTRODUCTION

Epidemiologic studies inevitably provide the most convincing tests of hypotheses that genetic traits modify the effects of environmental exposures on human disease risks. In this chapter we will describe the common epidemiologic approaches applied to investigate gene–environment interactions. We will focus primarily on sporadic (i.e., not known to have direct familial inheritance patterns), complex diseases. By "complex," we mean that a particular disease or physiologic dysfunction may have variable phenotypic expression patterns, and may also have multiple causal pathways that are likely to involve both genes and environmental exposures. Many cancers, neurodegenerative diseases, and diseases of the cardiovascular and respiratory systems, for example, fit this description. Also, we will not discuss genes for which a single mutation is a sufficient cause of disease, irrespective of other genetic mutations or environmental exposures.

We will describe the design features of various types of epidemiologic studies. In addition, we will explain the important biases that may arise in epidemiologic research, along with methods to avoid or minimize these biases. Methodological points will be illustrated with examples drawn from

Gene-Environment Interactions: Fundamentals of Ecogenetics, edited by Lucio G. Costa and David L. Eaton

the scientific literature. Readers seeking more comprehensive explanations of epidemiologic study techniques and principles are referred to standard textbooks (Rothman and Greenland 1998; Koepsell and Weiss 2003; Savitz 2003). We recommend the text by Khoury et al. (2004), which contains in-depth descriptions of genetic epidemiology methods. Excellent reviews of epidemiologic considerations in the study of gene–environment interactions are also available in scientific journal articles (Ottman 1996; Clayton and McKeigue 2001; Rothman et al. 2001; Kelada et al. 2002).

4.2. EPIDEMIOLOGIC FRAMEWORK FOR STUDYING GENE–ENVIRONMENT INTERACTIONS

It should be recognized that epidemiology is for the most part an observational rather than an experimental science. In the context of studying gene–environment interactions, this means that epidemiologists do not intentionally dose people with toxic agents, nor do epidemiologists manipulate people's genes. Instead, we observe the occurrence of disease among free-living human populations whose genotypes are determined by their ancestry (and the evolutional influences on their ancestors' gene pools).

As a starting point, we can consider the following simplified, and admittedly idealized, hypothetical example that would allow for an explicit and highly specific test of a gene–environment interaction. A particular environmental toxicant, such as a workplace chemical, is established as a potent risk factor for a certain disease. It is also known that this chemical is detoxified very efficiently to an innocuous form by a single enzyme, and that the gene encoding this enzyme has a clearly defined functional polymorphism; persons harboring the "variant" allele do not detoxify the compound, whereas persons who are homozygous for the wildtype allele metabolize the chemical to its nontoxic form. Given this scenario, we can readily imagine an epidemiologic study that would establish whether the genetic variant does indeed modify risk from this chemical. The conclusive epidemiologic study would define two groups of people who at the beginning of the study are free from the disease of interest: those with the variant allele (either heterozygotes or homozygous variant) and those who are wildtype homozygotes. Included in these two groups would be people who are exposed to the chemical under study and people who are not exposed. Let's also assume that exposure assessment and disease occurrence are both determined accurately and with no bias. Then, we would follow study subjects classified by genotype and exposure to determine their relative incidence rates of disease. According to prediction based on genotype, we would expect that there would be no effect of exposure on disease risk among persons who are homozygous wildtype, as contrasted with an elevated disease risk in exposed persons who harbor the variant allele.

4.3. STUDY DESIGNS

The various epidemiologic study designs have relative strengths and limitations that determine their suitability for investigating particular exposures and health outcomes. It is customary to subdivide study designs on the basis of subject selection methods (based on either exposure or health status) and temporality (prospective vs. cross-sectional vs. retrospective). Before describing each study design, we should point out that the logical inference of all designs is identical; every approach addresses the basic question of whether a particular exposure affects disease risk. The differences among designs are dictated primarily by logistics. In other words, some designs are more appropriate (practical) than others for studying a specific exposure–disease relation.

4.3.1. Cohort Studies

The idealized "definitive" study described in the preceding section is a cohort study. In this design, an index group, defined on the basis of some exposure or other (e.g., genetic) trait, is followed for detection of disease incidence, and its disease experience is compared with a reference group composed of people who are not exposed or do not have the trait in question. Among epidemiologic study designs, cohort studies are often considered the "gold standard" because cohort studies most closely mimic the standard experimental paradigm of following two or more groups who differ in exposure status. This design is also akin to therapeutic clinical trials that follow patients, given different treatments, for subsequent health outcomes.

The relevant contrasts in a cohort study are depicted in Table 4.1, where results are expressed in terms of relative risks associated with exposure (the rate of disease in exposed divided by the disease rate in nonexposed), separately for each of the genotype groups. Table 4.1 is the most basic, simplified data layout. An expanded version of this table would be necessary if we were comparing risks among multiple exposure groups, as in a dose–response analysis, or according to multiple genotypes, such as homozygous wildtype, heterozygous, and homozygous variant.

Table 4.1 Data Layout for a Cohort Study of Disease Risk[a] in Relation to Exposure and Genotype

	Variant Allele		Wildtype Allele	
	Cases	Number of People	Cases	Number of People
Exposed	a	N_1	e	P_1
Nonexposed	c	N_0	g	P_0

[a] *Key:*

Relative risk associated with exposure among persons with variant allele = $(a/N_1)/(c/N_0)$

Relative risk associated with exposure among persons with wildtype allele = $(e/P_1)/(g/P_0)$

Relative risk associated with the *interactive* effect of exposure and the variant allele = $(a/N_1)/g/P_0)$. [*Note*: This relative risk is compared with independent relative risk for exposure, $(e/P_1)/(g/P_0)$, and the independent relative risk for the variant allele, $(c/N_0)/g/P_0)$.]

Cohort studies can be classified on the basis of their temporal sequencing as either prospective or retrospective (also known as "historical"). In a *prospective* cohort study, groups with and without the trait or exposure of interest are followed forward in time to determine relative rates of disease incidence. This approach is especially well suited for relatively common outcomes that have short induction and latency periods, such as the acute onset of asthma-like symptoms among persons exposed to air pollution. However, prospective cohort studies are seldom practical for studies of rare diseases or chronic conditions that develop after many years of exposure, such as most environmentally related adult cancers.

The *retrospective* cohort design was devised as an alternative to prospective cohort studies. In a retrospective cohort study, disease incidence in the index and comparison (reference) groups are compared beginning from a timepoint in the past and followed through historical time to the present. Thus, even rare outcomes with long induction and latent periods can be investigated in a retrospective cohort study. However, a disadvantage of the retrospective design is that exposures assessed retrospectively may be unmeasureable or subject to recall bias.

Although the cohort design offers the clearest evidence for causation in epidemiology, prospective and retrospective cohort studies have seldom been applied to investigate gene–environment interactions. Prospective cohort studies usually entail large costs in time and resources; exceptions are short-term outcomes. Retrospective cohort studies reduce these cost and time inefficiencies. Also, it was rare in past cohort studies that biological specimens were archived for genetic investigations. A notable exception is the Nurses Health Study, which is a comprehensive prospective cohort study of numerous exposures and health endpoints in over 100,000 women in which blood samples have been obtained for a sizable fraction of subjects (Colditz et al. 1997). Because of the keen interest in genotyping and biomarkers, many cohorts that are currently being established include banking of biologic samples, such as sources of DNA. Such cohorts include the European Prospective Investigation into Cancer and Nutrition (EPIC) study (Riboli et al. 2002), the multiethnic cohort in Hawaii and Los Angeles (Kolonel et al. 2000), and the Women's Health Initiative (WHI Study Group 1998). These are large cohort studies, with more than 100,000 subjects each, which will provide increasing opportunities to study gene/environment interactions as the cohorts mature.

4.3.2. Case–Control Studies

An alternative to the cohort design is the case–control method, in which persons with the index disease (cases) are compared with persons free of the condition (controls) with respect to past exposure and other risk factors (e.g., genetic traits). The distinction between a case–control and cohort study is that subject selection is based on health status (disease present or absent) in a case–control study, whereas in a cohort study, selection is based on exposure status. The exposure studied can be an environmental factor, genetic trait, or the combination of the two. A major advantage of the case–control approach is that it enables efficient accumulation of large numbers of cases, even for relatively rare diseases. Cases can be identified from clinical settings, population disease registries, or other sources. In contrast, a cohort study would need to include tens of thousands of subjects in order to generate a sufficient number of cases of an uncommon disease for a meaningful study.

Table 4.2 shows the basic data layout for a case–control study. It should be noted that the cell entries for cases in Table 4.2 are the same as those from Table 4.1, but the controls in Table 4.2 represent a sample of the unaffected people in the source population. The odds ratio, which is the ratio of the odds of a case having been exposed divided by the odds of a control having been exposed, provides a valid estimate of relative risk in case–control studies.

There are two main types of case–control study, depending on their source population. The first, most common, type is the *community-based case–control study*, in which cases are identified from the population at large (e.g., from hospitals or disease registries) and controls are also selected from same source population. Community-based studies are most effective for studying associations with environmental exposures that occur widely in the population (e.g., air pollution), rather than exposures that are highly spe-

Table 4.2 Data Layout for a Case–Control Study of Gene–Environment Interaction

	Variant Allele		Wildtype Allele	
	Cases	Controls	Cases	Controls
Exposed	*a*	*b*	*e*	*f*
Nonexposed	*c*	*d*	*g*	*h*

[a] *Key:*

Relative risk (odds ratio) associated with exposure among persons with variant allele = *ad*/*bc*.

Relative risk (odds ratio) associated with exposure among persons with wildtype allele = *eh*/*fg*.

Relative risk (odds ratio) for the *interactive* effect of exposure and the variant allele = *ah*/*bg*. [*Note*: This odds ratio is compared with the independent odds ratios for exposure, *eh*/*fg*, and the odds ratio for the variant allele, *ch*/*dg*.]

cific to certain subgroups of the population (e.g., specific workplace chemicals). The second type is a *nested case–control study*, in which both cases and controls are members of a clearly defined cohort, such as an occupational group or some specially enumerated study group (e.g., the Nurses Health Study cohort). A nested case–control study can be considered as an efficient shortcut for investigating an entire study cohort. The following examples illustrate the two types of case-control studies.

Nelson et al. (2002) examined the modifying role of the Arg399Gln polymorphism of the XRCC DNA repair gene on squamous cell skin cancer risk associated with sunlight exposure. In total, 246 cases of squamous cell carcinoma diagnosed during 1993–1995 were identified from a population-wide survey in New Hampshire, and 431 age- and sex-matched controls were recruited from enrollment lists of the New Hampshire Department of Transportation and the Health Care Financing Administration. All subjects were Caucasians. As anticipated, sunlight exposure, inferred from a self-reported lifetime number of severe sunburns, was significantly associated with skin cancer. The most striking finding from the study was a strong interaction between genotype and sunlight exposure. The estimated relative risk of contracting squamous cell carcinoma associated with 3 or more sunburns, compared to 0–2 sunburns, was 1.5 and 1.6 in wildtype homozygotes and heterozygyotes, respectively, but increased more than fourfold to 6.8 in homozygotes of the variant allele (Table 4.3). This example exemplifies the community-based case–control method in that study subjects were identified from the general population, and the exposure of interest was widespread and relatively common.

Table 4.3 Case–Control Study of Squamous Cell Skin Cancer, Lifetime Sunburn History, and XRCC1 Genotype

Lifetime sunburns	Wildtype Homozygous (arg/arg)	Heterozygous (arg/gln)	Variant Homozygous (gln/gln)
0–2[a]	1.0[b]	1.0	1.0
≥3	1.5	1.6	6.8

[a] Reference category.
[b] Odds ratio.
Source: Adapted from Nelson et al. (2002), with permission of the American Association for Cancer Research, Philadelphia, PA.

An example of a case–control study nested within a defined cohort is the study of the interrelations between polychlorinated biphenyls (PCBs), variations in the cytochrome P450 1A1 (CYP1A1) gene, and breast cancer conducted among members of the Nurses Health Study (Laden et al. 2002). CYP1A1 was considered a plausible candidate gene because of prior experimental evidence that the enzyme produced by the gene plays a role in the formation of PCB-induced adducts on DNA. The cohort was composed of 121,700 female registered nurses originally enrolled in 1976 and who have been followed since then. Breast cancer cases diagnosed during 1976–1994 and matched controls were identified from a subset of 32,826 women who provided blood samples during 1989–1990. Breast cancer diagnoses were self-reported and corroborated by subsequent reviews of medical records. A matched control without breast cancer was selected for each case; matching factors were year of birth, menopausal status, month and time of blood draw, fasting status at blood draw, and use of postmenopausal hormonal replacement. Blood plasma concentrations of PCBs and DNA specimens were available for 378 cases and their matched controls. Two genetic variants of CYP1A1 were investigated as potential modifiers of associations with PCBs: an A4889G single nucleotide polymorphism (SNP) in exon 7, and the MspI T6235C transition. There was no evidence in this study that PCB exposures alone were related to increased breast cancer risk, nor were the CYP1A1 polymorphisms independently associated with breast cancer. However, there did appear to be an interaction between the exon 7 polymorphism and PCB exposure among postmenopausal women. As shown in Table 4.4, there was a striking dose–response gradient for exposures to PCBs among carriers of the variant allele, but no association with PCB exposures among women with the homozygous wildtype genotype. This

Table 4.4 Associations of PCBs with Breast Cancer Risk in Postmenopausal Women, by Cytochrome P450 1A1 exon 7 polymorphism

PCB	Wildtype Homozygous			Variant Carrier		
Tertile	Cases	Controls	Odds Ratio	Cases	Controls	Odds Ratio
1[a]	84	81	1.0	9	16	1.0
2	82	87	0.9	15	12	2.2
3	84	90	0.9	19	7	4.8

[a] Reference category.

Source: Adapted from Laden et al. (2002), with permission of the American Association for Cancer Research, Philadelphia, PA.

study illustrates some of the efficiency advantages of the nested case–control design compared to a full cohort analysis of gene–environment interaction. The investigators were able to assess the effects of PCBs and CYP1A1 genotypes on breast cancer risk by assaying blood samples for all of the cases but just for a sample of the noncases (controls). Conducting PCB measurements and performing genotyping for all 32,826 women with available blood samples would have been substantially more expensive and time-consuming, with only marginal gains in statistical precision.

4.3.3. Cross-Sectional Studies

This design involves the comparison of disease prevalence, the number of persons with the index condition at a given time, among groups classified with respect to exposure and other risk factors. In contrast, cohort and case–control designs usually involve the study of new-onset (incident) disease. The distinguishing feature of cross-sectional studies is that, typically, exposure and health outcome are determined at the same time. Subjects in cross-sectional studies are selected on the basis of exposure status, as with a cohort study, but in a cross-sectional study there is no follow-up over time.

Cross-sectional studies are especially suitable for investigating slowly developing diseases that do not have sharp onset times (e.g., the pneumoconioses) and conditions that are not overt diseases, such as disease symptoms or variability in physiological function (e.g., blood pressure). An illustrative example is the study by Richeldi et al. (1997) of berylliosis, a granulomatous lung disease that is triggered by immunity-mediated beryllium sensitization. Berylliosis most often is a chronic condition with a slow,

Table 4.5 Berylliosis Prevalence by Job Type (Beryllium Exposure) and HLA-DPD1 Allele Status

HLA-DPB1Glu69 Allele Status	Machinists (High Exposure)			Nonmachinists (Low Exposure)		
	Cases	Number of Workers	Prevalence (%)	Cases	Number of Workers	Prevalence (%)
Negative	1	39	3.2	0	55	0
Positive	4	16	25.0	1	25	4.0
Total	5	47	10.6	1	80	1.3

Source: Adapted from Richeldi et al. (1997), with permission of John Wiley & Sons, Hoboken, NJ.

insidious onset. Thus, prevalence is more readily determined than incidence. The investigators compared berylliosis prevalence in machinists (heavily exposed) and nonmachinists (minimally exposed) employed at a beryllium ceramics plant. A putative berylliosis susceptibility genetic polymorphism in the HLA-DP gene, HLA-DPB1Glu69, was examined as a modifier of exposure. As shown in Table 4.5, the excess prevalence of berylliosis in machinists was higher than in nonmachinists overall (10.6% vs. 1.3%), which was due largely to the pronounced difference in berylliosis prevalence among workers who carried the variant allele. Prevalence was 25.0% among machinists with the variant allele compared to just 3.2% among those without the variant. Among nonmachinists, prevalence was also higher among workers with the allele (4.0%) than those without (0%).

4.4. METHODOLOGICAL CONSIDERATIONS

The validity of all epidemiologic research depends on the extent to which bias is avoided or minimized. Bias can stem from inappropriate selection of study subjects (selection bias), misclassification of exposure or health outcome (information bias), confounding by unmeasured risk factors, or some combination of these. We will consider each of these types of bias in turn. Population-based case–control studies are far and away the most common design for gene–environment interaction investigations; thus, we will focus most subsequent methodological discussion on this design. It should be appreciated, however, that all study designs can suffer from bias, with some forms of bias more likely to arise in certain designs than in others.

4.4.1. Selection Bias

Subject selection in a cohort or cross-sectional study should be made independently of health outcome, which means that the groups compared should, on a priori grounds, have similar baseline disease risks. Thus, at least theoretically, differences in risk observed should be due to exposures and genetic factors of concern, or the interplay between these factors. Analogously, in a case–control study, subject selection is based on health outcome status and should be made independently of exposure status.

There are some notable methodological challenges in case–control studies, particularly in regard to the choice of controls. Ideally, the case group represents all subjects with the disease of interest in a defined population, sometimes referred to as the "source population" or "study base." In order to sharpen any causal inferences that are ultimately reached, it is highly desirable that the case group be a homogeneous clinical entity with a high degree of diagnostic accuracy. Thus, for example, in a study of non-Hodgkin's lymphoma, it is preferable to include only cases who have had appropriate pathological diagnostic confirmation and disease subtyping.

Controls should be representative of the study base in terms of their exposure experience and genetic composition. Another requirement for controls is that, had they developed the disease of interest, they would have been identified as cases, in the same way that the index cases were identified. To understand these requirements, consider a case–control study of a particular disease, say, bladder cancer, conducted in 2004 in which cases were identified from the patient rosters of a cancer referral and treatment hospital for the years 1997–2003, and controls were identified by random digit dialing (a method akin to opinion polling that is often used to identify controls) in the community where the hospital is located. In this instance, there are several potential sources of bias that might be introduced by the subject selection procedures. First, cases referred to a treatment center may come from many different communities, and thus may not represent a well-defined study base. The controls, however, were identified from the local community, and thus their exposure experience might not be representative of the study base from which the cases were drawn. There is also no assurance that any cases that might have occurred among controls would have been identified in the same manner as the cases. All that is known about the controls is that they resided in this community in 2004, so in- or outmigrations during the years when the cases came to the hospital would not be known. Finally, the fact that controls were selected by random digit dialing means that they had working telephones and were agreeable to participate in the study. Telephone ownership and willingness to participate in research can both be related to socioeconomic status and ethnicity, which in turn may be associated with environmental exposures (e.g., job), lifestyle factors (e.g., smoking, diet), and genetic background (via ethnicity).

When cases are selected from hospitals, it is sometime convenient to choose controls from among other hospitalized patients who do not have the disease under investigation. This strategy can balance the study base representation of cases and controls, provided that the controls' diseases have similar referral patterns as the index cases (note that this would not be true if the hospital specialized in certain diseases). Care should be taken to avoid selecting hospitalized controls whose conditions are related to either the exposures or genes of interest. For example, patients with chronic obstructive pulmonary disease would be a poor choice of controls for a case–control study of lung cancer because cigarette smoking is the predominant cause of both conditions. In less obvious situations, the absence of such associations can be a difficult requirement to verify. Choosing controls from among patients with several conditions thought to be unrelated to the index case disease can safeguard against inadvertently selecting an inappropriate control disease (Wacholder et al. 2002).

The selection biases just described in somewhat hypothetical terms can have important distorting influences on epidemiologic research. Nonetheless, bias can still be minimized if the sources of bias are identified and controlled. (We will discuss control of bias in the context of confounding later in this chapter.) For instance, if we had reason to be concerned that socioeconomic status and ethnicity were biases introduced by the subject selection process, we could adjust for these factors in the analysis to the extent that we have relevant data on these factors. Additionally, we might perform sensitivity analyses to quantify the potential magnitude of these biases on the results (Rothman and Greenland 1998; Lash and Fink 2003). For example, consider a case–control study in which cases were identified from a large referral hospital that provides services mainly to patients of higher socioeconomic status, and that controls were selected from the local community, which included residents with a wide range of socioeconomic status. We might be concerned that socioeconomic status can influence subjects' responses regarding exposure, either overreporting or denying (by some percentage) true exposure. We could therefore make calculations of relative risks associated with exposure that take the presumed level of reporting bias into account, and compare these with our observed relative risk estimates to see the possible magnitude of selection bias.

Study participation rates are often considered potential sources of selection bias. It is quite common to find that participation rates for cases exceed those for controls, probably because cases have greater inherent motivations to contribute to scientific research. This is not always true, however, as cases themselves, their families, or healthcare providers may feel that study participation is too disruptive or invasive. Complete participation of all subjects, although ideal, is seldom realized. The general rule regarding participation rates and bias is that unequal participation rates will produce bias only if participation is related to health status *and* exposure. This applies to any study design. Thus, for example, if 85% of cases and 60% of controls participate, yet participation is not dependent on exposure history, there will not be a bias. Of course, establishing that participation is unrelated to exposure history may be difficult or impossible to verify, and becomes more complicated when we are interested in studying exposures, genetic factors, and their interactions. The potential for bias might be inferred from ancillary information about participants and nonparticipants (Savitz 2003), such as from comparisons of demographic characteristics that are likely to be correlated with or are surrogates of exposure (e.g., urban or rural residence, ethnicity).

4.4.2. Information Bias

Misclassification of disease or environmental and genetic risk factors, which are the primary sources of information bias, can reduce research validity. It is generally, although not always, the case that misclassification that is nondifferential will result in a reduced ability to identify a true association. By *nondifferential*, we mean that the misclassification of disease is equivalent for persons with and without the risk factor, and similarly, that misclassification of the risk factor is equivalent for persons with and without the disease of interest. To illustrate, consider a case–control study of the congenital malformation microcephaly (abnormally small head) and maternal exposure to lead during pregnancy. If the diagnosis of microcephaly were perfectly accurate in every case, and a diagnosis of microcephaly could be ruled out in all of the controls, but exposure determination were randomly inaccurate in 20% of cases and 20% of controls (i.e., equivalent exposure misclassification by disease status), then the ability to observe an association, if one indeed exists, would be diminished. This phenomenon is often referred to as "bias toward the null." A similar bias toward the null occurs when disease status is misclassified equivalently in exposed and nonexposed persons. To see this, consider a cohort study in which pregnant women were classified accurately according to lead exposure, yet microcephaly was incorrectly diagnosed in 5% of offspring of exposed women and in 5% of offspring of nonexposed women (i.e., nondifferential misclassification of disease by exposure status). Here again, a true association would be underestimated or masked because of bias toward the null. We might think of this as a problem of distinguishing the signal from the noise.

There may also be situations in which misclassification of risk factors differs according to disease status, or conversely, accuracy of disease classification differs by risk factor status. This situation might arise if exposure assessment were performed more rigorously among cases than among controls, or if disease incidence were determined more accurately in the exposed group than in the nonexposed. Differential misclassification can bias observed findings either toward or away from the null, thus either underestimating or exaggerating true associations. The direction of bias will depend on the configuration of misclassification. An extreme scenario that would lead to bias away from the null would be a situation where exposures are determined more accurately among cases than among controls, and where the cases who had been exposed underwent additional procedures to confirm disease diagnoses .

Epidemiologists make concerted efforts to minimize misclassification of disease status and risk factors, although some misclassification is inevitable. In the absence of a gold standard for disease or risk factor determination, which obviously would be adopted, were one available, the default assumption is that any misclassification is nondifferential—unless there is evidence to the contrary.

In the context of ecogenetic epidemiology research, the main concern is misclassification of environmental exposure, which typically is the most difficult factor to ascertain. Certainly, accurate disease diagnosis is crucial to validity, as is valid determination of genetic markers. (The latter topic is addressed elsewhere in this volume.) In view of the challenges and implications for validity posed by exposure assessment, the topic of exposure assessment approaches deserves some extended discussion.

Exposures can be determined in epidemiologic research in numerous ways. In some situations, exposures may be measured directly in environmental media, such as from air, water, or soil sampling. These measurements can then be linked to individuals for investigation of exposure–disease relations. However, levels of contemporaneous environmental toxicants may not be the most relevant exposures, particularly for health outcomes that require prolonged exposure periods to be effective in inducing physiological impairment or disease. Most of the so-called chronic diseases (e.g., cancer) fall into this category. Instead, subjects' past exposures need to be reconstructed to estimate lifetime temporal exposure profiles and doses. The principal sources that epidemiologists rely on for exposure reconstruction include past workplace or ambient (community) environmental monitoring data and questionnaire responses provided by study subjects. Biological monitoring can be conducted to infer past exposures for some persistent toxicants; blood levels of dioxin (Salvan et al. 2001) and bone concentrations of lead measured by X-ray fluorescence (Chettle et al. 1991) are two well-known examples. In addition, some clinical measures may be used to infer past exposures, such as chest X-ray opacities that indicate a history of high-dose silica dust exposure (Dosemeci et al. 1994).

Questionnaires are the mainstay source of exposure information in population-based case–control studies. Subjects may be asked detailed questions about past environmental exposures, including those experienced occupationally, along with numerous other questions on other potential disease risk factors (demographic factors, medical history, smoking, alcohol use, diet, etc.). The level of detail sought for environmental exposures varies from study to study. For example, some questionnaires seek information only on whether and for how long a person worked in a particular industry, whereas other, more elaborate questionnaires elicit extensive detail about specific job tasks, materials handled, and use of personal protective equipment. The amount of detail included in a questionnaire will be dictated by the study hypotheses and logistics. Generally, the level of detail for a given exposure will be in direct proportion to its anticipated priority as a risk factor. It is always essential that questions be asked in a consistent, unbiased manner. Ideally, interviewers should not be aware of subjects' health status, although blinding is often difficult to achieve.

In case–control studies, the quality and validity of questionnaire response information may vary between cases and controls, especially when proxy respondents (e.g., next of kin) are required for deceased or severely ill subjects (Nelson et al. 1990). Information bias may also arise because cases, due to their concerns about the causes of their illness, at times provide more accurate and detailed exposure history responses than do controls, who are not ill. There is also some evidence that volunteered information, such as from open-ended questions, may be more subject to recall bias than information from closed-ended questions (Teschke et al. 2000). Including some presumably irrelevant ("bogus") questions can indicate subjects' propensities for erroneous reporting, and thus may help distinguish whether self-reporting bias is at play.

4.4.3. Confounding

Confounding factors are those that are related to the environmental and genetic risk factors of concern, *and* are associated with risk of the disease of interest independent from the environmental and genetic risk factors of concern. Figure 4.1 is a schematic representation of a confounding factor. The solid line between the confounder (C) and the disease (D) indicates that the C is a determinant of disease risk, even in the absence of the exposure (E) under study. The solid line between E and C denotes the correlation between the two factors. The dashed line between E and D indicates

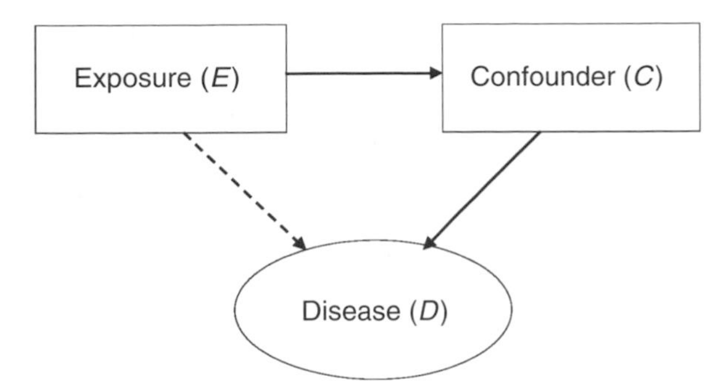

Figure 4.1. Depiction of a spurious association between exposure (E) and disease (D) due to a confounding factor (C).

a spurious association between the two. Thus, we might observe an elevated risk for D that appears to be related to E, but it could be the result of the correlation between C and E. For example, C may be a factor such as age or cigarette smoking, both of which are risk factors for many diseases. Failure to account for C in evaluating the relation between E and D (assuming that C and E are indeed correlated) would lead to an erroneous conclusion about risk related to E. Potential confounders that are frequently of concern in epidemiologic studies, and fortunately are amenable to control, are age, sex, race, and education.

Confounding can be controlled in several ways, including matching cases and controls on presumed potential confounders, limiting the study to a homogeneous confounder category (e.g., selecting only cases and controls who are African-American women aged less than 50 who had never smoked cigarettes), or by statistical adjustment in the data analysis. These approaches are described in detail in epidemiologic textbooks [see Rothman and Greenland (1998)]. An example of careful confounder control is in the study of PCBs, CYP1A1 genetic polymorphisms, and breast cancer described earlier in this chapter, in which breast cancer cases and controls were matched on year of birth (age), menopausal status, and other factors that are potentially correlated with PCB exposure and are known to influence breast cancer risk (Laden et al. 2002).

Assessment and control of confounding in studies of gene–environment interaction is somewhat more complex than the typical epidemiologic study that chiefly (or exclusively) addresses exposure–disease associations. The reason is that, in a gene/environment interaction study, we need to be concerned with extraneous risk factors that may be correlated with environmental exposures, genetic factors, or with both.

Studies of gene–environment interaction can be confounded by other, unmeasured genetic variants. This is often due to linkage disequilibrium of the two variants, a coexistence of two variants in the population more often than would be expected by chance, usually in two alleles that are in close proximity on the DNA strand. Thus, any observed association between one variant and disease may actually result from confounding by its linkage disequilibrium with another variant. Statistical methods aimed at analyzing haplotypes (allele combinations on the same or related genes) instead of single genotypes may overcome some of this difficulty, since haplotypes may be more specific markers of common inherited genetic variation than any single polymorphism (Johnson et al. 2001).

A particularly vexing type of confounder in gene–environment studies is what is known as *population stratification* which is due to imbalances in ethnicity between cases and controls that result from genetic mixing among populations. For example, we might decide to control for race and ethnicity in a case–control study, and accordingly, we could stratify cases and controls into various categories of non-Hispanic white, Hispanic white, nonwhite Hispanic, African-American, Pacific Islander, Asian, and "other". Within any one of these groups, there is likely to be considerable ethnic heterogeneity that may be associated with both the genetic risk factors and the index disease under study. How to control for population stratification is a matter of some dispute in ecogenetics (Wacholder et al. 2000; Cardon and Palmer 2003). Standard techniques, such as matching on or statistically adjusting for self-reported ethnicity, may not be adequate when there is severe population admixture. The crux of the problem is the ambiguity of defining ethnicity. What subjects report as their ethnicity, their parents' ethnicities, and so forth will not completely account for genetic heterogeneity. However, the alternative of characterizing ethnicity from intensive genomic screening for racial markers is not at present a practical option.

It should be appreciated that no epidemiologic study is ever completely free from confounding (or, for that matter, all other biases). It is simply not possible to identify and measure, in either the design or analysis of a study, every imaginable confounder. The importance of a confounder on causal inference will depend on how strong a risk factor the presumed confounder is, based on prior knowledge or on empirical findings from one's own study, and the degree to which the confounder is associated with the exposure of interest. For instance, linkage disequilibrium between measured and unmeasured (or unknown) genetic alleles can pose a serious confounding bias if the unmeasured allele is a strong disease risk factor. Small levels of confounding can be tolerated as inevitable consequences of observational research, although their influence will be more damaging if the anticipated effects of the exposures and genes of interest are modest.

4.4.4. Choice of Controls in Population-Based Case–Control Studies

There is no ideal choice for control selection in gene–environment case–control studies. The most common control selection strategy is to choose unaffected (by the disease of interest) persons who presumably come from the same study base as the cases. Matching the groups on relevant factors that might relate to disease risk and exposure (i.e., potential confounders) is a standard strategy to balance the case and control groups. These factors are typically age, sex, race, and ethnicity. As mentioned earlier, study validity can be improved to the extent that we can verify that the case and control groups emanate from the same source population. Except in certain instances, controls are ordinarily selected from among persons unrelated to the cases.

There are some alternative control selection choices to unaffected and unrelated. One is to select spouse controls. Spouses have the advantage of being relatively convenient study participants, as compared with controls selected from the community at large. Spouses (in most instances) are not blood relatives, and therefore can provide valuable comparisons for genetic factors except, of course, those that are X-linked. The disadvantage of spouse controls is that environmental history, especially workplace exposures, will be decidedly different between spouses in most instances, thus nullifying the ability to study certain risk factors. Also, spouses typically share the same home and lifestyle-related environments during adulthood, which will diminish the ability to distinguish differences in some relevant exposures.

Other choices of controls have been devised specifically for gene–environment studies [see Umbach and Weinberg (2000), Botto and Khoury (2004), and Thomas (2004) for reviews]. These include blood relatives, especially siblings and parents. Selection of unaffected sibling controls, other than identical twins, can facilitate the study of certain genes while controlling for other genetic traits. However, this advantage may be offset by limitations that result from potential overmatching on genetic factors and from cases' and controls' having shared environments that preclude studies of early-life exposures. On the other hand, selection of unaffected nontwin siblings or fraternal twins as controls can in theory provide a good, but not exact, match on genetic traits, and in most instances would allow for study of associations with postchildhood exposures and their interactions with genotypes. A difficulty can arise if the unaffected sibling is younger than the index case, and has thus not had an equivalent opportunity to manifest the disease under study. This approach is always best when controls are as old as or older than cases.

Another version of the family-based control approach for genetic case–control studies is to construct expected control genotypes from the case parents' genotypes, sometimes known as the "case–parent triad" design (Witte et al. 1999). Limited availability of living parents can be a major logistical constraint for studies of adult-onset diseases, but this design may have good utility for studies of childhood diseases. This approach mitigates the problem of selecting as controls younger, unaffected siblings. The case–parent triad design is valuable for determining genetic risk factors, particularly for diseases in which both the cases' and their parents' genes may be governing risk. One drawback is that the main effects of exposure cannot be estimated, for lack of a valid reference group. However, gene–environment interactions can be evaluated indirectly by comparing an association with a genetic variant between exposure groups. Such an approach was used in an investigation of effects of variants in the methylenetetrahydrofolate reductase gene on the risk of orofacial clefts in children (Jugessur et al. 2003). In this study, children carrying the C677T variant allele had a twofold increased risk of cleft palate only (without cleft lip), and the magnitude of increased risk was higher (RR = 4.3) in children whose mothers reported using folic acid than among children of nonusers (RR = 1.4). The direction of this association was unexpected and may be due to chance. Nonetheless, this example illustrates approaches to investigating gene–environment interaction in case–parent triad studies.

Yet another alternative method is the "case-only" design, which has been developed as a relatively simple approach to investigate gene–environment interactions (Khoury and Flanders 1996). The case-only design permits an evaluation of gene–environment interaction from the estimate of the association (odds ratio) between the exposure factor and genotype among cases. The validity of this estimate as an indicator of interaction requires that the two factors, environment and genotype, are independent (uncorrelated) among persons without the disease of interest in the source population (from which controls would be selected). Because this design does not include controls, the individual effects of exposure and genotype on disease risk cannot be determined, even though their relative interaction can be estimated. An example where the case-only method was applied was the Infante-Rivard et al. (2002) case–control study of exposure to byproducts of the drinking water disinfectant trihalomethane (THM), genetic polymorphisms, and childhood acute lymphoblastic leukemia. The analysis was limited to a subset of 170 of the total 491 cases for whom THM data and DNA samples were available. A very strong association (odds ratio 9.7) was observed for the relative interaction between prenatal THM exposure above the 75th percentile and a variant of the cytochrome P4502E1 (CYP2E1) gene. This result was in sharp contrast to the weak relation with THM found previously in the entire study of 491 cases and 491 controls (Infante-Rivard et al. 2001), suggesting that that the CYP2E1 gene variant may be a predisposing factor to a THM-associated leukemogenic effect. The chief advantage of the case-only approach is that it is a relatively cost-efficient way to examine gene–environment interactions since it requires data only for cases. This circumvents the difficulties in selecting and recruiting a suitable control group, and can take advantage of situations in which biologic materials exist only for cases (e.g., biopsy specimens). Findings from case-only studies may then suggest future, larger-scale research initiatives. However, enthusiasm for this design is limited by the inability to estimate the independent effects of exposures and genes, which are generally of paramount interest in epidemiologic studies (assessing interactions is often a secondary research goal). Also, verifying the assumption of independence between the candidate genes and exposures requires controls, and thus is precluded.

4.5. SUMMARY AND CONCLUSIONS

The decisions as to which diseases, exposures, and genes deserve study will inevitably dictate the choice of epidemiologic study design. As we have described in this chapter, certain study designs are more appropriate than others for investigating associations between environmental exposures and disease. The putative gold-standard design, prospective cohort study, provides the clearest evidence for causality, but is generally feasible only for relatively short-term health outcomes. Alternative approaches, including case–control and cross-sectional designs, are more suited to epidemiologic investigations of chronic health outcomes.

Irrespective of the study design, accurate and valid exposure assessment plays a critical role in determining the informativeness and validity of any epidemiologic investigation that attempts to address gene–environment interactions. This is not to say that epidemiologic studies of the main effects of candidate genes, and of interactions among genes on disease risk, are not valuable. Indeed, such studies can offer valuable research guidance, and may in fact reveal causal genes or genetic pathways when genetic factors predominate disease etiology. In addition, because certain gene products have multiple substrates, thus interacting with many exogenous and endogenous chemicals, the main effect of a polymorphism can be useful in reflecting the average effect of the genotype in the context of multiple substrates, indicating the importance of a certain biologic pathway on disease risk. Since most complex diseases are influenced by multiple environmental and genetic factors, careful planning will be necessary to obtain relevant data from numerous sources. It is also fair to say that in most instances the expected contributions to risk of complex diseases of any one environmental factor or genetic trait will be modest. Either the relative risk associated with a given exposure or genetic trait will be small (e.g., less than 1.5), or the frequency of the exposure or trait in the population will be low (e.g., less than 10%). Consequently, individual epidemiologic studies will need to have very large sample sizes, particularly when gene–environment interactive effects, even strong associations, are of primary concern (Garcia-Closas and Lubin 1999). Larger epidemiologic studies are clearly preferable to smaller ones, all things being equal, although resource limitations will constrain the size and scope of any individual study.

The role of chance in producing spurious findings always needs to be considered in epidemiologic research. Although false-positive and false-negative findings are by no means unique to epidemiologic studies of gene–environment interaction, their recognition has received considerable attention as technological advances have increasingly facilitated analyses of multiple candidate genes (Colhoun et al. 2003). Independent replication of findings is clearly essential to confirm genetic associations (Ioannidis et al. 2001), although even replication cannot ensure freedom from bias, such as linkage disequilibrium.

Systematic literature reviews and metaanalyses in which study findings from multiple similar studies are combined and summarized are often undertaken to clarify epidemiologic associations. An alternative approach is to pool the actual data from multiple studies and to perform a combined analysis of the aggregated datasets. Data pooling sometimes involves new data collection, such as additional genotyping.

Throughout this chapter, we have limited our discussion of genetic risk factors to genotypes that are readily determined from any tissues containing DNA. (We have also made the implicit assumption that nuclear DNA is of most interest, although mitochondrial genes may also be important in some diseases.) Other genomic techniques, such as gene expression array studies, have promise for providing further insights complementary to what can be derived from genotype data. Data collection for gene expression investigations imposes some complexities, particularly in obtaining tissue specimens from affected organs. Even when such tissue can be obtained, interpretation of gene expression data can be much more complex than interpretation of genotype data because of the possible distorting influences of recent exposures and disease state on expression profiles. By way of illustration, consider a study of gene products indicative of oxidative stress in

Parkinson's disease using brain samples from persons who died from Parkinson's disease and persons who died from other, presumably unrelated conditions. A feasible study would compare gene expression profiles in samples from persons with Parkinson's disease of regions of the brain affected by the disease (e.g., substantia nigra, striatum) and unaffected regions (e.g., cortex) from the same group, as well as comparable data from brains of persons who did not die from Parkinson's disease. It should be evident that what is measured in dead brain may not necessarily reflect typical conditions before disease onset or even during the course of disease. Also, the choice of controls (other deaths) may be based on convenience rather than on prior epidemiologic design considerations. Adherence to the basic epidemiologic principles, in terms of subject selection, exposure assessment, and avoidance of bias, will be necessary even in the most specialized circumstances.

5

Statistical Issues in Ecogenetic Studies

Stephanie A. Monks
University of Washington, Seattle, WA

5.1. INTRODUCTION

Since the mid-1980s, numerous genes have been identified for phenotypic variation in humans, such as the genes for cystic fibrosis and Huntington disease. In fact, as of April 2004, 1519 phenotypes are listed by the *Online Mendelian Inheritance in Man* as phenotypes with known molecular basis (OMIM 2004). Most of these phenotypes are disease-related, due to genetic variation in a single gene and represent simple modes of inheritance. Despite such apparent success, finding genes for diseases, and their associated risk traits, that are of public health interest has proved difficult.

There are several reasons for this lack of success. First, although it is well known that genetic variation produces differences in how an individual responds to the environment, the resulting phenotype is due, only in part, to genetic variation. Phenotypes of interest to public health, such as asthma, diabetes, and obesity, are due to the interactions among many genetic and environmental factors; that is, there are likely to be gene–gene interactions as well as gene–environment interactions. For disease phenotypes, this dictates that some individuals will carry the genetic variant for disease without exhibiting characteristics of the diseased state. This phenomenon is due to *reduced penetrance*. Others will be diseased without the presence of the genetic variant, that is, representing a *phenocopy*. In addition, some individuals may be diseased because of the *interaction* of an environmental factor and a genetic variant. Others will be diseased as a result of genetic

Gene-Environment Interactions: Fundamentals of Ecogenetics, edited by Lucio G. Costa and David L. Eaton

variation in a different gene; that is, there is likely to be *genetic heterogeneity* for the disease and its associated risk factors. The complexity introduced by reduced penetrance, phenocopies, gene–gene interaction, gene–environment interaction, and genetic heterogeneity has provided a challenge for scientists seeking to understand determinants of human health.

Fortunately, since the mid-1990s there has been an explosion of technology allowing scientists to study and understand phenotypic variation on a level never before possible. For instance, technology now allows for obtaining an individual's genetic makeup for over a million genetic locations and measuring expression for over 30,000 genes. These data are in addition to the increasingly detailed characterizations for the phenotype of interest and potentially associated environmental factors. With such a wealth of data, we are now challenged with not only deciphering this information but also sifting through to obtain what is relevant to our phenotype. Inherent in this process is the need to utilize appropriate experimental design, and the associated statistical methods, to answer our questions of interest. The basic statistics for genetic studies are discussed below. This is followed by information on the various stages of a statistical genetics study and includes

1. Aggregation and segregation studies
2. Family-based studies for gene mapping
3. Population-based studies for gene mapping

Throughout, the goal is to provide the best analysis so that the interactions can be understood between genetics, human health, and the environment.

5.2. BASIC CONCEPTS IN STATISTICS AND GENETICS

It is well established that there are wide differences in allele frequency distributions among racial populations for numerous genetic polymorphisms. Since disease prevalence can also vary by race, this creates the potential for *confounding* in studies aimed at associating genetic variation with a particular phenotype. Thus, the relationship between a genetic variant and disease risk can be distorted because of the differences in allele frequency by race, which are also related to differences in disease risk. This becomes even more problematic in genetic studies since racial background for numerous individuals is difficult to quantify. It is possible to detect confounding by determining whether the genetic variant under consideration is in *Hardy–Weinberg equilibrium* (HWE). Consider a diallelic locus with alleles

B and b, where the frequency of allele B (b) is represented by p_B (p_b). This locus is in HWE if the genotype frequencies in the population are equal to p_B^2, $2p_Bp_b$ and p_b^2 for genotypes BB, Bb, and bb, respectively. Although this doesn't provide a formal test of confounding, it can indicate the presence of potential confounders. In addition, errors due to genotyping assays often exhibit themselves through departure from Hardy–Weinberg proportions. It is noted that allele frequencies can always be estimated from genotype counts or frequencies:

$$\hat{p}_B = \frac{2N_{BB} + N_{Bb}}{2N} = \hat{p}_{BB} + \frac{\hat{p}_{Bb}}{2} \tag{5.1}$$

$$\hat{p}_b = \frac{2N_{bb} + N_{Bb}}{2N} = \hat{p}_{bb} + \frac{\hat{p}_{Bb}}{2} \tag{5.2}$$

where N_G are the number of genotypes of type G in a sample of N individuals and $\hat{p}_G = N_G/N$. However, genotype frequencies can be validly estimated from allele frequencies only when the population is in HWE: $\hat{p}_{BB} = \hat{p}_B^2$, $\hat{p}_{Bb} = 2\hat{p}_B\hat{p}_b$ and $\hat{p}_{bb} = \hat{p}_b^2$. Without the assumption of HWE, genotype frequencies must be estimated by direct counting of the number of genotypes in the sample as in Equations (5.1) and (5.2).

Most statistical genetics tests rely on the assumption of HWE, thus making it important to confirm before proceeding to further analysis. The two most utilized tests of HWE are a goodness-of-fit chi-square test

$$\chi^2 = \frac{[N_{BB} - (N \times \hat{p}_B^2)]}{N \times \hat{p}_B^2} + \frac{[N_{Bb} - (N \times 2\hat{p}_B\hat{p}_b)]}{N \times 2\hat{p}_B\hat{p}_b} + \frac{[N_{bb} - (N \times \hat{p}_b^2)]}{N \times \hat{p}_b^2}$$

and a likelihood ratio test:

$$\Lambda = -2\ln\left[\frac{(\hat{p}_B^2)^{N_{BB}}(2\hat{p}_B\hat{p}_b)^{N_{Bb}}(\hat{p}_b^2)^{N_{bb}}}{(\hat{p}_{BB})^{N_{BB}}(\hat{p}_{Bb})^{N_{Bb}}(\hat{p}_{bb})^{N_{bb}}}\right]$$

Note that the numerator contains estimates for genotype frequencies under the assumption of HWE while the denominator contains estimates for genotype frequencies without this assumption, through direct count of the number of observed genotypes. Each statistic provides a statistical test of the null hypothesis of HWE versus the alternative hypothesis of departure from HWE. Departure from HWE is also referred to as *Hardy–Weinberg disequilibrium* (HWD). For a diallelic locus, statistical significance for both tests can be assessed with the chi-squared distribution with one degree of freedom; however, for small samples or markers with rare alleles, an exact test should be used in order to ensure appropriate type I error for the test. Weir (1996) provides details on both types of tests, including extensions to multiallelic markers, as well as for assessing statistical significance with the chi-squared distribution or an exact test.

When considering more than one genetic variant, there are generally two phenomena of interest in statistical genetics studies. Both of these relate to deviations from Mendel's second law, which states that "During gamete formation, the segregation of one gene pair is independent of other gene pairs." There are two reasons why loci would deviate from Mendel's second law: linkage and linkage disequilibrium. When two loci follow Mendel's second law, recombinants and nonrecombinants are produced with equal frequency. When loci are physically located close to one another on a chromosome, there is a deviation from this relationship, termed *linkage*, which is characterized by the recombination fraction. Generally, the *recombination fraction* is denoted by θ where $\theta = \Pr(\text{recombinant gamete})$. When loci are unlinked, $\theta = \frac{1}{2}$. When loci are completely linked, $\theta = 0$. For θ in between 0 and $\frac{1}{2}$, the loci are said to be linked or in genetic linkage. Family-based methods for detecting genetic linkage between a genetic marker and a trait are discussed below.

We have discussed the possibility of dependence between alleles within a locus (HWD). Regardless of linkage, dependence can exist for alleles from separate loci as well. We will begin by considering gamete frequencies for two diallelic loci where locus 1 has alleles A/a and locus 2 has alleles B/b. If the allele at locus 1 occurs independently of the allele at locus 2, then we would expect the probability of a *haplotype* with alleles A and B to be the product of the frequencies for A and B. When alleles occur together more or less often than expected by chance, the alleles are said to be in *linkage disequilibrium*. Dependence can be caused by many factors such as natural selection, founder effects, migration, mutation, and random drift. A measure of this dependence is the *linkage disequilibrium coefficient* or the coefficient of gametic phase disequilibrium, which is defined to be

$$D_{\mathrm{AB}} = \Pr(\mathrm{AB}) - \Pr(\mathrm{A})\Pr(\mathrm{B})$$

In other words, the linkage disequilibrium coefficient measures the deviation of the haplotype frequency from what is expected by the random pairing of alleles at the two loci. In most studies, the linkage disequilibrium coefficient cannot be directly estimated since the phase cannot be determined for double heterozygotes. If it is reasonable to assume random mating, an unbiased estimator of D_{AB} can be found by (Lynch and Walsh 1998)

$$\hat{D}_{AB} = \frac{N}{N-1}\left[\frac{4N_{AABB} + 2(N_{AABb} + N_{AaBB}) + N_{AaBb}}{2N} - 2\hat{p}_A\hat{p}_B\right]$$

where

$$\hat{\text{Var}}(\hat{D}_{AB}) = \frac{\hat{p}_A\hat{p}_a\hat{p}_B\hat{p}_b}{N-1} + \frac{(2\hat{p}_A - 1)(2\hat{p}_B - 1)\hat{D}_{AB}}{2N} + \frac{\hat{D}_{AB}^2}{N(N-1)}$$

Here, N_G represents the number of individuals with the two locus genotype G. Alternatively, the expectation maximization algorithm can be used to estimate the haplotype frequency, Pr(AB), which can then be used to estimate D_{AB} (see Section 5.5, below).

5.3. AGGREGATION AND SEGREGATION STUDIES

Aggregation and segregation studies are generally the first step when studying genetic influences of a trait. While it is expected that genetic variation has an effect on most, if not all, traits, aggregation and segregation studies seek to determine the extent to which a trait is affected by genetic variation. In particular, *aggregation studies* determine whether there is familial aggregation of the trait. For instance, are relatives of diseased individuals more likely to be diseased than the general population? Is the clustering of disease in families different from what is expected on the basis of the prevalence in the general population? *Segregation studies* moves beyond aggregation of disease and seek to discriminate the factors responsible for familial aggregation. For instance, is the aggregation due to environmental, cultural, or genetic factors? What proportion of the trait variation is due to genetic factors? What mode of inheritance best represents the genetic factors? Does there appear to be genetic heterogeneity? The data required for such studies consist of the pedigree structure, measures of the trait of interest, and any environmental variables that might influence the trait.

Standard statistical measures are used in aggregation studies with a measure estimated for each type of relative of a diseased individual. In particular, the *relative risk ratio* represents the increased risk of disease given an individual is related to a diseased person. This is often denoted by λ_R and is equal to

$$\lambda_R = \frac{\text{Pr(relative of type R is diseased} \mid \text{diseased individual)}}{\text{Pr(diseased individual)}}$$

The numerator is often called the *relative recurrence risk*. A general framework has been developed for computing the relative recurrence risk of disease as a function of the underlying genetic parameters and the relative type. This framework has proved useful for studying patterns of risk across relatives as an indicator for the underlying mode of inheritance. In particular, there are two different modes of inheritance, additive and multiplicative, which provide a simple relationship among the relative risk ratios of first-, second-, and third-degree relatives. For an additive model, the Pr(disease given genotype *i* at locus 1 and genotype *j* at locus 2) = Pr(disease given genotype *i* at locus 1) + Pr(disease given genotype *j* at locus 2). Risch (1990) showed, for a model with additive risk, the adjusted risk ratio, $\lambda_R - 1$, decays by $\frac{1}{2}$ for each degree of relationship. For example, the grandchild of a diseased individual will have an adjusted risk ratio that is $\frac{1}{2}$ of the adjusted risk ratio for the child of a diseased individual. It is of note that this relationship provides information on how multiple loci interact; however, it does not provide information regarding the number of loci influencing the trait. For a multiplicative model, Pr(disease given genotype *i* at locus 1 and genotype *j* at locus 2) = Pr(disease given genotype *i* at locus 1) × Pr(disease given genotype *j* at locus 2). The decay of risk per degree of relationship is much more rapid than with the additive model. Risch (1990) provided an example for schizophrenia where observed relative risk ratios were compared across many types of relatives with what is expected from seven different disease models. The example showed a clear departure from an additive mode of inheritance for risk.

Once there is evidence for a genetic component for a trait, a logical next step is to determine what this genetic component looks like. Our discussion of relative risk ratios allowed us to informally evaluate what models were more consistent with an observed pattern of relative risk ratios. However, it is possible to use maximum-likelihood techniques to test hypotheses representing different sources of genetic influence and then provide estimates for the corresponding model parameters. For instance, is there a single major gene (oligogene)? Are there many genes of small effect that influence our trait (polygenes)? Could there be two major genes that are interacting to cause variation in our trait? *Segregation analysis* can be used to answer these questions. In particular, likelihood models are used to provide estimates of

- Single-locus parameters consisting of allele frequencies, or genotype frequencies if HWE is not assumed
- Polygenic background, representing the cumulative effect on the trait due to many genes

- Transmission parameters that allow testing of consistency with Mendelian segregation for major loci
- Penetrance parameters, for disease traits, which correspond to the probability an individual develops the disease for each disease genotype
- Mean parameters, for quantitative traits, representing a shift of the mean trait value for a given major locus genotype

For hypotheses that are nested within each other, likelihood ratio tests can be used to assess statistical significance. For instance, a model that includes a single locus and a polygenic effect on disease risk can be compared to a model with only a single-locus effect. This provides evidence, or the lack there of, for additional genetic influences beyond the single locus. Information gained from segregation studies can be used to advise further studies of genetic linkage and/or association in terms of sample size planning, minimum detectable risk, and power. In addition, some gene-mapping methods require values for the parameters specifying the mode of inheritance.

5.4. FAMILY-BASED METHODS IN GENE MAPPING

Once aggregation and/or segregation studies have established a genetic component for the phenotype of interest, we are next confronted with finding the underlying genes. Two general mechanisms could be used at this stage of study: designs and methods based on detection of (1) linkage and (2) association. There is still much debate regarding which of these should come first, and the answer will certainly be context-dependent. We will discuss studies aimed at detection of linkage first and then follow with studies utilizing linkage disequilibrium between trait and marker alleles. In addition to the data needed for aggregation/segregation studies, linkage analysis requires genetic marker data on families.

In its simplest form, *linkage analysis* consists of modeling the number of recombination events between a marker and a trait locus. If we are able to directly count the number of nonrecombinants, N, and recombinants, R. Then we would have a likelihood of

$$L(\theta) = \Pr(\text{data} \mid \theta) = \binom{N+R}{R}(1-\theta)^N \theta^R$$

The maximum-likelihood estimate for the recombination fraction, θ, is $R/(N + R)$. Likelihood ratio tests can be used to assess linkage between the marker and trait locus.

Morton (1955) suggested the use of a sequential testing procedure for testing $H_0: \theta = \frac{1}{2}$ versus $H_1: \theta = \theta_1$ for some fixed value θ_1 with $\theta_1 < \frac{1}{2}$. On the basis of Wald's sequential probability ratio test, he proposed sequentially collecting families until LOD > 3 or LOD < −2 where

$$\text{LOD} = \log_{10}\left(\frac{L(\theta = \theta_1 \mid \text{data})}{L\left(\theta = \frac{1}{2} \mid \text{data}\right)}\right)$$

Here, $L(\theta = \theta_1 \mid \text{data})$ is the likelihood for the data under the alternative hypothesis, $\theta = \theta_1$ and $L(\theta = \frac{1}{2} \mid \text{data})$ is the likelihood for the data under the null hypothesis, $\theta = \frac{1}{2}$. If the LOD score were greater than 3, then no additional families would have been collected, and it would be concluded that there was significant support for linkage between the marker and trait locus. If the LOD score were less than −2, then it would have been concluded that there was not significant evidence for linkage. Under the assumption of a marker and disease locus with Mendelian segregation, each with completely known mode of inheritance, these cutoffs corresponds to a power of 0.99 and a significance level of 0.001 for the fixed alternative.

Of course, we are seldom able to determine the number of recombinants with absolute certainty. Instead, we must

1. Determine the genotypes at the marker and disease locus, along with phase information, when possible.
2. List the different scenarios and their probability of being observed in the population when it is not possible to determine these with certainty.
3. Count the number of recombinants for each scenario.

The likelihood is then a weighted sum of the likelihoods conditional on the different scenarios. Penetrance parameters can be included in this likelihood, allowing for estimation of the underlying mode of inheritance.

The arguments for justifying a LOD cutoff of 3 came with many assumptions: (1) the only unknown parameter was the recombination fraction; (2) testing is performed sequentially, that is, the sample size is not fixed; (3) there is a gene of interest somewhere in the genome; (4) the only factor perturbing the relationship between the two Mendelian loci is linkage; and (5) testing involves a simple hypothesis of H_1: $\theta = \theta_1$. For complex traits, none of these assumptions will generally hold. Let's consider a general statistic, Z, which has been derived for testing linkage between a position in the genome and an underlying trait locus. We'll assume that, under the null hypothesis, Z is asymptotically standard normal. The genomewide significance level is then the probability that the statistic Z exceeds some threshold T. Two key points will help us determine an appropriate Z cutoff for assessing statistical significance: (1) for nonlinked positions in the genome, the statistics will be uncorrelated; and (2) for linked positions, the statistics will be correlated. Further, for a given statistic, this correlation can be computed. Lander and Kruglyak (1995) derived the *genomewide significance* for the statistic Z exceeding a threshold T to be approximately $\Pr(|Z| > T) = 1 - \exp\{-(C + 2\rho GT^2)\alpha_T\} \approx (C + 2\rho GT^2)\alpha_T$ when this quantity is small. Here C is the number of chromosomes, G is the genome length in Morgans, α_T is the pointwise significance level for exceeding T, and ρ is a parameter describing the correlation of the statistic across the genome. For LOD score analysis in humans, a LOD score cutoff of 3.3 provides an adjustment for multiple testing incurred by a genomewide linkage scan at a level of 0.05.

To this point, we have discussed *"model-based" linkage analyses*, namely, methods of linkage analysis that require parameterization of the underlying genetic model. For a disease trait, these methods require the parameterization of the frequency of a disease allele and the associated penetrances for a diallelic disease locus. Clearly, this is not an accurate characterization of the underlying etiology of a complex trait. The true model for a complex trait is likely not caused by a single gene, but rather depends on multiple genetic and environmental factors, many of which are interacting to produce phenotypic variation. One approach to this problem is to abandon the "model-based" or "parametric" framework for linkage analysis and focus on the sharing of marker alleles by sets of relatives and how this relates to similarities or dissimilarities in their phenotypes.

Two alleles are *identical by descent* (IBD) if they have descended from the same allele in a common ancestor. We are interested in studying how IBD sharing at genetic markers relates to phenotypic sharing. IBD values will be independent for markers that are not linked but positively correlated for linked markers with the correlation increasing for markers increasingly close.

For instance, a common design in gene-mapping studies is to sample sibling pairs, where both siblings are diseased. The idea behind affected sibling pair designs is to determine locations in the genome where the IBD sharing between siblings is consistent with a region in linkage to a complex trait locus. If a marker is unlinked, then the siblings will share one allele IBD in expectation. One test of linkage, the means test, uses this information to test for linkage by comparing the average number of alleles shared IBD in the sample with the expectation of one (Blackwelder and Elston 1985). For markers linked to a trait locus, the average number of alleles shared IBD should exceed one.

The same ideas can be applied to quantitative traits. Haseman–Elston regression is based on comparing allele sharing to similarity of a quantitative trait between relatives (Haseman and Elston 1972). For a marker that is linked to a *quantitative trait locus* (QTL), pairs of relatives that share marker alleles IBD will share QTL alleles IBD and therefore the quantitative traits for these relatives should be similar; that is, the squared difference, $Y_j = (X_{1j} - X_{2j})^2$, between trait values for two relatives, X_{1j} and X_{2j}, is expected to decrease as they share more marker alleles IBD. In fact, for any relationship between two individuals, we have $E(Y_j \mid \pi_{j,\mathrm{m}}) = \alpha + \beta\pi_{j,\mathrm{m}}$ where $\pi_{j,\mathrm{m}}$ is the proportion of alleles shared IBD for the two individuals at the marker. The coefficients α and β depend on the type of relative pair; however, in all cases a test of linkage is obtained by performing a one-sided test of H_0: $\beta = 0$ versus H_1: $\beta < 0$. Standard statistical software packages can be used to conduct this test.

We have discussed the basics of linkage analysis for both (trait) model-based and non-model-based methods. Several other issues that affect the power of a linkage study are discussed below. While we don't have complete control over the types of pedigrees available, we will generally have some flexibility with regard to pedigree structure. It is important to consider which relatives provide more information for linkage than others. Chapter 5 of Ott (1999) provides guidance for simple family structures and model-based methods. New methods in statistical genetics are generally published with information on power. This provides some data regarding power for a proposed study; however, each genetic study is unique. In addition, it is rare that all families will have the same structure. Simulation studies provide a flexible framework for evaluating the power of various pedigree structures and methods. Section 9.7 of Ott (1999) provides a discussion of practical issues for conducting simulations for genetic studies. For disease studies, sampling must be done through diseased individuals in order to obtain enough data to assess segregation of disease within families. This ascertainment must be accounted for when generalizing estimates of parameters from the study sample to the overall population. For studies of quantitative traits, it may be of interest to study natural variation in the population. The power of any segregation or linkage study is greatly affected by the homogeneity of the phenotype under study. Unfortunately, we seldom have a good handle on how to best characterize complex traits, so that additional phenotypic characterization could lead to better defined subgroups of disease that yield greater power than would analyzing the overall heterogeneous sample. The power for any method of linkage analysis will depend on the informativeness of the marker or marker set; i.e., in order to test for linkage, it is necessary to see the segregation in the genetic region under study. In addition, markers need to be spaced close enough to detect a trait locus.

5.5. POPULATION-BASED METHODS IN GENE MAPPING

While linkage methods allow for the identification of an area of a chromosome that is harboring a disease locus or a quantitative trait locus, the scope of resolution is typically limited by the number of informative meioses in the data. For example, Hastbacka et al. (1992) was able to localize a gene for diastrophic dysplasia to within a 1.6-cM (centimorgan) region of chromosome 5 using linkage methods. This is too large of a region for identifying the specific causative gene. *Population-based association methods* rely on the association between a marker allele and a disease susceptibility allele at the population level. This association is dependent on the number of meioses that have occurred since the initial disease mutation was introduced to the population. With such a large number of meioses, association methods are able to provide fine localization for susceptibility loci. Hastbacka et al. (1992) followed up the significant linkage to chromosome 5 by genotyping additional markers in the linked region and conducting association analyses. This allowed for localization of the causative variant to within 60 kbp. As with linkage methods, the data required for such studies consist of marker and phenotype data for a sample of individuals. While association tests can be conducted in family-based and/or population-based designs, we will discuss only population-based designs below.

Any of the standard statistical tests can be used to determine whether there is any association between a marker and trait. For dichotomous traits, this includes chi-square tests of independence and logistic regression. For continuous traits, this includes analysis of variance (ANOVA) and linear regression. A more interesting issue is how we deal with testing association between a set of tightly linked markers and a phenotype. In other words, what if we are interested in testing whether a specific *haplotype*, or any of the possible haplotypes in a gene, are associated with a phenotype?

Suppose that we have genotypes for multiple loci such that the haplotypes are not known with certainty. In the discussion below, this is referred to as the *marker phenotype*. We're interested in determining the particular combination of two multilocus haplotypes into what will be referred to as the *multilocus genotype*. In other words, the phenotype provides the genotypes across the markers but does not provide phase information whereas the genotype describes the two haplotypes that make up our marker phenotype. Generally, we will not be able to explicitly determine an individual's multilocus genotype; however, we can use likelihood methods to estimate the probability that the individual has each of the different possible multilocus genotypes that are consistent with their marker phenotype. Excoffier and Slatkin (1995) proposed using the expectation–maximization (EM) algorithm to obtain estimates of haplotype frequencies. Here, the likelihood L is defined as a function of the phenotype frequencies P_i:

$$L(P_1,\ldots,P_m) = P(\text{sample}|P_1,\ldots,P_m) = \frac{n!}{n_1!n_2!\cdots n_m!} \times P_1^{n1} \times P_2^{n2} \times\cdots\times P_m^{nm}$$

where there are m different observed phenotypes with counts $n_1, n_2, \ldots, n_m$. If we assume random mating at the haplotype level, then the probability of the jth phenotype is the sum of the probabilities of each of the possible c_j multilocus genotypes. The c_j genotypes are those consistent with the observed marker phenotype

$$P_j = \sum_{i=1}^{c_j} P(\text{genotype } i) = \sum_{i=1}^{c_j} P(h_k h_l)$$

where $P(h_k h_l) = p_k^2$ if $k = l$ and $P(h_k h_l) = 2p_k p_l$ if $k \neq l$, where p_k is the frequency of haplotype k represented by h_k above; that is, the genotype probabilities assume that the haplotypes are in Hardy–Weinberg equilibrium. The EM algorithm starts with an initial set of guesses for the haplotype frequencies and proceeds as follows. For each iteration of the algorithm, the current estimates of the haplotype frequencies are used to estimate the genotype frequencies. This splits the observed phenotypes into the unobserved genotypic classes. Next, the new genotype frequencies are used to estimate haplotype frequencies. These two steps are continued until changes in the haplotype frequencies in consecutive iterations are small.

If marker genotypes are available for a case–control study, then it is of interest to determine whether the haplotype frequencies differ between cases and controls. Thus, the following two hypotheses are compared:

H_0: the haplotype frequency distribution, $(p_1, \ldots, p_n)$, is equal for cases and controls.

H_1: the haplotype frequency distribution for cases $(q_1, \ldots, q_n)$ differs from the haplotype frequency distribution for the controls $(r_1, \ldots, r_n)$.

In other words, the parameters under the null hypothesis represent n haplotype frequencies when there is no difference between cases and controls. Under the alternative hypothesis, two different sets of n haplotype frequency estimates are needed: one set for cases and one set for controls. A likelihood ratio test can be used to compare these hypotheses:

$$\Lambda = -2\ln\left(\frac{L(p_1,\ldots,p_n\,|\,\text{data})}{L(q_1,\ldots,q_n\,|\,\text{data})L(r_1,\ldots,r_n\,|\,\text{data})}\right)$$

The test of association based on a comparison of the haplotype frequency distributions does not require one to assume a model for relating haplotypes to the phenotype. This will allow for a test that is robust to model misspecification and can determine which haplotype(s) is (are) responsible for the deviation between the haplotype frequency distributions. However, it requires one to maximize three separate likelihoods, which is computationally more expensive than other tests and does not provide information on what type of effect each haplotype has on the trait.

Schaid et al. (2002) provided a flexible group of methods, and associated software, that allows for tests of association for dichotomous, quantitative, and categorical traits. The methods entail constructing a likelihood assuming that an individual's multilocus genotype is known. The uncertainty in the individual's genotype is taken into account by averaging over all possible genotypes that are consistent with an individual's marker phenotype. The association tests are based on a score test for this likelihood. While this does not provide information regarding the haplotype effect sizes, it requires only one maximization step and so has less computational cost. This can become an issue if a large number of tests are being conducted.

Epstein and Satten (2003) proposed a retrospective likelihood for estimating the effects of haplotypes on disease risk in case–control studies. The model is general so that various relationships can be modeled for the effect of haplotype on disease risk. These include modeling the effect of a haplotype on the log(odds of disease) as additive, dominant, recessive, or general (no imposed structure between genotype risks). For instance, the additive model is commonly used and relates the log(odds of disease) to the number of haplotypes that an individual possesses. For haplotype i, this provides the following model:

$$\log\left[\frac{\Pr(\text{diseased}\,|\,x\text{ copies of haplotype }i)}{\Pr(\text{not diseased}\,|\,x\text{ copies of haplotype }i)}\right] = \alpha + \beta_i x$$

The additive model, while generally an oversimplification of the true disease etiology, has proved to be a robust model for association tests. In fact, Pfeiffer and Gail (2003) found that for many scenarios the additive model is more efficient than using dominant scores when the true underlying mode of inheritance is dominant or using recessive scores when the true underlying mode of inheritance is recessive. This is because the scores are applied to a marker in linkage disequilibrium with the disease locus rather than the disease locus itself.

When the overall sample size is small and/or the haplotype probabilities are small, p values based on an asymptotic distribution can be inaccurate. A common procedure that is used is to estimate p values in this situation relies on *permutations* of the original dataset to empirically estimate the p value. Recall that a *p value* is the probability of observing something as extreme or more extreme then the observed test statistic if the null hypothesis is true. In genetic tests of association, the null hypothesis is either (1) the haplotype frequency distribution is equal between cases and controls or (2) a measure of disease risk [e.g., log(odds of disease)], does not depend on haplotype. In either case, under the null hypothesis, the assignment of case and control is random with respect to marker genotypes. An estimate of statistical significance can be obtained by comparing the observed statistic T_{OBS} to the empirical null distribution generated by randomly shuffling case status as follows. For a dataset with N_{cases} and N_{controls}, a large set of permuted datasets are formed, and for each dataset the statistic is recomputed. Each permuted dataset is obtained by randomly assigning case status to N_{cases} individuals from the full sample. All other individuals are assigned to the control group. The p value is estimated to be the proportion of permutation-based statistics that are as extreme as or more extreme that T_{OBS}. Using a permutation distribution provides a nice framework for testing among a large number of haplotypes. In particular, one may want to base significance on whether any haplotype shows a deviation between the cases and controls. If a chi-square goodness-of-fit statistic was used to compare the proportion of haplotype i in cases with the proportion in controls, then this corresponds to computing a test statistic for each haplotype, χ_i^2, and basing significance on the maximum of those statistics. It is difficult to access significance for such a statistic using asymptotic distributions; however, assessing significance through a permutation distribution is simple. It is noted that this type of max statistic appropriately adjusts for the multiple tests introduced by considering many haplotypes. For more information, see Fallin et al. (2001) and/or Zhao et al. (2000).

5.6. DISCUSSION

Inclusion of genetic data in human studies is becoming increasingly common. From a scientific point of view, genetic studies can utilize inherited differences among individuals to provide an unbiased approach for detecting causes, initially unknown and unsuspected, of phenotypic variation. Other reasons for pursuing genetic studies include the promise of genetic testing, although a great deal of thought is needed as to (1) how genetic testing will benefit an individual and/or society, (2) how such information will be used by the public health system, and (3) how to keep such information from being misused for financial gain. Another reason for pursuing genetic studies is drug development. Finding genes that contribute to a disease, or even a subtype of disease, provides valuable insight into how pathways could be targeted for drug development.

Studies that utilize genetic information are becoming routine. With the increasing amount of in-depth data available for such studies comes both great promise and responsibility. We have discussed the various types of designs and methods that are available to the researcher along with what questions these designs can answer and the type of data required; however, we must continue to look ahead for new tools for understanding complex traits. At the most basic level, efforts should be directed to define and quantify accurate environmental and phenotypic measures. New technologies are providing such information. For example, microarrays, consisting of probes that have been synthesized or attached to a device, allow for the detection and quantification of abundance of the corresponding nucleic acids. This allows for a snapshot of expression for a large set of genes for the tissue under study for each individual (see Chapter 3, on tools of ecogenetics). This type of data is providing insights into the molecular level of disease etiology that has never before been possible. Further, current efforts are underway to harness the high-throughput capacity of microarrays to genotype individuals for hundreds of thousands of single-nucleotide polymorphisms. In addition to new tools, new statistical methodologies are being developed. These include Markov chain Monte Carlo (MCMC) methods that allow for a more general parameterization of the trait model. For example, MCMC methods have been developed to combine linkage and segregation analyses for quantitative traits (Heath 1997; Daw et al. 1999). These methods do not require one to assume, a priori, the number of QTLs influencing the complex trait and instead provide an estimate for the number of underlying QTL along with estimates of their effect sizes.

Inherent in large-scale studies is the problem of multiple testing. Thus, if a large number of tests are conducted, say, at the often used type I error rate of 0.05, then positive results will occur as a result of chance alone. More recent advances in multiple testing methods are providing new Bayesian techniques. In particular, false discovery rate (FDR) methods are constructed to control the number of significant results that are incorrectly called "significant" (Benjamini and Hochberg 1995; Storey and Tibshirani 2003).

With technology and statistical methods providing more and more sophisticated tools, it may well be possible to crack the puzzle that is a complex trait. Studies will need to continue to evaluate the most appropriate and efficient designs and the most powerful statistical methods. For regardless of how refined our technology becomes, it will always be essential to apply the proper design and method to answer the primary questions of interest.

ACKNOWLEDGMENTS

This project was supported under a cooperative agreement (U36/CCU300430-23) from the Centers for Disease Control and Prevention (CDC) through the Association of Schools of Public Health. The contents of this article are the responsibility of the author and do not necessarily represent the official views of CDC or ASPH.

Part II

6

Overview of Section II

Lucio G. Costa and David L. Eaton
University of Washington, Seattle, WA

In section I we provided an introduction to the field of ecogenetics by discussing its evolution from pharmacogenetics, and from studies of phenotypic determinations to molecular approaches. The new molecular tools that are allowing the rapid progress in the field, and the general epidemiologic and statistical strategies used in molecular epidemiology studies of gene–environment interaction, have also been discussed. This second section provides a more in-depth discussion of a limited number of genetic polymorphisms. The list of genes discussed is by no means exhaustive, and is meant only to provide selected examples in areas that have received substantial study. Functional polymorphisms in these genes can lead to significant changes in the pharmacokinetics and pharmacodynamics of important exogenous, as well as endogenous, chemicals.

Three of the five chapters in this section focus on biotransformation enzymes, an emphasis that reflects the fact that exploration of the roles of these enzymes in processing toxic substances has been the major area of focus in pharmacogenetics and ecogenetics research to date. An entire series of enzymes may be involved in the biotransformation of a given chemical. Chapter 7, on polymorphisms in cytochrome P450 and flavin-containing monooxygenase genes, discusses so-called phase 1 biotransformation enzymes, which are activated early in the biotransformation process, while Chapter 8, on polymorphisms in xenobiotic conjugation, and Chapter

Gene-Environment Interactions: Fundamentals of Ecogenetics, edited by Lucio G. Costa and David L. Eaton

9, on paraoxonase, butyrylcholinesterase, and epoxide hydrolase, focuses on phase II biotransformation enzymes. The remaining two chapters in this section cover polymorphisms in DNA repair enzymes (Chapter 10) and receptors and ion channels (Chapter 11).

Decades of research have shown that individual differences in the biotransformation enzymes that bioactivate or detoxify pharmaceutical drugs can limit their efficacy, or lead to severe adverse reactions. Indeed, given the recent (as of 2005) recalls of several widely utilized drugs due to unexpected severe toxic effects, the focus of some pharmacogenetic research is now toward the definition of individualized therapies that take into account individuals' genetic makeup. In addition, evidence is emerging that these same biotransformation enzymes can also modulate an individual's susceptibility to environmental and occupational toxicants, the specific area of ecogenetics research.

Chapter 7 describes the family of biotransformation enzymes responsible for the oxidative metabolism of a host of drugs and other chemicals foreign to the body (xenobiotics), also known as *phase I biotransformation enzymes*. The most widely studied class of enzymes that participate in oxidation of xenobiotics is the cytochromes P450 (CYP), which are involved in the metabolism of almost every known chemical. Several CYPs display common functional polymorphisms that significantly affect biotransformation, particularly the activation to toxic intermediates, of a large number of chemicals. Of particular interest is the CYP-mediated oxidation of a wide variety of chemicals that, when oxidized by a CYP, generate reactive forms of the molecule that can bind to and damage DNA, potentially increasing the risk of cancer. Indeed, over 90% of all known chemical carcinogens are not directly carcinogenic, but require metabolic activation (usually via one or more CYPs) to exert their carcinogenic effects. Thus, it is perhaps not surprising that subtle genetic differences in the rates at which one enzyme activates a "procarcinogen" to the "ultimate carcinogen" might contribute to important differences in susceptibility to cancer-causing chemicals. Thus, several of the chapters in Section III, on genetic polymorphisms and disease, that address gene–environment interactions in the development of particular types of cancer, also discuss CYP polymorphisms.

Although CYPs are responsible for the oxidation of most xenobiotics, the flavin-dependent monooxygenases (FMOs), a related, much smaller family of enzymes, also participate in oxidation of xenobiotics that contain a nitrogen or sulfur atom. Several of the FMOs are also polymorphic, and these genetic differences can contribute to variability in how people respond to certain drugs and chemicals.

Although the rate of activation of a relatively nontoxic chemical to a highly reactive, toxic intermediate via a CYP or FMO enzyme is certainly important, fortunately the body possesses other enzymes that are able to intercept and detoxify reactive intermediates. These phase II biotransformation enzymes provide an important means of protecting cells from both endogenous and exogenous chemicals. Chapters 8 and 9 discusses polymorphisms in phase II enzymes that are involved primarily in detoxication reactions. These phase II enzymes include the family of glutathione *S*-transferases, glucuronosyl transferases, sulfotransferases, and various esterases and hydrolases, in particular paraoxonase. The latter enzyme, discussed in Chapter 9, is of interest, because of its diverse enzymatic activities toward certain toxic organophosphorus insecticides, oxidized lipids, and pharmaceutical drugs. When considering the importance of a polymorphism in a specific xenobiotic biotransformation enzyme, it is important to keep in mind that it is the overall *ratio* of the rates of activation to detoxification of a given chemical that is most relevant. Thus, it is important to understand all of the biotransformation pathways that might be acting on that chemical. The field of pharmacokinetics (or toxicokinetics) provides mathematical descriptions of the rates of enzymatic reactions, and can be extremely useful in helping to determine the relative importance of a polymorphism in a biotransformation enzyme.

The toxic effects of many chemicals result from the ability of the activated chemical to damage DNA. DNA damage from both endogenous and exogenous sources occurs frequently in every living cell in our bodies. Fortunately, our cells possess remarkably efficient repair processes that remove and correct such DNA damage. Enzymes involved in the repair of damaged DNA are discussed in Chapter 10. We know from studies of rare genetic mutations in certain DNA repair enzymes that loss of DNA repair capacity is associated with a variety of chronic diseases, especially some forms of cancer. There is much interest now in identifying common genetic polymorphisms in DNA repair capacity that might cause slight alterations in the efficiency and/or accuracy of DNA repair. It is intuitively obvious that mutations in such enzymes that lead to inefficient DNA repair may represent a significant genetic susceptibility risk factor for exposure to mutagens and carcinogens.

Chapter 11, the final chapter in this section, deals with genetic polymorphisms found in receptors for neurotransmitters and hormones, as well as ion channels. Again, only selected examples are discussed to highlight the concept that, although genetic polymorphisms leading to pharmacokinetic differences have been most widely investigated, one should not forget the pharmacodymanic aspects of gene–environment interactions; indeed, mutations in target proteins can also significantly alter the response to drugs and environmental chemicals, in addition to predisposing individuals to disease.

7

Polymorphisms in Cytochrome P450 and Flavin-Containing Monooxygenase Genes

Catherine K. Yeung, Allan E. Rettie, and Kenneth E. Thummel
University of Washington, Seattle, WA

Mutations in the genes encoding the cytochrome P450 and FMO family of enzymes contribute to interindividual differences in the biological effects of xenobiotics by affecting the peak concentration and duration of exposure of the parent molecule or "active" metabolites in the body. However, the phenotype of interest—risk of disease, for example—often involves a complex interplay among multiple genetic and environmental factors that affect the primary or secondary activation and detoxification processes and excretion of the parent molecule and metabolites as well as factors affecting the cellular/tissue response to potentially toxic chemical species. Although it may be difficult to characterize the contribution of individual components such as P450/FMO mutations, the exercise should ultimately lead to a more useful understanding of our inherent differences in xenobiotic disposition and disease risk.

7.1. INTRODUCTION

7.1.1. Basic Principles

Biotransformation is a critical step in the pathway of xenobiotic detoxification and elimination. Passive or active excretion of highly lipophilic

---

Gene-Environment Interactions: Fundamentals of Ecogenetics, edited by Lucio G. Costa and David L. Eaton

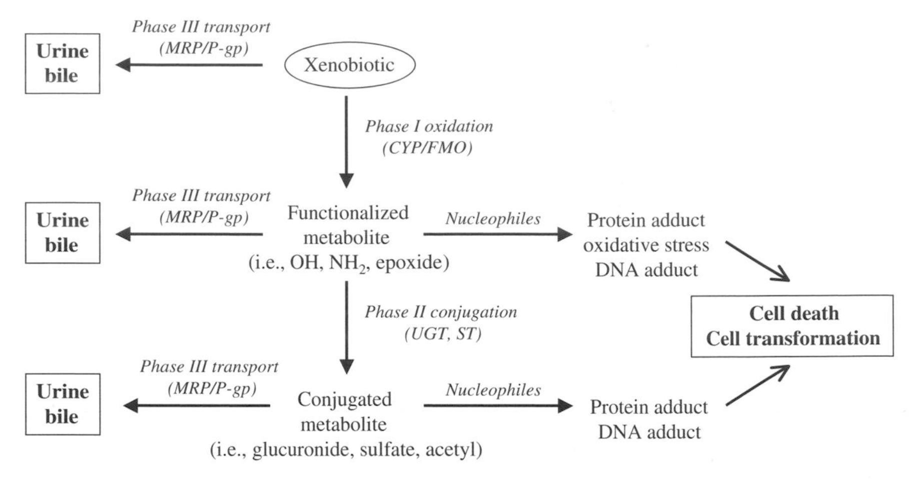

Figure 7.1. Duality of xenobiotic detoxification and bioactivation initiated through cytochrome P450 and flavin monooxygenase-catalyzed oxidations.

molecules from the body is generally inefficient as a result of extensive binding to plasma proteins that limits both glomerular filtration and biliary excretion, and thermodynamically favored reabsorption from either the renal tubular lumen or gastrointestinal tract. For most foreign molecules, an oxidation reaction catalyzed by the microsomal monooxygenases—cytochromes P450 (CYP) or flavin monooxygensases (FMO)—permits either further conjugative or oxidative reactions or export of the metabolite into bile and urine with less favorable competing reabsorption. In addition, for some molecules, the initial oxidation reaction represents a toxification or bioactivation step. This generally involves the formation of a chemically reactive primary or secondary molecule (e.g., arene oxides, carbenes, aryl hydroxylamines) that can modify cellular macromolecules, disrupt cellular homeostasis, and result in cell death or cancerous transformation (Figure 7.1). Thus, the time course and severity of toxicological effects attributable to a xenobiotic are often determined by the efficiency of the initial monooxygenase-catalyzed metabolic process. Accordingly, interindividual differences in P450/FMO expression and function influence the risk of adverse effects following xenobiotic exposure.

7.1.2. Genetic Basis of Interindividual Variability

There are multiple sources of variability in the efficiency of P450- and FMO-catalyzed reactions that are both genetic and environmental in origin. One useful framework for understanding this relationship is to consider

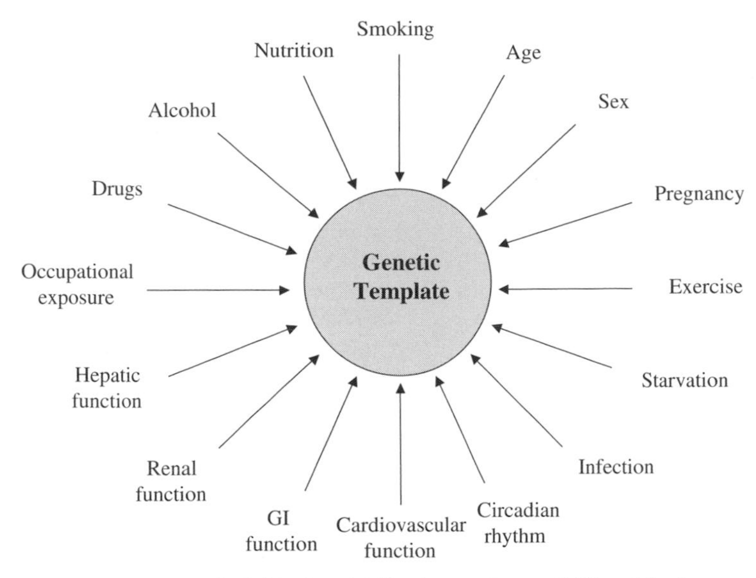

Figure 7.2. Sources of interindividual variability in cytochrome P450 and flavin monooxygenase protein expression. [Reproduced from unpublished figure with permission from William Evans, St. Jude Children's Research Hospital, Memphis, TN. Adapted from Vesell (1981).]

that environmental factors, which can vary among individuals and over time, modify an invariant but unique genetic template (Figure 7.2). Mutations in the genes that code for the P450/FMO proteins are numerous and sometimes common (polymorphic, ≥1% allele frequency). There are a total of 57 CYP proteins that are encoded in the human genome. These can be grouped into 18 distinct gene families, most of which play a role in the formation and elimination of various endogenous molecules with important physiological functions (Guengerich 2003a), including steroids, bile acids, eicosanoids, prostaglandins, vitamins, and even byproducts of sugar metabolism (e.g., acetone). At least 17 different, individual human CYP enzymes, most of them located in the CYP1–CYP3 gene families, are involved in the metabolism of xenobiotic molecules (Table 7.1). In addition, at least two human FMOs, namely, FMO1 and FMO3, may play a significant role in xenobiotic metabolism. Websites cataloging mutations for these CYPs and for FMO(3) are maintained by the Human Cytochrome P450 (CYP) Allele Nomenclature Committee, http://www.imm.ki.se/CYPalleles/htm (HCPANC 2004) and the Medical Research Council Human Genetics Unit (MRCHGU 2003), respectively.

Most gene mutations do not alter the rate of metabolism of a xenobiotic molecule significantly. Mutations may arise in a noncoding region and be

Table 7.1 Representative Substrates and Major Sites of Tissue Expression of Human Cytochrome 450 Enzymes (CYPs) Involved in Xenobiotic Metabolism

CYP Enzyme	Important Sites of Protein Expression	Representative Substrates[a]
1A1	Lung, placenta, small intestine	Benzo[*a*]pyrene, PAHs, 7-ethoxyresorufin
1A2	Liver	Caffeine, theophylline, heterocyclic amines
1B1	Breast, lung	DMBA, nitroarenes, heterocyclic amines
2A6	Liver	Nicotine, coumarin, MTBE
2A13	Nasal mucosa, lung	NNK, *N*,*N*-dimethylaniline, hexamethylphosphoramide
2B6	Liver, brain, kidney, small intestine	Buproprion, cyclophosphamide, 6-aminochrysene
2C8	Liver	Paclitaxel, carbamazepine, retinoic acid
2C9	Liver, small intestine	*S*-Warfarin, ibuprofen, losartan, phenytoin
2C19	Liver, small intestine	*S*-Mephenytoin, diazepam, proguanil
2D6	Liver, brain	Metoprolol, nortriptyline, flecainide, perphenazine
2E1	Liver, brain	Aniline, enflurane, *n*-butanol, benzene, acetaminophen, chlorzoxazone
2F12	Small intestine, kidney	Terfenadine, LTB_4, arachadonic acid
2J2	Small intestine, heart, kidney, lung, liver	Astemizole, ebastine, terfenadine
3A4	liver, small intestine	Midazolam, alfentanil, quinadine, verapamil, felodipine, simvastatin, cyclosporine, aflatoxin B_1
3A5	Liver, small intestine, kidney, prostate	Tacrolimus, midazolam, carbamazepine, nifedipine
3A7	Liver	DHEA–sulfate, aflatoxin B_1, heterocyclic amines
4A11	Liver, kidney	Sulindac, 3-methylindole, leukotrienes

[a] Identification of a substrate molecule with a given enzyme does not imply that no other enzyme can metabolize the same molecule; indeed, overlapping substrate specificity is common [PAHs, polycyclic aromatic hydrocarbons; DMBA, dimethylbenz[*a*]anthracene; MTBE, methyl *ter*-butyl ether; NNK, 4-(methylnitrosoamino)-1-(3-pyridyl)-1-butanone; LTB_4, leukotriene B_4; DHEA–sulfate, dehydroepiandrosterone 3-sulfate]. For additional substrates, see Omiecinski et al. (2000).

irrelevant for transcription, processing, and stability of the mRNA, or they may reside in a coding region but elicit no change in primary amino acid sequence (synonymous mutation). Alternatively, a mutation may result in an amino acid alteration (nonsynonymous mutation), but function is conserved despite the structural change. However, as illustrated in the relationship presented in Equation (7.1) (below), some mutations can alter the maximum rate of xenobiotic metabolism (V_{max}) by decreasing the steady-state enzyme level [E_{ss}]. This occurs as either a result of a change in the zero-order synthesis rate (R_0) or a change in the first-order degradation rate constant. In this scheme, which is derived from an assumption of Michaelis-Menten kinetics, k_{cat} represents a first-order rate constant for the irreversible catalytic step in the reaction:

$$V_{max} = [E_{ss}] \cdot k_{cat} \qquad \text{where} \qquad [E_{ss}] = \frac{R_0}{k_{deg}} \tag{7.1}$$

The molecular mechanism underlying change in enzyme content may involve mutation in the 5′-flanking region of the gene that affects the binding of regulatory proteins (promoters, enhancers, or repressors) and gene transcription, duplication of an intact functional gene, altered mRNA splicing or stability, or altered protein synthesis or stability. In addition, some mutations (gene deletions or frameshifts) cause a complete loss of functional transcript and enzyme (null mutation), resulting in profound reductions in xenobiotic metabolism. Even when mRNA production is unaffected, a genetic mutation may change the structure of the enzyme (amino acid substitution or deletion) such that one or more of the steps of the respective catalytic cycle (e.g., substrate binding, oxidation, or product release) are altered. These changes can affect both the K_m (a measure of the stability of the enzyme–substrate complex) and k_{cat} for a reaction [see Equation (7.2)], ultimately changing the maximum efficiency (CL_{int}, intrinsic clearance) for a particular metabolic reaction:

$$CL_{int} = \frac{V_{max}}{K_m} \tag{7.2}$$

The overall phenotype for xenobiotic clearance in a population may take the shape of a clear biomodal (or multimodal) distribution if the penetrance of a polymorphic mutation is pronounced (e.g., a null mutation), or it may appear as a skewed but unimodal distribution when the penetrance of one or more polymorphisms is low (Figure 7.3).

7.1.3. Gene–Environment Interactions

The hepatic and extrahepatic expression of P450s and FMOs appears to be regulated by a variety of different cell signaling pathways and endogenous ligands, including corticosteroids (Gerbal-Chaloin et al. 2002), growth

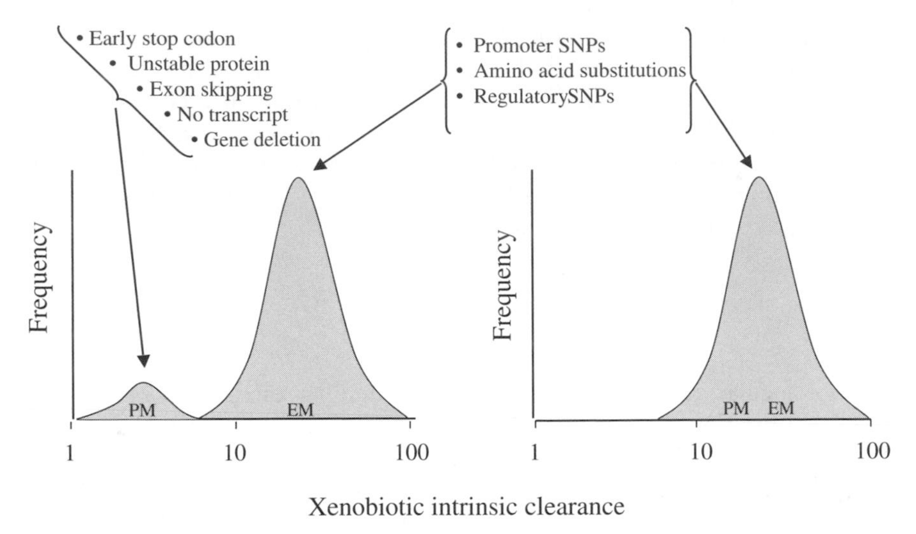

Figure 7.3. High- and low-penetrance genotype–intrinsic clearance phenotype relationships for xenobiotic metabolizing enzymes. [Reproduced from unpublished figure with permission from William Evans, St. Jude Children's Research Hospital, Memphis, TN. Adapted from Relling and Dervieux (2001).]

hormone (Liddle et al. 1998), thyroxine (Liddle et al. 1998), 1,25-dihydroxy vitamin D_3 (Thummel et al. 2001), lipophilic bile acids (Makishima et al. 2002) and fatty acids/eicosanoids (Zhang et al. 2000) for the P450s, and sex steroids for several of the FMOs. Presumably, "normal" variability in hormone (ligand) concentrations and cell target receptors [GR, glucocorticoid receptor; GHR, growth hormone receptor; VDR, vitamin D receptor; PXR (also known as SXR), pregnane X receptor; PPARα, peroxisome proliferator-activated receptor] accounts for some of the differences in *constitutive* gene transcription and steady-state enzyme levels. There are also well-defined pathways for P450 enzyme *induction* (FMOs do not appear to be "inducible") in response to exogenous molecules, such as drugs and drug-like xenobiotics. The best characterized involve activation of intracellular receptors [e.g., arylhydrocarbon hydroxylase receptor, or AHR; PXR; and CAR. For more details, see discussion below and in Goodwin et al. (2002) and Nebert et al. (2004)] that bind to nuclear DNA and result in enhanced gene transcription and enzyme synthesis (Table 7.2). In addition, a reduction in tissue enzyme content in response to xenobiotics or pathophysiological states can occur through suppressed expression of key regulatory elements, including receptors such as PXR (Pascussi et al. 2000).

Table 7.2 Xenobiotic Responsive Transcription Factors that Enhance Cytochrome P450 Transcription[a]

Ligand	Receptor Complex	CYP Enzymes Affected
PAHs, antiestrogens	AHR–ARNT	CYP1A1, 1A2, 1B1
TCPOBOP, PB, androstanes	CAR–RXRα	CYP2B6, 3A4
Hyperforin, rifampin, pregnanes	PXR–RXRα	CYP3A4, 2C9, 2C19, 2B6
Fibrates, glitazones, PUFA	PPARα–RXRα	4A11

[a] *Key:* PAH, polyclic aromatic hydrocarbons; TCPOBOP, 1,4-bis-[2-(3,5-dichloropyridyloxy)]benzene; PB, phenobarbital; PUFA, polyunsaturated fatty acids.

It is important to recognize that these "environmental" sources of interindividual variability may also have genetic components, although the observed phenotypic differences may be the result of multiple mutations of weak individual penetrance and mechanisms that are indirect and not readily apparent. For example, mutations in AHR and PXR may alter the transcription of *CYP1A1* and *CYP3A4*, respectively, in response to receptor agonists (Zhang et al. 2001; Harper et al. 2002).

On the basis of the discussion above, it should be apparent that a complete understanding of individual risk to an adverse event or disease following exposure to a xenobiotic molecule depends on a complex interplay between genetic and environmental factors. In addition, a lifetime of environmental exposures and disease history can modulate the penetrance of a given set of gene mutations as a result of chronic pathophysiological changes in cellular and organ homeostasis. Thus, individual response to a xenobiotic is a function of a dynamic gene–environment interaction that may change over a lifetime. However, this chapter focuses primarily on the direct effects of inherited (genomic) mutations in the cytochrome P450 and FMO genes that either enhance the bioactivation or reduce the metabolic elimination of environmental toxicants and carcinogens.

7.2. CYTOCHROMES P450

7.2.1. CYP1 Family

The *CYP1A* subfamily is often cited as a critical player in defining the risk of toxicity from many xenobiotics commonly found in the environment (Nebert et al. 2004). It is comprised of three gene products, CYP1A1, CYP1A2, and CYP1B1, which exhibit overlapping substrate specificities and a common pathway of induction involving AHR.

7.2.2. AH Receptor—ARNT and CYP1 Induction

It is a paradox of nature that an enzyme such as CYP1A1, which can initiate carcinogenesis through the formation of reactive metabolites, can be readily induced by the same family of polycyclic aromatic hydrocarbons (PAHs), especially halogenated hydrocarbons, that the gene or enzyme presumably has evolved to protect against. Indeed, detection of CYP1A1 induction in body tissues is presumptive evidence for exposure to this class of cancer-causing molecules and is associated with increased cancer risk [reviewed in Ma and Lu (2003) and Nebert et al. (2004)].

In humans, *CYP1A1* expression is generally low and is restricted largely to extrahepatic organs, even after exposure to enzyme-inducing agents. For example, inhalation of tobacco smoke and consumption of charbroiled meats is associated with increased CYP1A1 content in the lung (Mollerup et al. 1999) and small intestine (Fontana et al. 1999), respectively. Placentae from women who had smoked during pregnancy likewise show induced levels of CYP1A1 (Pasanen 1999), suggesting a protective function at this and other barrier sites. There appears to be significant interindividual differences (up to 100-fold) in the extent of extrahepatic CYP1A1 induction following exposure of humans to xenobiotics, some of which has been attributed to genetic variability (Ma and Lu 2003).

The molecular mechanism of CYP1A1 induction is well established (Sogawa and Fujii-Kuriyama 1997; Ma and Lu 2003). *CYP1A1* gene transcription is controlled by both positive and negative regulatory elements as well as multiple transcription factors (Figure 7.4). In the absence of a ligand for the AHR, transcription of the *CYP1A1* gene appears to be silent; the *CYP1A1* promoter is inaccessible because of higher order folding of the DNA into a nucleosome. Binding of an inducing agent to the AHR triggers a cascade of cytosolic and nuclear events that ultimately permit binding of RNA polymerase and gene transcription. Formation of the AHR/ARNT complex will also activate the transcription of a "battery" of genes, including *CYP1A2, CYP1B1*, GST, GT, NQO, and ALDH (Nebert et al. 2000).

The AHR-ARNT dimer is a positive *trans*-acting factor that binds to a xenobiotic response element on the 5′-flanking region of the gene. There are multiple tandem XREs on the human *CYP1A1* gene. It is believed that occupation of one or more XREs by the AHR/ARNT complex causes the basal *CYP1A1* promoter (i.e., GC box) to be accessible to its own transcription factors (e.g., Sp1) (Sogawa and Fujii-Kuriyama 1997).

Given this mechanism of induction, one could anticipate that mutations in the genes coding for AHR, ARNT, and CYP1A1 and other undefined ancillary transcription factors (e.g., Sp1) could influence CYP1A1 inducibility. With respect to the AHR, a number of coding mutations have been identified and studied in vivo and in vitro [reviewed by Ma and Lu (2003)].

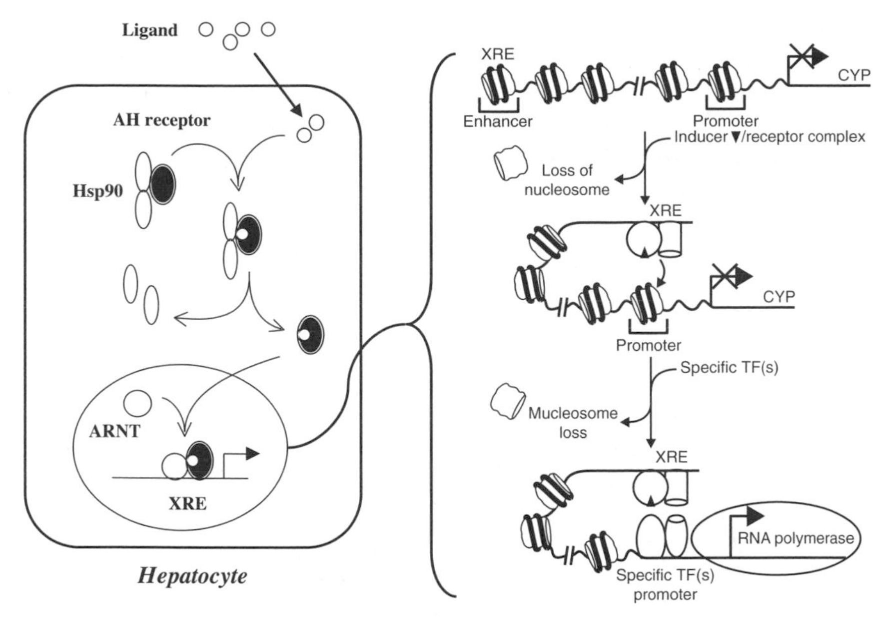

Figure 7.4. Mechanism of CYP1A1 induction. [Adapted from published figures, with permission of authors (Sogawa 1997).]

Investigations of the Arg554Lys coding variant have yielded mixed results. Some investigators have reported greater inducibility for the Lys554 variant compared to wildtype, whereas others have failed to find a significant association. However, it was reported that a combination (haplotype) of coding variants—Lys554 and Ile570, but not Lys554 alone—resulted in a reduced inductive response (Harper et al. 2002), suggesting the need for comprehensive genomic testing and large study populations to elucidate an informative genotype–phenotype association. Mutations in the promoter region of the AHR gene have also been identified, but they do not appear to affect AHR expression or CYP1A1 inducibility (Racky et al. 2004). Coding and noncoding mutations in the ARNT gene have been identified, but they too have not been associated with an altered CYP1A inductive response (Scheel et al. 2002).

7.2.3. CYP1A1

7.2.3.1. CYP1A1 Function. Aromatic hydrocarbons are common contaminants of our industrial environment. Most are highly lipophilic and thus will accumulate in lipoidal matrices of all multicellular organisms, including humans. Some can cause overt toxicity through direct mechanisms, whereas others cause toxicity after bioactivation to a reactive metabolite. CYP1A1 catalyzes both types of functionalization reactions, and its activity can reduce or enhance the risk of toxicity accordingly, depending on the structure of the substrate. For example, aromatic hydroxylation of benzo(a)pyrene followed by sulfation or glucuronidation of the primary phenol can facilitate the urinary or biliary excretion of the potential carcinogen. Alternatively, formation of a nucleophilic arene oxide can lead to covalent addition to electrophilic sites in DNA and cellular proteins and initiate the development of cancer.

A single exposure to benzo(a)pyrene may not elicit a carcinogenic event, although a lifetime of repeated exposures will increase the probability. Likewise, mutations in the *CYP1A1* gene that alter the rate of protein synthesis or degradation or catalytic steps (K_m and k_{cat}), even those with limited penetrance, could increase an individual's lifetime risk of cancer from exposure to the procarcinogen. Although enhanced CYP1A1 activity due to either genetic predisposition or enzyme induction could be viewed as a protective process in the short term, it is generally considered to be a liability from the perspective of long-term survival.

7.2.3.2. CYP1A1 Mutations. Multiple single-nucleotide polymorphisms (SNPs) in the *CYP1A1* gene have been identified [see Committee (2004)]. Two relatively common point mutations, T3801C (MspI, downstream of 3′-polyA tail) and A2455G (Ile462Val, in the heme-binding domain), have been linked with increased CYP1A1 protein expression (MspI) and catalytic activity (Val462), although these effects are controversial (Wu et al. 1998). Presence of the C3801 SNP alone, C3801 and G2455, and G2455 alone have been designated as the *CYP1A1*2A, *2B*, and **2C* alleles, respectively.

Numerous associative investigations of the common *CYP1A1* SNPs and various solid organ cancers (e.g., lung, bladder, breast, and colon) have yielded mixed results. For example, a relatively large metaanalysis failed to find a significant association between the MspI (*CYP1A1*2A/2B*) polymorphism and the risk of lung cancer (Houlston 2000). However, a more recent report of another large pooled analysis of original data from multiple studies revealed a significant odds ratio of 2.36 for the occurrence of lung cancer among Caucasians, but not Asians, after adjustment for age, gender, and smoking habit (Vineis et al. 2003). Additionally, other investigators are exploring the possibility that characterizing a combination of mutations in the *CYP1A1* gene (or its regulatory genes) and genes coding for proteins that detoxify products of CYP1A1 activity, such as GSTs or epoxide hydrolase (EH), could be a better predictor of cancer risk than a single gene mutation. (see Chapter 8—Polymorphisms in Xenobiotic

Conjugation). For example, an analysis of DNA adduct formation in lymphocytes from individuals living in an urban environment revealed a trend for a higher level of adduct formation with combinations of *CYP1A1*2A* and reduced activity GSTP1, GSTM1, microsomal ETT (mEH) mutations after stratification for exposure to environmental toxicants (Georgiadis et al. 2004).

7.2.4. CYP1A2

7.2.4.1. CYP1A2 Function. CYP1A2 is predominantly a hepatic enzyme and thus plays an important role in systemic xenobiotic elimination. With respect to toxicants, it will metabolize many heterocyclic amines found in our industrial environment and in charbroiled foods (Eaton et al. 1995; Kim and Guengerich 2005). Substrates of CYP1A2 include aromatic amines (2-aminofluorene, 2-aminoanthracene, 2-amino-3-methylimidazo[4,5-*f*] quinoline, 2-aminobiphenyl, and 2-amino-1-methyl-6-phenylimidazo[4,5-*b*] pyridine or PhIP), the neurotoxin MPTP (1-methyl-4-phenyl-1,2,3,6-tetrahydropyridine), and the food contaminant, aflatoxin B_1.

Like CYP1A1, CYP1A2 expression can be induced by exposure to the PAHs found in cigarette smoke, an effect mediated by the AHR. There is large interindividual variability (>60-fold) in the expression of hepatic CYP1A2, and some of the functional differences appear to be a result of genetic mutations (Nebert et al. 2004), although constituents of the diet (e.g., glucosinolates) and pathophysiological states (e.g., cirrhosis) can up- and downregulate *CYP1A2* gene transcription (George et al. 1995; Lampe and Peterson 2002).

7.2.4.2. CYP1A2 Mutations. More than 20 different mutations in the *CYP1A2* gene have been identified to date [see Committee (2004)]. Some of the noncoding mutations, such as the *CYP1A2*1F* and *CYP1A2*1C* variants, have been linked to an increased or decreased response, respectively, to enzyme inducers. The −163C > A mutation within intron-1 (*CYP1A2*1F*) putatively affects gene transcription and has been associated with enhanced induction of in vivo CYP1A2 activity (caffeine metabolic ratio) in subjects exposed to cigarette smoking (Sachse et al. 1999). In contrast, the *CYP1A2*1C* SNP (−3860G > A) reduced enzyme inducibility in response to cigarette smoke exposure (Nakajima et al. 1999).

The *CYP1A2*1F* and *CYP1A2*1C* alleles are potentially of clinical importance, as 80% of patients with schizophrenia are smokers and CYP1A2 metabolizes some important antipsychotic drugs. However, results from different pharmacogenetic studies relating the CYP1A2 polymorphism to drug toxicity or efficacy have been conflicting (Mihara et al. 2001; Ozdemir et al. 2001; Schulze et al. 2001). These discrepancies could be the result of differences in drug therapy and study design, or they could suggest an incomplete genetic picture. Indeed, data from a publication in 2003 indicate a more complicated genotype–phenotype relationship in which a gene haplotype that includes the **1F* SNP and other noncoding SNPs (*CYP1A2*1K*) exhibits reduced transcription and in vivo catalytic activity (Aklillu et al. 2003). Resolution of apparent discrepancies such as this will require comprehensive testing of all potentially relevant noncoding *CYP1A2* SNPs in a large population of subjects with careful control of all potential environmental confounding factors.

Interestingly, a more recent study of subjects receiving a diet both low and high in heterocyclic amines revealed a significant association between the *CYP1A2*1F* polymorphism and the urinary excretion of PhIP conjugates, which represent nontoxic metabolic pathways that compete with CYP1A2-catalyzed oxidation to a reactive metabolite (Moonen et al. 2004). The authors reported greater PhIP (unchanged + direct conjugate) excretion in subjects homozygous for the less inducible (–163C) allele, consistent with reduced oxidative metabolism and, by inference, lower DNA adduct formation. Although interesting, these results must be viewed with caution, as the authors also reported poor overall urine recovery of the calculated exogenous PhIP dose in the diet and a failure to detect N^2-hydroxy products, which are known major urinary metabolites of PhIP.

Multiple mutations within coding regions of the *CYP1A2* gene have also been identified. Of these, the product of the *CYP1A2*11* variant (558C > A, F186L) exhibits an intrinsic enzymatic function (determined by ethoxyresorufin-*O*-deethylase activity) that is 20% that of the wildtype enzyme. An intronic variant, *CYP1A2*7*, is associated with reduced in vivo activity, presumably as a result of improper mRNA splicing at the exon-6/intron-6 boundary (Allorge et al. 2003). Both variant alleles are relatively rare (<1%) and thus are not likely to be a major source of population variability in CYP1A2 activity. No other functionally significant variant alleles have been reported.

Only a limited number of *CYP1A2* genotype–cancer association studies have been reported to date. Most often, investigators have chosen to determine CYP1A2 activity phenotype (determined by urinary caffeine metabolite ratios) rather than genotype, but such data have generally not suggested an association between *CYP1A2* genotype (as reflected by catalytic activity) and cancer risk (Eaton et al. 1995). However, it is becoming increasingly clear that, for heterocyclic amines such as PhIP, the status of conjugating enzymes such as the SULTs, UGTs, and NATs is an equally important variable for determining disease risk at the oxidative (CYP1A) bioactivation pathway. Indeed, multigenetic contributions to disease risk are probably the general rule rather than an exception.

7.2.5. CYP1B1

7.2.5.1. CYP1B1 Function. *CYP1B1* transcript and protein have been detected in many organs of the body, including parenchymal and stromal tissue of the brain, kidney, breast, cervix, uterus, ovary, prostate, and lymph nodes (Muskhelishvili et al. 2001). The enzyme is expressed constitutively, but it can also be induced via activation of the AHR (Nebert et al. 2004). CYP1B1 exhibits a relatively broad substrate specificity, metabolizing many of the same substrates as CYP1A1 (e.g., polycyclic aromatic hydrocarbons) and CYP1A2 (e.g., heterocyclic amines) but often with a different efficiency and product regioselectivity (Guengerich et al. 2003). One area of current intensive study involves the bioactivation of endogeous estrogenic steroids to reactive catechol metabolites. Both CYP1A1 and CYP1B1 catalyze the hydroxylation of 17β-estradiol (E_2) to a pair of regioisomeric catechols. Both products have carcinogenic properties, but the favored product of CYP1B1, 4-OH-E_2, exhibits greater potency (Hanna et al. 2000).

7.2.5.2. CYP1B1 Mutations. More than 20 variants of the *CYP1B1* gene have been identified [see Committee (2004)], some of which result in altered catalytic function. The most highly penetrant alleles, a series of rare mutations that are predicted to result in a truncated transcript and inactive protein, were first identified as a major cause of primary congenital glaucoma (Stoilov et al. 1997; Bejjani et al. 1998). Some of the other missense mutations (*CYP1B1*2–*7*) are also found in the general population and display relatively modest differences (two- to three-fold) in catalytic efficiency toward E_2 and benzo(a)pyrene in comparison to the wildtype allele (Hanna et al. 2000).

A number of epidemiologic studies have been conducted to determine whether inheritance of the common coding variants affects cancer risk. Both positive and negative associations between the presence of coding variants and breast cancer have been reported (Miyoshi and Noguchi 2003), and all indicate a small relative risk (<2.5 odds ratio). The kidney expresses CYP1B1 constitutively, and coding mutations in the *CYP1B1* gene have been identified as a risk factor for renal cell carcinoma (Sasaki et al. 2004); more recent investigations have pointed out the importance of a multigene–environment study design in elucidating single gene effects (Listgarten et al. 2004). In this regard, a study of bladder cancer failed to identify *CYP1B1* polymorphisms as a risk factor but did reveal a significant association with occupational exposure to aromatic amines and GSTM1, GSTT1, and NAT2 genotypes (Hung et al. 2004).

7.2.6. CYP2A6 and CYP2A13

7.2.6.1. CYP2A Function. CYP2A6 is a hepatic and nasally expressed P450 (Koskela et al. 1999) that catalyzes the metabolism of few drugs, but many environmental toxicants including procarcinogenic nitrosamines. The expression of CYP2A6 in liver is highly variable, which has been attributed in significant part to genetic polymorphisms in the human population (Xu et al. 2002; Tricker 2003). The *CYP2A6* gene is part of a complicated gene locus containing pseudogenes and a structurally related gene, *CYP2A13*, which is also polymorphic and metabolically active towards nitrosamines (Su et al. 2000).

7.2.6.2. CYP2A6 Mutations. With respect to *CYP2A6*, 15 different gene variants have been reported [see HCPANC (2004)]. A complete loss of enzyme activity is observed with *CYP2A6*2* (no heme incorporation), the *CYP2A6*4A–D* gene deletion alleles, and *CYP2A6*5* (putatively unstable protein). A previously reported null variant, *CYP2A6*3*, appears to be an artifact from early nonspecific genotyping methods (Xu et al. 2002). Several other coding and noncoding mutations lead to reduced metabolic activity and gene duplication that results in an ultrarapid metabolizer status (similar to *CYP2D6*). The population frequency of known *CYP2A6* variant alleles varies widely throughout the world, and this may contribute to interethnic differences in addictive behaviors (e.g., smoking) and disease (e.g., lung cancer) (Xu et al. 2002).

CYP2A6 is the principal catalyst of the oxidation of nicotine to its inactive metabolite, cotinine (Nakajima et al. 1996). This represents the major route of nicotine elimination in vivo; thus, there has been considerable interest in defining the relationship between smoking behavior and the *CYP2A6* polymorphisms. Unfortunately, the epidemiologic data have produced conflicting conclusions; some studies suggest a significant association between the slow metabolizer genotype and tobacco dependence and self-reported consumption rate (Sellers et al. 2003), whereas others indicate no association between *CYP2A6* genotype and smoking behavior (Tricker 2003). As noted previously, direct comparison between studies is difficult because of differences in methodologies employed to assess smoking behavior. Further studies are warranted given the health consequences of smoking.

In the case of tobacco-related cancers, the hypothesis that genetically determined variability in CYP2A6-catalyzed activation of procarcinogenic nitrosamines, such as 4-(methylnitrosamino)-1-(3-pyridyl)-1-butanone (NNK) and *N′*-nitrosonornicotine (NNN), affects the risk of disease is quite plausible. In this context, a slow metabolizer may exhibit reduced exposure to NNK and NNN due to reduced consumption of cigarettes and reduced bioactivation of the procarcinogens. Once again, the experimental data available are mixed, with some groups reporting no difference between subjects homozygous for the deletion allele (*CYP2A6*4*) and those homozygous for the wildtype allele (*CYP2A6*1A*) in terms of cigarette consumption and lung cancer risk (Loriot et al. 2001; Wang et al. 2003a) and others reporting a significant difference for both traits (Ariyoshi et al. 2002).

7.2.6.3. CYP2A13 Function and Mutations. The emerging picture for the *CYP2A13* polymorphism is also inconsistent. One group has reported a reduced risk of lung adenocarcinoma for subjects carrying the *CYP2A13*2* allele, compared to those homozygous wildtype, particularly after stratification for smoking status (Wang et al. 2003a). This variant allele (Arg257Cys) was associated with a reduced intrinsic metabolic clearance for NNK metabolism (Zhang and Robertus 2002). In contrast, the presence of one copy of a putatively null allele (*CYP2A13*7*), and presumably reduced metabolism, was associated with an increased risk of small cell lung cancer (Cauffiez et al. 2004). In this case, the authors argued that in this context, CYP2A13 plays an important detoxification role and protects against the development of this type of lung cancer. Clearly, there is a need for additional work to resolve the experimental dichotomy.

7.2.7. CYP2E1

7.2.7.1. CYP2E1 Function. CYP2E1 is found predominantly in the liver, but the enzyme has also been detected in human brain, lung, and nasal mucosa (Lieber 1997). Although expression in extrahepatic tissues is much lower than that in the liver, there is considerable interest in determining its role in the bioactivation of xenobiotics from the environment that may initiate disease processes such as cancer. Substrates for CYP2E1 are generally low-molecular-weight, polar molecules, such as primary (e.g., ethanol) and secondary alcohols, amines, ethers, halogenated alkanes (e.g., chloroform, halothane) and alkenes, simple aromatic molecules (e.g., benzene, pyridine), and nitrosamines (e.g., *N,N*-dimethylnitrosamine), many of which undergo CYP2E1-dependent metabolism to yield toxic or carcinogenic products (Lieber 1997). In addition to these metabolic processes, uncoupled activation of the CYP2E1 catalytic cycle results in an increased formation of reactive oxygen species (e.g., superoxide anion, hydroxyl radical) and thus is thought to contribute to oxidative stress and chronic disease in the organs in which CYP2E1 is expressed (Lieber 2004). In this regard, CYP2E1 has been implicated in the development of alcoholic liver disease and non-alcoholic steatohepatitis (Lieber 1997, 2004).

Although basal expression of hepatic CYP2E1 is relatively constant in healthy people, the enzyme is readily induced by many of its substrates, including ethanol, acetone, and pyridine. The mechanism of induction can be complex, involving both transcriptional activation and protein stabilization (Lieber 1997). The latter process results in a complicated time course of in vivo activity that oscillates between reduced and enhanced metabolic clearance as the inducer/inhibitor molecule is absorbed and then cleared (Chien et al. 1997). Chronic exposure to CYP2E1-inducing agents (e.g., ethanol and nicotine) is thought to increase the risk of organ toxicities resulting from xenobiotic exposures (Lieber 2004).

7.2.7.2. CYP2E1 Mutations. Multiple noncoding mutations in the *CYP2E1* gene have been identified [see HCPANC (2004)]. The most extensively studied are SNPs in the 5′-regulatory region (RsaI and PstI) and a SNP within intron-6 (DraI), all causing restriction fragment length polymorphisms (RFLPs). A standardized nomenclature for different combinations of SNP inheritance has been assigned [e.g., *CYP2E1*5B* (PstI+, RsaI–); *CYP2E1*6* (DraI–); *CYP2E1*5A* (PstI+, RsaI–, DraI–)], although much of the older literature describes testing for one or the other RFLP that can be a composite of multiple gene alleles. Some studies of white populations suggest little or no difference in CYP2E1 expression or in vivo metabolic activity for the DraI and RsaI mutations (Carriere et al. 1996), whereas other research has indicated higher gene expression for the variant PstI+/RsaI– allele (Watanabe et al. 1994). Part of this discrepancy may stem from limited sample size and a difference in variant allele frequency (higher in Chinese, Japanese, and Korean populations than in white populations).

Two additional mutations in the 5′-flanking region of the CYP2E1 gene have been described since the mid-1990s. A variable tandem repeat (*CYP2E1*1D*) and a promoter SNP (*CYP2E1*7B*) have both been linked to altered gene transcription (Fairbrother et al. 1998; McCarver et al. 1998). In the case of *CYP2E1*7B*, current evidence suggests that it modifies the repressive effects of inflammatory cytokines (Qiu et al. 2004), whereas an increase in tandem repeats (from 6 to 8) appears to enhance the inducibility of *CYP2E1*1D* (McCarver et al. 1998). The predicted phenotype of these variant alleles is one of enhanced metabolic bioactivation and toxicity risk, compared to wildtype, under the appropriate exposure or disease condition.

A number of epidemiological studies have been conducted to determine the relationship between *CYP2E1* polymorphisms and disease risk, including lung, gastric, and esophageal cancer. Data relating to the DraI and RsaI SNPs and lung cancer have been inconsistent, possibly a result of racial differences in allele frequency and inadequate sample size, failure in early studies to evaluate specific *CYP2E1* haplotypes, differences in lung cancer cell type, and differences in CYP2E1 induction status (Oyama et al. 2003). It is also likely that genetically based differences in phase II metabolism and the degree of environmental exposure to protoxic molecules and enzyme inducers need to be factored in to achieve full resolution of discrepancies. For example, a publication involving Chinese subjects indicated a significantly higher risk of adenocarcinoma for individuals homozygous for the wildtype RsaI allele, and the risk increased further when coupled with a GSTM1 null genotype (Wang et al. 2003b).

A complex multivariate association was also reported for subjects suffering from benzene poisoning. These data suggested a combined effect from *CYP2E1* PstI and NQO1 c.609 C > T SNPs, GSTT1 null status, and ethanol consumption (Wan et al. 2002). Benzene is metabolized by CYP2E1 to phenol and subsequently to *o*- and *p*-hydroquinones (Figure 7.5). Ethanol is an inducer of CYP2E1. The phenolic metabolites can be eliminated after sulfation or glucuronidation, or they can be oxidized further to benzoquinones. The final bioactivation step can occur enzymatically (e.g., myeloperoxidase) or nonenzymatically. The electrophilic quinone metabolites can either be detoxified through the actions of NQO1 and GSTs, or initiate cellular injury, presumably in lymphocyte progenitor cells that are a critical site of leukemogenesis. In agreement with this mechanism and potential points of genetic penetrance, the risk of blood chromosomal aberrations in workers exposed to benzene was linked to a combination of CYP2E1, GSTM1, GSTT1, and NQO1 genotypes (Kim et al. 2004). Mutations in NQO1 and myeloperoxidase genes associated with reduced white cell counts have also been observed in workers exposed to benzene, even at relatively low (<1 ppm) concentrations (Lan et al. 2004).

The development of alcoholic liver disease has been associated with CYP2E1 metabolic activity. Uncoupled CYP2E1 catalytic cycle turnover contributes to hepatocellular oxidative stress (Caro and Cederbaum 2004); thus, mutations that affect gene transcription might be expected to affect the risk of disease. However, once again, the literature is conflicting on this point, probably reflective of the very complicated etiology of alcoholic liver

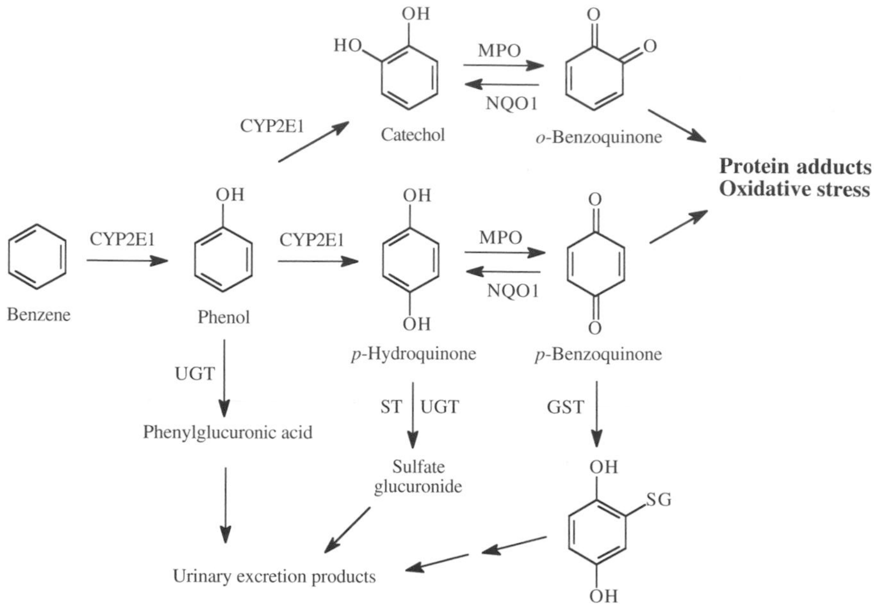

Figure 7.5. Metabolic pathways of benzene bioactivation and detoxification.

injury, including the possible contribution of genetic polymorphisms in *ADH2/3* and *ALDH2* genes. An association between *CYP2E1* polymorphisms (principally *CYP2E1*5B* and *1D* studied) and alcohol dependence is equally conflicted (Howard et al. 2002), most likely because it is also a polygenic behavioral trait. A clear *CYP2E1* genotype–phenotype association may also be confounded by the complex relationship between smoking and alcohol consumption and the inductive (CYP2E1) effects of nicotine and ethanol. Indeed, a more recent study (Howard et al. 2003) reports a significant association between the *CYP2E1*1D* genotype and both alcohol and nicotine dependence in Canadian Native Indians.

7.2.8. Other Members of the CYP2 Family

CYP2B6, CYP2C8, CYP2C9, CYP2C19, and CYP2D6 are all found in the liver, where they play a critical role in metabolic drug elimination. CYP2C9 and CYP2C19 are also expressed at appreciable levels in mucosal epithelia of the small intestine (Paine and Thummel 2003) and, thus, may contribute to first pass drug metabolism. Polymorphisms in the *CYP2B6*, *CYP2C8*, *CYP2C9*, *CYP2C19*, and *CYP2D6* genes contribute importantly to the high degree of interindividual variability in enzyme expression and function that is observed in vivo, in terms of both drug clearance and risk of adverse drug toxicity (Daly 2003; Pirmohamed and Park 2003). Important examples include the well-known *CYP2D6* polymorphism that affects the clearance and possibly the toxicity of some psychotropic drugs (Bertilsson et al. 2002) and the *CYP2C9* polymorphism that affects warfarin dosing and toxicity (Higashi et al. 2002). In the case of warfarin, a highly penetrant coding mutation (*CYP2C9*3*) causes reduced warfarin clearance and excessive accumulation under standard dosing conditions. Although doses are adjusted through therapeutic monitoring, the risk of an adverse event is increased in the poor metabolizers.

There has been considerable speculation over the years as to whether these drug-metabolizing enzyme polymorphisms also affect the risk of diseases that are thought to develop in part from long-term exposure to environmental toxicants. Examples include the *CYP2D6* polymorphism and the risk of Parkinson's disease (Riedl et al. 1998) and lung cancer (London et al. 1997), but the available epidemiological data are conflicting, and very few detailed mechanistic insights (i.e., substrate–toxic product) have ever been developed. There has also been interest in determining whether the *CYP2D6* polymorphisms (and other P450 gene mutations) affect the risk of addiction to psychotropic drugs that are cleared or activated by the enzyme, including nicotine, dextromethorphan, codeine, and amphetamines [reviewed in Howard et al. (2002)]. The available data are interesting but preliminary, and more definitive studies are needed.

7.2.9. CYP3A4/5/7

7.2.9.1. CYP3A Function. The metabolically active CYP3A gene products CYP3A4, CYP3A5, CYP3A7, and CYP3A43 are perhaps the most important drug-metabolizing enzymes in the body because of their relatively high level of expression in the major metabolic clearance organs (liver and small intestine) and broad substrate selectivity. Thus, there has been considerable interest in defining the relationship between CYP3A allelic variants and the efficiency of enzymatic activity and its contribution to interindividual differences in the bioactivation or detoxification of protoxic or procarcinogenic molecules.

Of the human CYP3A isoforms, CYP3A7 and CYP3A5 are expressed in the fetal liver, whereas CYP3A4 and CYP3A5 are the dominant forms in the adult liver (Stevens et al. 2003). CYP3A43 also appears in the adult, primarily in extrahepatic tissues (Westlind et al. 2001), but its role in xenobiotic disposition remains to be defined, as expression levels are low, and catalytic activity is largely unknown. In contrast, CYP3A4/5/7 enzymes contribute importantly to the hepatic and extrahepatic metabolism of numerous xenobiotics.

7.2.9.2. CYP3A7 Function and Mutations. CYP3A7 is the dominant P450 enzyme in fetal liver. It is replaced by CYP3A4 postpartum in response to ontogenic changes in the expression of hepatic transcription factors. However, this developmental pattern is not absolute, as some adult livers exhibit significant levels of CYP3A7 that appear to be related in part to a genetic polymorphism in the 5′-flanking region of the gene (Kuehl et al. 2001; Burk et al. 2002). The contribution of CYP3A7 to the formation or inactivation of toxic molecules during fetal life has been the subject of much speculation. Unfortunately, there is only a limited knowledge base available, most pertaining to CYP3A7 metabolic activity toward potential toxicants and nothing related to genetic variation and disease risk. One example involves CYP3A7-catalyzed activation of various heterocyclic amines, a process associated with increased mutagenicity in model cell culture systems (Hashimoto et al. 1995).

7.2.9.3. CYP3A4/5 Mutations. Although CYP3A4 can be found in the liver of most (if not all) adults, there is considerable interindividual variability in protein accumulation and in vivo activity (Lamba et al. 2002), and much of this variability has been attributed to genetic factors (Ozdemir et al. 2000). Although numerous SNPs in the CYP3A4 gene have been identified [see HCPANC (2004)], those that result in significantly altered metabolic activity (Dai et al. 2001; Lamba et al. 2002) or gene transcription (Matsumura et al. 2004) appear to be present at relatively low population allele frequencies (<5%), whereas the more common CYP3A4 polymorphisms do not appear to exhibit a clear phenotype (Lamba et al. 2002).

The most frequent and penetrant CYP3A polymorphism described to date is that associated with variable CYP3A5 expression. It is well established that CYP3A5 protein can be readily detected in both fetal and adult liver, but only in a subset of the population. Multiple mutations in the CYP3A5 gene (*CYP3A5*3*, *6*, and *7*) elicit an essentially null poor metabolizer (PM) phenotype, with very little accumulation of the functional protein (Kuehl et al. 2001). The frequency of CYP3A5 PM variant alleles varies widely throughout the world population. With respect to *CYP3A5*3*, the allele frequency is as high as 96% in some white European populations and as low as 6–7% in sub-Saharan Africa (Namibia, Congo, and Nigeria) (Thompson et al. 2004). The **3*, **6*, and **7* alleles of CYP3A5 result in an improperly spliced mRNA and a truncated, inactive protein. Although the penetrance of these mutations is not complete (i.e., a small amount of properly spliced mRNA is generated in homozygous mutants), characterization of liver, small intestine, and kidney tissues from genotyped donors has revealed a strong association between the specific content of CYP3A5 protein and inheritance of the *CYP3A5*1* allele (Lin et al. 2002; Givens et al. 2003).

7.2.9.4. CYP3A5 Genetic Variants and Altered Xenobiotic Metabolism. Compared to tissues with a homozygous, *CYP3A5*3/*3* genotype, those containing the active *CYP3A5*1* allele exhibit higher metabolic activity, on average, toward substrates of both CYP3A4 and CYP3A5, including the drugs nifedipine, midazolam, diltiazem, carbamazepine, and tacrolimus (Gillam et al. 1995; Bader et al. 2000; Huang et al. 2004; Yamaori et al. 2004). In the case of tacrolimus, the *CYP3A5*3* allele appeared to explain 18% of the total interindividual variability in tacrolimus blood concentration : dose ratios in kidney transplant patients receiving the drug (Haufroid et al. 2004).

It is not surprising to find that CYP3A5 may also play an important role in the metabolism of several environmental toxins. Substrates for CYP3A include several heterocyclic aromatic amines (e.g., 1-nitropyrene and pyridoindoles) and other polycyclic aromatic hydrocarbons such as benzo(a)pyrene and aflatoxin B_1 (AFB_1) (Kim and Guengerich 2005). AFB_1 is a fungus-derived contaminant of several plant foods that is associated with an increased risk of hepatocellular carcinoma. As seen in Figure 7.6, AFB_1 is oxidized by CYP3A4/5 and CYP1A2 to *exo*- and *endo*-epoxides (Guengerich 2003b). The *exo* epoxide, and to a lesser extent the *endo*-epoxide, can react with N^7-guanine residues of DNA, or it can be detoxified through the formation of a GSH conjugate or hydrolysis or the corresponding dihydrodiol. The dihydrodiol can undergo further transformation and bind to cellular proteins, but this is not thought to contribute to carcinogenic risk. CYP3A4 catalyzes stereoselective epoxide formation, generating only the more genotoxic *exo*-isomer, in contrast to a racemic mixture of epoxides produced by CYP1A2. CYP3A5 can also catalyze AFB_1 8,9-epoxidation (Gillam et al. 1995), although the stereochemistry for the reaction is currently undefined.

On the basis of the available in vitro data, one might anticipate that carriers of the functional *CYP3A5*1* allele would exhibit, on average, a higher degree of reactive product formation than those homozygous for inactivating alleles. Indeed, Wojnowski et al. (2004) reported the detection of significantly higher (16–23%) aflatoxin–albumin adducts in the blood of *CYP3A5*1* carriers from Gambia, West Africa. Although the effect is modest, it illustrates the potential contribution from one gene polymor-

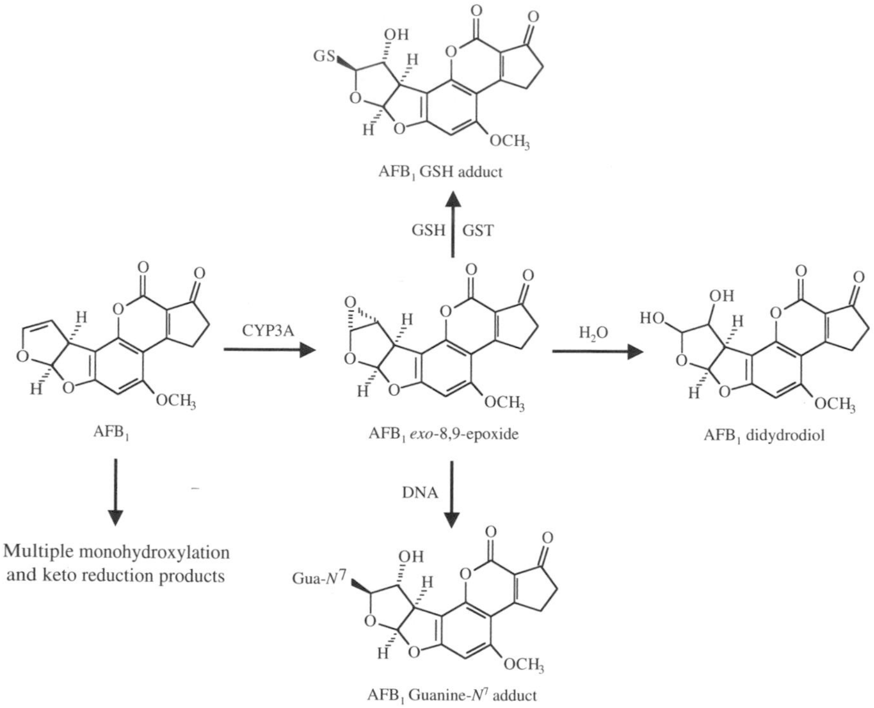

Figure 7.6. Metabolic pathways of aflatoxin bioactivation and detoxification.

phism to what is likely to be a multi-faceted (genes and environment) disease risk.

7.2.9.5. CYP3A5 Genetic Variants and Disease Risk. With the discovery of the genetic basis for polymorphic CYP3A5 expression, Kuehl et al. (2001) postulated that marked difference in *CYP3A5*3* allele frequency among white and African-American populations may reflect an evolutionary pressure associated with sodium retention in the kidney. CYP3A enzymes (including CYP3A5) catalyze the 6β-hydroxylation of cortisol, and the metabolite stimulates sodium retention and hypertension in experimental models systems (Duncan et al. 1988; Watlington et al. 1992; Ghosh et al. 1995). CYP3A5, not CYP3A4, is the major renal CYP3A enzyme (Schuetz et al. 1992; Haehner et al. 1996), and it is found at the highest levels in the kidneys of individuals carrying the wildtype *CYP3A5*1* allele (Givens et al. 2003). Thus, high CYP3A5 activity and elevated intrarenal 6β-hydroxycortisol production may have provided a selective advantage in geographical areas prone to drought. In support of this hypothesis, Thompson et al. (2004) showed that the frequency of the *CYP3A5*3* variant was positively associated with distance from the equator. Presumably, in contemporary societies, renal expression of the wildtype CYP3A5 allele with their associated sodium-sparing effects may adversely enhance the risk of salt-sensitive hypertension.

CYP3A5 also is an efficient catalyst of estrogen oxidation, including estrone and estradiol (E_2) (Huang et al. 1998; Lee et al. 2001). These reactions are quite intriguing, as some of the products (e.g., 4-hydroxy-E_2) have been implicated in breast carcinogenesis. However, no data defining the relationship between the CYP3A5 genotype and breast cancer risk have been reported.

Another example of the potential contribution from CYP3A genetic variability to disease risk involves prostate cancer. A mutation in the 5′-flanking region of the CYP3A4 gene (*CYP3A4*1B*) has been associated repeatedly with the incidence and severity of prostate cancer (Keshava et al. 2004). The molecular basis for this association has been attributed to altered interactions between nuclear proteins and the putative nifedipine response element (NFSE) and gene transcription. However, demonstration of an appreciable functional effect of the mutation on gene transcription (Ando et al. 1999) and hepatic protein expression (Westlind et al. 1999) is lacking. This raises the possibility that the NFSE mutation is simply a reporter for another causal gene allele that is in linkage disequilibrium. One possibility is *CYP3A5*1*, which has been shown to be strongly linked to the *CYP3A4*1B* allele (Kuehl et al. 2001). Because CYP3A5 mRNA is the dominant CYP3A transcript in human prostate (Wojnowski et al. 2002), higher prostatic metabolism of androgens or protoxic molecules could conceivably affect disease risk. Indeed, Plummer et al. (2003) reported a reduced risk of prostate cancer among white individuals carrying the wild-type *CYP3A5*1* allele, particularly compared to those with less aggressive disease. Interestingly, the investigators did not observe the same association for African-Americans, but that finding may be due to a failure to test for the inactivating **6* and *7 alleles, which are quite common in that population. Clearly, for future evaluations of the association between CYP3A5 SNPs and disease risk of any kind, it will be essential to conduct complete characterization of all polymorphic alleles.

7.2.10. Other P450 Enzymes

A number of other microsomal P450 enzymes may also contribute to interindividual differences in both disease risk and drug-induced toxicities. For example, mutations in the genes that code for enzymes that metabolize arachidonic acid (e.g., *CYP2J2*, *CYP4A11*, and *CYP4F2*) may affect the risk of cardiovascular disease, renal hypertension, and some inflammatory diseases (Kroetz and Xu 2004). CYP4A11 and CYP4F2 catalyze the ω-hydroxylation of arachidonic acid to 20-HETE. In contrast, CYP2J2 catalyzes an *endo*-epoxidation of arachidonic acid into epoxyeicosatrienoic acids (EETs), which can be further metabolized to corresponding dihydroxy products (DHETs). In the kidney, these molecules appear to exert powerful effects on vascular endothelium and tubular epithelium, altering regional blood flow, salt reabsorption, and vascular cell growth. Although largely unexplored, one can anticipate that mutations in the *CYP4A11*, *4F2*, and *2J2* genes that affect catalytic function might alter the risk of disease.

7.3. FLAVIN MONOOXYGENASES

The flavin-containing monooxygenases (FMOs) are a family of microsomal, NADPH- and oxygen-dependent enzymes that are present in all mammalian species studied to date. A total of five functional isoforms have been characterized in humans and designated with Arabic numerals (FMO1–FMO5) corresponding to the chronologic order of discovery (Lawton et al. 1994). A sixth FMO (FMO6) has been determined to be a pseudogene prone to exon skipping and does not encode a functional FMO enzyme (Hines et al. 2002). The various FMO enzymes exhibit 50–55% identity with ~87% homology between orthologs from other mammalian species (Phillips et al. 1995). All genes remain as a single gene cluster (Phillips et al. 1995) and have been localized to human chromosome 1, with *FMO1-4* located within the 1q23–q25 region and *FMO5* at the 1q21.1 locus (McCombie et al. 1996). The evolutionary divergence of these isoforms likely occurred some 250–300 million years ago (Phillips et al. 1995). The substrate selectivity of FMOs is quite broad, encompassing xenobiotics from a wide range of pharmacologic and environmental classes. All substrates are oxidized at a heteroatom, typically a soft nucleophile with an electron-rich center, commonly nitrogen, sulfur, phosphorus, or selenium (Ziegler 1993; Lawton et al. 1994).

7.3.1. FMO1

7.3.1.1. FMO1 Function. FMO1 is present in fetal liver and adult kidney at levels of 47 ± 9 and 25 ± 10 pmol/mg of microsomal protein, respectively (Yeung et al. 2000). Therefore, it seems to be a quantitatively significant monooxygenase and may play a role in fetal and renal metabolism or bioactivation of susceptible environmental agents (Figure 7.7). Extremely low

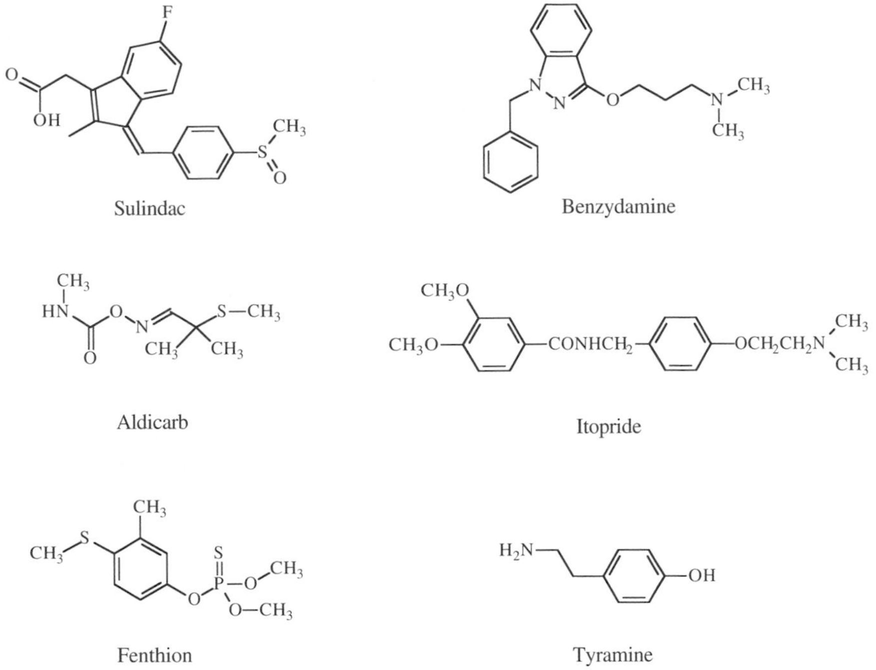

Sulindac

Benzydamine

Aldicarb

Itopride

Fenthion

Tyramine

Figure 7.7. Representative substrates of FMO1 and FMO3.

levels (2.9 ± 1.9 pmol/mg) of FMO1 have also been detected in human intestinal microsomes (Yeung et al. 2000), but the physiologic significance of such low enzyme levels is unclear. Interestingly, FMO1 is not expressed appreciably in adult liver, but is the dominant hepatic form of the enzyme in most experimental animals (Rettie and Fisher 1999). Therefore, metabolism and toxicity outcomes that reflect hepatic FMO activity may not correlate well between humans and animals as a consequence of this difference in hepatic FMO complement.

Aldicarb [2-methyl-2-(methylthio) propanal *O*-(methylamino) carbonyl oxime; Temik], a heavily used carbamate insecticide and nematocide, acts as a reversible inhibitor of acetyl and butyryl cholinesterases (Tomlin 1994). It is bioactivated by human FMO1 to its more toxic sulfoxide form (Figure 7.7), possibly in the intestine prior to transport into the blood, where it could inactivate red blood cell and plasma cholinesterases (Schlenk et al. 2002). FMO3 and the cytochrome P450 enzymes do not oxidize aldicarb to any significant extent, further implicating FMO1 as a primary contributor to aldicarb toxicity (Schlenk et al. 2002). FMO1 is also involved in the metabolism of fenthion (*O*,*O*-dimethyl *O*-[3-methyl-4-(methylthio)phenyl] phosphorothioate), another organophosphate insecticide, but in this example, the sulfoxide is less toxic than the parent compound, and FMO1 may act here as part of a detoxification pathway (Furnes and Schlenk 2004).

Sulindac sulfide (SID), a nonsteroidal antiinflammatory drug (NSAID), is proposed to cause fewer adverse renal side effects than other available NSAIDs, including increased sodium and potassium retention, nephrotic syndrome, and decreased glomerular filtration (Bunning and Barth 1982; Abraham and Keane 1984). This "renal-sparing" effect might result in part because of the high levels of FMO1 present in kidney tissue because FMO1 rapidly metabolizes SID to the less pharmacologically active sulindac sulfoxide (SOX) (Hamman et al. 2000). Normal renal prostaglandin synthesis is thereby maintained, and renal tissue is protected from prolonged homeostatic disruption. Conceivably, the presence of polymorphic variation in the *FMO1* gene that compromises either catalytic activity or protein expression could predispose the individual to increased renal damage with exposure to pharmacologic or environmental toxins, but this has yet to be tested.

As noted earlier, human FMO1 is subject to ontogenic regulation, and within 3 days postpartum, FMO1 expression is suppressed in the liver. Conversely, the onset of expression of FMO3 begins between 1 and 2 years of age (Koukouritaki et al. 2002). The exact control mechanisms underlying this ontological "switch" have not yet been elucidated. Conceivably, polymorphisms in regulatory domains that control expression of FMO1 may alter fetal or neonatal response to either therapeutic agents or environmental toxicants.

7.3.1.2. FMO1 Mutations. Several rare (<2% population frequency) *FMO1* variants have been identified within the coding region of the *FMO1* gene in a population of African-Americans. These include H97Q, I303V, I303T, and R502X (Furnes and Schlenk 2004). Compared with the wildtype protein, the effect of these mutations on enzymatic activity is relatively modest, with the exception of R502X, which exhibits significant reduction in the intrinsic clearance for methimazole and imipramine *N*-oxidation. Additionally, limited studies suggest that African-American kidney samples contain higher levels of renal FMO1 protein expression than do Caucasian samples (Krause et al. 2003). Extensive polymorphism also exists in the upstream promoter regions (Hines et al. 2003), which may influence protein expression levels of FMO1. At least, 10 common SNPs exist upstream of the *FMO1* coding regions (Hines et al. 2003). One of these polymorphisms, *FMO1*6* (found in approximately 13% of African-Americans, 29.6% of Hispanic-Americans, and 10.8% of northern European-Americans), lies within a highly conserved YY1 regulatory domain. The *FMO1*6* mutation (–9536C > A transversion) impedes the binding of the YY1 transcription factor, resulting in a two- to three-fold loss of FMO1 promoter activity in HepG2 cells (Hines et al. 2003). However, two homozygous *FMO1*6* individuals were found to have FMO1 protein levels in the upper quartile of their age bracket, suggesting that variations in the YY1 domain are not solely responsible for regulating FMO1 expression (Hines et al. 2003).

7.3.2. FMO3

7.3.2.1. FMO3 Function. FMO3, the major hepatic FMO in humans, is present at levels of ≤100 pmol/mg microsomal protein (Overby et al. 1995). FMO3 mRNA has been detected at low levels in fetal liver, kidney, and lung and at higher levels in adult liver and kidney (Dolphin et al. 1996); however, mRNA levels do not always correlate with protein expression. A great deal of interindividual variation in hepatic FMO3 protein levels and activity has been reported, with as much as 20-fold variation in protein concentration (Koukouritaki et al. 2002) and similar variability in functional activity measurements (Stormer et al. 2000). FMO3 substrates include the antipsychotic drug clozapine (Tugnait et al. 1997), the H_2-receptor blockers cimetidine (Ziegler 1990) and ranitidine (Kang et al. 2000), nonsteroidal antiinflammatory agents benzydamine (Ubeaud et al. 1999) and sulindac sulfide (Hamman et al. 2000), (*S*)-nicotine (Cashman et al. 1992), and the primary biogenic amines tyramine and phenylethylamine (Lin and Cashman 1997a, 1997b) (Figure 7.7). FMO3 is the primary route of metabolism for the gastroprokinetic agent itopride (Mushiroda et al. 2000).

FMO3 displays a marked preference for tertiary amine substrates, such as the endogenous compounds tyramine and trimethylamine, but it can also accommodate nucleophilic primary and secondary amines (Figure 7.7) as well as nucleophilic sulfides, often resulting in the generation of stereoselective sulfoxides (Cashman 1998). Tyramine is an indirectly acting sympathomimetic that can be either formed via endogenous biosynthetic pathways or ingested in elevated amounts in foods such as cheese, yeast products, fermented foods, beer, and wine (Lin and Cashman 1997a). Additionally, tyramine may also potentiate histamine toxicity (Taylor 1986) and trigger migraine headaches (Crook 1981). It is possible that FMO3 polymorphisms that result in decreased FMO3 activity may predispose susceptible individuals to variable tyramine tolerance and associated symptoms.

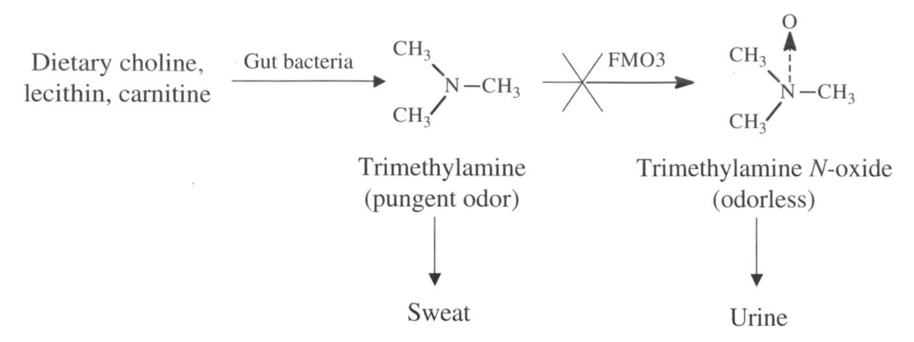

Figure 7.8. Mechanism of trimethylaminuria caused by FMO3 deficiency.

7.3.2.2. Trimethylaminuria (TMAU). Trimethylaminura (fish-odor syndrome) is an autosomal recessive disorder caused by the inability to convert trimethylamine, a malodorous compound, to its innocuous *N*-oxide metabolite (Figure 7.8). Trimethylamine is generated in vivo from bacterial degradation of dietary choline, carnitine, and lecithin (Holmes et al. 1997), which are found in foods such as egg yolk, liver, kidney, legumes, soybeans, peas, shellfish, and saltwater fish (Ruocco et al. 1989). In addition to the "fishy" body odor, TMAU often presents with accompanying constellation of psychosocial dysfunction, including social isolation, clinical depression, and increased risk of suicide (Ayesh et al. 1993). Case reports of individuals with trimethylaminuria have also reported concurrent seizures and behavioral disturbances (McConnell et al. 1997), hypertension (Treacy et al. 1998), transient trimethylaminuria (Mayatepek and Kohlmuller 1998), perimenstrual exacerbations (Zhang et al. 1995), dietary exacerbations (precipitated by cow's milk) (Rothschild and Hansen 1985), urticaria and sulfur drug intolerance (Treacy et al. 1998), and migraine possibly due to altered tyramine metabolism (Akerman et al. 1999).

TMAU has been documented in the white Caucasian population for almost 300 years and was rumored to be a major cause of suicide among concubines in the Thai Sukhothai era more than 700 years ago (Thithapandha 1997). About 1% of the white British population and up to 11% of the New Guinean population are thought to be carriers of TMAU (Zhang et al. 1996; Mitchell et al. 1997). Normal individuals metabolize TMA to trimethylamine *N*-oxide (TMANO) with a ratio of TMA:TMANO greater than 95:5 and excrete less than 18 μmol of TMA/μmol creatinine (Treacy et al. 1995; Zhang et al. 1995), whereas affected individuals excrete more than 46 μmol TMA/μmol creatinine (Akerman et al. 1999), with a TMA:TMANO ratio of less than 80:20 (Hadidi et al. 1995).

Primary TMAU is a consequence of metabolically defective genetic variants of FMO3.

7.3.2.3. FMO3 Mutations. No fewer than 27 allelic variants have been reported for FMO3 (see Table 7.3), many of which have been determined to compromise catalytic activity. Most were derived from patients exhibiting TMAU. This latter group includes P153L (Dolphin et al. 1997), E305X

Table 7.3 FMO3 Genetic Variants and Trimethylaminuria Phenotype

Accession Number	Systematic Name	Trivial Name	Phenotype
1	c.472G > A	E158K	Normal
2	c.198G > T	M66I	TMAU
3	c.458C > T	P153L	TMAU
5	c.769G > A	V257M	Normal
6	c.154G > A	A52T	TMAU
7	c.923A > G	E308G	Normal
8	c.940G > T	E314X	TMAU
9	c.1160G > T	R387L	TMAU
10	c.1474C > T	R492W	TMAU
11	c.428T > A	V143E	TMAU
12	c.569T > C	I199T	TMAU
13	c.1424G > A	G475D	TMAU
14	c.442G > T	G148X	TMAU
15	c.182A > G	N61S	TMAU
16	c.539G > T	G180V	Normal
17	c.1302G > A	M434I	TMAU
18	c.245T > C	M82T	TMAU
19	2092_10145del		TMAU
20	c.394G > C	D132H	Normal
21	c.668G > A	R223Q	TMAU
22	c.1079T > C	L360P	Normal
23	c.1084G > C	E362Q	Normal
24	c.1507G > A	G503R	Normal
25	c.172G > A	V58I	TMAU
26	c.594T > A	D198E	Unknown
27	c.613C > T	R205C	TMAU
28	c.94G > A	E32K	TMAU
29	c.191A > del	E64R M65D M66X	TMAU

Source: http://human-fmo3.biochem.ucl.ac.uk/Human_FMO3/.

(Treacy et al. 1998), E314X (Akerman et al. 1999), M434I (Dolphin et al. 2000), M661I, and R492W (Akerman et al. 1999).

Although TMAU is most often caused by a single mutation in the *FMO3* gene (primary TMAU) that results in a dysfunctional enzyme, this metabolic syndrome can also be caused by nongenetic factors, including adult hepatitis or hepatic pathology (Marks et al. 1978; Ruocco et al. 1989; Mitchell et al. 1999) and dietary precursor overload (Marks et al. 1976). An international database has been created to catalog novel SNPs of *FMO3*, with the majority of the current entries derived from individuals affected by TMAU (see http://human-fmo3.biochem.ucl.ac.uk/Human_FMO3/). African-Americans represent a higher percentage of patients (33%) than found in the general population (12%) (Cashman et al. 2003), but white, Hispanic, and Asian individuals are also affected. Throughout all examined ethnic groups, women make up a significantly greater fraction (87%) of the affected population than do their male counterparts (Cashman et al. 2003). However, in some instances, individuals with either FMO3 SNPs thought to be benign (e.g., E158K/E308G and V277A) or lacking FMO3 SNPs altogether exhibited severe symptoms, suggesting that *FMO3* coding sequence variation is not the only factor involved in development of TMAU (Cashman et al. 2003). Clearly, TMAU is a multifactorial disorder that can be caused by both genetic variation and dietary/environmental and gender influences.

7.3.3. FMO2

FMO2 is expressed predominantly in human lung tissue, but it is present in most humans as a nonfunctional, truncated protein (designated hFMO2*2) lacking 64 amino acids from the *C* terminus (Dolphin et al. 1998). All other mammals examined to date, including nonhuman primates, express a full-length, functional protein (Dolphin et al. 1998). However, ethnically diverse genotyping has shown that 26% of African-Americans and a smaller percentage of Mexican-Americans possess at least one copy of the allele encoding full-length protein (designated hFMO2*1), which is able to catalyze the bioactivation of thioureas, thioacetanilide, and thiobenzamide (Krueger et al. 2002). Because the lung is a major route of entry of many environmental toxicants, it is possible that the presence of catalytically active FMO2 may predispose polymorphic individuals to increased risk of adverse events caused by exposure to inhaled toxicants.

7.3.4. FMO4

FMO4 protein is expressed at extremely low levels in many animal tissues and appears to be the predominant isoform expressed in the rat brain

(Furnes et al. 2003). However, its significance in human brain (and other) tissues is unknown. FMO4 is structurally unique amongst the FMOs because of a *C*-terminal extension of 20–25 amino acids downstream of the "normal" stop codon found in the other enzymes (Itagaki et al. 1996). Also, some FMO4 transcripts have a premature start codon that is eliminated from a fraction of the transcript population by alternative splicing (Itagaki et al. 1996). Little is known about the substrate preferences of FMO4, but preliminary studies suggest it is catalytically active against some tricyclic antidepressants and nonsteroidal antiinflammatory agents (Lang and Rettie 1998). More recently, several polymorphisms were identified in the *FMO4* gene—I37T, V323A, and E339Q present in 1 in 100, 27 in 100, and 1 in 100 samples, respectively (Furnes et al. 2003). However, nothing is known about the functional consequences (if any) of these polymorphisms.

7.3.5. FMO5

FMO5 is a minor component of human liver microsomes, present at approximately one third the concentration of hepatic FMO3 and also at low levels in fetal liver microsomes (Overby et al. 1995). Recombinantly expressed human FMO5 does not catalyze the oxidation of methimazole, a classic FMO substrate, but does catalyze the oxidation of the straight-chain primary amine, *n*-octylamine (Overby et al. 1995). More recent reports suggest that FMO5 mRNA is upregulated by progesterone in cultured breast cancer cells, thereby enhancing the FMO-mediated carcinogenicity of tamoxifen in target tissues. However, this work was performed in subcultured breast cancer cell lines, so its application to in vivo situations is unclear (Miller et al. 1997). Despite its ability to oxidize *n*-octylamine and its possible link to tamoxifen toxicity, FMO5 exhibits severely restricted substrate specificity for most xenobiotics tested to date. It is not currently considered to be a significant xenobiotic metabolizing enzyme.

7.4. SUMMARY AND FUTURE DIRECTIONS

Polymorphisms in the genes encoding the cytochrome P450 and FMO family of enzymes contribute to interindividual differences in the biological effects of xenobiotics by affecting the peak concentration and duration of exposure of the parent molecules or active metabolites in the body. The most penetrant mutations are associated with the metabolic clearance of drugs (e.g., warfarin–CYP2C9; nortrityline, CYP2D6; tacrolimus, CYP3A5), and these may be the targets of prospective pharmacogenetic testing in order to improve drug safety and efficacy.

In theory, the very same mutations could affect the metabolic clearance of nontherapeutic xenobiotics and their associated toxicity. However, this does not appear to be widespread, primarily because molecules with serious toxicity are not metabolized exclusively by the most penetrant and polymorphmic CYP/FMO enzymes. In addition, common mutations in the genes coding for enzymes that often play a much greater role in the bioactivation or elimination of toxic molecules (e.g., CYP1A1, CYP1A2, CYP1B1, CYP2E1) do not appear to elicit profound changes in enzyme function. These are interesting observations from a teleological perspective, but the practical reality is that genetic contributions to interindividual differences in the risk of toxicity from exposure to foreign molecules will be multifactorial, often involving multiple enzymes that can catalyze the same critical reaction(s) and genes encoding proteins that regulate *CYP* and *FMO* gene expression (i.e., transcription factors) and protein degradation. Risk of toxic effects may also depend on the relative efficiency of phase II (conjugation) or phase III (transport) processes that work to eliminate many products of oxidative metabolism. For example, the toxicity of benzene appears to depend on polymorphisms affecting the bioactivation step(s) as well as the mutations in the *NQO1* and *GST* genes that are involved in the detoxification of the chemically reactive metabolites.

Environmental factors will also modify the impact of many high- and low-penetrance gene mutations by affecting the synthesis, degradation, or function of the polymorphic enzyme as well as secondary enzymes that catalyze the same or parallel metabolic event. For example, induction of CYP1A2 activity by cigarette smoking can mask the penetrance of a CYP2A6-catalyzed reaction if both enzymes contribute to the elimination of the molecule of interest. Conversely, enhanced degradation of CYP1A2 in response to viral infection may enhance the penetrance of a parallel polymorphic enzyme process. Numerous molecules found in the diet and industrial environment and normal and pathophysiological disease processes have been shown to affect steady-state CYP and FMO expression and function. Thus, a complete understanding of individual risk to xenobiotic-induced disease will require careful assessment of all genetic and environmental factors that impinge on the efficiency of the relevant metabolic processes. Although this may be quite daunting to implement experimentally, it should ultimately lead to a more satisfying and useful understanding of our inherent differences in xenobiotic metabolism and disease risk.

8

Polymorphisms in Xenobiotic Conjugation

Helen E. Smith, David L. Eaton, and Theo K. Bammler
University of Washington, Seattle, WA

8.1. INTRODUCTION

There are many enzymes in the body that participate in the metabolism and elimination of endogenous compounds and xenobiotics. These biotransformation reactions usually aid in the ultimate elimination of such compounds, although in some cases the reactions bioactivate parent compounds. Typically, biotransformation reactions are categorized into three processes: phase I, phase II, and phase III. Phase I reactions are generally oxidative, and usually increase by a small amount the hydrophilicity of the original compound by the introduction of —OH, —NH_2, —SH, or —COOH functional groups. Phase II reactions usually greatly increase the hydrophilicity of the original compound or the compound modified by a phase I reaction, promoting the excretion of the compound in urine or feces (via the bile). Phase III processes represent carrier-mediated transport that facilitates the movement across biological membranes of the water-soluble products formed by phase I and phase II reactions.

This chapter focuses on phase II reactions including glucuronidation, sulfation, acetylation, methylation, conjugation with glutathione, and reduction of quinones. These reactions are carried out by UDP–glucuronosyltransferases (UGTs), sulfotransferases (STa; SULTs), *N*-acetyltransferases (NATs), catechol *O*-methyltransferase (COMT),

Gene-Environment Interactions: Fundamentals of Ecogenetics, edited by Lucio G. Costa and David L. Eaton

glutathione *S*-transferases (GSTs), and NAD(P)H:quinone oxidoreductases (NQOs), respectively. Each of these enzyme classes is coded by multiple genes that are often polymorphic in humans, resulting in variable biotransformation capabilities among people. These differences may increase, or decrease, the susceptibility of some individuals to endogenous compounds or exposures to environmental contaminants or medications, or alter the efficacy of medications (Parkinson 1996).

8.2. UDP-GLUCURONOSYLTRANSFERASES

The conjugation of compounds with glucuronic acid is a major biotransformation reaction known as *glucuronidation*. This reaction is carried out by the UDP–glucuronosyltransferases (UGT) family of microsomal enzymes using uridine diphosphoglucuronic acid (UDP–glucuronic acid) as a cofactor. Glucuronidation enhances the clearance of a wide variety of compounds, including drugs, dietary chemicals, environmental pollutants, chemical carcinogens, phase I oxidation products, and endogenous compounds such as bilirubin, bile acids, steroid hormones, and thyroid hormones. The substrates for these enzymes are aliphatic alcohols and phenols, carboxylic acids, primary and secondary aromatic and aliphatic amines, free sulfhydryl groups, and nucleophilic carbon atoms, all which have the nucleophilic heteroatoms necessary for conjugation (Parkinson 1996; Miners et al. 2002). Figure 8.1 illustrates an example of an UGT-mediated

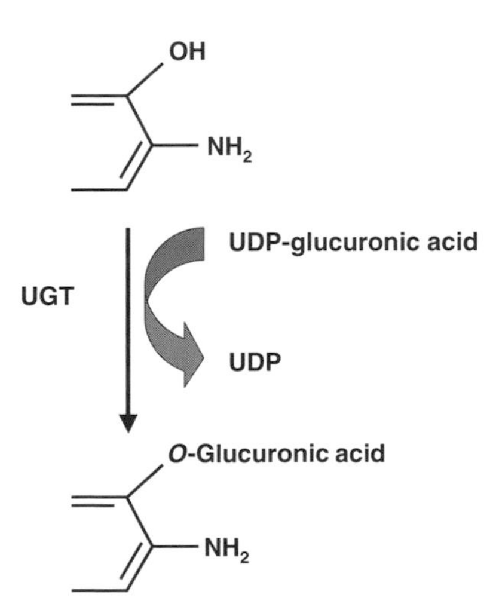

Figure 8.1. Example of UGT glucuronidation reaction: glucuronidation of aromatic amine.

Table 8.1 UGTs Tissue Distribution and Substrates

Isoforms	Tissue Distribution	Example Substrates
1A1	Liver[a]	Menthol[a] Tolbutamine[a] Rifamycin[a] Acetaminophen[a] Irinotecan[a] Indinavir[a] Bilirubin[a] Estrogens[a]
1A3		
1A4		
1A5		
1A6		Xenobiotic phenols[b]
1A7	Upper GI tract (orolaryngeal mucosa,[b] esophagus,[b] stomach,[b] lung); not expressed in liver[c]	Benzo(a)pyrene phenol[b]
1A8	Small intestine and colon mucosa[b]	
1A9		
1A10	Small intestine and colon mucosa[b]	
2A1		
2B4		
2B7		Wide range of xenobiotics[c] Wide range of hydroxysteroids[c] Zidovudine[c]
2B10		
2B11		
2B15		Many plant-derived chemicals[c] 17B-hydroxyandrogens[c]
217		

[a] Bosma (2003).
[b] Bock (2003).
[c] Miners et al. (2002).

conjugation reaction. Table 8.1 lists examples of UGT substrates by UGT isoform.

Most glucuronide products are polar and water-soluble, allowing rapid and efficient excretion in urine or bile. As a result, glucuronidation is usually considered to be a detoxification reaction. There is evidence, however, that the acyl glucuronides produced from glucuronidation of carboxylic acid substrates may be reactive intermediates that can covalently bind to endogenous proteins and other macromolecules, potentially contributing to toxic responses (Bailey and Dickinson 2003).

The UGTs are a super family of enzymes grouped into two families: UGT1 and UGT2. More than 16 functional human UGTs have been characterized (Bock 2003). The UGT1 members are formed by alternative splicing of one gene locus on chromosome 2q37, while the UGT2 members are products of distinct genes clustered on chromosome 4q13. Table 8.1 lists the locations where the UGT isoforms are known to be expressed.

The UGT1 gene consists of five exons, but is unusual in that there are 13 different variants of exon 1, most with their own promoter region. The remainder of the coding region (exons 2–5) does not vary, and this region includes the cofactor binding site and the membrane insertion portion. A mutation in the UGT1 gene common region (exons 2–5) can result in the elimination of expression of this entire family. In contrast, a mutation in exon 1 of a UGT1 or a mutation in any of the UGT2 genes only affects that particular gene product (Bosma 2003). The different gene products of the UGT1 family arise via the splicing of different "versions" of exon 1 to the common region (exons 2–5). Although there are 13 variants of exon 1, and therefore 13 possible versions of this gene family, UGT1 produces only nine functional proteins, as two of the exon 1 variants are pseudogenes, another contains a frameshift mutation that produces a truncated and inactive protein, and yet another contains a G–C clamp that inhibits strand separation and transcription (Bosma 2003). Figure 8.2 illustrates the complex structure of the UGT1 gene.

UGT isoforms 1A1, 1A6, 1A7, 2B4, 2B7, and 2B15 are polymorphic in humans (see Table 8.2). Because of inadequate information regarding substrate specificity and substrate binding regions, the functional significance of these polymorphism variants is not clear (Miners et al. 2002). To complicate matters further, expression of these enzymes changes during

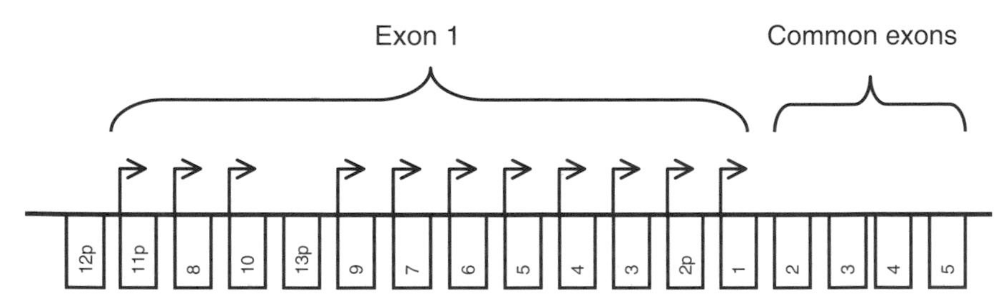

Figure 8.2. Structure of the human UGT1 gene. Exon 1 consists of 13 unique exons. Except for exons 12p and 13p, each has its own promoter region, indicated by an arrow. Exon 11p is not transcribed, while exon 2p produces a truncated, inactive protein.

Table 8.2 UGT Polymorphisms

Isoform	Allele	Polymorphism	Significance	Allelic Frequency
1A1	1A1*28[a]	(TA)7TAA[a]	Gilbert syndrome in Caucasians[a]	Caucasians, 35–40%; other, varies[a]
	1A1*27[a]	211G > A[a]	Possibly associated with Gilbert syndrome in Asians[a]	Not known[a]
	Not named[a]	1456T > G[a]	Possibly associated with Gilbert syndrome in Asians[a]	Not known[a]
1A6	1A6*2[a]	Thr181Ala and Arg184Ser[b]	Lower rate of phenol glucuronidation in vitro[b]	17–29%[b]
	NA	Arg184Ser[b]	Lower rate of phenol glucuronidation in vitro[b]	0.5–2%[b]
1A7	1A7*1[b]	Asn129, Arg131, Trp208[b]	Wildtype	36%
	1A7*2[b]	Lys129,131; Trp208[b]	Lower V_{max} toward phenolic derivatives of benzo(a)pyrene glucuronidation in vitro[b,c]	26%
	1A7*3[b]	Lys129,131; Arg208[b]	Lower V_{max} toward phenolic derivatives of benzo(a)pyrene glucuronidation in vitro[b,c]	36%
	1A7*4[b]	Asn129; Arg131,208[b]	Lower V_{max} toward phenolic derivatives of benzo[*a*]pyrene glucuronidation in vitro[b,c]	2%
2B4	NA	Asp458Glu[b]	Does not alter substrate specificity[b]	Asp^{458} threefold higher than the Glu^{458} variant in Caucasians[b]
2B7	2B7*1[b] 2B7*2[b]	His268[b] Tyr268[b]	No marked difference between the two variants toward various substrates[b]	Allelic frequency about equal in Caucasians; *1 allele threefold more prevalent in Japanese[b]
2B15		Asp85[b] Tyr85[b]	V_{max} for dihydrotestosterone glucuronidation may be lower for Asp85[b]	Approximately equally distributed[b]

[a] Bosma (2003).
[b] Miners et al. (2002).
[c] Bock (2003).

development (de Wildt et al. 1999), and there are substantial interindividual differences in the levels of expression in liver and other organs (Mackenzie et al. 2003).

It is known that various polymorphisms in 1A6, 1A7, and 2B15 can affect the catalytic activity of the enzymes toward some substrates (Miners et al. 2002). The UGT1A1 polymorphisms are the only ones whose functional significance has been clearly demonstrated. Individuals with altered expression and activity of this gene may suffer from hyperbilirubinemia in the form of Crigler–Najjar syndromes type I and type II, which may result in serious mental retardation, or Gilbert syndrome, which is idiopathic (e.g., lacks any clear pathology). The two Crigler–Najjar syndromes result from the rare absence (type I) or very low expression (type 2) of UGT1A1, due to mutations in coding or promoter regions. As of 2003, over 35 mutations have been identified in Crigler–Najjar syndrome type I patients and 18 in type II patients (Bosma 2003). Gilbert syndrome is a milder hyperbilirubinemia inherited as an autosomal recessive trait. It usually occurs as a result of a polymorphism in the promoter where a TA repeat occurs 7 or 8 times, rather than the normal 6 times. The syndrome may also result from missense mutations in the coding region. In individuals with Gilbert's syndrome, the activity of bilirubin glucuronosyltransferase is about 30% of normal. In addition to causing mild hyperbilirubinemia (generally recognized only after fasting), these polymorphisms could conceivably put persons at risk for adverse drug reactions by causing an altered metabolism that decreases clearance, therefore increasing drug toxicity. For example, these polymorphisms may contribute to adverse reactions to the chemotherapeutic agent irinotecan, and the anticonvulsant lamotrigine (Miners et al. 2002; Bosma 2003; Bock 2003).

8.3. SULFOTRANSFERASES

The sulfotransferase family of enzymes (STs) is responsible for the sulfonate (SO_3^-) conjugation of a variety of xenobiotics and endogenous compounds. This reaction is often referred to as *sulfuryl transfer*, *sulfation*, or *sulfonation*. The conjugation reaction occurs via nucleophilic attack by the substrate on the electrophilic sulfur in 3′-phosphoadenosine-5′-phosphosulfate (PAPS), the cofactor used in the reaction. This cleaves the phosphosulfate bond in PAPS and transfers the sulfonate (SO_3^-) to the substrate. Sulfonation reactions are classified by the acceptor involved in the reaction. The presence of an alcohol functional group on a substrate leads to an *O*-sulfonation reaction producing an ester. *N*-Sulfonation reactions produce amides, whereas *S*-sulfonations produce thioesters. Of the

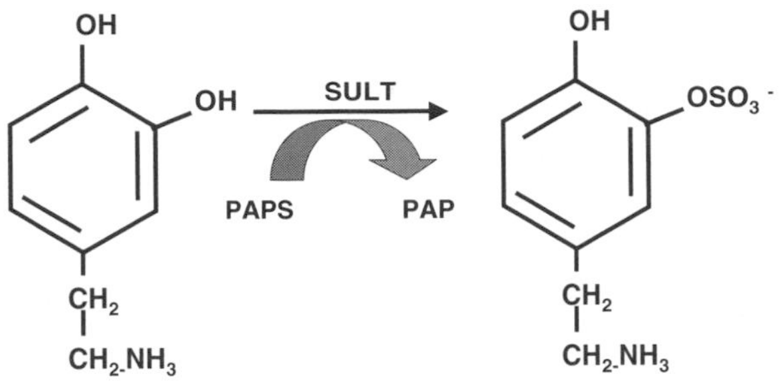

Figure 8.3. Example of SULT conjugation reaction: sulfonation of dopamine.

various types of reactions, *O*-sulfonations predominate. Figure 8.3 illustrates an example of a ST-mediated conjugation reaction. ST substrates include a large array of compounds. Some example substrates are listed in Table 8.3.

Physiologically, the STs are important for activating endogenous compounds needed for cell growth and/or development, the biosynthesis of heparin, and the inactivation of endogenous signaling molecules such as steroids, thyroid hormones, and neurotransmitters, including the catecholamines dopamine and norepinephrine. STs also biotransform many xenobiotics, such as phenolic drugs (Negishi et al. 2001; Glatt et al. 2000, Strott 2002). These biotransformation reactions increase the water solubility of most substrates, allowing them to be excreted, predominantly in the urine and sometimes in bile. Lipophilic substrates are even converted to amphiphiles that, because of the low pK_a of sulfonates, will remain ionized at physiological pH.

However, in some cases sulfonation reactions bioactivate compounds. For example, minoxidil, a potent vasodilator, is sulfonated to its active form by ST enzymes (Strott 2002). Other conjugates are strong electrophiles and may damage cellular macromolecules such as DNA and proteins by covalently binding to them. The sulfo conjugates of benzylic and allylic alcohols and of aromatic hydroxylamines may cause cellular damage due to their electrophilicity (Glatt et al. 2001).

STs are classified according to whether they are soluble or membrane-bound. Membrane-bound STs are embedded in the *trans*-Golgi complex where they posttranslationally modify cellular carbohydrates, peptides, and proteins. They have not been found to metabolize xenobiotics (Strott 2002). Therefore, this chapter will focus on soluble STs.

Soluble STs are now referred to as "SULTs" followed by numbers and letters indicating family and subfamily categories. SULT substrates are usually relatively low-molecular-weight compounds (Strott 2002;

Table 8.3 ST Tissue Distribution and Substrates

Isoform	Chromosome Location	Tissue Distribution	Example Substrates[a]
SULT1A1	16p11.2–12.1[b]	Liver—highest SULT1 expressed hepatically;[b] present in other tissues—platelets, placenta, adrenal gland, endometrium, colon, jejunum, brain, leukocytes[b,c]	Phenols—most important SULT for biotransformation of these compounds[b,d] Benzylic alcohols[b] Secondary nitroalkane 2-nitropropane[b] Aromatic amines[b] Hydroxylamines[b] Hydroxamic acids[b] *N*-Oxide minoxidil[b] Phenolic steroids[b] Catecholamines[d] 4-Nitrophenol (standard substrate)[b] Catecholamines[d]
SULT1A2	16p11.2–12.1[b]	Liver[b,c] Bladder tumors[c] Other tissues[b]	Most efficient human enzyme for sulfonate ion of several aromatic hydroxylamines and hydroxamic acids[b] Phenolic drugs[d] Catecholamines[d] Substrate specificity not well understood[b]
SULT1A3	16p11.2–12.1[b]	Very high in jejunum and colon[b,c] Present in other tissues: platelets, placenta, brain, leukocytes[c] Negligible in adult liver[b,c]	Phenols Dopamine-selective substrate[b] Catecholamines[d]
SULT1B1	4q13.1[b]	Highest in colon and GI tract[b] Lower levels in liver and leukocytes[b,c]	Thyroid hormones[b,d] Phenols[b] Benzylic alcohols[b]
SULT1C1	2q11.2[b]	Fetal lung and kidney[b] Kidney[c] Stomach[c] Thyroid gland[c]	Specific substrate has not been idenitified[b] Some thyroid hormomes[b]

Table 8.3 *Continued*

Isoform	Chromosome Location	Tissue Distribution	Example Substrates[a]
SULT1C2	2q11.2[b]	Some fetal tissues[b] Kidney[c] Ovary[c] Spinal cord[c]	4-Nitrophenol[b] *N*-Hydroxy-2-acetylaminofluorene[b] 17B-estrone[b] Bisphenol A[b] 4-Octylphenol[b] Nonylphenol[b] Diethylstilbestrol[b] 1-Hydroxymethylpyrene[b]
SULT1E1		Liver, endometrium, jejunum, mammary epithelial cells, adrenal gland, liver, small intestine, fetal kidney, fetal lung, fetal liver[b,c]	17B-estrone[b] 17B-estradiol[b] Thyroid hormomes[b] 4-Hydroxylonazolac[b] Pregnenolone[b] DHEA[b] Diethylstilbestrol[b] 1-Naphthol[b] Naringenin[b]
SULT2A1		Fetal adrenal gland[b] Adult liver, adrenal gland, jejunum[b,c]	Prototype substrate: DHEA[b,d] 17B-estradiol[b,d] Androsterone[d] Bile acids[d] Pregnenolone[d] Testosterone[d] Estrone[d]
SULT2B1a,b	19q13.4	Placenta, prostate, trachea[c]	3B-hydroxysteroids[b] Pregnenolone—2B1a preferential substrate[b,c] Cholesterol—2B1b preferential substrate[b,c]
SULT3**			
SULT4A1	22q13.1–2[b]	Brain[b–d]	No substrates yet found[b]
SULT5***			
Membrane bound			

[a] The SULTs biotransform a wide array of endogenous compounds including hormones and neurotransmitters, as well as drugs and xenobiotics. Their substrates specificity overlaps (Glatt et al. 2001; Strott 2002).
[b] Glatt and Meinl (2004).
[c] Glatt et al. (2001).
[d] Strott (2002).
[e] Not yet detected in humans. Expressed in rabbits. Murine sequence deposited in Genbank (Glatt and Meinl 2004).
[f] Not yet detected in humans. Murine sequence deposited in Genbank (Glatt and Meinl 2004).

Glatt et al. 2001). In mammals, 44 soluble SULTs have been identified that make up five families. The first two families are the largest and best characterized. Of these SULTs, 11 distinct proteins, encoded by 10 genes located on various chromosomes, have been identified in humans (Glatt and Meinl 2004) (see Table 8.3). Of these, SULTs 1A1, 1A3, 2A1, and 1E1 have been purified from human tissues. In humans, gene family members share amino acid sequence identity of at least 60% for SULTs, compared to 45% for STs in general, with the exception of SULT4A1, which has less than 36% amino acid homology to the other SULTs. The SULTs are expressed primarily in the liver, kidney, intestinal tract, lung, platelets and brain, and have a large variety of overlapping substrates (Strott 2002; Glatt et al. 2001).

Polymorphisms have been identified for SULTs 1A1, 1A2, 1A3, 1A4, 1B1, 1C1, 1C2, and 2A1 (Glatt and Meinl 2004). Table 8.4 lists the known polymorphisms for these enzymes, along with their allelic frequency and significance of the polymorphisms with regard to activity toward substrates.

SULT1A1 has only 1 variant that is common (1A1*2, allele frequency ~30%). SULT1A2 also has two relatively common nonsynonymous SNP variants (1A2*2, allele frequency ~30%, and 1A2*3, allele frequency about 18%), but neither seem to change catalytic function significantly (Glatt and Meinl 2004).

For SULT1A3, 9 SNPs have been identified, but only one, designated SULT1A3*2, resulted in an amino acid change, (Lys234Asn), and was found only in African-Americans. The change in amino acid sequence didn't change the kinetic characteristics of the enzyme, but did appear to increase the rate of protein degradation (Thomae et al. 2003). It also appears that some individuals have two copies (gene duplication) of SULT1A3, with the second copy of the gene identified as SULT1A4, both of which appear to be transcriptionally active (Hildebrandt et al. 2004). Several variants in SULT1A4 were identified that appeared to have reduced activity because of a decrease in protein levels. As with the SULT1A3*2 variant, these were seen only in African-American populations in that study. Additional variants in SULT1A3 were also identified, including two cSNPs (Cys302Thr, and Cys302Ala) that seemed to decrease immunoreactive protein but had little effect on catalytic activity (Hildebrandt et al. 2004).

More recent studies are providing a better understanding of the role of SULT polymorphisms. The 1A1 variants affect the enzyme function in platelets; however, their significance in liver function is not clear. The 1A2 variants may have differential activity toward promutagens such as *N*-hydroxy-2-acetylaminofluorene, potentially making some people more susceptible to the adverse affects of exposure to such compounds (Glatt et al. 2001; Glatt and Meinl 2004). Polymorphisms found in SULT1C1 alter catalytic activity in COS-1 cells; however, it is not known whether these

Table 8.4 ST Polymorphisms

Isoform	Allele[a]	Polymorphism	Significance	Allelic Frequency[b]
SULT1A1	1A1*1	Reference allele	The variant forms of SULT1A1	0.67
	1A1*2	Arg213His	may have	0.32
	1A1*3	Met223Val	altered activity	0.01
	1A1*4	Arg37Gly	in platelets, but	0.003
	1A1*V	Ala147Thr, Glu181Gly,	do not appear to affect activity in	NA
	1A1*VI	Arg213His Pro90Leu, Val243Ala	liver	NA
SULT1A2	1A2*1	Reference allele	May alter enzymatic activity toward	0.51
	1A2*2	Ile7Thr, Asn235Thr	procarcinogens	0.29
	1A2*3	Pro19Leu		0.18
	1A2*4	Ile7Thr, Arg184Cys, Ans235Thr		<0.01
	1A2*5	Ile7Thr		<0.01
	1A2*6	Asn235Thr		<0.01
SULT1A3[c]	1A1*1A-K		9 SNPs, resulting in 11 haplotypes (*2 below)	
	1A3*2	Lys234Asp	Only nonsynonimous SNP; seen only in African-Americans; protein half-life reduced	0.045[c]
SULT1A4[d]	1A4*1		Gene duplication of SULT1A3	
	1A3/1A4*3	Pro101Leu	80% decrease in activity; seen only in African-Americans	0.025[d]
	1A3/1A4*4	Pro101His	50% decrease in activity; seen only in African-Americans	<0.01[d]
	1A3/1A4*5	Arg144Cys	Modest change in V_{max}	0.025[d]

Table 8.4 *Continued*

Isoform	Allele[a]	Polymorphism	Significance	Allelic Frequency[b]
SULT1C1	1C1*1	Reference allele	These polymorphisms probably will not have any physiological significance since other SULTs have higher activity toward the 4-nitrophenol substrate evaluated and have greater expression	0.905
	1C1*2	Ser193Ala		0.067
	1C1*3	Asp60Ala		0.011
	1C1*4	Arg73Gln		0.011
SULT1C2	1C2*I	Reference allele	NA	NA
	1C2*II	Asp5Glu		NA
SULT2A1	2A1*I	Reference allele	NA	NA
	2A1*II	Met57Thr		
	2A1*III	Glu186Val		
	2A1*IV	Met57Thr, Glu186Val		
	2A1*VI	Leu159Val		
	2A1*VII	Ala64Pro		
	2A1*VIII	Lys227Glu		
	2A1*IX	Ala261Thr		
	2A1*X	Ala64Pro, Ala261Thr		

[a] Roman numerals were polymorphism identifiers assigned by Glatt and Meinl (2004).
[b] Allelic frequency in Caucasians unless otherwise stated (NA—information not available).
[c] From Thomae et al. (2003).
[d] From Hildebrandt et al. (2004).

variations have any physiological significance. Unless SULT1C-specific substrates are identified, the importance of SULT1C polymorphisms may be overshadowed by the fact that other SULTs have greater expression and higher activity toward the 4-nitrophenol substrate (Glatt and Meinl 2004).

Many 2A1 variants have been found but as yet these polymorphisms have not been fully characterized nor some of their sequences verified (Thomae et al. 2002).

Of the various polymorphisms in SULTs, potential disease outcomes have been studied to any extent only for the SULT1A1 Arg213His variant. Colon and breast cancers are the most common disease states investigated for an association with this polymorphism, and no clear associations has been identified for either of these (Langsenlehner et al. 2004; Moreno et al. 2005).

8.4. *N*-ACETYLTRANSFERASES

N-Acetyltransferases (NATs) are cytosolic enzymes that catalyze the *N*- or *O*-acetylation of compounds with aromatic or heterocyclic amines or hydrazines, or their *N*-hydroxylated metabolites. These enzymes are highly expressed in the liver, and are also expressed in most other tissues in mammals. Acetylation reactions require acetyl–coenzyme A (acetyl–CoA) as a cofactor (Daly 2003; Hein 2002; Pompeo et al. 2002). These reactions are carried out in two sequential steps. First, the acetyl group from the cofactor is transferred to an active-site cysteine within the NAT, releasing coenzyme A. Next, the acetyl group is transferred from the acetylated enzyme to the amino group of a substrate (Parkinson 1996). Figure 8.4 illustrates an example of *N*-acetylation by NATs.

Acetylation of drugs and other chemicals may be either bioactivation or detoxification reactions, depending on the substrate. For example, *N*-acetylation may be a detoxification step for aromatic amines that cause urinary bladder cancer. In contrast, *O*-acetylation of aromatic and heterocyclic amines bioactivates these compounds to electrophilic intermediates potentially involved in initiation of colon cancer (Hein 2002; Miller et al. 2001).

There are only two distinct NAT genes in the human genome: *NAT1* and *NAT2*. Both are intronless genes on chromosome 8p22, but are independently regulated. *NAT1* is expressed in most tissues, whereas NAT2 expression appears limited to the liver and gastrointestinal tract. The two enzymes have overlapping substrate profiles, and no substrate is exclusively acetylated by one enzyme or the other (see Table 8.5). The widespread tissue distribution and expression very early in development suggests that NAT1 may have a developmental function (Sim et al. 2000).

Numerous polymorphisms have been identified in both NAT genes. Many of the NAT1 polymorphisms are synonymous (do not change amino

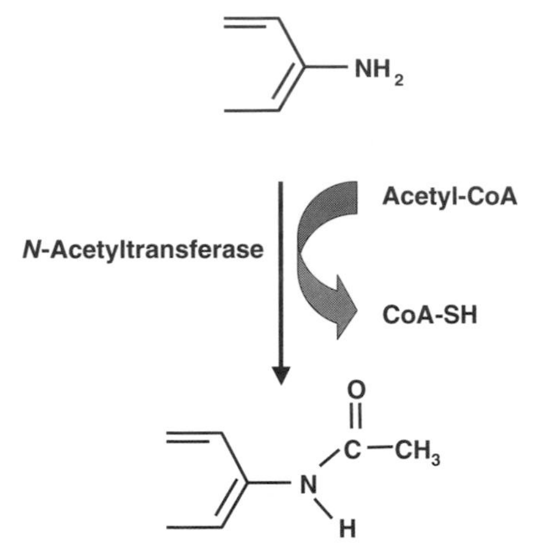

Figure 8.4. Example of NAT conjugation reaction: *N*-acetylation of aromatic amine.

Table 8.5 NAT Tissue Distribution and Substrates

Isoforms	Tissue Distribution	Example Substrates
NAT1	Liver[a] Gut[a] Leukocytes[a] Erythrocytes[a] Other[a]	*p*-Aminosalicylate[a] *p*-Aminobenzoic acid (*p*-ABA)[a] *p*-Aminobenzoyl glutamate[a]
NAT2	Liver[a] Intestinal[a] Epithelium[a]	Isoniazide[a,b] Sulfonamides[a] (sulfamethazine, sulfamethoxazole[b]) Dapsone[a] Arylamine carcinogens (aminofluorene)[a] Caffeine[b]

[a] Sim et al. (2000).
[b] Daly (2003).

acid sequence) (Hein 2002). However, in the NAT1*15 and NAT1*19 alleles, point mutations create a premature stop codon resulting in a non-functional protein (Sim et al. 2000; Daly 2003). These variant alleles are rare, as less than 1% of Caucasians are homozygous for them (Daly 2003). The functional activity of other NAT1 alleles may not be controlled entirely by

Table 8.6 NAT1 Polymorphisms

Allele[a]	Polymorphism	Significance	Allelic Frequency
NAT1*5	Arg117Thr, Arg166Thr, Glu167Gln[b]	NA	NA
NAT1*10	Reference allele[b]	Rapid acetylation[c]	NA
NAT1*11A	Val149Ile, Ser214Ala[b]	Slow acetylation[c]	NA
NAT1*11B	Val149Ile, Ser214Ala[b]	Slow acetylation[c]	NA
NAT1*11C	Sedr214Ala[b]	Slow acetylation[c]	NA
NAT1*14A	Arg187Gln[b]	Slow acetylation[c,d]	NA
NAT1*14B	Arg187Gln[b]	Slow acetylation[c,d]	NA
NAT1*15	Arg187Stop[b]	Protein not produced[b]	NA
NAT1*17	Arg64Trp[b]	NA	NA
NAT1*19	Arg33Stop[b]	Protein not produced[b]	NA
NAT1*21	Met205Val[b]	NA	NA
NAT1*22	Asp251Val[b]	NA	NA
NAT1*24	Glu261Lys[b]	NA	NA
NAT1*25	Ile263Val[b]	NA	NA

[a] There is linkage disequilibrium between NAT1*10 and NAT2*4. This combination occurs 3.5 times more frequently than expected by chance (Sim et al. 2000). There is also linkage disequilibrium between NAT1*15 and NAT2*4. This combination occurs more frequently than expected by chance (Sim et al. 2000).
[b] Hein (2002).
[c] Sim et al. (2000).
[b] Daly (2003).

genotype, as expression of the gene may be modified by the presence of certain substrates (Sim et al. 2000). Tables 8.6 and 8.7, respectively, list the NAT1 and NAT2 polymorphisms that affect the amino acid sequence of these enzymes.

The significance of the NAT1 polymorphisms is not fully understood and needs to be characterized further (Sim et al. 2000; Daly 2003). In contrast, the genotype–phenotype relationship for the NAT2 isoforms is well established, and represents one of the earliest examples that gave rise to the field of pharmacogenetics (Sim et al. 2000) (see also Chapter 2, on historical perspectives of ecogenetics). Numerous NAT2 polymorphisms, mostly cSNPs, change the rate of acetylation of compounds, giving rise to either "slow" or "rapid" acetylator phenotypes. Slow acetylators have various mutations in the NAT2 gene that decrease activity or enzyme stability. The incidence of slow acetylation varies by population. About 50% of people with European ancestry exhibit the NAT2 slow acetylator phenotype (Miller et al. 2001). The percentage of slow acetylators in northern Africans is 90%, but it is

Table 8.7 NAT2 Polymorphisms

Allele[a]	Polymorphism	Significance	Allelic Frequency
NAT2*4	Reference allele	Rapid acetylator[b]	NA
NAT2*5A	T341C, C481T	Slowest acetylator; high urinary bladder cancer risk with aromatic amines[b]	NA
NAT2*5B	T341C, C481T, A803G	Slowest acetylator; high urinary bladder cancer risk with aromatic amines[b]	NA
NAT2*5C	T341C, A803G	Slowest acetylator; high urinary bladder cancer risk with aromatic amines[b]	NA
NAT2*5D	T341C	Slowest acetylator; high urinary bladder cancer risk with aromatic amines[b]	NA
NAT2*5E	T341C, G590A	Slowest acetylator; high urinary bladder cancer risk with aromatic amines[b]	NA
NAT2*5F	T341C, C481T, C759T, A803G	Slowest acetylator; high urinary bladder cancer risk with aromatic amines[b]	NA
NAT2*6A	C282T, G590A	Slow acetylation[c]	NA
NAT2*6B	G590A	Slow acetylation[c]	NA
NAT2*6C	C282T, G590A, A803G	Slow acetylation[c]	NA
NAT2*6D	T111C, C282T, G590A	Slow acetylation[c]	NA
NAT2*7A	G857A	Slow acetylation[c]	NA
NAT2*7B	C282T, G857A	Slow acetylation[c]	NA
NAT2*10	G499A	NA	NA
NAT2*11	C481T	NA	NA
NAT2*12A	A803G	NA	NA
NAT2*12B	C282T, A803G	NA	NA
NAT2*12C	C481T, A803G	NA	NA
NAT2*13	C282T	NA	NA
NAT2*14A	G191A	NA	NA
NAT2*14B	G191A, C282T	NA	NA
NAT2*14C	G191A, T341C, C481T, A803G	NA	NA
NAT2*14D	G191A, C282T, G590A	NA	NA
NAT2*14E	G191A, A803G	NA	NA
NAT2*14F	G191A, T341C, A803G	NA	NA
NAT2*14G	G191AC282T, A803G	NA	NA
NAT2*17	A434C	NA	NA
NAT2*18	A845C	NA	NA
NAT2*19	C190T	NA	NA

[a] There is linkage disequilibrium between NAT1*10 and NAT2*4. This combination occurs 3.5 times more frequently than expected by chance (Sim et al. 2000). There is also linkage disequilibrium between NAT1*15 and NAT2*4. This combination occurs more frequently than expected by chance (Sim et al. 2000).
[b] Hein (2002).
[c] Sim et al. (2000).
[d] Daly (2003).

less than 10% in many Asian populations. Screening for three variant alleles of the NAT2 gene (NAT2*5, NAT2*6, and NAT2*7) detects the majority of slow acetylators in populations with northern European ancestry, although screening for 11 NAT2 SNPs will identify additional slow acetyaltors, and thus decrease misclassification and the population size needed for statistical analysis in association studies (Deitz et al. 2004).

Because of the large number of NAT polymorphisms and the common exposure to aromatic and heterocyclic amines that are activated to carcinogens by NATs, it is likely that the polymorphisms in NATs modify cancer risk (Hein 2002). The significance of the alterations in activity across the different NAT alleles varies depending on the disease state being evaluated. For example, the polymorphisms in NAT2 may modulate the risk of lung, bladder, breast, and colon cancer due to the acetylation of aromatic amines in tobacco smoke and cooked foods (Daly 2003).

More recent studies suggest that rapid acetylation by NAT2 increases risk for colon cancer but decreases risk for bladder cancer (Hein 2002; Miller et al. 2001; Thier et al. 2003). Rapidly acetylating NAT2 variants may modulate the risk of bladder cancer from exposure to aromatic amines (e.g., benzidine, 4-aminobiphenyl, and 1-naphthylamine) because NAT2 acetylation of these amines is considered a detoxification reaction. In general, slow acetylators appear to be at somewhat greater risk for developing bladder cancer (Hung et al. 2004), although this was not seen in a Chinese population (Ma et al. 2004). On the other hand, rapid acetylation of aromatic and heterocyclic amines that must be bioactivated to become carcinogenic increases the risk of developing colon cancer caused by these compounds.

There has also been some interest in determining whether the NAT genes are related to risk of stomach cancer. The study results have been somewhat mixed. A study by Lan et al. (2003) determined that the weight of evidence from previous studies and their own suggested that there is likely no association between the NAT1*10 or the rapid acetylator phenotype NAT2 genotypes and stomach cancer risk (Lan et al. 2003).

Many drugs acetylated by NAT2 are not used in modern medicine, except for isoniazide and the sulfonamides. Slow acetylators are more likely to suffer side effects from isoniazide than are normal acetylators.

8.5. METHYLTRANSFERASES

Methylation is generally a minor xenobiotic biotransformation pathway. It differs from the other phase II reactions because it does not result in products that are more water-soluble than the parent compound. Several enzymes carry out the methylation reaction. One of these is catechol *o*-

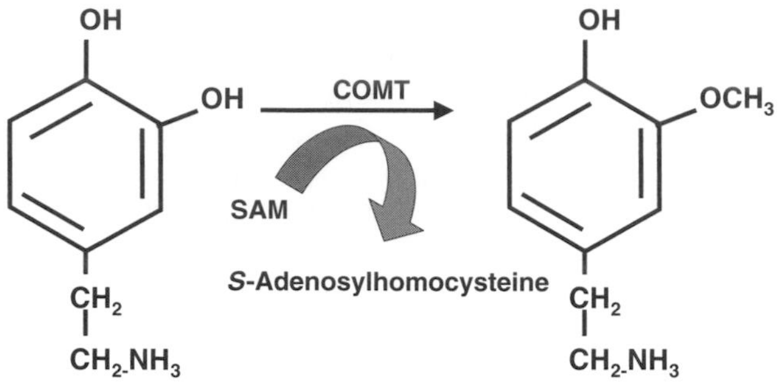

Figure 8.5. Example of methylation reaction: *O*-methylation of dopamine.

Table 8.8 COMT Tissue Distribution and Substrates

Tissue Distribution[a]	Example Substrates
Brain[b]	L-Dopa[c]
Endometrium[b]	Catechol estrogens[c]
Liver[b]	Dopamine[c]
Kidney[b]	Noradrenaline[c]
Adrenal gland[b]	3,4-Dihydroxybenzoic acid[c]
Lungs[b]	
Heart[b]	
Mammary gland[b]	
Erythrocytes[b]	

[a] COMT is thought to be ubiquitously expressed. These are examples of tissues in which COMT has been detected and should not be considered an exhaustive list.
[b] Tenhunen et al. (1994).
[c] DeMille et al. (2002).

methyltransferase (COMT). This enzyme, which is expressed throughout the body, catalyzes the methylation of hydroxylated compounds using *S*-adenosyl methionine (SAM) as a cofactor (Lundstrom et al. 1995). Figure 8.5 illustrates an example of a COMT-mediated methylation reaction. Table 8.8 lists the tissue distribution and example substrates for this enzyme.

The COMT gene is located on chromosome 22 and codes for an enzyme that has both a soluble (S-COMT) form and a membrane-bound (MB-COMT) form (Lundstrom et al. 1995). The relative activities of S-COMT and MB-COMT toward dopamine, noradrenaline, L-dopa, and 3,4-dihy-

Table 8.9 COMT Polymorphisms

Allele	Polymorphism	Significance	Allelic Frequency
WTCOMT	Reference allele		50%[a]
LACOMT	Val108 to Met108 in S-COMT; Val158 to Met in MB-COMT[a]	Decreased protein activity and increased thermolability[c]	50%[a]
NA	6–13 repeats of TAAA (tetranucleotide short tandem repeat) in intron 1[b]	NA	NA
NA	*Hind*III restriction site polymorphism in P2 promoter at 5′ end[b]	NA	NA
NA	*Bgl*I restriction site polymorphism at 3′ end in exon 6[b]	NA	NA

[a] Scanlon et al. (1979).
[b] DeMille et al. (2002).
[c] Lotta et al. (1995).

droxybenzoic acid were evaluated in lysates from *Escherichia coli* and baculovirus-infected insect cells expressing the enzyme. The V_{max} of *S*-COMT was higher than that of MB-COMT for all compounds studied, while the K_m values varied depending on the substrate and enzyme form. *S*-COMT had a K_m about 15 times higher than that of MB-COMT for catecholamines. MB-COMT also had a higher affinity for SAM, the coenzyme needed by COMT (Lotta 1995).

COMT is polymorphic (see Table 8.9). One common SNP results in a variant with lower catalytic activity and greater thermolability (LACOMT) compared to the wildtype (WTCOMT) (Scanlon et al. 1979). This SNP is a substitution of Val108 to Met in S-COMT and Val158 to Met in MB-COMT. The LACOMT variant was discovered when a trimodal distribution of COMT activity in red blood cells was identified (Assicot 1971). Fifty percent of the population carries this allele, and therefore 25% is homozygous for this low-activity variant (Scanlon et al. 1979).

Interestingly, various studies have found that the differences in LACOMT and WTCOMT activity may be attributed to the stability of LACOMT at physiological temperatures. When their relative activities were compared after heating to 37°C, the activity of the Met108 variant was reduced compared to the Val108 form. This indicates that the Met108 form is thermolabile even at physiological temperatures. The thermolability of the variant was stabilized when bound to SAM, indicating the activity of the thermolabile form may be controlled by the physiological availability of SAM (Lotta et al. 1995).

Other COMT polymorphisms have been identified. These include a tetranucleotide short tandem repeat in intron 1, a HindIII restriction site polymorphism (RSP) in the 5′ region of the gene, and a BglI RSP in the 3′ region of the gene. Because of a lack of effect on the protein sequence, these polymorphisms have not been evaluated for their effect on COMT activity in vivo (DeMille et al. 2002). Resequencing of the human COMT gene in DNA samples from 60 African-American and 60 Caucasian-American subjects identified 23 SNPs, including a novel nonsynonymous cSNP present only in DNA from African-American subjects, and one insertion/deletion (Shield et al. 2004), but only the previously identified Met108 variant seemed to have functional significance.

The significance of low COMT activity in various diseases has been evaluated by a number of authors, in particular influences on schizophrenia and estrogen-sensitive cancers. Conclusions vary thus far. A metaanalysis of studies evaluating the association between the COMT Val158/108Met SNP and schizophrenia by Bray et al. (2003) concluded that existing case–control studies showed no association between either allele and the disease, while family-based studies found a modest association between the Val allele and schizophrenia (Glatt et al. 2003). Results regarding the influence of the Val158/108Met SNP in estrogen-sensitive cancers have varied depending on the population studied and other risk factors evaluated.

8.6. GLUTATHIONE *S*-TRANSFERASES

The glutathione *S*-transferases (GSTs) are a large, multigene family of enzymes that catalyze the conjugation of electrophilic compounds with glutathione (GSH). The reaction occurs via the nucleophilic attack of GSH on an electrophilic substrate. Figure 8.6 illustrates an example of a GST-mediated glutathione conjugation reaction. The result of the reaction is

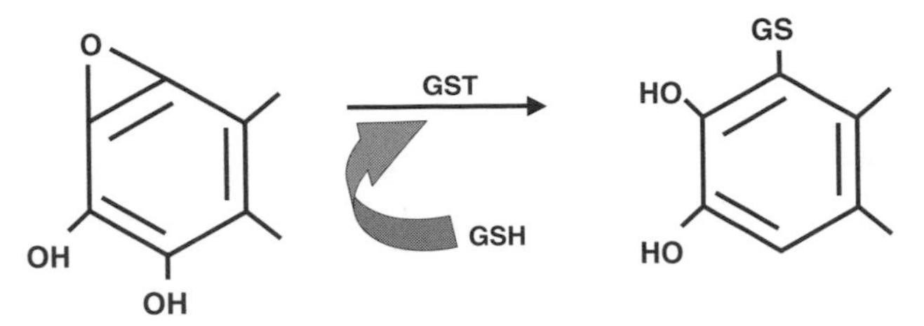

Figure 8.6. Example of glutathione conjugation: conjugation of a DNA reactive moiety.

usually the reduction of the substrate reactivity toward cellular macromolecules and an increase in water solubility. Although these are generally considered to be detoxification reactions, some haloalka(e)nes, are activated by GSTs to highly reactive episulfonium ion intermediates that can bind to cellular macromolecules, resulting in toxicity.

GST-mediated reactions can result in the opening of epoxide rings, nucleophilic aromatic substitutions, reversible Michael additions to aldehydes and ketones, isomerizations, and peroxidations. The electrophilic functional groups involved in these reactions may be carbon, nitrogen, or sulfur atoms (Eaton and Bammler 1999). One of these enzymes, GSTP1 (see discussion below) has been found capable of inhibiting Jun *N*-terminal kinase (JNK) activity, thereby protecting cells from H_2O_2-induced cell death (Sheehan et al. 2001) as well as regulating the constitutive expression of specific downstream molecular targets of the JNK signaling pathway (Elsby et al. 2003).

GSTS are found in bacteria, fungi, insects, helminths, and plants, and have been well characterized in mammals. Many of the three-dimensional structures of these enzymes have been determined in nonmammalian species, and a few novel functions first found in these species were later found to exist in mammals (Sheehan et al. 2001).

Two supergene families code for GST enzymes. They are grouped according to whether they are soluble or membrane-bound. In humans, 16 genes have been found for soluble GSTs and six for membrane-bound GSTs (Eaton, 1999; Nebert and Vasiliou 2004). The 16 soluble GSTs are classified into six classes: alpha (GSTA), mu (GSTM), theta (GSTT), pi (GSTP), zeta (GSTZ), and omega (GSTO, also called *chi*). GST kappa and one alpha class (GSTA4) are expressed in mitochondria (Strange et al. 2001; Gallagher and Gardner 2002).

The soluble GSTs have been well characterized. Members within a class of soluble GSTs share 40–50% of their sequence, but only 25–30% of their sequence with members of other classes. In most of these classes, several genes are designated as subfamilies. The nomenclature for these genes is as follows. A lowercase letter denotes the species the gene has been found in (e.g., "h" for human, "m" for mouse, "r" for rat), an uppercase letter denotes the class ("A" for alpha class), and an Arabic numeral identifies the subfamily. For the genes with polymorphism variants, the different alleles were initially designated using lowercase letters following the numeral denoting subfamily (Eaton and Bammler 1999), although more recent nomenclature utilizes the "*" designation, followed by a capital letter assigned in the order of discovery.

Soluble GSTs function as dimeric proteins. Only subunits from the same class may pair to form functional enzymes (Eaton and Bammler 1999). These enzymes have a very large number of substrates that often overlap across classes. Substrates include both endogenous and exogenous compounds (see Table 8.10).

Polymorphisms have been described for many genes in the superfamily of soluble GSTs, although most attention has been focused on the mu, theta, and pi families (see Table 8.11). Polymorphisms have been identified in hGSTA1, hGSTA2, hGSTM1, hGSTM3, hGSTT1, and hGSTP1. Certainly the most studied polymorphism in the GSTs, and probably in all biotransformation enzymes, is the GSTM1 null polymorphism (GSTM1*0). This polymorphism arose through a complete gene deletion, and is very common. Approximately 50% of populations of northern European heritage are homozygous for the null allele, and thus express no functional protein (Eaton and Bammler 1999). There is also a variant allele of GSTM1, which is the result of a one basepair variation in exon 7. The variant is GSTM1*b, while the wildtype allele is called GSTM1a. The catalytic efficiency of these two variants is similar (Strange et al. 2001).

The polymorphism identified in the GSTM3 gene is due to a 3-bp deletion in intron 6 that creates a recognition site for the YY1 negative transcription factor. The enzyme's expression may be reduced in individuals with this variant allele (Eaton and Bammler 1999; Strange et al. 2001; Daly 2003). Interestingly, the GSTM3*b variant appears to be in linkage disequilibrium with GSTM1*a (Strange et al. 2001).

A deletion polymorphism of the GSTT1 gene is also fairly common, and individuals homozygous for the GSTT1*0 allele also lack expression of functional enzyme. The frequency of this polymorphism varies with ethnicity from approximately 15–25% in Caucasians to over 60% in some Asian populations (Eaton and Bammler 1999; Strange et al. 2001).

There are four known allelic variants of GSTP1. GSTP1*a is the most common. GSTP1*b is the result of a change in codon 104 (Ile to Val). GSTP1*c has the same codon change as GSTP1*b combined with a change of Ala to Val in codon 113. This latter change alone is the rare GSTP1*d variant. The variant alleles exhibit different kinetics compared to the

Table 8.10 GST Tissue Distribution and Substrates

Isoform	Tissue Distribution[a]	Example Substrates[b]
A1	Liver	Environmental carcinogens
	Testis	Benzo(a)pyrene7,8-dihydrodiol-9,10-epoxide
	Kidney	AFB-8,9-epoxide
	Adrenal	Styrene oxide
	Pancreas	5-Hydroxymethyl-chrysene sulfate
	Lung	7-Hydroxymethylbenz(a)anthracene sulfate
	Brain	4-Nitroquinoline oxide
	Heart	Acrolein
A2	Liver	Hexachlorobutadiene
	Pancreas	Butadiene
	Testis	Trichloroethylene
	Kidney	Methylene chloride
	Adrenal	Ethylene oxide
	Brain	PhiP (2-amino-1-methyl-6-phenylimidazo[4,5*b*]
	Lung	pyridine)
	Heart	Pesticides
A3	Placenta	Lindane
A4	Small intestine	Alachlor
	Spleen	Atrazine
M1	Liver	DDT
	Testis	Methyl parathion
	Brain	EPN (*O*-ethyl-*O*-4 nitrophenylphenyl
	Adrenal	phosphonothione)
	Kidney	Drugs
	Pancreas	*cis*-Platin
	Lung	Chlorambucil
	Heart	Cyclophosphamide
M2	Brain	BCNU
	Testis	Thiotepa
	Heart	Fosfomycin
	Pancreas	Ethacynic acid
	Kidney	Nitroglycerine
	Adrenal	Menadione
	Lung	Acetaminophen
	Liver	Mitozantrone
M3	Testis	Adriamycin
	Brain	Endogenous molecules
	Spleen	4-Hydroxy-2-nonenal
	Other	Cholestrol-5,6-oxide

Table 8.10 *Continued*

Isoform	Tissue Distribution[a]	Example Substrates[b]
M4	Liver	Adenine propenal
	Skeletal muscle	9-Hydroperoxylinoleic acid
	Heart	Dopaminochrome
	Brain	Aminochrome
	Pancreas	Catechol estrogen quinones
	Lung	Malelylacetoacetate
	Kidney	
	Placenta	
M5	Brain	
	Testis	
	Lung	
P1	Brain	
	Lung	
	Heart	
	Testis	
	Adrenal	
	Kidney	
	Pancreas	
	Liver	
T1	Kidney	
	Liver	
	Small intestine	
	Brain	
	Spleen	
	Prostate	
	Pancreas	
	Testis	
	Heart	
	Lung	
T2	Liver	
Zeta	NA	
Sigma	NA	
Kappa	NA	
Chi	NA	

[a] Information modified from Eaton and Bammler (1999), Table 1.
[b] Information modified from Eaton and Bammler (1999), Table 2.

wildtype, but the physiological or pharmacological significance of these changes may not be great (Daly 2003).

Polymorphisms upstream of GSTA1 have been reported (Coles et al. 2001). These variations appear to affect levels of hepatic expression for both

Table 8.11 GST Polymorphisms

Isoform	Allele	Polymorphism	Significance	Allelic Frequency
GSTM1	GSTM1*0	Deletion mutation[a]	Small increased risk for lung and bladder cancer[a]	40–50% homozygosity in various ethnic groups;[c] 50% homozygosity in Caucasians[a]
	GSTM1*a	Wildtype[b]	—	—
	GSTM1*b	Asn172Lys	No significant difference in activty compared to GSTM1a[b]	NA
GSTM3	GSTM3*b	3-basepair deletion in intron 6 introducing the negative transcription factor[a,b]	Downregulation of expression;[c] may increase risk for basal cell carcinoma[a,b]	16% allelic frequendy in Caucasians[a]
GSTT1	GSTT1*0	Deletion mutation[a]	Increased risk for some brain tumors[a]	12–62% homozygosity;[a] 20% homozygosity in Caucasians[b] or Europeans[c]
GSTP1	GSTP1*a	Wildtype[a]	—	
	GSTP1*b	Ile104Val[a]	Increased risk for testicular, oral pharyngeal, bladder cancers, and teratomas[a]	NA
	GSTP1*c	Ile104Val and Ala113Val[a]	Increased risk for testicular, oral pharyngeal, bladder cancers, and teratomas[a]	NA
	GSTP1*d	Ala113Val[a]	Increased risks for testicular, oral pharyngeal, bladder cancers, and teratomas,[a] decreased risk for asthma[b]	NA

Table 8.11 *Continued*

Isoform	Allele	Polymorphism	Significance	Allelic Frequency
GSTA1	GSTA*A GSTA*B	Upstream polymorphisms that affect expression of GSTA1 and GSTA2; the GSTA*A haplotype expresses more GSTA1 and less GSTA2 than does the GSTA*B haplotype[c]	NA	NA

[a] Eaton and Bammler (1999).
[b] Strange et al. (2001).
[c] Daly (2003).

GSTA1 and GSTA2. Two haplotypes have been described: GSTA*A and GSTA*B. The GSTA*A results in higher expression of GSTA1 but lower expression of GSTA2 than the GSTA*B haplotype (Daly 2003).

Literally hundreds of studies have examined whether the GSTM1 deletion polymorphism is associated with increased risk for a variety of diseases with known or suspected environmental etiology. The GSTM1 polymorphism has been extensively studied for the following reasons:

1. The homozygous null variant is very common in the population (~50%), allowing for relatively good statistical power to see an effect in modest sample sizes.
2. The relationship between genotype (GSTM1*0/*0) and phenotype (GSTM1 enzymatic activity) is straightforward; for instance, there is no question that homozygous null people lack activity, and that heterozygotes (or, more rare, homozygotes) have catalytic activity (assuming, of course, that there are no promoter polymorphisms that would block expression of the functional allele).
3. There is reasonably good "biological plausibility," for the theory that GSTM1 deletion is associated with increased disease risk, since

GSTM1 activity has been shown in a number of studies to detoxify a variety of drugs and chemicals.

4. The genotyping assay is simple and inexpensive.

Among the most frequently studied outcomes is that of cancer in smokers. A Medline search of "GSTM1 polymorphism, lung cancer, smoking" yields more than 100 papers since 2004. Many of these studies have been discussed in Chapter 13, on lung cancer, so a detailed discussion will not be presented here. Suffice it to say that the majority of studies reveal a modest but usually statistically significant increase in lung cancer risk among those homozygous for the null allele, compared to those who have one or two copies of the functional allele. Early studies, including a meta-analysis (McWilliams et al. 1995), suggested that the relative risks (or odds ratios) for the GSTM1 null allele were on the order of 40–50%, but more recent meta-analyses find only a very small (OR 1.17) increase in risk (Benhamou et al. 2002; Houlston 1999). Numerous studies have also reported a significant association between bladder cancer risk and the GSTM1 null genotype among smokers, with increased relative risks (or odds ratios) on the order of 30–40% (Johns and Houlston 2000). Many other studies have been conducted to identify possible associations with the GSTM1 deletion polymorphism with various other cancers, but associations have been mixed and generally weak. For example, over 20 different studies have examined the association between GSTM1 genotype and breast cancer. A more recent meta-analysis of many of these studies (Sull et al. 2004) found a nonsignificant increase in overall breast cancer risk (OR 1.06, CI 0.99–1.14), and a small but significant increase in postmenopausal women (OR 1.19, CI 1.08–1.34) who were GSTM1-null.

The GSTM3*b variant has been associated with an increased risk of basal cell carcinoma (Strange et al. 2001) and more recently with prostate cancer (Medeiros et al. 2004).

Although GSTP1 is expressed in relatively high levels in the lung and has high catalytic activity toward carcinogenic arene epoxides generated from CYP oxidation of PAHs in tobacco smoke, relatively few studies have found an association between any of the GSTP1 cSNP variants and increased risk of lung cancer (Schneider et al. 2004). However, the GSTP1313GG polymorphism was found to modify the relationship between pack years of smoking and lung cancer risk by a factor of nearly 2 (Miller et al. 2003). The GSTP1 variants have also been associated with increased risks for testicular, oral pharyngeal, and bladder cancers, and teratomas in various studies [reviewed in Eaton and Bammler (1999)].

Many studies have also examined the relationship between GSTT1 deletion polymorphism and various disease outcomes. There is some evidence that the lack of expression of GSTT1 is associated with increased risks for the development of myelodysplastic syndrome and acute myeloid leukemia, diseases that are linked to occupational and environmental exposures (Arruda et al. 2001).

The significance of GST polymorphisms in drug response, including response to and toxicity from chemotherapeutics, is unclear (Innocenti and Ratain 2002; Daly 2003; Townsend and Tew 2003).

8.7. NQO1 AND NQO2

Two enzymes are involved in the two-electron reduction of quinones to dihydroquinones: NAD(P)H:quinone oxidoreductase 1 (NQO1), also referred to as DT dipahorase; and NRH:quinone reductase (NQO2). The hydroquinones produced by these enzymes may be further metabolized and excreted. This process is considered a protective mechanisms because it removes dangerous quinones that may redox-cycle to produce reactive oxygen species (Foster et al. 2000).

NQO2 is a truncated homolog of NQO1, with 49% of its protein identical to that of NQO1. Both proteins are cytosolic. The enzymes are encoded by separate genes on different chromosomes, although they are expressed in many of the same tissues. NQO1 is an unusual enzyme in that it can utilize both NADH and NADPH as a cofactor. NQO2 uses nonphosphorylated nicotinamides such as *N*-ribosyl and *N*-alkyldihydronicotinamide as cofactors (Foster et al. 2000; Chen et al. 2000). Figure 8.7 illustrates the NQO1 reduction of a quinone.

The two-electron reduction of quinones carried out by these enzymes is thought to proceed by two direct hydride transfers. One is from NAD(P)H to FAD and the other, from $FADH^2$ to the quinone. Two sequential one-electron transfers may occur in this reaction by these enzymes, but this is

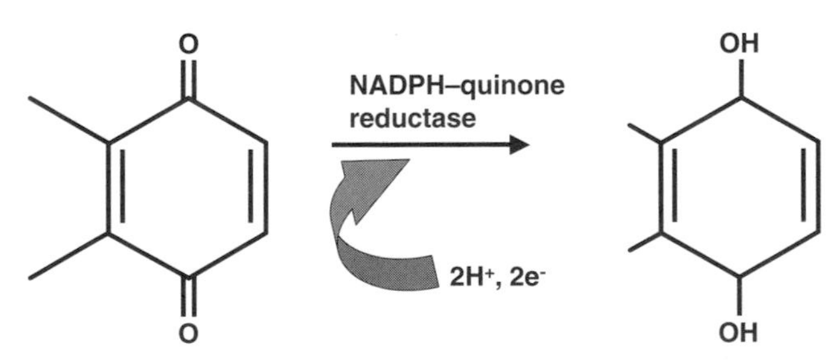

Figure 8.7. Example of quinone reduction by NQOs.

Table 8.12 NQO Tissue Distribution and Substrates

Isoform	Tissue Distribution	Example Substrates
NQO1	Ubiquitous; highest in kidney > skeletal muscle > lung[a]	Environmental carcinogens/ toxicants PAH quinones[a] Drugs Mitomycin[b] Anthracyclines[b] Aziridinyl benzoquinones[b] Nitrobenzamides[b] Endogenous molecules Vitamin K hydroquinone[b]
NQO2	Similar to NQO1[a]	Endogenous molecules Vitamin K hydroquinone[b]

[a] Jaiswal (2000).
[b] Chen et al. (2000).

not thought to be likely as no semiquinone intermediate has been identified (Foster et al. 2000; Chen et al. 2000).

NQO1, and possibly NQO2, have physiological roles in the reduction of vitamin K during production of blood coagulation factors (Chen et al. 2000). NQO1 is clinically important because of its ability to reductively activate cytotoxic antitumor quinone prodrugs such as mitomycin (Chen et al. 2000; Gutierrez 2000). The enzyme is also thought to be important for its ability to protect tissues against the mutagenic, carcinogenic, and cytotoxic effects of xenobiotic semiquinones or those produced from metabolism of endogenously produced quinones (Chen et al. 2000; Dinkova-Kostova and Talalay 2000). Table 8.12 lists example substrates for these enzymes and their expression location.

There are four gene loci that code for different NAD(P)H quinone reductases in humans. NQO1 is located at 16q22, while NQO2 is found at 6p25. NQO1 is expressed in all tissues. Its expression is highest in kidney, then skeletal muscle and lung. NQO2 appears to have a similar tissue expression. There is, however, large variability in activity across individuals, across tissues within an individual, and between normal and tumor tissues. Tumor tissues and cells of hepatic and colon origin express more of the protein than do normal tissues and cells of similar origins. NQO1 is inducible by some xenobiotics, antioxidants, oxidants, heavy metals, and radiation (Jaiswal 2000). This induction may be via the activity of the antioxidant response elements (AREs) of the NQO1 gene. AREs are DNA sequences to which intracellular signaling molecules activated by exposure to conditions of oxidative stress bind. This binding of signaling molecules to the ARE stimulates the expression of enzymes that are protective of oxidative stress. The exact mechanisms for how NQO1 induction may occur during oxidative stress are not fully understood (Jaiswal 2000).

Two polymorphic variants of NQO1 have been found (see Table 8.13). The *2 allele is the result of a C > T SNP at nucleotide 609 (Pro187Ser), while the *3 variant is the result of a C > T SNP at nucleotide 465 (Arg139Try). Both polymorphisms were identified in human tumor cell lines and in the general population. The *2 allele causes reduced activity of the NQO1 protein (Chen et al. 2000). People homozygous for this allele exhibit less than 5% of the normal NQO1 activity. As a result, the protein is either nondetectable or apparent only at trace levels using Western blotting. In addition, any variant protein expressed may be unstable. The *3 allele does not appear to affect enzymatic activity, although it appears to have decreased expression compared to the wild type. How the *3 allele causes decreased expression is unclear (Chen et al. 2000; Pan et al. 2002). An altered NQO1 mRNA has also been found. It is the product of a deletion of exon 4. However, no protein corresponding to this mRNA has been detected (Pan et al. 2002). Table 8.13 lists the polymorphisms of NQO1, including their significance and frequency.

The *2 allele may have significance if its decreased activity is low enough to affect a cell's protective measures against cancer. Persons with kidney tumors have been found to have a higher rate of this mutation in comparison to the general population (4% rather than 1%). This polymorphism may also have implications for cancer treatment if it reduces activation of anticancer prodrugs (Chen et al. 2000). A large epidemiology study found that persons homozygous for the *2 allele had a sevenfold increase in the risk of bone marrow toxicity with benzene exposure. Such bone marrow toxicity may lead to diseases such as aplastic anemia and leukemia (Nebert et al. 2002).

8.8. CONCLUSIONS

Although many gene–environment interaction studies of phase II biotransformation enzymes in human populations have been completed since the mid-1990s, the number of examples demonstrating important and consistent positive relationships is disappointingly small. One reason for this may be that the penetrance (e.g., functional contribution to phenotype) of variations in phase II metabolism is likely to be highly dependent on other pathways involved in the disposition of a particular chemical. For example,

Table 8.13 NQO Polymorphisms

Isoform	Allele	Polymorphism	Significance	Allelic Frequency
NQO1	NQO1*1	Reference allele	—	—
	NQO1*2	C > T substitution at nucleotide 609 resulting in Pro187Ser[a,e]	Could reduce cellular protection against reactive oxygen species generated during quinone redox cycling;[a,b] could reduce efficacy of antitumor quinone prodrugs that require activation via 2 electron reduction;[a,c] sevenfold increased risk of bone marrow toxicity with benzene exposure[d]	Allelic frequency —22% Caucasians, 45% Asians;[d] frequency of homozygosity—5% Caucasians, 16% Mexican-Americans, 20% Asians[e]
	NQO1*3	C > T substitution at nucleotide 465 resulting in Arg139Try[a,e]	Could reduce cellular protection against reactive oxygen species generated during quinone redox cycling;[a,b] could reduce efficacy of antitumor quinone prodrugs that require activation via 2 electron reduction[a,c]	Allelic frequency: 0–5% for general population[e]

[a] Chen et al. (2000).
[b] Dinkova-Kostova and Talalay (2000).
[c] Gutierrez (2000).
[d] Nebert et al. (2002).
[e] Pan et al. (2002).

the significance of a variant in a GST enzyme that detoxifies a particular reactive intermediate may depend on the rate at which the intermediate is formed. That, in turn, may vary greatly between individuals because of genetic variation (e.g., CYP polymorphisms) or because of different levels of environmental influence on the expression of phase I genes (e.g., enzyme induction or inhibition from drugs, diet, or other environmental factors).

Thus, future population-based studies will require consideration of multiple genetic variants and more detailed assessment of environmental exposures in order to unravel possible gene–gene and gene–environment interactions associated with phase II biotransformation. This will, in turn, require large samples sizes, making such studies both difficult and expensive.

8.9. SUGGESTED READING

Additional information on polymorphisms in xenobiotic conjugation is available in the literature. Online sources include publications by Demille et al. (2002) and Glatt and Meinl (2003). The following journal articles are also recommended:

Bosma (2003)
Dawling et al. (2001)
Eaton and Bammler (1999)
Lundstrom et al. (1995)
Miners et al. (2002)
Negishi et al. (2001)
Pan et al. (2002)
Strange et al. (2001)
Strott (2002)
Townsend and Tea (2003)

9

Paraoxonase, Butyrylcholinesterase, and Epoxide Hydrolase

Lucio G. Costa, Toby B. Cole, Gary K. Geiss, and Clement E. Furlong
University of Washington, Seattle, WA

9.1. INTRODUCTION

Paraoxonase, butyrylcholinesterase, and epoxide hydrolase are examples of hydrolytic enzymes that are involved in the metabolism of both exogenous and endogenous chemicals. The genes encoding these proteins contain coding region polymorphisms that result in altered enzymatic activity, as well as regulatory region polymorphisms that affect how much enzyme is produced, via effects on transcription or RNA splicing. The case of paraoxonase illustrates the importance of a concept that is often overlooked in human genetic studies: that the ultimate goal of genotyping polymorphisms is to predict differences in protein function. For paraoxonase, a high-throughput functional enzyme assay is available that detects differences not only in enzymatic activity but also in plasma enzyme levels. Because it is a functional assay, any genetic variation that affects total enzymatic activity is taken into account. The examples given demonstrate how even a single nucleotide difference in DNA can result in a large difference in enzymatic activity for specific substrates. Such polymorphisms can be quite common in human populations, with striking differences in prevalence among ethnic groups.

Gene-Environment Interactions: Fundamentals of Ecogenetics, edited by Lucio G. Costa and David L. Eaton

9.2. PARAOXONASE (PON1)

9.2.1. Background

The plasma enzyme paraoxonase (PON1, EC 3.1.1.2 and EC 3.1.8.1) was so named for its ability to hydrolyze paraoxon, the highly toxic metabolite of the insecticide parathion (Figure 9.1). PON1 is a member of a family of

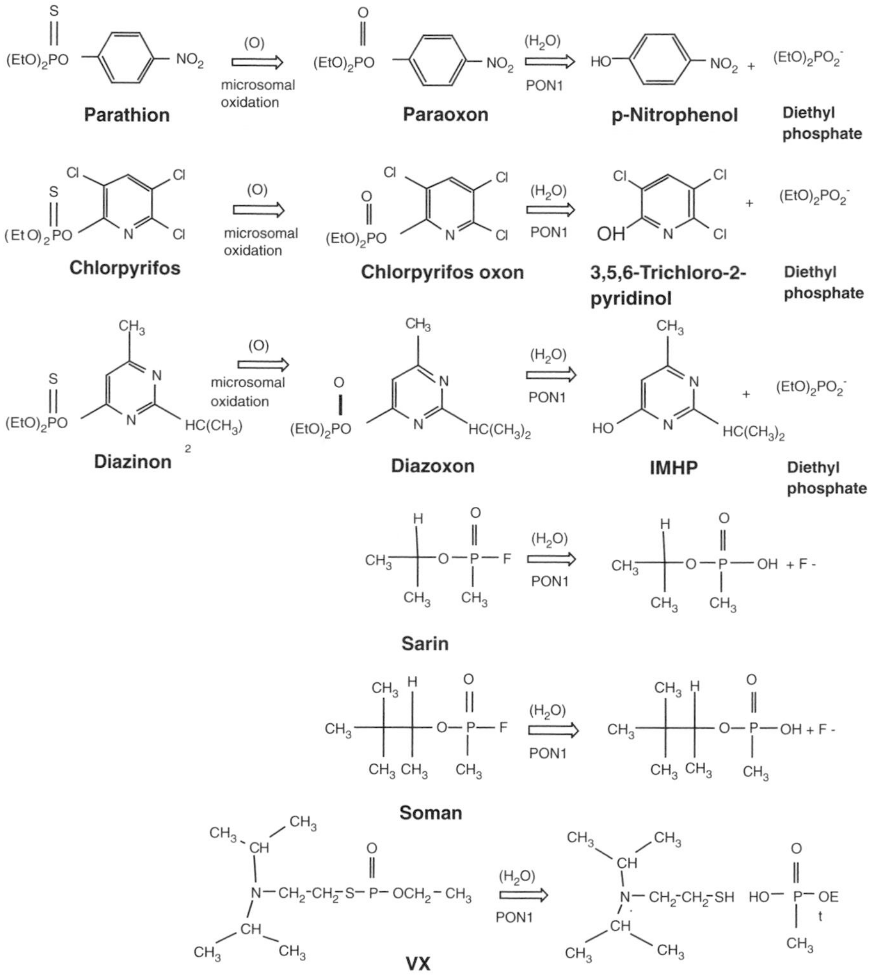

Figure 9.1. Cytochrome P450/PON1 pathway for organophosphorus compound detoxication. Note that the nerve agents do not require bioactivation and can be hydrolyzed directly by PON1.

proteins that also includes PON2 and PON3, the genes for which are clustered in tandem on the long arm of chromosome 7 (q21, 22). PON1 is synthesized primarily in the liver and a portion of PON1 is secreted into the plasma, where it is associated with high-density lipoprotein (HDL) particles.

PON1 has both endogenous and exogenous substrates. The natural physiological function of PON1 appears to be the metabolism of oxidized lipids in both low-density lipoprotein (LDL) particles and HDL particles. It is often stated that PON1 prevents oxidation of lipids in lipoprotein particles; however, it is probably more accurate to state that PON1 metabolizes toxic oxidized lipids, converting them to nontoxic molecules. While a primary function of PON1 in lipid metabolism has been elucidated only relatively recently, its role in the metabolism or aromatic esters and toxic organophosphorus compounds has been studied for more than five decades. Figure 9.1 shows the pathways for bioactivation and detoxication of three common organophosphorus insecticides, parathion, diazinon, and chlorpyrifos, as well as the nerve agents sarin, soman, and VX. The insecticides are manufactured as organophosphorothioates (or parent compounds) that contain a sulfur atom attached to the phosphorus atom, and need to be bioactivated to a more toxic oxon form by microsomal cytochrome P450s, which remove the sulfur and replace it with oxygen (Figure 9.1). In addition to bioactivating the parent compounds, the cytochrome P450s can also detoxify them, with varying ratios of activation versus detoxication depending on the specific cytochrome P450. The oxon forms of the organophosphorothioates are potent inhibitors of acetylcholinesterases, enzymes crucial for controlling cholinergic neurotransmission at the neuromuscular junction and in the central nervous system. In contrast, the parent compounds are poor inhibitors of acetylcholinesterases; for example, chlorpyrifos oxon inhibits acetylcholinesterase at 1000 times the rate of chlorpyrifos. Given the large difference in toxicities between the parent compounds and their oxon metabolites, it is possible even that the experimentally observed inhibition of acetylcholinesterase by parent organophosphorothioates in vitro is actually due to the presence of trace quantities of their oxon forms.

9.2.2. PON1 Polymorphisms

More than 160 polymorphisms have been described in *PON1*, some in coding regions and others in regulatory regions of the gene. Even before the discovery of DNA polymorphisms in *PON1*, it was found that the plasma paraoxonase activity of PON1 in human populations exhibited a polymorphic distribution. Human populations could be split into two groups, those with high paraoxonase activity (high metabolizers) and those

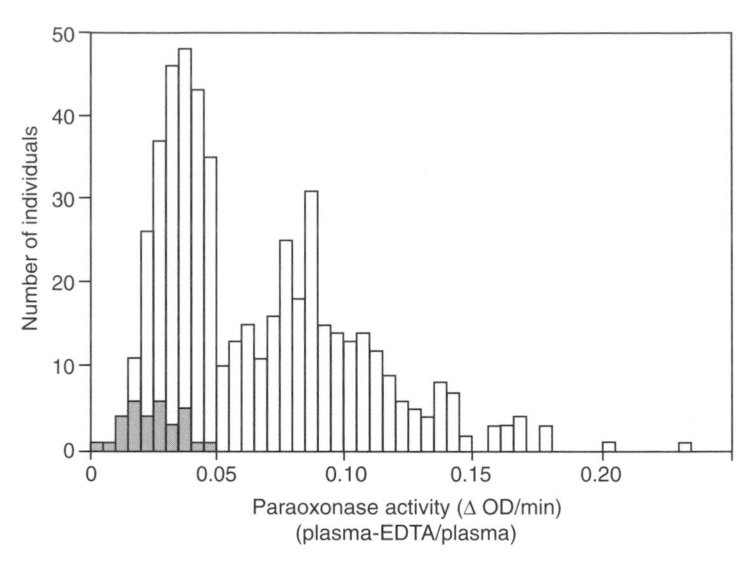

Figure 9.2. Serum paraoxonase activity in 521 random Seattle Caucasian blood donors and 31 cord blood samples (hatched bars). [Data from Mueller et al. (1983).]

with low paraoxonase activity (low metabolizers). Gene frequencies for high metabolizers and low metabolizers varied among groups of different geographic origin. The first analyses were carried out using histograms of paraoxonase activity similar to that shown in Figure 9.2. For these types of studies, many different assays were developed for measuring rates of paraoxon hydrolysis in human plasma. A much more useful analysis of PON1 functional genomics is described in the section on PON1 status below.

The molecular basis of the paraoxonase activity polymorphism turned out to be a single-nucleotide polymorphism in the coding region of *PON1*, resulting in a glutamine (Q)/arginine (R) substitution at codon 192. Elucidating the molecular basis of the paraoxonase activity polymorphism required purification of rabbit and human PON1. Rabbit PON1 is very stable, allowing for its purification to homogeneity, sequencing, and the design of probes that were used to isolate rabbit cDNA from a rabbit liver library. The rabbit cDNA clone was used in turn to isolate clones of human cDNA from a human liver cDNA library. Sequencing of the isolated human PON1 cDNA clones revealed two coding region polymorphisms, resulting in amino acid substitutions at L55M and Q192R. Further research showed that the Q192R amino acid substitution was responsible for the activity polymorphism. The $PON1_{Q192}$ alloform hydrolyzes paraoxon much less efficiently than does $PON1_{R192}$. The effect of the polymorphism is reversed for

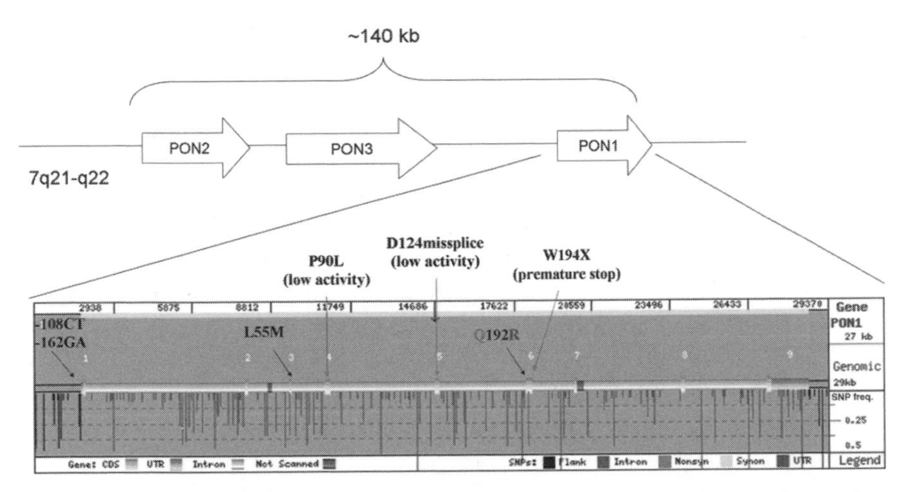

Figure 9.3. The PON1 gene is located on human chromosome 7, near the genes encoding PON2 and PON3. The expanded PON1 gene diagram shown below the flowchart illustrates the locations and frequencies of SNPs that have been identified in the PON1 gene, with SNP locations indicated by vertical lines below the gene structure diagram. The numbered boxes correspond to the 9 exons found in *PON1*. The length of each line indicates the frequency of that SNP in the population. In addition to the previously identified SNPs that result in amino acid substitutions at L55M and Q192R, 166 new SNPs were identified by the UW/FHCRC Variation Discovery Resource (Seattle SNPs). The PON1 SNP diagram at bottom was created at the SeattleSNPs Website (http://pga.gs.washington.edu/), using the GeneSNPs public Web resource (http://www.genome.utah.edu/genesnps) available from the Utah Genome Center.

other OP compounds such as the nerve agents soman and sarin. $PON1_{M55}$ has been associated with low plasma PON1 protein levels; however, as noted below, this is primarily a result of linkage disequilibrium with the low efficiency-108T allele of the C-108T promoter region polymorphism.

Five different promoter region polymorphisms were also identified and characterized by three research groups. Of these, the C-108T polymorphism had the most significant effect on plasma PON1 levels, with the -108C allele providing levels of PON1 about twice as high as those seen with the -108T allele. With the more recent advent of high-throughput technologies for resequencing genes from multiple individuals for the Environmental Genome Project, more than 160 new polymorphisms have been identified in PON1 (Figure 9.3).

9.2.3. PON1 Genotyping and Disease

Approximately 100 studies have been carried out to date that have examined the association of PON1 polymorphisms with various diseases. Some studies have reported an association of $PON1_{R192}$ with risk for cardiovascular disease, while other studies have found no association. Virtually all of these studies examined only the nucleotide polymorphisms via PCR assays. While it would be possible to genotype an individual for all of the known PON1 polymorphisms, this analysis would not provide the level of plasma PON1 nor the phase of polymorphisms (i.e., which polymorphisms are on each of an individual's two chromosomes). Further, it is not yet known whether any of the as-yet-uncharacterized polymorphisms are important in the splicing efficiency required to provide the final PON1 mRNA. More recent studies have shown that polymorphisms in both introns and exons can affect splicing efficiency, which can result in low levels of mRNA, and in turn low levels of the encoded protein product. Polymorphisms in 3′ untranslated regions (3′UTR) can also affect message stability and efficiency of polyadenylation.

9.2.4. PON1 Status

A functional genomic analysis provides the most informative approach for examining the relationship of PON1 genetic variability to disease risk, because measurement of an individual's PON1 function automatically takes all polymorphisms that might affect activity into account. Fortunately, the functional genomics of PON1 can be examined easily through the use of a high-throughput enzyme assay involving two PON1 substrates. By directly measuring PON1 hydrolytic activities in human plasma, one can determine the sum of all factors that affect PON1 levels or activity. An example of this analysis is shown in Figure 9.4. Rates of hydrolysis of diazoxon are plotted against rates of paraoxon hydrolysis for each plasma sample in a population. As a result of different catalytic efficiencies of the rates of hydrolysis of these two substrates by $PON1_{Q192}$ and $PON1_{R192}$, this two-substrate assay separates the population into three distinct groups: individuals homozygous for $PON1_{Q192}$, heterozygotes, and individuals homozygous for $PON1_{R192}$. It is important to note that this analysis, in addition to providing the functional expression of the plasma $PON1_{192}$ alloforms, also provides the plasma level of PON1 for each sample. We refer to this type of analysis as a determination of "PON1 status" for an individual. Figure 9.4 also illustrates the power of this type of approach over conventional PCR genotyping. Several individuals were identified whose PON1 status did not match what would be predicted solely on the basis of their Q192R genotype (i.e., PON1 activity was lower than expected). The discovery of new *PON1* polymorphisms by resequencing the PON1 genes of many individuals revealed why this "discrepancy" occurred—previously unknown polymorphisms were resulting in at least one allele that produced inactive

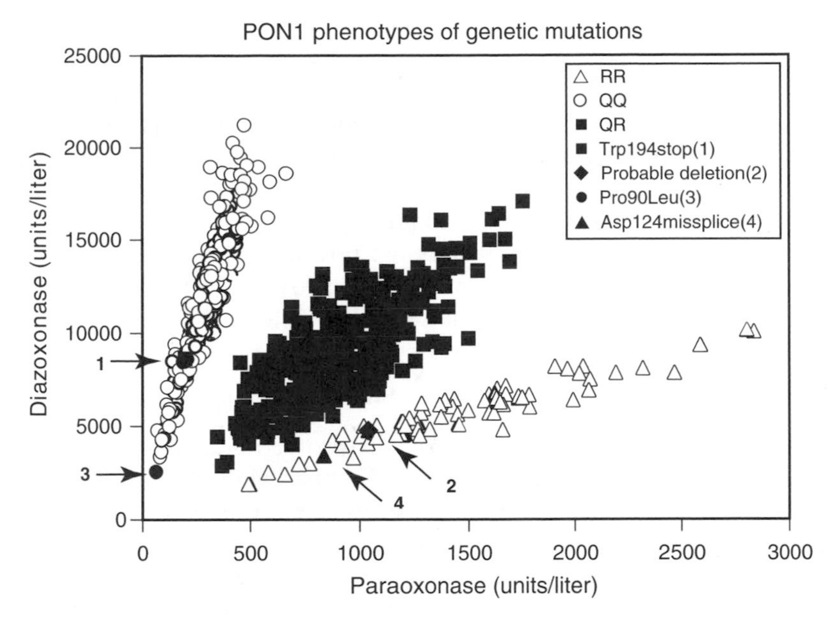

Figure 9.4. Determination of PON1 status. Plot of diazoxonase versus paraoxonase activities in plasma of carotid artery disease cases and controls, coded for PON1Q192R genotype (determined by PCR) (Richter and Furlong 1999; Jarvik et al. 2003) Note that the two-substrate assay provides an accurate inference of $PON1_{192}$ genotype as well as the level of plasma PON1 activities (PON1 status). Because it is a functional analysis, it provides a 100% accurate determination of the functional genomics of PON1 status. Newly discovered SNPs explain why some individuals have lower PON1 activity than would be predicted by Q192R genotype alone. Individuals 1–4 genotyped as heterozygotes (Q/R192); however, their enzyme analysis indicated homozygosity for Q or R. Complete sequencing of their *PON1* genes revealed mutations in one allele, resulting in only one alloform of PON1 in their serum. The mutations revealed are noted in the figure. [Note the large variability in PON1 levels, even among individuals of the same Q192R genotype (Jarvik et al. 2003).]

PON1. Whereas Q192R genotyping alone did not predict this reduction in PON1 activity, measurement of PON1 status detected these genotype–activity discrepancies.

9.2.5. PON1 Levels Are Variable Among Individuals

Once plasma PON1 reaches mature levels, between 6 and 15 months of age, it remains fairly constant throughout an individual's lifetime. For a given Q192R genotype, there is a large variability of PON1 levels among individuals. Fundamental biochemical considerations dictate that it is the catalytic efficiency with which PON1 degrades toxic OP compounds and metabolizes oxidized lipids that determines the degree of protection provided by PON1 against these insults. In addition, higher concentrations of PON1 provide better protection. Thus, for adequate risk assessment, it is important to know PON1 levels. Also, for substrates affected by the Q192R polymorphism, one Q192R alloform may protect better than the other. In cases where substrate hydrolysis is affected by the Q192R polymorphism, it is important to know the individual's genotype at Q192R as well as plasma PON1 levels. Knowing simply the position Q192R genotype alone or those of the other PON1 polymorphisms provides insufficient information for proper epidemiological analyses.

9.2.6. PON1 Status and Organophosphorus Compounds

What is the evidence that PON1 provides protection against exposure to organophosphorous (OP) compounds? The first evidence that high PON1 levels were protective against OP exposures came from experiments where purified PON1 from rabbit was injected into rats or mice to raise their plasma PON1 levels. These experiments provided clear evidence that high PON1 levels were protective against exposure to chlorpyrifos/chlorpyrifos oxon. What about the consequence of low levels of plasma PON1? This question was addressed by exposing mice that lacked PON1 (due to removal of part of the *PON1* gene) to OP compounds. Dramatic increases in sensitivity were observed with exposures to chlorpyrifos oxon and diazoxon, with less dramatic increases in sensitivity to the respective organophosphorothioate parent compounds. Surprisingly, these mice were not more sensitive to paraoxon exposure.

Measurement of the catalytic efficiencies of PON1 for hydrolyzing paraoxon, diazoxon, and chlorpyrifos oxon provided an explanation for these observations (Table 9.1). While there was nearly a tenfold difference in the catalytic efficiency for paraoxon hydrolysis between $PON1_{Q192}$ and $PON1_{R192}$ (0.7 vs. 6.7), the catalytic efficiency of $PON1_{R192}$ was not sufficient to provide protection against paraoxon exposures. However, a tenfold increase in catalytic efficiency for diazoxon hydrolysis (~77) was sufficient to provide significant protection against diazoxon exposures. Both alloforms had equivalent catalytic efficiencies and provided equivalent protection against diazoxon exposures. For chlorpyrifos oxon exposures, the $PON1_{R192}$ alloform had a higher catalytic efficiency (250) and provided significantly better protection than did the $PON1_{Q192}$ alloform, which had a catalytic efficiency of 150 (Table 9.1).

Thus, simply knowing that PON1 hydrolyzes a specific substrate does not enable one to determine whether it actually provides protection against exposure to the compound in question. While PON1 hydrolyzes the nerve

Table 9.1 Catalytic Efficiency Determines the in vivo Efficacy of PON1 for Detoxifying OP Dompounds

	Paraoxon		Diazoxon		Chlorpyrifos-oxon	
	$PON1_{Q192}$	$PON1_{R192}$	$PON1_{Q192}$	$PON1_{R192}$	$PON1_{Q192}$	$PON1_{R192}$
Catalytic efficiency $(V_{max}/K_m)^a$	0.71	6.27	75	77	152	256
Provides protection in vivo[b]	–	–	+	+	+	++

[a] Paraoxonase, diazoxonase, and chlorpyrifos oxonase activities were determined in vitro at physiologically relevant salt concentrations, using purified human plasma $PON1_{192Q}$ or $PON1_{192R}$ isoforms [data from Li et al. (2000b)].

[b] PON1 knockout mice were injected with purified human plasma $PON1_{Q192}$ or $PON1_{R192}$ 4 h prior to OP exposure (0.3 mg/kg, paraoxon; 1 mg/kg, diazoxon; 2 mg/kg, chlorpyrifos-oxon), and cholinesterase inhibition was measured in the brain and diaphragm 2 h following OP exposure. Neither PON1 isoform protected against paraoxon toxicity. The two isoforms provided equivalent protection against diazoxon toxicity. $PON1_{R192}$ afforded twice the protection as $PON1_{Q192}$ against chlorpyrifos oxon toxicity (Li et al. 2000b).

agents soman and sarin, the catalytic efficiencies for hydrolysis are not sufficient to protect against exposure to these OP compounds. However, it is possible that site-directed mutagenesis of PON1 in vitro could be used to generate engineered recombinant variants of PON1 that will be useful as therapeutics for humans exposed to nerve agents or other OP compounds.

9.2.7. PON1 Status and Disease

A similar approach was used, instead of genotyping alone, to determine whether PON1 status is important in protecting against specific diseases. An examination of PON1 status of individuals with carotid artery disease compared with disease-free control subjects showed that low PON1 levels were a risk factor for this disease among $PON1_{Q192}$ homozygotes and heterozygotes. In the population studied, there were too few $PON1_{R192}$ homozygotes from which to obtain meaningful data. Figure 9.5 illustrates the application of the analysis of PON1 status to this study.

9.2.8. PON1 and Drug Metabolism

PON1 has been shown to metabolize a number of drugs and prodrugs, inactivating some while activating others. It is likely that the metabolism of at least some of these drugs will be affected in a pharmacokinetically significant manner by the widespread differences in PON1 status among

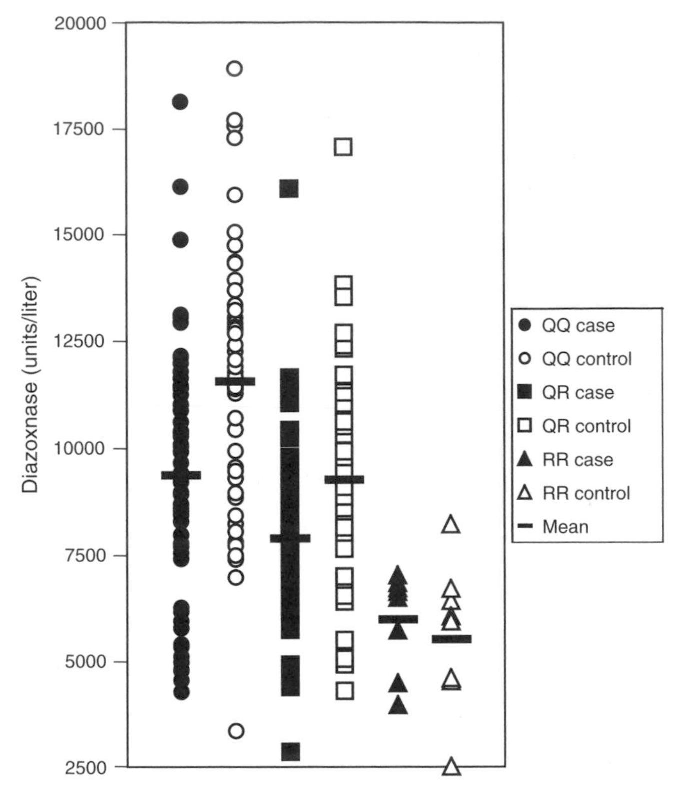

Figure 9.5. Diazoxonase levels of CAAD (carotid artery disease) patients and controls separated by position 192 genotypes as indicated. Note the abundance of individuals with low diazoxonase activity among the PON1QQ cases and the sparsity of cases with high diazoxonase activity. Diazoxonase activity is a relative measure of plasma PON1 protein. Means are indicated by horizontal bars. [Reproduced from Jarvik et al. (2000).]

individuals. For such drugs, it will be important to tailor drug dose to fit an individual's specific rates of activation or inactivation.

9.3. BUTYRYLCHOLINESTERASE

9.3.1. Historical Background

Butyrylcholinesterase was discovered in the 1930s by biochemists isolating enzymes from horse serum that hydrolyzed esters of choline. Two different forms of cholinesterases were identified in humans, based on their biochemical properties and in which fraction of blood they were found (plasma or red blood cells). The first enzyme was found to hydrolyze acetylcholine specifically and was therefore called *specific cholinesterase*, later to be renamed *acetylcholinesterase* (AChE, EC 3.1.1.7). AChE is now known to have a critical role in modulating the transmission of neuronal signals. The other enzyme metabolized a wider variety of ester-containing substrates and was named *nonspecific* or *pseudocholinesterase*. This enzyme's activity, localized primarily to the plasma fraction of serum, has been officially named *butyrylcholinesterase* (BChE, EC 3.1.1.8), and its biochemistry and genetics have been the subject of decades of study. BChE's function is less definitive than that of AChE, but the majority of evidence suggests that it removes toxins from blood that would otherwise inhibit AChE and perturb neuronal signaling.

Butyrylcholinesterase is produced in liver and secreted into serum, where it is relatively abundant in plasma (~5 mg/L). However, it is also present at significant levels in the lung, intestines, and brain. No doubt partially because of its accessibility and abundance in serum, it was purified biochemically in the 1930s. Although the exact physiological role of BChE is still a matter of debate, the enzyme clearly plays a role in metabolizing poisonous toxins that enter the bloodstream. Since BChE hydrolyzes a broad range of ester-containing compounds, it probably evolved as protection against natural poisons that are found in some plant species, in particular the glycoalkaloids. These compounds are poisonous because they inhibit AChE, thus interfering with normal nerve and brain cell signaling, causing adverse effects to the organisms that ingest them. Although we now know which plants to avoid eating, there are toxins in our present-day environment that are capable of inhibiting AChE. For example, compounds used in medicine, insecticides, and biochemical warfare agents all, either deliberately or inadvertently, inhibit AChE. Thus, the amount of active BChE present in serum and its functional relationship to AChE makes it an excellent model for human ecogenetics studies.

The first evidence for the importance of BChE as an ecogenetic marker was obtained in the 1950s. Some patients who received the muscle relaxant succinylcholine suffered from a potentially fatal prolonged apnea lasting several hours. Succinylcholine works by binding to nicotinic receptors, and usually has a short duration of action due to degradation by BChE in the bloodstream, effectively lowering the concentration of active drug. Follow-up studies indicated that the affected patients had lower levels of plasma BChE activity than did normal unaffected individuals. It was later discovered that healthy family members of affected individuals also exhibited lower BChE activity, suggesting that the trait was heritable. This classic example of a polymorphism associated with drug sensitivity represents the first description of a BChE polymorphism that could have adverse consequences when an individual is exposed to drugs metabolized by BChE. As discussed below, several other mutations in BChE were also discovered that resulted in altered biochemical activity. These results suggest the possibility that individuals with certain BChE genotypes may exhibit increased sensitivity to exposure to cholinesterase inhibitors that are present in today's environment.

9.3.2. Gene and Protein Structure

The cDNA encoding BChE was cloned in 1987 and encodes a protein consisting of 574 amino acids. BChE usually consists of four identical subunits, each of which is heavily glycosylated. The human BChE gene is on chromosome 3 (3q26) and consists of four exons and three large introns that span a region greater than 70 kbp. As expected from its biochemical properties, BChE is structurally related to AChE, with approximately 50% amino acid identity shared between the two proteins. The crystal structure of BChE has just recently (as of 2005) been obtained and should reveal important insights into the mechanisms of BChE action, its similarity with AChE, and how individual coding mutations alter BChE structure/function.

9.3.3. BChE Polymorphisms

As mentioned above, the first evidence of a functional genetic mutation in plasma cholinesterase, or BChE, dates back to the 1950s in patients suffering from succinylcholine-induced apnea. Termed the "atypical" enzyme, we now know this mutation is due to an aspartic acid to glycine point mutation at amino acid position 70 that lies near the active site and alters the binding affinity for positively charged substrates such as succinylcholine. The atypical form of the enzyme represents one of the more frequent BChE alleles, occurring in at least one copy in approximately 1 in 25 individuals. However, there is an ethnic component to BChE polymorphisms. For example, this mutation was rarely found in the Japanese populations studied. Today, we know that there are approximately 40 mutations that alter BChE activity, a likely underestimate considering that new mutations are usually identified as a result of functional differences in rates of hydrolysis of ester compounds (Table 9.2). The high-throughput sequencing efforts associated with the Environmental Genome Project will undoubtedly elucidate many more polymorphisms across the entire BChE gene as it has done in the case of the *PON1* gene.

The majority of BChE mutations occur in the coding sequence of the gene and result in altered enzymatic activity. In fact, 30 of the 40 known mutations cause silent BChE, resulting in less than 10% of BChE activity.

Table 9.2 Major Genetic Variants of Butyrylcholinesterase

Genetic Variant	Mutation	Activity	Allele Frequency
Wildtype	None	100	0.87
Atypical[a]	Asp70Gly	70	0.02
K variant	Ala530Thr	50	0.12
Silent	>30 mutations	0–10	0.003
C5+	Unknown	125	Unknown

[a] Altered response to succinylcholine.

Source: Adapted from Lockridge and Masson (2000).

Of those silent BChE mutations whose molecular basis is known, nearly all are contained within BChE's coding sequence. Although silent BChE mutations are extremely rare in most human populations, individuals void of all detectable BChE activity (homozygous "silent") have been identified and have added uncertainty to the discussions regarding BChE's natural role. Another class of coding sequence mutations includes the "fluoride alleles" that are resistant to the inhibitory effects of sodium fluoride. These mutations alter the binding of fluoride to a noncatalytic site on BChE that probably changes the conformation of active enzyme, again altering its biochemical activity. The fluoride variants, like most BChE mutations, occur at low frequencies in most populations. Other mutations that alter the protein sequence and subsequent activity of BChE are the H, J, and K variants. The K variant represents the most widely distributed BChE polymorphism with approximately 1 in 4 persons being a carrier, while the H and J variants occur at low frequency. With a high frequency of K variants with lower enzymatic activity, the K variants represent likely candidates for increased risk of consequences following toxin exposures.

Several silent forms of BChE have been identified that have no mutations in their coding regions, suggesting that there are mutations in noncoding regions of BChE. One of the silent alleles has been identified as a splice mutation in intron 2, resulting in a nonfunctional protein. In addition, there are likely to be as-yet-undiscovered mutations in other intronic sequences, promoter regions and 5′ and 3′ untranslated regions that will alter BChE activity by raising or lowering the amount of enzyme produced. For instance, the J and K variants of BChE have been postulated to have regulatory region mutations, supported by the fact that the K-variant reduction in activity is due to lower levels of BChE protein in the blood. However, the source of this mutation is not presently known. In addition, the Johannesburg variant, one of two high-activity alleles, has been reported to have no coding sequence mutation, also suggesting a regulatory mutation. Now that the sequence of the human genome is complete, it should be possible to identify and characterize additional BChE mutations that alter the overall activity of the enzyme.

9.3.4. BChE and Environmental Toxins

The routine use of organophosphates and carbamates as pesticides (designed to inhibit insect flight muscle cholinesterase) has raised concerns that people with low-activity BChE mutations might be more susceptible to these agents. These compounds, or their bioactivated derivatives, exert their effects via the inhibition of AChE. As a stoichiometric scavenger such as carboxylesterases, BChE, along with other plasma catalytic scavengers like PON1, assists in the first line of defense for preventing AChE inhibition. Consistent with this role, human plasma BChE activity decreases in response to exposure to pesticides and precedes a drop in AChE activity, even in cases where no symptoms are apparent. Evidence that low-activity BChE mutations result in a higher risk of poisoning comes from a study of pesticide workers that revealed that 11% (8 in 72) of mildly poisoned workers (exposed to either organophosphate or carbamate compounds) had at least one copy of a mutated gene compared to 1.6% (1 in 62) of the control group. Although the results of this study are intriguing, additional research on larger groups will be required to assess the overall risk due to BChE mutations.

More recent events have raised concerns regarding exposure of civilians or military personnel to chemical weapons such as the nerve agents soman and sarin. Both agents are ester-containing compounds whose mechanism of action is the inhibition of AChE and both can be inactivated by covalent binding to BChE. Catalytic inactivation by PON1 is probably not efficient enough to provide significant protection. As in the case with pesticides, this suggests that a low plasma BChE concentration may result in increased susceptibility and more severe effects in response to agent exposure. Support for this hypothesis is found in a series of studies in rats, mice, and monkeys demonstrating that injections of purified BChE before exposure to lethal doses of sarin and soman will protect the animal from death. In humans, there has been one suggestive case of a Gulf War veteran who was known to carry the atypical BChE allele and suffered adverse effects in response to the carbamate pyridostigmine, a nerve agent prophylactic that reversibly inhibits cholinesterases. This example is similar to the situation with succinylcholine-induced apnea and indicates that individuals with low-activity BChE alleles may suffer adverse effects during treatment with prophylactics or drugs designed to inhibit cholinesterases (see discussion below).

More recently, BChE has also gained attention for its ability to metabolize ester-containing drugs, such as cocaine and its derivatives. Although the rate of cocaine hydrolysis by BChE is rather slow, site-directed mutagenesis of the active site in vitro has resulted in more efficient cocaine hydrolysis. This suggests the possibility that an engineered version of the enzyme could be used as therapy for cocaine induced toxicity. If a similar approach were successful for organophosphate compounds or nerve agents, purified BChE could eventually become a treatment for exposure to a variety of toxic compounds.

9.4. ROLE OF BChE IN NEURODEGENERATIVE DISEASES

For many years, the focus in neuronal studies was understandably on AChE, due to its prominent role in the metabolism of the neurotransmitter acetylcholine. However, in more recent years, BChE has been implicated in neuronal development as well as neurodegenerative diseases, primarily Alzheimer's disease (AD). The evidence for such a correlation is found in a number of studies indicating that neuronal BChE levels are increased in patients with AD and are associated with plaques and neurofibrillary tangles, one of the hallmarks of AD. On the other hand, studies aimed at investigating the relationship between BChE polymorphisms and AD have not been conclusive. A positive correlation between risk of AD and carriers of the K variant was reported initially but subsequent work failed to corroborate this finding and one study even showed the K variant was protective. Regardless, further research on the role of BChE in AD is important, since therapeutic treatments employed against AD are cholinesterase inhibitors designed to increase the amount of available acetylcholine, thereby improving cognitive abilities. Since BChE is also capable of metabolizing acetylcholine, and some drugs are more specific inhibitors of BChE, targeting both enzymes might improve results. Thus it is likely that BChE will remain a topic of investigation in Alzheimer's disease–related research.

9.5. EPOXIDE HYDROLASES

9.5.1. Background

Epoxide intermediates are frequently generated in situ through oxidative metabolic processes. These intermediates may function as critical initiators of cellular damage that may include adduction of proteins, RNA and DNA.

The epoxide hydrolases (EHs; EC 3.3.2.3) are a family of enzymes that function to hydrate simple epoxides to vicinal diols and arene oxides to *trans*-dihydrodiols. These enzymes thus represent one category of the broader group of hydrolytic enzymes that include esterases, proteases, dehalogenases, and lipases. EHs are associated with the metabolism of exogenous and endogenous chemicals. In most cases, they catalyze a detoxication reaction, but in some instances they play a role in bioactivation, as illustrated by the metabolism of the procarcinogenic polyaromatic hydrocarbons to highly reactive and mutagenic bay region diolepoxides (Figure 9.6).

9.5.2. Soluble Epoxide Hydrolase (sEH)

Five classes of mammalian EH have been characterized (Table 9.3). Of these, the soluble and microsomal EHs are of most interest with regard to xenobiotic metabolism. The soluble, or cytosolic, form of EH (sEH) metabolizes various *trans*-epoxides, such as *trans*-stilbene oxide, as well as various endogenous substrates (e.g., arachidonate and prostaglandin-derived epoxide intermediates). In humans, the sEH protein is expressed in most tissues from the gene localized on chromosome 8p21-p12. So far, a total of seven polymorphisms have been identified in sEH; four of these have significant effects on enzymatic activity in vitro, and one of the latter (the double mutant Arg287Gln/Arg103Cys) has significant effects on apparent K_m, V_{max}, and enzyme stability. The sEH has been suggested to regulate blood pressure via regulation of renal *cis*-epoxyeicosatrieonic acids (metabolites of arachidonic acid). In this respect, it is significant that a higher level of polymorphisms have been found in a black population, which has a higher incidence of hypertension.

9.5.3. Microsomal Epoxide Hydrolase (mEH)

The microsomal form of EH (mEH) was the initial member of the EH family characterized. Although, as mentioned, mEH is involved in bioactivation processes, it plays a pivotal role mostly in deactivating epoxide intermediates, and is thus considered a major detoxifying enzyme. The mEH protein comprises 455 amino acids, has an apparent molecular mass of 49 kDa (kilodaltons), is highly conserved among species, and is expressed in most tissues and cell types. The gene for human mEH has been mapped to chromosome 1q42.1.

Microsomal EH gene expression in humans and rodents appears to be developmentally regulated. Low levels of mEH functional activity in early-stage fetal liver tissues may represent a risk factor contributing to

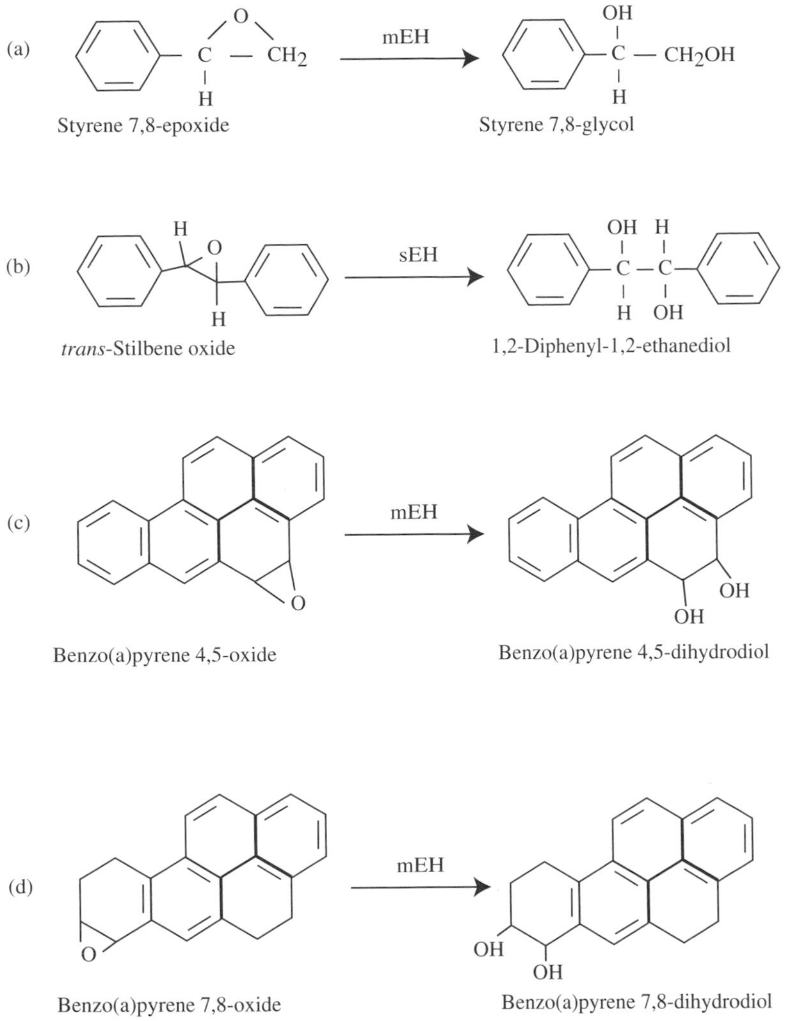

(a) Styrene 7,8-epoxide —mEH→ Styrene 7,8-glycol

(b) *trans*-Stilbene oxide —sEH→ 1,2-Diphenyl-1,2-ethanediol

(c) Benzo(a)pyrene 4,5-oxide —mEH→ Benzo(a)pyrene 4,5-dihydrodiol

(d) Benzo(a)pyrene 7,8-oxide —mEH→ Benzo(a)pyrene 7,8-dihydrodiol

Figure 9.6. Representative substrates of epoxide hydrolases: (a) hydrolysis of styrene oxide by microsomal EH (detoxication reaction); (b) hydrolysis of *trans*-stilbene oxide by soluble EH (detoxication reaction); (c) hydrolysis of benzo(a)pyrene 4,5-oxide by mEH (detoxication reaction); (d) hydrolysis of benzo(a)pyrene 7,8-oxide by mEH. The resulting benzo(a)pyrene 7,8-dihydrodiol is further oxidized by cytochrome P450 or prostaglandin H synthase to benzo(a)pyrene 7,8-dihydrodiol-9,10-epoxide, which is resistant to hydrolysis by EH and binds covalently to DNA (activation reaction).

Table 9.3 Classes of Mammalian Epoxide Hydrolase

Cholesterol oxide hydrolase	Converts cholesterol 5,6-oxide to cholestane; microsomal enzyme
Hepoxilin A_3 hydrolase	Involved in arachidonic acid metabolism; cytosolic enzyme
Leukotriene A_4 hydrolase	Converts leukotriene A_4 (LTA4) to LTB4, a proinflammatory mediator; cytosolic enzyme
Soluble epoxide hydrolase	Metabolizes *trans*-epoxides and several arachidonic and prostaglandin-derived epoxides
Microsomal epoxide hydrolase	Detoxifies several toxic epoxide intermediates; activates polyaromatic hydrocarbons to diolepoxides

Table 9.4 Substrates of Microsomal Epoxide Hydrolase

Benzo(a)pyrene 4,5-oxide
Styrene oxide 7,8-epoxide
cis-Stilbene oxide
Carbamazepine 10,11-epoxide
1,2-Epoxy-3 butene (epoxide of 1,3-butadiene)
Naphthalene 1,2-oxide
Benzene oxide
Nitropyrene oxide
Benzo(a)pyrene 7.8-oxide[a]

[a] Part of activation of benzo(a)pyrene to benzo(a)pyrene 7,8-dihydrodiol–9,10-epoxide, which is resistant to EH.

developmental abnormalities consequent to toxicant exposures. For example, the antiepileptic agent carbamazepine, which is a known teratogen, is oxidatively metabolized through the cytochrome P450 monooxygenase system to an epoxide intermediate, carbamazepine 10,11-epoxide, which is a substrate for mEH.

Unlike cytochromes P450, available evidence suggests that human mEH is not modulated to any major extent by chemical exposures. In rodents, prototypic inducers such as phenobarbital, β-naphthoflavone or polyaromatic hydrocarbons demonstrate modest increases of mEH expression (6.5-fold induction), while imidazole compounds have a slightly greater effect (tenfold induction).

Microsomal EH exhibits broad substrate specificity, and its major function is related to the hydrolysis of epoxides derived from xenobiotics, resulting in the formation of dihydrodiol products. A partial list of substrates of mEH is given in Table 9.4.

9.5.4. Polymorphisms in mEH

The existence of genetic polymorphisms for mEH has been established. Two genetic polymorphisms have been characterized in the gene's coding region: one in exon 3 (Tyr113His) and one in exon 4 (His139Arg), with Tyr and His as the predominant amino acid at positions 113 and 139, respectively. The allelic distributions exist in Hardy–Weinberg equilibrium. The individual haplotypes, expressed in vitro from cDNA constructs and assayed for mEH activity using benzo(a)pyrene-4,5-oxide as substrate, resulted in activities varying by approximately twofold. The Tyr_{113}/Arg_{139} construct exhibited the highest activity, while the His_{113}/His_{139} allele exhibited the lowest. In addition to the coding region polymorphisms, seven additional 5′-region noncoding polymorphisms have also been identified, with evidence that they may affect expression of mEH. Both coding and noncoding region polymorphisms contribute to the population variation in mEH activities, which has been reported to range from two- to eightfold.

Studies on the role of mEH polymorphisms and disease susceptibility have focused on lung cancer. Overall, the polymorphism in exon 3 was associated with a decreased risk in lung cancer (His113/His113 homozygotes, displaying lower EH enzymatic activity), suggesting that mEH may indeed play a role in activating polyaromatic hydrocarbons to genotoxic metabolites. Associations of mEH polymorphisms with ovarian cancer and polycistic ovary syndrome have also been reported.

Acknowledgments

The research described in this chapter was supported by the following grants: ES09883, ES07033, ES-09601/EPA-R826886, ES04696, ES11387, and HL67406.

9.6. SUGGESTED READING

Additional literature on *paraoxonase* (PON1) is available from Costa et al. (2003a, 2003b), Costa and Furlong (2002), and Jarvik et al. (2003).

The following sources on *butyrylcholinesterase* (BChE) are recommended: Darvesh et al. (2003), Lockridge et al. (2000), Schwarz et al. (1995), Soreq and Zakut (1993), and Whittaker (1986).

Further information on *epoxide hydrolase* (EH) can be found in the journal articles by Armstrong and Cassidy (2000), Fretland and Omiecinski (2000), Lee et al. (2002), and Omiecinski et al. (2000).

10

DNA Repair Enzymes

Jon P. Anderson
LI-COR Biosciences, Lincoln, NE

Lawrence A. Loeb
University of Washington, Seattle, WA

10.1. INTRODUCTION

Ever since the origin of DNA-based life, genomes have been subjected to environmental stresses. Every second, the genome of each of our cells is altered, broken, and reassembled. The survival of cells, humans, and species depends on mechanisms to repair this damage and reconstitute genomes. Evolution has been at work for billions of years to produce exquisite DNA repair systems that patrol the genome, fixing or replacing damaged, altered, and miscoded nucleotides. The ability of the cell to maintain its genetic integrity is crucial; without this ability there would ensue a cascading of mutations reaching an error threshhold at which point the wildtype genetic information could no longer be maintained (Eigen 2002). Unrepaired DNA damage can reduce the overall fitness of a cell, triggering cell cycle arrest, apoptosis, unchecked growth, or other diminished functionality. Therefore, loss of any of these repair pathways in humans can result in mutations, cancer, and death, and variability in the ability of these repair systems to perform may ultimately lead to an increase in mutations in somatic cells (any cells that are not egg or sperm) and to a higher risk for disease. In this chapter we will analyze the role of DNA repair enzymes and focus on how polymorphisms in these enzymes may increase the risk of disease. We will

Gene-Environment Interactions: Fundamentals of Ecogenetics, edited by Lucio G. Costa and David L. Eaton

first summarize the types of DNA damage occurring *in vivo*. We will then focus on different DNA repair pathways and effects of the known single-nucleotide polymorphisms (SNPs) in genes that encode components of these pathways. Finally, we will review human diseases linked to deficits in DNA repair.

10.2. DNA DAMAGE

DNA damage can be grouped into two main categories, exogenous (environmental) and endogenous (spontaneous) damage. Exogenous DNA damage can be caused by many environmental agents, including natural chemicals found in food, such as aflatoxins; synthetic (human-made) chemicals, including benzopyrene, which is found in cigarette smoke; and chemicals used in chemotherapy of cancer, such as cisplatin. Other sources of exogenous DNA damage include exposure to UV radiation produced naturally by the sun or artificially by tanning booth lamps, as well as ionizing radiation, such as γ rays and X rays.

Sources of endogenous DNA damage include the chemical instability of DNA and the production of reactive molecules by normal cellular processes. Chemical instability of DNA can manifest as depurination and depyrimidation events, resulting in the loss of a base from the DNA strand. Elegant experiments by Lindahl first demonstrated the rapid loss of purines and pyrimidines from DNA in an aqueous environment (Lindahl and Nyberg 1972). From these data it can be estimated that 10,000 bases per cell per day are lost spontaneously and subsequently repaired (Loeb et al. 1986). Damage by forms of reactive oxygen (e.g., hydroxyl radicals, superoxide anion) in cells also occurs at a high rate, with an estimated 10,000 events per cell per day (Helbock et al. 1998). Other types of damage are caused by chemical alterations, such as methylation, and incorporation of incorrect bases during DNA synthesis. With all of this damage occurring within the cell, it is no wonder that drugs that reduce the damage load on cells, such as antioxidants, are being promulgated for cancer prevention (Weiss and Landauer 2003).

If the tens to hundreds of thousands of damaging alterations that occur in each cell each day are not repaired, these altered, damaged, and miscoded nucleotides can result in somatic mutations. An important example of damage leading to mutation is the deamination of cytosine to uracil, which causes a code change from CG to TA. Other mutations, including (6–4) photoproducts and cyclobutane dimers caused by UV light exposure, block both DNA replication and RNA transcription by terminating synthesis by DNA and RNA polymerases, respectively. In order to correct this damage before it affects cellular functionality or triggers apoptosis (programmed cell death), the cell has evolved multiple repair mechanisms. In the following sections, we cover several of these systems including reversal of damage, base excision repair, nucleotide excision repair, mismatch repair, recombination with restoration of DNA sequences, and the bypass of lesions by special DNA polymerases. We will also consider genetic variations in the human population that affect the efficiency and accuracy of these repair mechanisms and can lead to greater disease susceptibility.

10.3. REPAIR MECHANISMS

10.3.1. Reversal of Damage

In principle, the direct repair of DNA damage (i.e., chemically correcting the change in a damaged base, without removing it) is error-free. Damaged DNA is returned to its original state without replacement of the damaged bases. Since this approach would appear to be the most accurate and efficient way to repair damaged DNA, why is it not the only approach used? Most DNA damage that takes place within a cell cannot be reversed as a result of kinetic barriers, so other repair mechanisms need to be implemented to maintain genetic integrity. In instances where the damage can be successfully reversed, the cell has developed mechanisms to exploit this feature. Reversal of damage includes several processes ranging from photoreactivation; to the removal of adducts (molecules which attach to DNA); and religation of nicked DNA.

Photoreactivation involves the direct reversal of UV damage that creates pyrimidine–pyrimidine cyclobutane dimers and (6–4) photoproducts. The repair is mediated by a class of enzymes called *photolyases* that bind to the site of UV damage. For example, to repair damage resulting in a dimer, the enzymes absorb blue light energy through chromophores and use this gathered energy to break the covalent bonds that link the bases' pyrimidine rings, thus converting the dimer back to the two separate bases. This pathway for DNA repair, although not conclusively demonstrated in humans, is present in bacteria, plants, fungi, insects, reptiles, amphibians, and marsupials.

Another DNA repair pathway targets alkylated bases. Alkylation damage is marked by the addition of carbon-containing alkyl groups to multiple sites on DNA molecules (Singer and Grunberger 1983). Most chemotherapeutic drugs, as well as other synthetic chemicals such as vinyl chloride and the chemical warfare agent mustard gas, are alkylating agents. These agents can produce oxygen and nitrogen alkylated products, including O^6-methylguanine and O^4-methylthymine. Once attached to the DNA, these compounds can cause errors, such as mispairing, during DNA synthesis, and ultimately lead to DNA mutations (Wood et al. 1992). Alkylated adducts can be directly removed by a class of enzymes called *alkyltransferases*, which are commonly referred to as "suicide enzymes." After removing an alkyl group from the damaged nucleotide, the alkyltransferase transfers the molecule to a cysteine residue in its own structure, inactivating itself as a result (Wani et al. 1990). Since these enzymes can undergo only one catalytic event, they might be classified as disposable repair proteins.

Why would such an energetically expensive system that uses a nonrenewable protein be used in the repair process? The answer may lie in the fact that this type of system is kinetically advantageous over other multiple-step processes that employ several gene products. Furthermore, evidence in *Escherichia coli* indicates that the inactivation of alkyltransferases may result in an adaptive response that triggers the production of more gene products that can protect against alkylation damage (Volkert 1988). In other words, when an *E. coli* alkyltransferase is inactivated because of the repair of a DNA adduct, it will induce the production of other repair proteins, increasing the overall ability of the cell to survive alkylation damage.

In humans, O^6-alkylguanine–DNA alkyltransferase (AGT) is used primarily for removing methyl groups from O^6-methylguanine and O^4-methylthymine. Transgenic mice overexpressing the human AGT protein have been shown to maintain a higher resistance to alkylation-induced tumor formation as compared to wildtype mice (Dumenco et al. 1993). Tumors with reduced AGT expression are particularly responsive to treatment with alkylating agents. Polymorphic sites have been identified in the human AGT protein, but it remains to be established whether these polymorphisms confer an increased risk for cancer (de Boer 2002).

Finally, certain single-strand breaks (nicks) in the DNA backbone can be directly repaired by DNA ligases. The frequency by which single-strand breaks occur spontaneously has not been established. However, these are one of the most frequent types of lesions caused by exposure of cells to X rays and γ rays. Most breaks in the DNA backbone result in a damaged 5′-phosphates or altered termini, rendering them unavailable for repair by ligases.

10.3.2. Base Excision Repair (BER)

As the name implies, base excision repair (BER) is a repair mechanism that removes and replaces damaged bases from a DNA strand. Primarily, BER

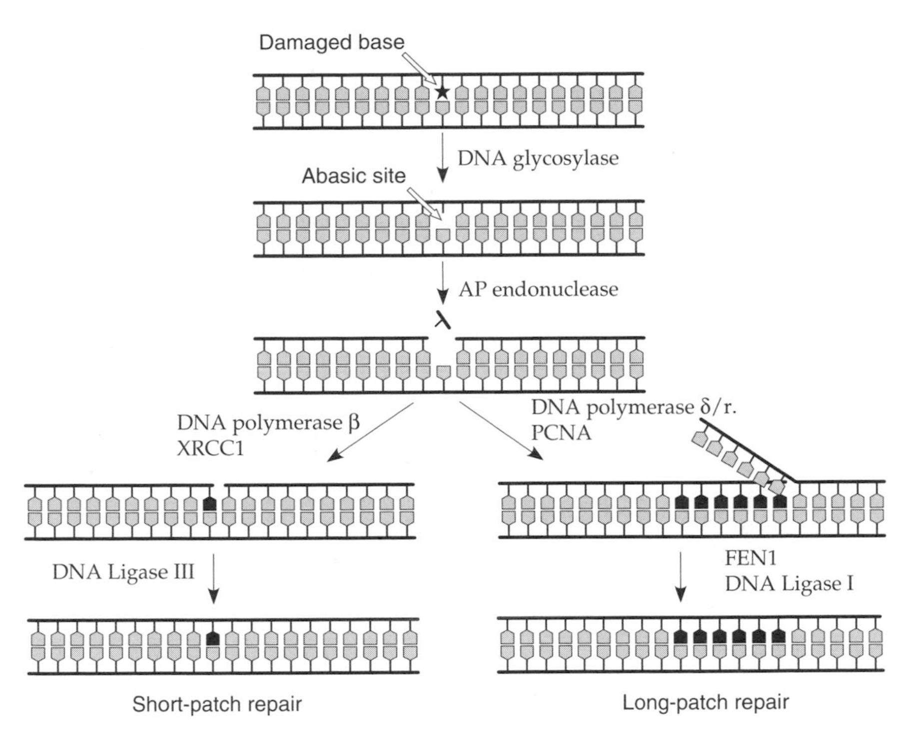

Figure 10.1. Schematic representation of human base excision repair showing removal and replacement of a single nucleotide (short-patch repair) or replacement of the damaged base along with 2–8 additional nucleotides (long-patch repair). The filled-in pentagons represent nucleotides incorporated during DNA repair.

replaces single bases in a process termed "short-patch repair." A second minor pathway for BER also exists, termed "long-patch repair," in which 2–10 nucleotides are displacing from the DNA strand.

Short-patch repair is the major BER pathway utilized in humans (Figure 10.1). DNA glycosylases are the major repair enzymes that initiate BER. These enzymes are believed to patrol DNA, recognizing and removing aberrant bases throughout the genome. DNA glycosylases are monomeric enzymes that first attach to sites of aberrant bases. After locating an aberrant base, a glycosylase attaches to the site and bends the DNA, causing the damaged base to flip out of the DNA strand into a cleft in the enzyme. The flipped-out base is then cleaved from the DNA sugar backbone by the DNA glycosylase. This action generates a gap, or abasic site, designated as apurinic or apyrimidinic (AP), depending on which type of base has been removed. The abasic site is further processed by an AP endonuclease that cuts one side of the deoxyribose (Mol et al. 1995) and results in a site lacking a single nucleotide. The DNA is then resynthesized by DNA polymerase β and XRCC1 (Singhal et al. 1995) and the nicked DNA is then sealed by DNA ligase III.

Although the short-patch BER pathway specifically targets damaged bases in DNA, it is also widely used to repair gaps, or AP sites, that occur spontaneously with an estimated frequency of 10,000 hits per cell per day. The human genome also sustains an additional 10,000 oxidative damaging hits per cell per day (Helbock et al. 1998), and most of these damaged sites are repaired by DNA glycosylases. There are more than eight known human DNA glycosylases that all function in specific situations, although some have overlapping specificities. Four of these remove uracil from the DNA strand. The major protein responsible for repairing AP sites in humans is the endonuclease APE1. Nonsynonymous polymorphisms in APE1 have been identified in the human population that exhibit a 40–60% reduction in incision activity (Hadi et al. 2000). These variants may be associated with an increase in disease susceptibility, as it has been reported that a 20–35% reduction in BER repair capacity leads to an increased risk for developing cancer (Hadi et al. 2000).

The most thoroughly studied glycosylase, OGG1, recognizes and removes the modified base 8-oxo-guanine from oxidatively damaged DNA. A reduction in the activity of this glycosylase results in an increase accumulation of oxidation-induced mutations, and a common polymorphism in OGG1 (S326C) has been associated with an increased risk of cancer (Xu et al. 2002). It has also been reported that smokers who develop lung cancer have decreased OGG1 activity in peripheral blood cells compared to smokers who do not develop lung cancer. The XRCC1 protein also takes part in BER by stimulating endonuclease activity following the excision of the damaged base.

A specific polymorphism in XRCC1 (R194W) has been linked to a reduction in cancer risk when this polymorphism is associated with other risk factors, including smoking and alcohol consumption (Goode et al. 2002). These types of studies, which link demographic risk factors and polymorphisms, may provide better insight into the biology of DNA repair mechanisms.

Long-patch repair is a minor component of the human BER pathway. Like short-patch repair, long-patch repair proceeds following removal of the aberrant base by a DNA glycosylase and nicking of the DNA strand by an AP endonuclease. However, unlike short-patch repair, Long-patch repair uses enzymes that are normally associated with DNA replication. PCNA and DNA polymerase-δ and/or -ε proceed to fill in the gap in the DNA strand. These enzymes synthesize a new stretch of DNA that displaces the original DNA strand and creates a single-stranded DNA flap of 2–10 bp in length. The flap endonuclease, FEN1, removes the DNA flap, and the nicked DNA strand is repaired with DNA ligase I (Figure 10.1).

10.3.3. Nucleotide Excision Repair (NER)

Many forms of DNA damage are not recognized by DNA glycosylases and thus cannot be repaired by BER. Furthermore, unlike many other organisms, humans do not appear to repair pyrimidine–pyrimidine dimers by the use of photolyases (see Section 10.3). A separate DNA repair system, nucleotide excision repair (NER), is therefore used to deal with these other DNA alterations, specifically those that generate large covalent additions to bases in DNA. NER differs from BER in that the excised product is a single-stranded oligonucleotide of some 30 bases in length, rather than a single base. The substrate specificity for NER is less stringent than that for base excision repair; NER recognizes extreme forms of DNA damage such as chemical crosslinks, pyrimidine dimers, (6–4) photoproducts, and large chemical adducts. Presumably the common intermediate in damage detection by NER is distortion of the double-helix DNA structure.

Nucleotide excision repair includes two different pathways. Global NER identifies and corrects damage throughout the genome, while transcription-coupled NER targets active genes engaged in transcription. The general steps in both pathways appear to be similar and include recognition of the DNA damage by repair proteins that bind at the site of damage, melting of the dsDNA at the site of damage, excision of a 24–32-nucleotide oligo that encompasses the damage, filling of the resulting gap by DNA polymerases, and ligation of the nicked DNA (Figure 10.2).

Approximately 20 proteins are required for either global or transcription-coupled NER. Global NER commences with the XPC/hHR23B heterodimer recognizing DNA damage, presumably by patrolling DNA and identifying distortions at the site of damage. Transcription-coupled NER does not require XPC to identify DNA damage, but relies on the stalling of RNA polymerase II at the site of damage during transcription (Hanawalt and Spivak 1999). The proteins CSA and CSB associate with the RNA polymerase complex and initiate transcription-coupled NER. For both systems, other proteins are then recruited to the site of damage, including the transcription factor IIH (TFIIH) proteins, the XPB and XPD helicases, the single-stranded binding protein RPA, and XPA. This protein complex proceeds to melt the dsDNA at the site of damage, allowing the endonucleases XPG and XPF/ERCC1 to cut out the region of damage. XPG cuts on the 3' side of the damage, 2–8 nucleotides from the lesion. The XPF/ERCC1

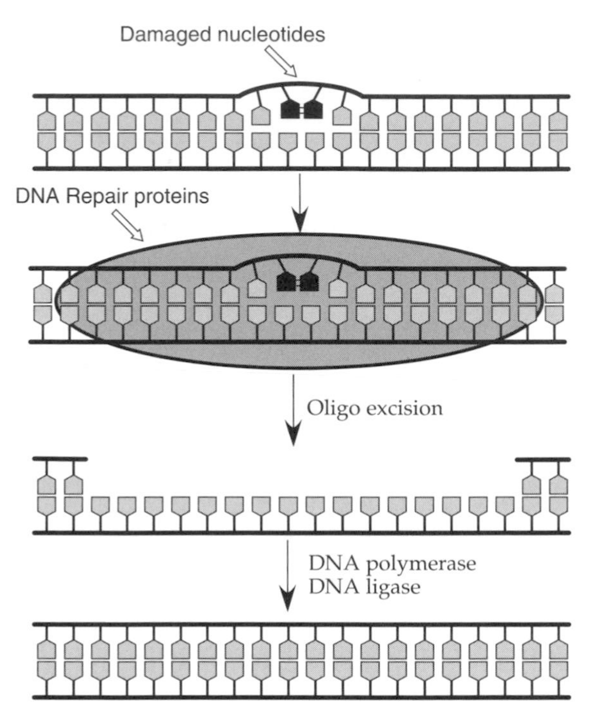

Figure 10.2. Schematic representation of human nucleotide excision repair. An oligonucleotide segment of 24–32 nucleotides, containing DNA damage, is excised by a set of approximately 20 proteins. The resulting gap is filled by DNA polymerase.

heterodimer cuts on the 5′ side of the damage, 15–24 nucleotides from the lesion. These endonucleases both cut ssDNA at junctions with dsDNA, creating an oligonucleotide 24–32 nucleotides in length harboring the damaged nucleotide. Following the removal of the oligonucleotide from the DNA strand, polymerase-δ or -ε along with PCNA proceeds to fill in the gap, and ligase I then seals the backbone of the DNA.

Several variants have been found for the genes of the NER pathway. In most cases, it is unknown whether these variants are associated with an increased risk for cancer. Many of the genes involved in the NER pathway are also involved in transcription initiation processes. Therefore, polymorphisms that affect these NER genes may influence multiple systems and can produce complex results.

10.3.4. Mismatch Repair

Throughout this chapter, we have outlined how our cells identify and repair various types of DNA damage. Each of these previous repair systems relies on the fact that the damaged DNA they target contains nucleotides that are different from those in native DNA and thus are easily identified. These repair systems do not appear to efficiently recognize DNA that contains unaltered nucleotides that have been incorrectly paired. If uncorrected, the mispairing of the DNA nucleotides leads to mutations following DNA replication. Mispairs usually occur because of replication errors in which the DNA polymerase incorporates the incorrect base. Mismatches can also occur via other reactions, such as the deamination of 5-methylcytosine to produce thymine, which is then mispaired with guanine.

Mismatch repair proteins (MMRs) are used to identify and correct base mismatches and small insertions or deletions throughout the genome. One of the main concerns with properly correcting a mismatch is identifying the incorrect nucleotide. Since both of the nucleotides present in a mismatch are potentially correct, MMR must repair only the newly synthesized daughter strand that contains the misincorporated base. For mammals, the signal that distinguishes daughter from parent strands has not been conclusively established, but is believed to involve the presence of a single-strand break.

However, it has been determined that in *E. coli*, repair enzymes rely on the methylation of DNA to distinguish daughter strands. During DNA replication in *E. coli*, the newly synthesized daughter strand remains unmethylated for several minutes. During this time MMRs can distinguish between the daughter and parental DNA strands, repairing any mismatches in the daughter strand. *E. coli* MMR uses three main proteins to identify mismatches: mutS, mutL, and mutH. DNA mismatches and small insertions/deletions are initially detected by mutS, which binds to the site of the mismatch and initiates the recruitment of mutL. MutL stabilizes the DNA protein complex and activates mutH, a protein that can identify a hemimethylated GATC site located hundreds of nucleotides away from the mismatch and nick the DNA strand opposite the methyl group. *E. coli* enzymes, including helicase II, ExoI, ExoVII, and RecJ, then unwind and degrade the daughter DNA strand starting from the hemimethylated GATC site and proceeding past the mismatched site. This degraded region of DNA can span more than 1 kbp and must be resynthesized by DNA polymerase III and sealed by DNA ligase (Figure 10.3). What an inefficient system for the replacement of a single mismatched nucleotide!

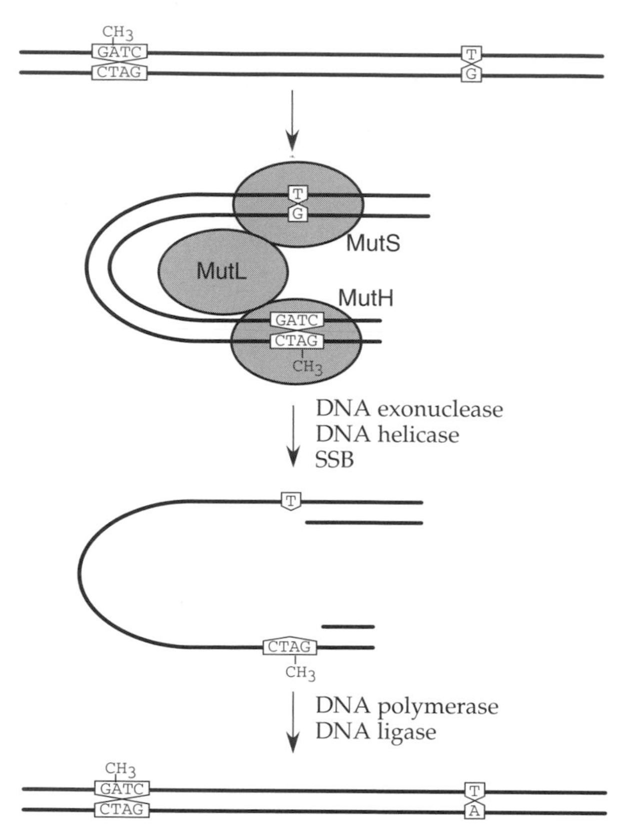

Figure 10.3. Representation of the mismatch repair pathway in *Escherichia coli*. Mismatched bases are bound by mutS, which recruits mutL and mutH proteins. MutH binds to a hemimethylated GATC site, discriminating the methylated parental strand from the unmethylated daughter strand. The daughter strand containing the mismatch is degraded, and DNA polymerase fills the resulting gap.

Several human homologs to the *E. coli* MMR genes have been identified and appear to function in a similar manner in humans (Modrich and Lahue 1996). Mismatched DNA, single insertions, and deletions are initially identified by MutSα, which is a heterodimer consisting of human mutS homolog 2 (hMSH2) and hMSH6. A second heterodimer, MutSβ, is responsible for identifying two to four nucleotide insertions and deletions, and is composed of hMSH2 and hMSH3. Several mutL homologs (MLH) also exist and function as heterodimers in humans, such as hMutLα (hMLH1 and hPMS2), and hMutLβ (hMLH1 and hPMS1). The DNA from the daughter strand is believed to be removed by exonuclease I and FEN1, and the newly formed gap is filled in by DNA polymerase-δ and/or -ε in association with PCNA. As mentioned previously, the mechanism that designates the daughter strand in human cells has not yet been definitively established.

Over 250 mutations have been identified in the hMLH1 and hMSH2 genes (de Boer 2002). Mutations in MSH3, MSH6, and MLH3 have been associated with microsatellite instability (Perucho 1996). Not surprisingly, most of these mutations were identified in cancerous cells but not in normal somatic cells from the same individuals. Reduced expression of several of these repair genes has also been linked to cancer, indicating that individual variations may increase the risk for disease (Wei et al. 1998).

10.3.5. Double-Strand Break Repair

Double-strand breaks are considered to be one of the most dangerous types of lesions produced in DNA; as few as one break per genome has the potential to kill the cell (Khanna and Jackson 2001). Double-strand breaks can arise both spontaneously and through exposure to ionizing radiation; spontaneous double-strand breaks are estimated to occur at a rate of 40 per cell per year in mammals (Stewart 1999). Unlike other repair systems, repairs of double-strand breaks cannot utilize the complementary, undamaged DNA strand to serve as a template for restoring the nucleotide sequence. Human cells employ two separate repair systems to combat these dangerous lesions. Nonhomologous end joining (NHEJ) is the predominant system used, while repair by homologous recombination occurs less frequently.

10.3.5.1. Non-homologous End Joining. Nonhomologous end joining directs the rejoining of the broken DNA strands without requiring the presence of complementary nucleotides at the site of the break. The proteins involved in NHEJ must first identify the ends of the fragmented DNA strand, and these ends must be brought into close proximity to one another. Any damage to the ends must be restored and the two ends ligated together to form an intact DNA strand. The major proteins responsible for repairing double-strand breaks through NHEJ are the Ku proteins, DNA-PK_{cs}, MRE11, Rad50, MBS1, XRCC4, and ligase IV.

The Ku proteins (Ku70 and Ku80) form a heterodimer that binds to the ends of the fragmented DNA. The Ku protein complex may also serve to bring the two fragmented ends near one another. DNA-PK_{cs} is a protein kinase that recognizes the DNA break and triggers a repair response by phosphorylating downstream proteins. MRE11, Rad50, and NBS1 are proteins thought to process the ends of the breaks, preparing them to be rejoined. XRCC1 and ligase IV rejoin the ends of the break, creating an intact DNA strand.

Polymorphisms have been identified in each of the genes responsible for NHEJ. Several of these SNPs have been linked to disease susceptibility. SNPs identified in ligase IV may modulate predisposition to multiple myeloma and breast cancer (Roddam et al. 2002). Polymorphisms in the Ku80 gene have been found in high frequency in groups of cancer patients (Price et al. 1997), and reduction of the Ku proteins by an autoimmune response has been shown to increase sensitivity to ionizing radiation (Harris et al. 1985). Finally, two polymorphisms (R2140C and M3844V) have been identified in the DNA-PK_{cs} gene of BALB/c mice that render the mice more susceptible to breast cancer and may explain why this strain is tenfold more sensitive to ionizing radiation than other mouse strains (Yu et al. 2001). Identification of such SNPs that increase disease susceptibility reiterates the importance of the NHEJ pathway in maintaining genetic integrity and may provide clues on genomic rearrangements that characterize certain cancers.

10.3.5.2. Homologous Recombination. Homologous recombination is only a minor component of the double-strand break repair mechanism in humans, but is the predominant system in other organisms, including yeast. Unlike NHEJ, which links the two broken strand ends together without any other information about the DNA being repaired, homologous recombination utilizes information from the homologous chromosome to aid in the repair (Figure 10.4).

Following a double-strand break, proteins possessing 5′ exonuclease activity "chew" back the ends of the DNA, producing single-stranded ends that can invade and recombine with the DNA of the homologous chromosome. The complementary DNA is then synthesized on the single-stranded sections using information from the sister chromosome. Finally, the two DNA strands are unwound and the nicks are ligated to form the intact DNA strand.

The proteins involved in homologous recombination include Rad50, MRE11, and MBS1, which are also involved in NHEJ. This pathway also uses the proteins Rad51, Rad52, Rad54, Rad55, Rad57, Rad59, XRCC1, BRCA1, BRCA2, and NbsI, independently from the NHEJ pathway. Polymorphisms have been found in the Rad51, XRCC1, BRCA1, and BRCA2 genes that confer an increased risk for lung and breast cancers (Levy-Lahad et al. 2001; Fu et al. 2003). Other SNPs have been identified in the genes of homologous recombination, but it is unclear whether these polymorphisms affect the risk for any disease.

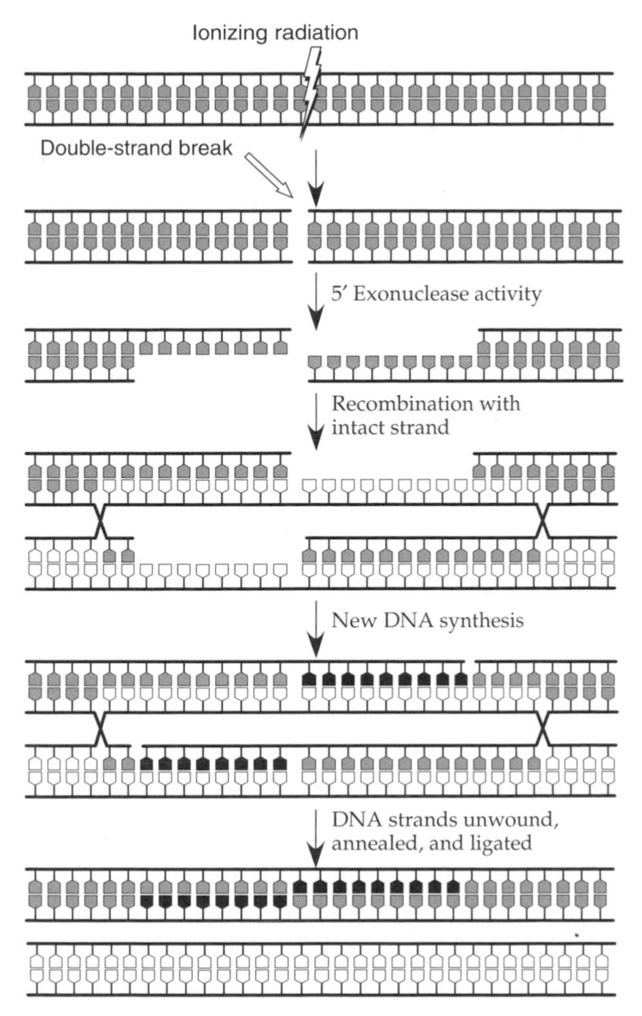

Figure 10.4. Schematic representation of homologous recombination. Broken DNA is resected by 5′ exonuclease activity. The resulting single-stranded DNA sections recombine with the sister chromosome using this intact DNA as a template for repair. Gaps in the DNA strands are filled and the strands are unwound, reannealed, and religated to form an intact DNA strand.

10.3.6. Damage Bypass

We have considered how cells use multiple repair systems to abnegate the effects of damage from both exogenous and endogenous sources. These repair mechanisms are able to overcome tens of thousands of DNA alterations and misincorporations per day, thereby restoring the nucleotide sequence and guaranteeing faithful passage to subsequent cell generations. Despite these strategies, many lesions escape DNA repair and are present at the time of DNA replication. Damage that is left unrepaired can block the major replicative DNA polymerases (Pol δ, ε, α), and thus halt replication. If the block is not bypassed, then the DNA will not be fully copied, leading to chromosomal truncation, cellular arrest, apoptosis, or altered functionality. Base excision repair and nucleotide excision repair pathways are no help in correcting this type of blocking damage because the advancing replication fork has separated the parental DNA strands, eliminating the opportunity to use the undamaged complementary strand as a template for repair.

Nevertheless, it has long been recognized that mammalian cells with thousands of blocking lesions in their DNA are able to replicate. The more recent discovery of special DNA polymerases able to copy past altered DNA templates provides an explanation for this enigma. Two separate bypass mechanisms, recombinational bypass and bypass synthesis, are thought to be used by the cell in order to overcome blocked replication. In both mechanisms, the damaged base is not repaired but simply traversed. This type of bypass allows the replication fork to continue while the damage is left for the other repair mechanisms to contend with at a later time. Recombinational bypass is thought to utilize template switching to bypass the DNA damage. As the replication machinery encounters DNA damage, one strand is impeded by the damage, while the complementary strand continues to be copied. The newly synthesized sister DNA from the complementary, unblocked strand now becomes the surrogate template for the replication that has stalled. By utilizing this template-switching scheme over a short distance, the DNA damage present in the parental strand can be bypassed, allowing replication to proceed. This form of lesion bypass is considered an error-free method that uses the normal replicative polymerases. It is also believed that the enzymes involved in homologous recombination play a role in recombinational bypass, aiding in the template-switching mechanism. The details of this method, however, remain to be delineated.

Synthesis by bypass DNA polymerases is an error-prone process; however, the cells are evidently willing to incur mutations as a price for survival. After the replicative polymerase(s) stalls (stall) at the site of damage, the replicative polymerase is replaced by a bypass polymerase that copies across the damaged bases. The bypass polymerase then dissociates from the DNA and the normal replicative machinery continues to synthesize the DNA. At least four mammalian bypass polymerase have been identified: ζ, η, ι, and κ (Friedberg et al. 2002). Each of these bypass polymerases seems to be specific in the type of DNA damage that it bypasses, and all share two main properties. They all have a lower fidelity and lower processivity than the replicative polymerases. The low fidelity is manifested by an increased rate of incorporation of noncomplementary nucleotides in copying of unaltered DNA, and has the potential to produce mutations throughout the genome. Bypass polymerases have an especially large binding pocket within the catalytic site that allows them to accommodate bulky template alterations in DNA. As a consequence they can also accommodate noncomplementary basepairs. The low processivity of these polymerases self-limits the stretch of DNA that they can copy during each binding event. Using polymerases with low processivity and low fidelity enables the cell to successfully bypass damaged nucleotides in DNA without incorporating significant amounts of error in the daughter strands.

10.4 DISEASES LINKED TO NONFUNCTIONAL REPAIR SYSTEMS

The findings by Cleaver regarding xeroderma pigmentosa (XP), a rare inherited DNA repair disease with a high incidence of skin cancers, have provided the strongest link thus far between cancer and deficits in DNA repair (Cleaver and Kraemer 1989). Individuals with XP have a 1000-fold increase in risk of skin cancer after exposure to sunlight. This specificity is mimicked in cell cultures; XP cells are sensitive uniquely to UV light. These cells also have mutations in the genes encoding nucleotide excision repair that presumably reduce or eliminate NER rates and so cause the cells' susceptibility to light. Other diseases where mutations in DNA repair enzymes have been linked to a high incidence of cancer include ataxia telangectasias, Fanconi's anemia, hereditary nonpolyposis coli, Bloom's syndrome, and inherited forms of breast cancer with mutations in BRCAI and BRCA2. The association between polymorphisms in DNA repair genes and cancer may also pertain in Werner syndrome, a disease of premature aging that is characterized by a high incidence of unusual cancers. Fortunately, most inherited DNA repair diseases are rare and require that the individual be homozygous for the polymorphisms. However, heterozygotes, who have a single mutated allele, are much more common in the population and also suffer from a reduction in expression that may render them susceptible to environmentally mediated DNA damage.

10.5. THE "MUTATOR PHENOTYPE"

Cells have evolved exquisite and redundant systems for maintaining genetic integrity. Every day, tens of thousands of DNA-damaging alterations and misincorporations are successfully repaired, thus preventing somatic mutations. This is not to say that no mutations occur within our genomes; the estimated mutation rate for human somatic cells is $\sim 2.0 \times 10^{-7}$ mutations per gene per division (Oller et al. 1989). Furthermore, chronic exposure to DNA-damaging agents, such as benzopyrene from cigarettes, increases mutagenesis, presumably by causing an amount of DNA damage that exceeds the cell's repair capacity. These somatic mutations can have dire consequences. It is the production of somatic mutations in oncogenes and tumor suppressor genes that is thought to trigger carcinogenesis, with an estimated 6–12 events required for tumor progression (Renan 1993). Since these events need to occur within specific genes, and within a particular class of proliferating cells (stem cells), the probability of an individual developing cancer under normal rates of somatic mutation is exceedingly low. Cancer, however, is frequently observed within the human population. Furthermore, at the chromosomal level, cancer cells contain many alterations involving thousands of nucleotides. In order to account for the high number of mutations found in most cancer cells, it is argued that a mutator phenotype must be expressed during the development of a tumor.

The mutator phenotype can arise through mutations in multiple genes whose normal function is to maintain genetic integrity. An acquired mutation in a DNA repair pathway that compromises efficiency or accuracy of repair will decrease the cell's ability to cope with subsequent genomic damage, resulting in an increase of mutations and a progression toward cancer. Mutations in genes that affect cellular replication—including the replicative polymerases, helicases, and DNA binding proteins—may also render them error-prone and produce a mutator phenotype.

The expression of a mutator phenotype is thought to be a relatively early event in the progression toward cancer. Cells that carry mutations in genes that regulate cellular growth, invasion, and metastasis have a selective advantage as tumors grow. This accumulation of mutations throughout the genome, along with multiple rounds of selection, leads to tumor progression and heterogeneity. It is estimated that it takes 20 years or more for DNA damage from exposure to a chemical carcinogen to develop into a clinically detectable tumor (Loeb 2001). It is hypothesized that during this time a mutator phenotype would be manifested in cancer cells, multiple rounds of selection would take place, and mutations within multiple onocogenes and tumor suppressor genes would produce a malignant phenotype (Figure 10.5).

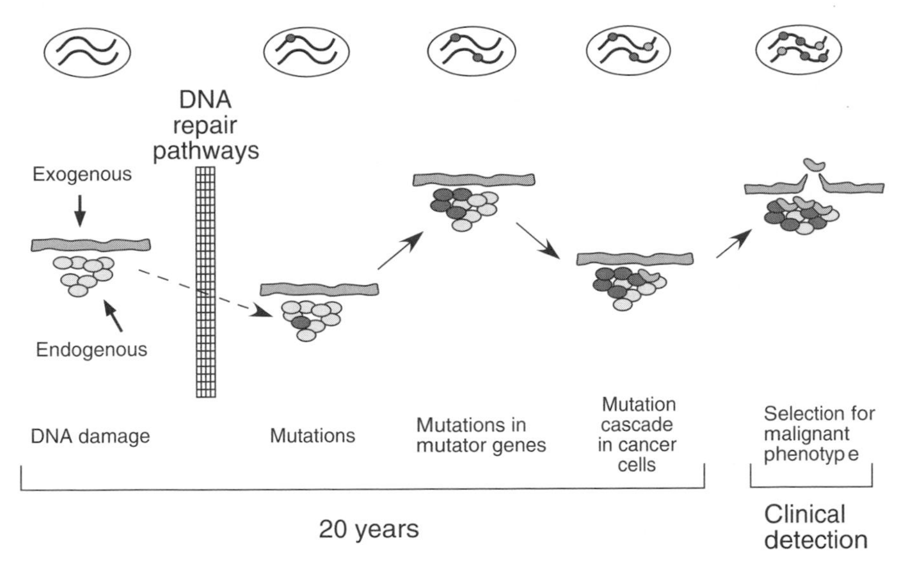

Figure 10.5. A mutator phenotype in cancer progression. Random mutations occur when endogenous and exogenous damage exceeds the capacity of the DNA repair mechanisms. Random mutations can both continue to expand clonally and result in mutations in mutator genes. Increase in the rate of mutation caused by mutations in mutator genes along with repetitive rounds of selection can result in additional mutations in mutator cells. Cancer cells expressing mutations in genes that regulate cellular growth, invasion, and metastasis are selected to produce a malignant phenotype.

The presence of a mutator phenotype may have some interesting consequences in the treatment and prevention of cancer. The overall effectiveness of chemotherapeutic agents may be diminished by preexisting drug-resistant mutants present within a tumor. On chemotherapy treatment, these drug-resistant mutants would have a replicative advantage and could repopulate the tumor. This provides one mechanism by which tumors routinely acquire drug resistance to a single chemotherapeutic agent, and provide a mechanistic explanation for the efficacy of, and frequent necessity for, multiple-drug therapy in cancer treatment. If the accumulation of mutations is directly correlated with tumor progression, then inhibiting mutation accumulation may delay carcinogenesis. Assuming a 20-year average progression to clinical cancer, even a twofold reduction in the rate of mutation accumulation would provide a significant clinical delay. By exploring ways to reduce the high amount of DNA damage occurring within the cell, we may successfully prevent cancer by delaying onset to the point that death from other causes occurs first. Mutations that reduce the fidelity of DNA polymerases could be exploited by using nucleotide analogs that are incorporated and terminate DNA synthesis preferentially in malignant cells. Finally, future treatments involving gene therapy may be able to specifically target repair system genes that are malfunctioning, enhancing their activities and reducing the mutational load.

Throughout this chapter we have investigated a series of repair mechanisms that are used by cells to maintain the integrity of the information contained within our DNA. By repairing the damage that is constantly occurring to our DNA, our cells reduce the amount of somatic mutations that take place, maintaining the cells' overall fitness. Destruction of one or more of these repair pathways can increase cancer risk by increasing the mutational load placed on the cell. Finally, polymorphisms in the genes of the repair systems may reduce the ability of these systems to maintain the genome. This reduction in repair activity, alone or in synergy with environmental factors, may place us at an increased risk for disease.

10.6. SUGGESTED READING

The journal articles by deBoer (2002), Goode et al. (2002), Loeb et al. (2003), Stoler et al. (1999), Valerie and Povirk (2003), and Wood et al. (2001) are recommended for further reading on DNA repair enzymes.

11

Receptors and Ion Channels

Lucio G. Costa
University of Washington, Seattle, WA

11.1. INTRODUCTION

Cells produce signaling molecules that communicate with and influence other cells. Some signaling molecules bind with high affinity to specific receptor proteins on the surface of target cells. Such is the case of neurotransmitters and of many growth factors, cytokines, and hormones. Once the ligand is bound to its receptor, the latter converts this interaction into one or more intracellular signals (Csaszar and Abel 2001). Another form of receptors, the intracellular type, is located in the cytosol or in the nucleus. They are usually in an inactive state, and the binding of a signal molecule gives rise to the translocation to the nucleus and/or binding to specific gene sequences that regulate transcription. Unlike intracellular receptors, cell surface receptors do not regulate gene expression directly, but do so through a series of various intracellular signal–transduction mechanisms (Csaszar and Abel 2001). On the basis of the transduction mechanisms, the receptors can be divided into three classes: channel-linked or ionotropic; G-protein-linked, or metabotropic; or catalytic (e.g., receptors with tyrosine kinase activity). Receptors play a crucial role in the regulation of cellular functioning, and small changes in their structure may influence, qualitatively or quantitatively, the nature of the response. Genetic polymorphisms have been reported for a large number of receptors, and some have been shown to be associated with specific diseases or with altered responses to endoge-

Gene-Environment Interactions: Fundamentals of Ecogenetics, edited by Lucio G. Costa and David L. Eaton

nous compounds, pharmaceutical drugs (55% of which target receptors), or environmental chemicals.

Cell membranes are usually impermeable to charged molecules, such as ions. This barrier is overcome by ion channels, which are proteins that form pores in the membrane, allowing selected ions to pass through. This selectivity allows classification of ion channels into Na^+, K^+, Ca^{2+}, and Cl^- channels, based on their preferred ion (Bockenhauer 2001). Cells actively maintain concentration gradients for ions; Na^+, Ca^{2+} and Cl^- are accumulated in the extracellular fluid, whereas K^+ is concentrated in the cytoplasm. Concentration gradients and the transmembrane voltage make up the electrochemical gradient, and serve as the driving force of ion movement (Bockenhauer 2001). Ion channels are also equipped with gates: mechanisms that open or close a pore in response to defined stimuli, such as voltage or binding of a ligand. Altogether, these ion channel systems allow the control of complex events such as heartbeat, muscle contraction, or neuronal functions. Polymorphisms of ion channels also exist, that have been associated with cardiac, CNS, and muscle diseases.

This chapter focuses on selected examples of genetic polymorphisms affecting receptors and ion channels. Examples discussed include the adrenergic receptors, a class of membrane receptors for a major neurotransmitter; the vitamin D receptors and the Ah receptors, as examples of cytosolic receptors for an important vitamin and for various classes of environmental contaminants (e.g. dioxins, polyaromatic hydrocarbons); and the nuclear estrogen receptors. In addition, polymorphisms of the sodium channel are also discussed.

11.2. ADRENERGIC RECEPTORS

Adrenergic receptors are the target for epinephrine and norepinephrine and represent a critical component of the sympathetic nervous system, which innervates most organs of the body and controls their functions. All adrenergic receptors are G-protein-coupled receptors and are the target of a multitude of drugs used to treat a wide variety of diseases ranging from hypertension, angina, and cardiac arrhythmia, to asthma, prostate hypertrophy, and glaucoma. The family of human adrenergic receptors consists of nine subtypes: α_{1A}, α_{1B} and α_{1C}; α_{2A}, α_{2B}, and α_{2C}; β_1, β_2, and β_3 (Small et al. 2003). In most cellular systems, α_1-, α_2-, and β-adrenergic receptors are coupled to G_q, G_i, and G_s proteins, respectively. This leads to stimulation of phospholipase C and increased function of inositol phosphates; inhibition of adenylyl cyclase, with a decreased formation of cyclic AMP; or stimulation of adenylyl cyclase.

It has been long recognized that physiologic responses and functions of adrenergic receptors, as well as response to adrenergic agonists or antagonists, display marked interindividual variations within the human population (Small et al. 2003). The role of genetic polymorphisms in at least some of the these interindividual variations has been extensively studied in most recent years. Seven of the nine adrenergic receptors (the exceptions are the α_{1B} and α_{1C} receptors) have polymorphisms that result in amino acid substitutions or deletions, which in some cases have clinical and/or pharmacological relevance (Table 11.1). Most polymorphisms result in alterations in the binding of agonists to the receptors with consequent changes in the coupling to their respective effector systems. Additionally, several missense mutations lead to altered ability of the receptor to desensitize or downregulate on exposure to an agonist (Table 11.1). Polymorphisms of the β_1 and β_2 receptors have been most extensively investigated.

Table 11.1 Polymorphisms of Adrenergic Receptors

Receptor	Polymorphism[a]	Biochemical/ Pharmacological Consequences	Clinical relevance
α_{1A}	Cys492Arg	None	None
α_{2A}	Asn251Lys	↑ inhibition of cAMP[b]	None
α_{2B}	Glu–Glu–Glu deletion	Lack of desensitization	Risk of acute coronary events?[c]
α_{2C}	Gly–Ala–Gly–Pro deletion	↓ inhibition of cAMP ↓ stimulation of PLC	Risk of heart failure
β_1	Ser49Gly	↑ downregulation	None certain
	Arg389Gly	↓ stimulation of cAMP	Reduced response of DBP to β blockers
β_2	Arg16Gly	↑ downregulation	↑ nocturnal asthma symptoms
	Gln27Glu	↓ desensitization	↓ response to β agonists
	Thr164Ile	↓ stimulation of cAMP	↑ risk of hypertension ↑ risk of heart failure
β_3	Trp64Arg	↓ stimulation of cAMP?	Obesity? Type 2 diabetes?

[a] The most common allele is given first, except for β_2 Arg16Gly, where the Gly16 is prevalent but the Arg16 has been traditionally considered the wildtype.

[b] *Abbreviations*: cAMP, cyclic adenosine monophosphate; PLC, phospholipase C; DBP, diastolic blood pressure.

[c] ? denotes the existence of contradictory studies.

Source: Adapted from Small et al. (2003).

11.2.1. β_1 Adrenergic Receptors

Beta$_1$ receptors are coupled through a G_s protein to adenylyl cyclase, and their activation causes an increase in cAMP (Liggett 2000). These receptors are expressed in many organs but appear to exert their major function in the cardiovascular systems; although β_2 adrenergic receptors are also expressed in the heart, the β_1 is the dominant subtype. Both agonists (e.g., xamoterol) and antagonists (e.g., atenolol) are used clinically to increase cardiac output, and to treat hypertension, angina, and cardiac arrhythmias, respectively.

The β_1-adrenergic receptors present two nonsynonimous coding region polymorphisms resulting in amino acid substitutions at position 49 (Ser/Gly) and 389 (Arg/Gly). The Arg389Gly polymorphism, which results in a receptor less responsive to the effect of agonists, is the most widely studied with regard to its potential involvement in cardiovascular disease and response to β-blockers. The allele frequency of Gly389 differs between Caucasians and African-Americans, with the latter ethnic group having a higher frequency (42% vs. 27%) (Small et al. 2003). Maximal stimulation of adenylyl cyclase of Gly389 is only about 30% that of the Arg389 receptor (Small et al. 2003). Genetic epidemiology studies have not yet provided strong evidence of an association between the Arg389Gly polymorphism and cardiovascular disease, although Arg389 has been associated with increased blood pressure. Interesting findings have, however, arisen from studies investigating a combination of polymorphisms in African-Americans. Individuals with the combination of dysfunctional α_{2C}-adrenergic receptor [with a Gly–Ala–Gly–Pro deletion; see Table 11.1 and Small and Liggett (2001)] and the Arg389 polymorphism, had a marked risk for heart failure (OR = 10.1) (Small et al. 2002). This may be explained by an increased norepinephrine release from cardiac presynaptic nerves (not antagonized by the dysfunctional α_{2c} receptor) and a hyperactive β_1-adrenergic receptor (the Arg389 type), that lead to a catecholamine-evoked cardiomyopathy (Small et al. 2003).

It is known from clinical practice that a large percentage of patients treated with β-blockers in monotherapy do not respond with an adequate decrease in blood pressure. More recent studies have shown that Gly389 homozygotes display a reduced sensitivity to β-blockers (Brodde and Stein 2003; Sofowora et al. 2003; Johnson et al. 2003). While the Arg389 genotype was associated with a significant reduction of diastolic blood pressure in response to metoprolol in comparison to that found in carriers of the Gly389 allele, the Ser49Gly polymorphism also appears to play a role. Indeed, the highest response was observed in individuals homozygous for the Ser49Arg389 haplotype (Johnson et al. 2003).

11.2.2. β_2 Adrenergic Receptors

The β_2-adrenergic receptor is also positively coupled to adenylyl cyclase through a G_s protein, and its most relevant role is in the lung, where its expression on bronchial smooth muscle serves to relax muscle, resulting in bronchodilation (Liggett 2000). Beta$_2$ agonists (e.g. albuterol) are thus utilized for treatment of bronchospasm in asthma and in chronic obstructive pulmonary disease (Fenech and Hall 2002). In the coding region of the human β_2-adrenergic receptor, three nonsynonimous polymorphisms have been identified (Table 11.1). Note that the Arg16 receptor is referred to as wildtype, since it was the first cloned, but is in fact the minor allelic variant (Small et al. 2003). All three polymorphisms (Arg16Gly, Gln27Glu, and Thr164Ile) result in altered response to agonists or altered downregulation (Leineweber and Brodde 2004). A total of 13 polymorphic sites have been identified in the β_2-adrenergic receptor, including nine in the 5′ promoter region that may influence the levels of receptor expression (Liggett 2000). Of the 8192 (2^{13}) possible combinations of polymorphisms, or haplotypes, only 12 were found, with 5 haplotypes representing 90% of those in the population, suggesting the existence of strong linkage disequilibrium between some alleles (Liggett 2000). This illustrates the importance of considering haplotypes in genetic epidemiology studies, and may explain some of the contrasting findings (e.g., weak or no associations) found in some studies (Fenech and Hall 2002; Leineweber and Brodde 2004). Indeed, while asthmatics with the Arg16 allele had a greater bronchodilating response compared to those with the Gly16 allele, individual β_2 adrenergic receptor SNPs were not predictive of asthma severity or bronchial hyperactivity. Beta$_2$-adrenergic receptor haplotypes, on the other hand, revealed associations with severity of asthma and protection against bronchial hyperactivity, and provided greater predictive values of the bronchodilator response to beta agonists (Fenech and Hall 2002; Small et al. 2003). Beta$_2$-adrenergic receptor polymorphisms have also been investigated in relationship to cardiovascular disease. In particular, the Gly16 and Gly27 alleles were associated with an increased risk of hypertension, and individuals with heart failure carrying the Ile164 allele had a more rapid progression to either death or heart transplantation (Bray et al. 2000; Liggett et al. 1998).

11.3. VITAMIN D RECEPTOR

Vitamin D plays a central role in calcium and phosphate homeostasis and is essential for the proper development and maintenance of the bone; deficiency of vitamin D leads to rickets and osteomalacia. Vitamin D can be obtained from the diet (fish oils, egg yolks, liver) as vitamin D_2 and D_3; many foods (e.g., milk) are also fortified with vitamin D. Vitamin D_3 is also produced in the skin by the action of sunlight. Vitamin D needs to be bioactivated to exert its biological effects. A first step is a 25-hydroxylation that occurs primarily in the liver by the action of cytochromes P450; the second and more important step is another hydroxylation, which occurs mainly in the kidney, to form 1,25-dihydroxy vitamin D_3 [1,25$(OH)_2D_3$] (Brown et al. 1999). Most of the biological activities of 1,25$(OH)_2D_3$ are mediated by a high-affinity cytosolic receptor (the vitamin D receptor, VDR), that acts as a ligand-activated transcription factor. The major steps involved in the control of gene transcription by the VDR include ligand binding, translocation to the nucleus, heterodimerization with the retinoid X receptor, binding of the heterodimer to vitamin D response elements in the DNA, and recruitment of other nuclear proteins into the transcriptional preinitiation complex (Brown et al. 1999). In addition to changes of gene expression mediated by VDR, 1,25$(OH)_2D_3$ also induces a variety of nongenomic effects (e.g., increased in intracellular calcium, activation of protein kinase C) that are mediated by one or more, as yet poorly characterized, membrane surface receptors, distinct from VDR (Brown et al. 1999). The VDR is localized in vitamin D target organs, such as the intestine, bone, kidney, and parathyroid glands, as well as in other tissues not involved in calcium homeostasis. Vitamin D is essential for the development and maintenance of a mineralized skeleton; 1,25$(OH)_2D_3$ induces bone formation by inducing the synthesis of bone matrix proteins and mineral apposition, also enhances the efficiency of the small intestine to absorb dietary calcium and phosphate, and modulates the maintenance of calcium balance by the parathyroid glands (Brown et al. 1999).

More than 25 different polymorphisms have been identified so far in the human VDR, mostly near the 3′ end of the gene, which is localized on chromosome 12q. The five most widely studied polymorphisms (Table 11.2) include three restriction fragment length polymorphisms (RFLP) of *BsmI*, *ApaI* and *TaqI*. The first two are located in intron 8, while *TaqI* is on exon 9 but is "silent"; that is, it does not result in amino acid changes. However, all may influence gene expression or mRNA stability (Csaszar and Abel 2001; Uitterlinden et al. 2002). A *FokI* restriction site located in the 5′ end of the VDR gene (exon 2) involves a single nucleotide substitution (T to C), which eliminates the first ATG translation initiation site, leading to two forms of VDR protein that differ by three amino acids in length (Uitterlinden et al. 2002). An additional genetic variation, in the 3′ untranslated region of the VDR gene, is the poly(A) polymorphisms, which may be important for mRNA stability. The number of polyA repeats varies between

Table 11.2 Major Vitamin D Receptor Polymorphisms

Polymorphism	Possible Clinical Relevance
BsmI	↑ risk of breast cancer?
	↑ risk of prostate cancer
	↑ blood pressure
	↑ blood and tibia lead levels
	↑ risk of lupus
TaqI	↑ risk of Crohn's disease
ApaI	↑ risk of diabetes mellitus
FokI	↑ risk of breast and prostate cancers
	↑ risk of malignant melanoma
	↓ bone mineral density
	↑ blood lead levels in children
Poly(A) repeat	↑ body mass index

11 and 23, and genotypes with <18 or >18 repeats are denoted as short (s) or long (L) (Grundberg et al. 2004).

VDR polymorphisms have been studied for associations with bone mineral density and hence risk of osteoporosis, various types of cancer, Crohn's disease, and possible susceptibility to lead toxicity (Zmuda et al. 2000). The *BsmI* polymorphism has been associated with a slight effect on bone mineral density, with the BB genotype resulting in a greater reduction compared to the bb genotype. The overall effect is small (2–3%) and would be equivalent to about a 2–3-year difference in age in postmenopausal women (Cooper and Umbach 1996). The FF genotype of the *FokI* polymorphism has been associated with increased bone mineral density in children (Ames et al. 1999). Several studies have found associations between VDR polymorphisms, particularly *BsmI* and *FokI*, and the risk of breast and prostate cancer, although contrasting findings have also been reported. Homozygosity for the wildtype (F) allele at the *FokI* restriction site was associated with a reduced risk of malignant melanoma (Csaszar and Abel 2001). The *TaqI* polymorphism has been shown to influence the development of Crohn's disease, an inflammatory bowel disease (Csaszar and Abel 2001). Finally, individuals with shorter polyA repeat (SS) and absence of the *BsmI* restriction site (BB) had higher muscle strength, body weight and fat mass (Grundberg et al. 2004).

Various studies have also investigated associations between VDR polymorphisms and pharmacokinetics of lead. The effects seen are small, and the polymorphisms may be effect modifiers or markers of other loci. For example, the *FokI* polymorphism was found to modify the relationship of lead exposure and blood lead levels during the first 2 years of life; as floor dust lead increased, children with the *FokI FF* genotype had a greater increase in blood lead concentration than did children with the *Ff* genotype (Haynes et al. 2003). Lead workers with the B allele of the *BsmI* polymorphism had higher blood and tibia lead levels than did those with the bb genotype after controlling for lead exposure (Schwartz et al. 2000). Overall, it is apparent that the interpretation of polymorphic variations in the VDR gene is severely hindered by the fact that several of the polymorphisms have no, or unknown, functional effects (Uitterlinden et al. 2002).

11.4. ESTROGEN RECEPTORS

The effects of estrogens are mediated by their binding to and activation of estrogen receptors (ER), which are members of the steroid/thyroid hormone superfamily of nuclear receptors. The basic structure of the ER protein contains a DNA-binding domain, a transcription-modulating domain, and steroid-hormone-binding domain (Rollerova and Urbancikova 2000). Two types of ER exist: ERα and ERβ. The ERα gene is located on human chromosome 6q25.1 and encodes for a 66-kDa protein of 595 amino acids. The ERβ gene is on chromosome 14q and codes for a protein with 530 amino acids and a mass of 59.2 kDa. The homology of ERβ and ERα is high in the DNA-binding domain, but lower in the ligand-binding domain. Both receptors bind estrodiol-17β with similar affinity, but the ERβ binds androgens and phytoestrogens with greater affinity (Kuiper et al. 1998). On binding to a ligand, ERs undergo a conformational change leading to dimerization, DNA binding, and activation of coactivators or corepressors and transcription factors. ERs dimerize as homodimers or heterodimers, and the hormone–ER complex binds to chromatin at estrogen response element in target genes (Rollerova and Urbancikova 2000). This leads to a cellular response, or to the stimulation or inhibition of target gene expression. After mRNA is produced by transcription of receptor genes, ERs migrate to the cytoplasm, where they are complexed to heat-shock proteins. ERs are expressed in a variety of tissues and are involved in the control of their proliferation, differentiation, and development. Important physiological roles are the development of reproductive organs, and modulation of nervous system and cardiovascular functions. In addition, ER can also mediate adverse effects of estrogens, such as breast or kidney cancers (Tanaka et al. 2003).

A large number of polymorphisms in the coding, regulatory, and intronic regions of the ERα gene have been identified. The most widely studied are two polymorphisms in intron 1 that give rise to *PvuII* and *XbaI* restriction sites. These two polymorphisms have been associated with breast cancer risk, although contrasting results have been reported (Cai et al. 2003a; Shin et al. 2003). Breast cancer is one of the hormone-dependent cancers, and expression of the ERα seems to be higher than ERβ in breast cancer tissue, while the opposite is true in normal breast tissue (Leclercq 2002). A GT dinucleotide repeat polymorphism upstream of the ERα gene was associated with breast cancer risk in Chinese women, with the $(GT)_{17}$ or $(GT)_{18}$ allele having a substantially decreased risk of breast cancer (Cai et al. 2003b), but other studies investigating associations between various coding region polymorphisms and breast cancer have been inconsistent. The *PvuII* and *XbaI* polymorphisms have also been associated with sporadic Alzheimer's disease in an Italian population (Brandi et al. 1999). In particular, the PPXX genotype had a much greater prevalence in Alzheimer's patients than in controls. The *PvuII* and the *XbaI* polymorphisms and a promoter region TA repeat polymorphism have been associated with cardiovascular disease (Herrington and Howard 2003), particularly in postmenopausal women undergoing hormone replacement therapy.

Several polymorphisms have also been identified in the ERβ gene, but again, association studies with breast cancer have provided contrasting findings. However, a more recent study has found a significant association between a dinucleotide CA repeat in intron 5 and bone mineral density in postmenopausal women (Scariano et al. 2004).

11.5. Ah RECEPTOR

The aryl hydrocarbon receptor (AhR) plays an important role in the toxic response to environmental pollutants. The response to environmental chemicals has been characterized, and binding of chlorinated dioxins, planar aromatic hydrocarbons (PAHs) and other halogenated hydrocarbons to the AhR can lead to cancer, thymic involution, chloracne, wasting, and birth defects. In the absence of ligand, the AhR is present in the cytosol in a complex with various proteins (Hsp90, XAP2, p23). On binding to a ligand, the AhR complex translocates to the nucleus where the AhR dissociates from the Hsp90 complex and forms an heterodimer with ARNT (Ah receptor nuclear translocator). The AhR/ARNT heterodimer recognizes an enhancer DNA element [known as *xenobiotic responsive element* (XRE)], resulting in the enhanced expression of a number of genes, exemplified by cytochrome P4501A1 (CYP1A1) (Mimura and Fujii-Kuriyama 2003) (see

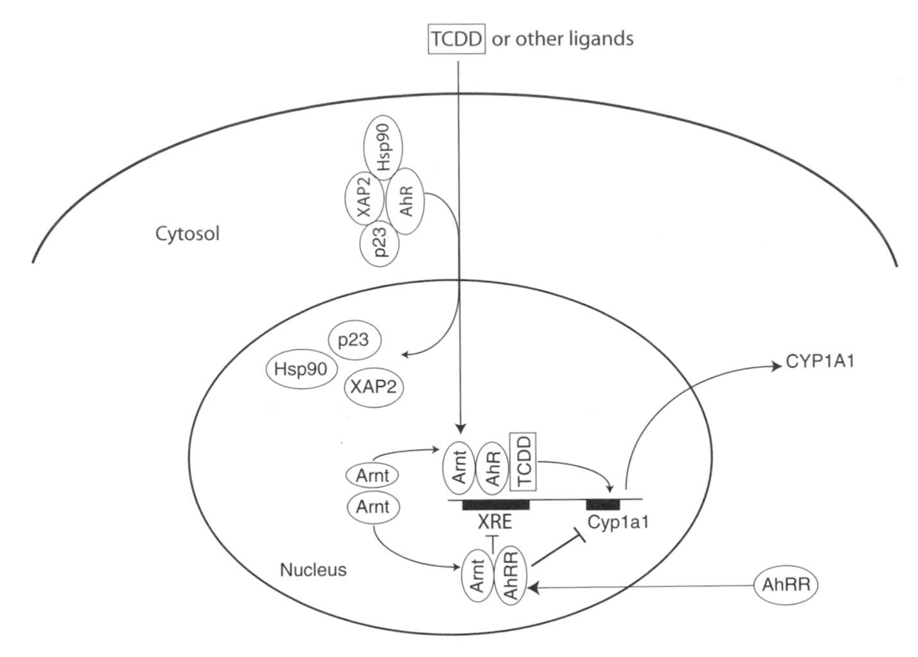

Figure 11.1. Mechanisms of transcriptional activation by the Ah receptor. [Adapted from Ma and Lu (2003) and Mimura and Fujii-Kuriyama (2003) Graphic adapted by J. Gill.]

also Figure 11.1). In addition to CYP1A1, several other genes involved in cell proliferation, cell cycle regulation, and apoptosis are induced by AhR ligands. Studies in AhR deficient mice indicate that most, if not all, toxic effects of TCDD (2,3,7,8-tetrachlorodibenzo-*p*-dioxin) are mediated by activation of the AhR. Engenous ligands have not been identified yet, although some have been suggested (Ma and Lu 2003).

It is known that there are marked species and strain differences in sensitivity to TCDD. For example, the LD_{50} for this compound ranges from 1 μg/kg for guinea pig, the most sensitive animal, to >5 mg/kg for hamster, the most resistant species. In mice, the DBA/2 strain is less sensitive to the toxic effects of TCDD and PAHs than is the C57BL/6 strain. This is due to a single-nucleotide difference at codon 375 of the AhR gene leading to an Ala-to-Val substitution, which in DBA/2 mice reduces the affinity of ligands for the AhR by tenfold (Wong et al. 2001). Polymorphisms in the AhR are also present in rats, with the Long–Evans strain proving 1000-fold more sensitive to the toxicity of TCDD than the Han–Wistar strain.

The human AhR, which is structurally more similar to the AhR from DBA/2 mice (Ma and Lu 2003), displays three of the most widely studied polymorphisms at codon 554 (Lys/Leu), codon 570 (Val/Ile), and codon 517 (Pro/Ser), in addition to several other mutations (Harper et al. 2002). In vitro studies in lymphocytes have shown that induction of CYP1A1 activity was higher in Lys554 homozygotes than in Leu554 homozygotes in a Caucasian population; however, other studies yielded very different results. Studies with codons 570 and 517 have provided similarly contrasting findings. Such inconsistencies in findings may be due to the presence of additional reported polymorphisms in ARNT or in the AhR repressor, which would influence AhR function (Harper et al. 2002). Furthermore, a tenfold variation in the binding affinity of the AhR for TCDD in a human population could not be explained by any specific polymorphism (Harper et al. 2002).

Genetic epidemiology studies in human population to investigate possible associations between AhR polymorphisms and disease (e.g., lung cancer in smokers, bladder cancer) have provided no clear indications. Also, no association between the codon 554 polymorphism and chloracne (a skin disorder associated with exposure to TCDD, which, however, occurs in only a few percentage of exposed individuals) has been found (Harper et al. 2002).

11.6. SODIUM CHANNELS

Sodium channels are transmembrane proteins responsible for the voltage–dependent increase in sodium permeability that initiates action potentials in neurons and other electrically excitable cells such as myocytes. When the cell membrane is depolarized by a few millivolts, sodium channels activate and inactivate within milliseconds. Influx of sodium further depolarizes the membrane and initiates the rising phase of the action potential (Catterall 2000; Yu and Catterall 2003). The sodium channel consists in a 260-kDa α subunit that contains the ion-conducting pore and the essential elements of sodium channel functions (opening, ion selectivity, rapid inactivation), and one or more auxiliary β subunits that modify the kinetics (opening and closing) of the channel (Yu and Catterall 2003). Nine α subunits have been functionally characterized, and a tenth related isoform (Na_x) may also function as a sodium channel. The sodium channels are identified by a nomenclature that takes into account the chemical symbol of the permeating ion (Na), the principal physiological regulator (voltage, subscript), a number indicating the gene subfamily (1 so far), and a decimal assigned to specific channel isoform (Goldin et al. 2000) (see also Table 11.3). Genes encoding different sodium channels are located on human chromosomes 2, 3, 11, and 15 (Table 11.3). Sodium channel isoforms have different tissue and cellular localization; for example, $Na_v1.1$ and $Na_v1.2$

Table 11.3 Mammalian Sodium Channel Subunits[a]

Type	Gene	Human Chromosome	Sensitivity to TTX	Primary Tissue	Relevance of Polymorphisms
$Na_v1.1$	SCN1A	2	+	Neurons (CNS,PNS)	Epilepsy
$Na_v1.2$	SCN2A	2	+	Neurons (CNS)	Epilepsy
$Na_v1.3$	SCN3A	2	+	Neurons (CNS)	
$Na_v1.4$	SCN4A	11	+	Skeletal muscle	Myasthenic syndrome
$Na_v1.5$	SCN5A	3	–	Heart	Cardiac arrhythmias[b]
$Na_v1.6$	SCN8A	15	+	CNS, PNS	
$Na_v1.7$	SCN9A	2	+	Neurons (PNS) Schwann cells	
$Na_v1.8$	SCN10A	3	–	Neurons (DRG)	
$Na_v1.9$	SCN11A	3	–	Sensory neurons (PNS)	
Na_x[c]	SCN7A	2	NA	Heart, uterus, astrocytes, skeletal muscle	
β_1	SCN1B	19	NA	CNS, heart	Epilepsy
β_2	SCN2B	11	NA	CNS	
β_3	SCN3B	11	NA	CNS, skeletal muscle	

[a] *Abbreviations*: TTX, tetrodotoxin; CNS, PNS, central, peripheral nervous system; DRG, dorsal root ganglia

[b] These include long QT syndrome and Brugada syndrome.

[c] Not voltage-regulated.

Source: Adapted from Goldin et al. (2000) and Yu and Catterall (2003).

channels are found primarily in CNS neurons, while $Na_v1.4$ and $Na_v1.5$ are expressed mostly in skeletal muscle and heart, respectively. Different sodium channel isoforms also have a differential sensitivity to tetrodotoxin (TTX), a toxin that binds to a specific site (receptor site 1) on the extracellular side of the membrane, and blocks sodium conductance (Cestele and Catterall 2000). Additional sites on the α subunit of the sodium channel are targeted by several other toxins, as well as by antiepileptic, local anesthetic, and antiarrhythmic drugs (Cestele and Catterall 2000; Catterall 2000). The three auxiliary β subunits identified (β_1, β_2, and β_3) are encoded by genes located on human chromosomes 11 and 19, and are also differentially expressed in tissues (Table 11.3).

Several different diseases have been associated with mutations in genes encoding ion channels, including sodium channels, and these diseases are thus defined as "channelopathies" (Bockenhauer 2001). Mutations that affect sodium channels often result in increased excitability, and are thus defined as "gain of function" mutations. Several mutations of the SCN5A gene (encoding the cardiac $Na_v1.5$ sodium channel; Table 11.3), interfere with the rapid inactivation of the channel after opening, or change its voltage dependence, thus prolonging cardiac depolarization (Bockenhauer 2001). These mutations cause inherited long QT syndrome of the type 3 variant, and Brugada syndrome, both of which lead to an increased risk of ventricular arrhythmias (Chen et al. 1998; Napolitano et al. 2003). A common SCN5A variant in African-Americans (Ser1102Tyr) has been shown to cause a small risk of cardiac arrhythmia (Splawski et al. 2002). Multiple mutations in the SCN4A gene (encoding the $Na_v1.4$ skeletal muscle channel) have been associated with various myotonic disorders, including paramyotonia congenita and hyperkalemic periodic paralysis (Cannon 1996). Additionally, a single specific mutation (V1442E) of the SCN4A gene has been shown to cause a myasthenic syndrome (Tsujino et al. 2003). While myotonia is linked to gain-of-function mutations, loss-of-function defects, leading to enhancement of fast inactivation, are responsible for paralysis and myastenia (Tsujino et al. 2003). Mutations in the SCN1A, SCN2A, and SCN1B genes, encoding α and β subunits expressed in the CNS (Table 11.3), which enhance the persistent inward sodium current and thus lead to prolonged neuronal depolarization, increased firing frequency, and enhanced excitability, have been shown to cause generalized epilepsy with febrile seizure (GEFS+) (Wallace et al. 1998; Lossin et al. 2002).

11.7. CONCLUSIONS

It has once been stated that, in contrast to enzymes, where genetic variability is very high, few genetic defects can be found in the case of receptors (Dryer and Rüdiger 1988). Yet, as indicated by the few examples discussed in this chapter and by several additional examples (Table 11.4), a large number of receptors and ion channels display various genetic polymorphisms, which in some cases have been associated with specific disease conditions or altered responses to pharmacological agents or xenobiotics. Given the relevant role of receptors and channels in mediating a vast array of physiological responses, and their importance as targets for pharmaceutical drugs, this area of research, while still lagging behind that studying enzyme polymorphisms, will no doubt see further significant advancements.

Table 11.4 Additional Examples of Polymorphic Receptors and Ion Channels and Their Possible Clinical Significance

Receptor/Ion Channel	Phenotypic association
Dopamine D_2 receptor	Alcoholism Schizophrenia
Serotonin 2A receptor	Schizophrenia Altered response to clozapine (an antipsychotic drug)
Tumor necrosis factor α receptor 2	Systemic lupus erythematosus
Androgen receptor	Prostate cancer
Peroxisome proliferator-activated receptor γ (PPAR-γ)	Obesity
Thromboxane A1 receptor	Bronchial asthma
Insulin-like growth factor I (IGF-I) receptor	Intrauterine growth retardation
Low-density lipoprotein (LDL) receptor	Familial hypercholesterolemia
Potassium channels	Long QT syndrome (some types) Neonatal familial epilepsy Familial hyperinsulinism

11.8. SUGGESTED READING

For further reading on receptors and ion channels, the review by Rollerova and Urbancikova (2000) and the papers by Harper et al. (2002), Small et al. (2003), Uitterlinden et al. (2002), and Yu and Catterall (2003) are recommended.

Part III

12

Overview of Section III

Lucio G. Costa and David L. Eaton
University of Washington, Seattle, WA

This section of the book examines gene–environment interactions from a different perspective. The focus is on the health outcome; specifically, the disease, and the discussion revolves around genetic polymorphisms that may predispose individuals to disease. Genetic disorders that precisely link a certain defect to a specific outcome (e.g., a missense mutation in the β-globin gene leads to sickle cell anemia) are not the main focus of these chapters. Rather, information is presented on complex diseases, where knowledge on the role of gene–environment interaction in the etiology of the disease is beginning to emerge. Various chapters address types of cancer (e.g., lung, gastrointestinal), chronic neurodegenerative diseases, cardiovascular diseases, chronic obstructive pulmonary diseases, infectious diseases, diabetes, and obesity. All these represent major public health issues, which cause high morbidity and mortality in the population, and all of them clearly have both a genetic and an environmental component to their etiology.

In lay terms, the overall scope of this section may be seen as an attempt to answer the question "Why me, doc?" The intent is to present a general discussion and the most recent information (with no attempt to be exhaustive) on research that has investigated the possible role of certain polymorphisms, some discussed in previous sections, in conferring susceptibility to disease. The complexity of the diseases poses substantial difficulties in establishing clear-cut associations. Most often the endpoint is the result of an array of multifactorial aspects involving both the individual's genome

Gene-Environment Interactions: Fundamentals of Ecogenetics, edited by Lucio G. Costa and David L. Eaton

and the environment. A typical example that may illustrate this issue is represented by the studies on the risk of squamous cell carcinoma of the lung among Japanese reported in 1992 by Hayashi et al. (1992). Individuals with the CYP1A1 Val allele had a significant relative risk (RR) of 3.3, while those with the GSTM1 deletion a RR of 2.3. However, individuals with both variants had a RR of 9.1. A follow-up study by the same group (Nakachi et al. 1993) found that the CYP1A1-GSTM1 gene–gene interaction was even greater (odds ratio of 41) when the dose of cigarette smoke was considered, with the interaction greater in the lower dose of cigarette smoking. Interestingly, this gene–gene–environment interaction has not been confirmed in Caucasian populations, probably because the allele frequency of the CYP1A1 variant is very low (~1%), relative to that seen in Japanese populations (~10%). This example emphasizes the fact that when attempting to determine the causes of complex diseases, such as those discussed in this section, one has to consider possible gene–gene interactions, in addition to gene–environment interactions, and that the gene–gene interaction may be influenced by dose.

In addition to the chapters focusing on different diseases, the two final chapters in this section deal with the issue of diet and nutrition and their interactions with genetic background, a topic that has widespread relevance given that "Good health starts with good eating habits." But the genes also play their role, especially with the important issue of genetic determinants of addiction to drugs of abuse, including two of the most common, alcohol and tobacco.

13

Lung Cancer

Valle Nazar-Stewart
Oregon Health and Science University, Portland, OR

13.1. INTRODUCTION

Although tobacco use is very strongly related to lung cancer and accounts for most cases of lung cancer, few smokers (5 cases per 1000 person-years of cigarette use) develop this disease. An understanding of the biology underlying this paradox may help us identify susceptible individuals and determine why they are at risk. Family history is a risk factor for lung cancer (Blot and Fraumeni 1996), suggesting the presence of a genetic component to risk and offering a possible moderating influence on risk for cancer caused by tobacco. Early studies of gene–environment interactions focused on lung cancer, and lung cancer continues to be a good model for this type of work for several reasons:

1. The main environmental cause of lung cancer, tobacco, is well established as a very strong risk factor (current smokers experience 10–17 times the incidence among never-smokers) with a large attributable risk (79–90% of lung cancers are caused by tobacco smoking), so most cases of lung cancer are caused by tobacco.
2. Lung cancer is one of the most common cancers, allowing accrual of large study populations.
3. Tobacco exposure is easily and reliably assessed, even through subjects' self-report (Patrick 1994). Thus, we can accrue large study

Gene-Environment Interactions: Fundamentals of Ecogenetics, edited by Lucio G. Costa and David L. Eaton

populations with known exposure to the environmental cause of cancer, and concentrate our investigations on the genetic components of risk.

Even so, genetic contributions to lung cancer are complicated, with multiple genes involved in complex metabolic pathways for numerous carcinogens in tobacco smoke. Furthermore, it is increasingly apparent that genes do not act in isolation of other genes or exposures that influence their function. Large study populations with known exposure to the main environmental cause allow study of the complexities of gene–environment interactions including the combined effects of multiple genes, relationships between genes and intermediate markers for disease, and influences of ethnic background, gender, and both endogenous and exogenous exposures on gene expression and protein function.

Because tobacco is the overwhelming cause of lung cancer, research on gene–environment interactions has focused on genetic differences in the internal processing of carcinogens and reactive oxygen that occur in tobacco smoke. Because lung cancer was a focus of early work, many studies preceded a detailed understanding of the underlying biology, as well as standards for the conduct of gene–environment studies, and study methods varied greatly. For these reasons, and because it now appears that many of the polymorphic genes may each contribute modestly to overall risk, it is unsurprising that for most polymorphisms there are studies with conflicting results. As research and knowledge of the relevant biological processes have progressed, studies have become increasingly sophisticated with respect to study design, subject selection, collection of data on exposures and confounders, and analytical methods. That learning process is ongoing. An examination of studies of genetic polymorphisms and risk for lung cancer provides insight into the conduct, and the benefits and difficulties, of gene–environment studies.

Cigarette smoke contains thousands of compounds, including 30–50 known carcinogens as well as reactive oxygen compounds that can damage DNA. Most tobacco carcinogens are inhaled as less harmful procarcinogens, which are activated by phase I enzymes (primarily cytochrome P450s) to metabolites capable of reacting with DNA and beginning carcinogenesis. Carcinogenesis is prevented when reactive carcinogens are detoxified by phase II enzymes (e.g., glutathione *S*-transferases) or DNA damage is repaired by DNA repair enzymes (see Figure 13.1).

Two classes of carcinogens important in lung cancer are polycyclic aromatic hydrocarbons (PAHs) and tobacco-specific *N*-nitrosamines (TSNA). PAHs, such as benzo(a)pyrene, are oxidized by cytochrome P450s, particularly CYP1A1, into highly reactive epoxides that can form DNA adducts.

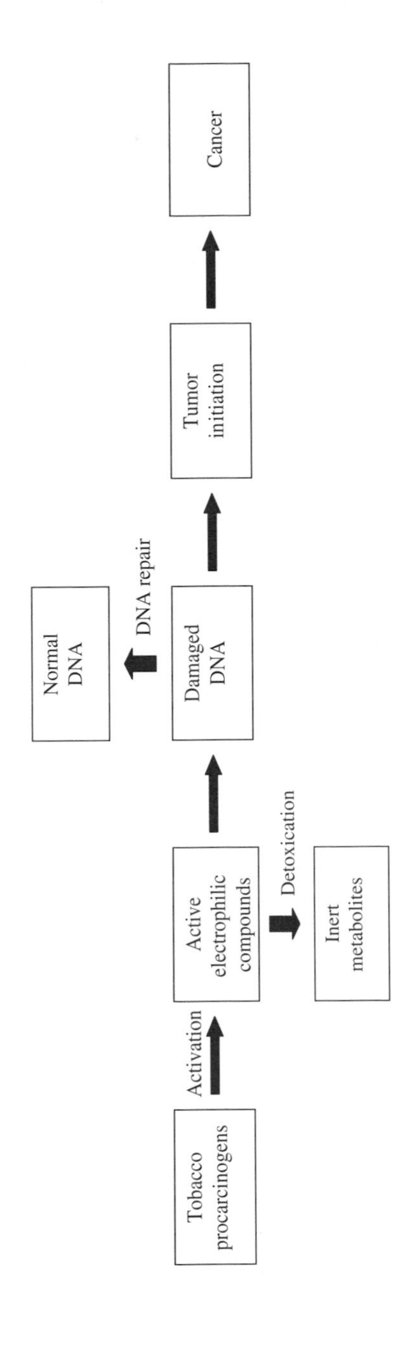

Figure 13.1. Biotransformation of tobacco carcinogens.

Alternatively, reactive PAHs can be detoxified by the glutathione *S*-transferases. For example, benzo(a)pyrene (BaP) is transformed into phenols and B(a)P-7,8-diol through reactions catalyzed by cytochrome P450 enzymes. These metabolites are then further transformed, via cytochrome P450s and microsomal epoxide hydrolase, to highly reactive (+)-*anti*-BPDE. BPDE can be detoxified by the GSTs, but (+)-*anti*-BPDE is thought to be the ultimate mutagenic form of benzo(a)pyrene and is a known direct cause of lung cancer (Bartsch et al. 2000).

Tobacco-specific *N*-nitrosamines (TSNA), including 4-methylnitrosoamino-1,3-pyridyl-1-butanone (NNK) and *N'*-nitrosonornicotine (NNN), are synthesized during smoking and are also metabolically activated, with some involvement of CYPs, to forms that bind with DNA (Bartsch et al. 2000).

Tobacco smoke also contains free oxygen as well as compounds that generate reactive oxygen species (ROSs) that can damage DNA. In particular, oxidatively damaged DNA base guanine is highly mutagenic, and it appears to be important in lung cancer. ROS levels may be reduced by GSTs, and oxidatively damaged DNA may be repaired by DNA repair enzymes.

The lung is an important line of defense against toxins. Several cell types in the lung are capable of activating inhaled chemicals, which need only cross intracellular membranes to reach DNA. Many inhaled lipophilic compounds (such as PAHs) have long retention times in the lung, so local lung metabolism, despite occasional low levels of the enzymes involved, can contribute significantly to transformation of carcinogens. However, pulmonary expression is not required for an enzyme to affect lung cancer risk because the entire cardiac output circulates through the lungs, so reactive metabolites generated in the liver will also pass through the lung. Many of the enzymes involved in these metabolic processes are genetically polymorphic, with some common genetic variants having differences in level of function. Many studies have sought to determine whether interindividual differences in ability to transform carcinogens and reactive oxygen in tobacco smoke determine susceptibility to tobacco-related lung cancer.

13.2. POLYMORPHISMS IN BIOTRANSFORMATION ENZYMES

13.2.1. Cytochrome P450s

Several cytochrome P450 (CYP) enzymes catalyze activation of procarcinogens in tobacco smoke, with overlapping substrate specificity. Early gene–environment studies focused on cytochrome P450s 1A1 (termed "aryl hydrocarbon hydroxylase") and CYP2D6 phenotypes. Levels of CYP1A1 and CYP2D6 function were known to vary across subjects, and these studies tested hypotheses that individuals with a high level of function would have excess risk for lung cancer. CYP1A1 activity toward specific substrates was measured in blood cells, and CY2D6 activity was measured as a function of urinary metabolites of orally administered debrisoquine. Early studies did report high CYP1A1 and CYP2D6 function among patients with lung cancer, but these results were difficult to replicate. These difficulties may have been due to the fact that the phenotyping methods were difficult and varied between labs; function of these enzymes can be induced (upregulated), but inducing exposures were not understood and could not be controlled for; variation in phenotypic activity was difficult to categorize for purposes of analysis; and although normal baseline levels of activity varied considerably, only small convenience study populations were used. Also confusing were questions of whether measurements of activity toward synthetic substrates (debrisoquine), or in blood cells (CYP1A1) were indicative of actual biology in the lung.

Most gene–environment studies now do direct DNA analyses, which are inexpensive, easier to replicate, representative of all cells in the body, and not subject to influence by inducing exposures. Genotyping, though, cannot reflect transcriptional or posttranslational influences on production and function of the gene product, and it may be level of enzyme and, ultimately, enzyme function, in which we are most interested. Thus, our understanding of how polymorphic genes influence risk may be served by both genotyping and phenotyping studies as well as data on how they relate to each other.

13.2.1.1. CYP1A1. CYP1A1 is the principal enzyme responsible for activation of polycyclic aromatic hydrocarbons (PAHs) including benzo(a)pyrene. CYP1A1 occurs in lung where expression is reported to vary up to 50-fold, is highly inducible by varied exposures, and is differentially inducible across subjects. Smokers have induced CYP1A1 activity, and levels of benzo(a)pyrene and total aromatic DNA adducts in lung tissue of smokers are directly correlated with pulmonary CYP1A1 expression.

Polymorphisms in the CYP1A1 gene affect transcription control and, consequently, enzyme expression and induction (Bartsch et al. 2000). Two gene polymorphisms occur among both Caucasian and Asian populations, one at a *Msp*I recognition site in the 3′ noncoding region (*m1*) and another involves a single base substitution Ile–Val in the exon 7 heme-binding region (*m2*). Genotype assignment is often not complete because the *Msp*I polymorphism and the Ile–Val variant are in incomplete linkage disequilibrium. Both Ile–Val (*m2*) and *MspI* (*m1*) variants have increased catalytic activity relative to the wildtype gene, and both have been linked to elevated levels of PAH-DNA adducts among smokers. Frequencies of *m1* and *m2* variants vary widely across genetically diverse ethnic populations. Both *m1* and *m2* rare alleles are more frequent among Japanese than among Caucasian and African-American populations. For example, *MspI m1* allele frequency is 0.33 in Japanese and 0.10 in Caucasians (Kiyohara et al. 2002). The rare *m2* allele occurs even more infrequently. Their roles in lung cancer have been examined in over 20 epidemiology studies involving a variety of ethnic populations.

The majority of studies among Japanese populations have reported an association of both *m1* and *m2* with lung cancer, with odds ratios ranging from 1.7 to 9.4, and some variation in risk according to lung cancer cell type and extent of smoking history (Hayashi et al. 1992). Results from Caucasian and mixed ethnic study populations are less conclusive, with many of the larger studies having positive but more modest results for *m1* (odds ratios of 1.5–3.1) and *m2*. Studies reporting no association between either *m1* or *m2* and lung cancer have included Scandanavian, European, French Caucasian, Chinese, and mixed study populations. Some of the negative studies had low statistical power to detect an effect, or design weaknesses such as an absence of data on tobacco use and ethnicity, and control study groups with genotypes not in Hardy–Weinberg equilibrium. Some data suggest that cancer risk associated with *m1* may be modified by smoking, with odds ratios stronger among light smokers than among heavy smokers and for squamous cell cancers. A meta analysis of 15 published studies with varied ethnic groups (Houlston 2000) found slight nonsignificant increases in risk for lung cancer among subjects with homozygous *m1* and *m2* variants, as might be expected when studies with larger and smaller estimates of risk are combined.

Because *m1* and *m2* are so rare, very large studies are required in order to achieve adequate power to study these alleles, especially among non-Japanese populations. The modest associations between *m1* and lung cancer measured in Caucasian populations suggest the possibility of some underlying ethnic difference in the biology. Alternatively, it is quite possible that cultural or genetic differences in inducing exposures, or genetic differences in function of related or competing tobacco metabolism pathways, or in DNA repair, are responsible for these differences in Japanese and Caucasians.

There is also some evidence for gender differences in CYP1A1 function. CYP1A1 mRNA expression in the lung has been reported to be twice as high in women as in men. The basis for this difference is unclear, but could be due to interactions between female steroid hormones or estrogen receptors and the aryl hydrocarbon receptor pathway (Mollerup et al. 1999,

Haugen 2000). This finding is consistent with evidence that women smokers also have higher frequencies of lung PAH adducts, adducts per pack-year of smoking (Mollerup et al. 1999), and p53 G-to-T transversions in lung tissue than do male smokers, as well as higher risk for lung cancer (Risch et al. 1993; Zang and Wynder 1996).

A third CYP1A1 polymorphism occurs only in people of African descent (*m3* or *4 RFLP) and has not been associated with lung cancer risk in several studies, except for risk for adenocarcinoma.

Design of CYP1A1 studies should recognize that expression and activity of this enzyme is determined by both genetics and inducing exposures (including exposure to tobacco). Phenotype may not correspond closely with genotype, and risk for lung cancer may not be related to CYP1A1 genotype among subjects who have had no exposure to PAHs or other inducers. CYP1A1 alone is not a sufficient cause for lung cancer. When nonsmokers are included in studies of CYP1A1, we are missing a piece of the causal pathway, and a true association of CYP1A1 with lung cancer risk may be undetected or attenuated, and this could account for some of the conflicting results above. Clearly, it is important in genotype studies to collect data on exogenous and endogenous exposures that may affect expression or activity of the corresponding protein. Also, the strength of the relationship between CYP1A1 and lung cancer appears to vary by gender, ethnicity, and diet. Study groups homogeneous for these factors will allow more precise measurement of the CYP1A1–lung cancer relationship. [For a review, see Bartsch (2000).]

13.2.1.2. CYP2D6. Poor ability to hydroxylate the drug debrisoquine (PM or poor metabolizer phenotype) is inherited as an autosomal recessive trait resulting from one of several defective CYP2D6 variants, three of which account for over 90% of PM phenotype persons. It has been unclear whether CYP2D6 is expressed in human lung, although this is not essential for involvement of CYP2D6 in lung cancer, and there is recent evidence that levels of the carcinogen-DNA adduct, 7-methyl-2′-deoxyguanosine-3′-monophosphate, are significantly higher in lung tissue of extensive metabolizers (EMs) than in that of poor metabolizers (PMs). However, the only tobacco substrates known to be activated by CYP2D6 are NNK and nicotine, and other P450s are more important in their activation. Absence of a known biological involvement in lung cancer would appear to limit interpretation of studies of CYP2D6 and lung cancer.

CYP2D6 has been studied exclusively in Caucasian populations. Early reports showed modest to strong excess risk (two- to three-fold) associated with the EM phenotype, although few reports achieved statistical significance, and some reported no association. A more recent study found an association between tobacco smoking and lung cancer that increased in strength with increasing CPY2D6 activity measured as a continuous variable, but no difference in risk when subjects were categorized according to PM or EM status. Unfortunately, distributions of CYP2D6 activity, documented in control subjects, seldom provide clear modes to suggest distinct phenotype categorizations.

Genotyping studies have had more mixed results, with most studies reporting an increase in risk (3–6.4-fold) associated with EM, and no modification of risk by either extent of smoking history or family history of lung cancer. Two meta analyses of phenotype and genotype studies together have found no association of CYP2D6 with lung cancer risk (Christensen et al. 1997) and an overall odds ratio of 0.7 (95% CI 0.5–0.9) associated with EM status (Rostami–Hodjegan et al. 1998). A meta analysis of genotype studies alone found the EM genotype to be associated with 2.3-fold excess risk (95% CI 1.6–3.4) (Amos et al. 1992).

There may be a modest association between CYP2D6 and lung cancer risk. However, the value of these findings is limited because the importance of CYP2D6 in tobacco metabolism, and a connection between the processes of debrisoquine metabolism and carcinogenesis, have not been well established. An understanding of underlying biology is important in establishing causality. Reports of possible relationships between CYP2D6 status, nicotine dependence, and smoking addiction further limit our understanding of CYP2D6 and cancer and complicate our ability to study CYP2D6.

13.2.1.3. CYP2E1. CYP2E1 is found in human lung and is known to metabolize many tobacco smoke procarcinogens, including NDMA, NNN, NNK, other volatile nitrosamines, and small organic chemicals (e.g., benzene). CYP2E1 also actively generates oxygen radicals. Regulation of CYP2E1 expression is complex, involving both transcriptional and post-transcriptional mechanisms. Known modulators of CYP2E1 activity include body weight (explaining perhaps half of variability), alcohol, diet (lettuce, black tea, and isothiocyanates), medications, and age (Le Marchand et al. 1999). Lettuce and isothiocyanate-containing cruciferous vegetables (including brussel sprouts, broccoli, and Chinese cabbage) inhibit CYP2E1. Alcohol induces CYP2E1.

There is wide interindividual variation in CYP2E1 activity in humans, explained partly by environmental influences. Activity in young, healthy Caucasians is reported to be unimodally distributed with a four- to five-fold range. Some ethnic variation in CYP2E1 function has also been reported: Japanese men have lower CYP2E1 activity than do Caucasians with similar smoking history and diet (Le Marchand et al. 1999). Caucasian and Japanese women have lower CYP2E1 activity than do their male counterparts, but this appears to be explained by women's lower body weight (Le Marchand et al. 1999).

Several polymorphic alleles have been identified: a *Dra*I RFLP (T-A substitution in intron 6), and two RFLPs occuring in the 5′-flanking transcription regulation region, *Rsa*I and *Pst*I (wildtype *c1*, rare variant *c2*), which appear to be in complete linkage disequilibrium. Additionally, *Dra*I may be partially linked to *Rsa*I. *Rsa*I affects binding of transcription factor and transcriptional activity of CYP2E1, and *Pst*I appears to have no effect on transcription (Watanabe 1994). For many genetically polymorphic enzymes, it is not known whether heterozygotes have function intermediate or similar to either the homozygote wildtype or homozygote variant. In this situation, it may be analytically prudent to try various genotype categorizations for comparison of risk. Indeed, such analyses could suggest level of function of the heterozygote. However, for CYP2E1 we have evidence that *Rsa*I c2 variant heterozygotes have activity similar to that of the homozygous wildtype (Le Marchand et al. 1999).

Variant *Dra*I allele has been associated with reduced risk for lung cancer in studies of Japanese, Hawaiian, African-American, Mexican-American, and mixed populations. Studies of the *Rsa*I *c2* allele suggest that variant c2 may have reduced CYP2E1 activity, perhaps more in Japanese subjects than in Caucasians, and may be less influenced by ethanol exposure than the wildtype allele. Large ethnic differences in allelic frequencies occur: *Rsa*I *c2* is rare (2–8%) among Caucasians and African-Americans, limiting studies in these populations, and more common among Japanese (24%) (Le Marchand et al. 1999). The *Pst*I/*Rsa*I *c2* allele (with apparently low activation capability) has been associated with reduced risk for lung cancer (with statistical significance) in Japanese (Uematsa 1991), as well as in Mexican- and African-American and Swedish study groups. Other studies have reported associations for only for squamous cell carcinoma and adenocarcinoma. Other studies have reported no association in Finnish, U.S. Caucasian and African-American, and Japanese study groups.

13.2.2. Myeloperoxidases

When tobacco smoke particulates (or asbestos fibers) reach the lungs, neutrophils are recruited in and release myeloperoxidase (MPO). MPO functions as an oxidative antimicrobial agent by producing hypochlorous activity, superoxides, and free radicals. MPO also activates a wide range of tobacco procarcinogens, including PAHs, to DNA-damaging metabolites. Importantly, MPO activates the 7,8-diol epoxide B(a)P intermediate to highly reactive B(a)P diol epoxide and increases binding of B(a)P to DNA. A G-to-A transition in the promoter region results in disruption of a SP1-binding site and a large reduction in MPO mRNA expression and activity. As expected for an enzyme that activates procarcinogens, the A/A low-activity genotype (and in some studies A/A and A/G combined or A/G alone) has been fairly consistently associated with reduced lung cancer risk (odds ratios of 0.10–0.84) in varied ethnic populations (London et al. 1997, Le Marchand et al. 2000; Feyler et al. 2002) but not in all studies.

13.2.3. *N*-Acetyltransferases

After *N*-oxidation of aromatic amine and heterocyclic amine carcinogens in tobacco smoke, these compounds are further activated catalytically by NATs (via *O*-acetylation) to active acetoxy compounds that can form DNA adducts. NATs also catalyze detoxification (via *N*-acetylation) of aromatic and heterocyclic amines. CYP1A2 is an integral part of these metabolic processes and may need to be considered with NAT2 in determining susceptibility to these compounds. Relationships between NATs and cancer risk may vary by organ, and may be dependent on tissue-specific expression (Hein et al. 2000). With regard to lung, NAT1 mRNA has been detected in both bronchial and peripheral tissues, and NAT2 has been found in peripheral lung but not in bronchus. Some histologic types of lung cancer tend to occur in the bronchial region (small cell), while others occur in the lung periphery (adenocarcinoma), and it would be interesting to determine whether NAT2 is associated with peripheral histologies more than others.

13.2.3.1. NAT2. Caffeine has been used successfully as a probe to determine NAT2 phenotype, but measurement of the urinary metabolites of caffeine may be affected by multiple genes and environmental influences (e.g., diet, disease, and drug use) (Hein et al. 2000). There are large differences in population frequencies of NAT2 slow acetylation phenotype (determined by caffeine administration), ranging from 5% among Japanese to 50–60% in Caucasians, and 90% among Egyptians. Rapid acetylators are either homozygous or heterozygous for the NAT2 rapid acetylator alleles, and there are at least 20 rare alleles, with two accounting for most of the slow acetylator phenotype in Caucasians and one in Asians (Hengstler et al. 1998). Slow acetylation is due, at least in part, to a reduction in NAT2 protein (Hein et al. 2000).

Most, but not all, phenotype and genotype studies have reported an increase in risk for lung cancer associated with NAT2 rapid acetylation phenotype. Finnish studies show increased risk for asbestos-associated malignant mesothelioma among NAT2 slow acetylators. Further evidence for a role for NAT2 in cancer includes data that NAT2 slow acetylators have high baseline frequencies of lymphocyte chromosomal aberrations, and a report that subjects with GSTM1 null/NAT2 slow acetylation status have very high DNA adduct levels (Hou et al. 1995).

13.2.3.2. NAT1. At least 24 different NAT1 alleles have been identified. Polymorphic variation in the polyadenylation signal of the gene results in variation in transcript half-life and quantity of NAT1. Approximately 15–20% of Caucasians and up to 50% of Asians are fast acetylators. There has been one finding of an association between low-activity NAT1 alleles (NAT1*14 and NAT1*15) and excess risk for lung cancer (Bouchardy et al. 1998).

13.2.4. Microsomal Epoxide Hydrolases

Microsomal epoxide hydrolase (mEH) is involved in both activation and detoxification reactions, generating reactive epoxide intermediates and detoxifying epoxides via hydrolysis. Whether the role of mEH is predominantly one of activation or detoxication, and whether functional variants might be expected to increase or reduce risk for lung cancer, has been in question. Microsomal EH is strongly expressed in bronchial epithelial cells and is believed to metabolize some components of tobacco smoke, including PAHs. Two genetic polymorphisms have been reported. A polymorphism on exon 3 results in a tyrosine-to-histidine change and reduced enzymatic activity. Another polymorphism on exon 4 results in a change from histidine to arginine, increased protein stability, and correspondingly increased enzymatic activity. Microsomal EH activity [toward (±)-benzo(a)pyrene-4,5-oxide] varies at least eight-fold, and is correlated with protein content but not with mEH RNA content. Thus, genotype alone may not explain differences in mEH activity.

Nonetheless, a few reports have suggested an association between genotypes predictive of low activity and reduced risk for all lung cancer, and among squamous cell lung cancers and African-Americans considered separately. A recent meta analysis of eight published studies found no relationship between mEH and lung cancer, but a pooled analysis (including raw data from four published and four unpublished datasets) found reduced risk for lung cancer (OR = 0.7, 95% CI 0.51–0.96) among subjects with exon 3 His/His genotype, especially among heavy smokers and among adenocarcinomas (Lee et al. 2002).

13.2.5. Glutathione *S*-Transferases

Glutathione *S*-transferases (GSTs) can detoxify mutagenically active forms of several carcinogens in tobacco smoke, including active electrophilic benzo(a)pyrenes, by conjugating them with cellular glutathione. The active benzo(a)pyrene diol epoxide metabolite of cigarette smoke, benzo(a)-pyrene-7β,α-dihydrodiol-9α,10α-epoxide (BPDE) is a good substrate for each of the GSTs, particularly GSTP1. Low or no GST activity is generally presumed to predispose an individual to cancer. However, certain GSTs, particularly GST theta, may have a more complex role as they appear to be involved in the activation of certain xenobiotics (e.g., some alkane and alkene procarcinogens) (Hengstler et al. 1998). GSTM1, GSTT1, and GSTP1 have overlapping substrate specificities and are polymorphic in humans, with level of expression influenced by both the genetic polymorphisms and exposures that induce production of the enzyme. GSTM1 is expressed at a low level, and GSTP1 at a high level, in human lung. GST "high risk" genotypes are much more common than CYP450 "high risk" alleles, so even a modest role for GSTs in determining risk may be equated with a large population attributable risk.

13.2.5.1. GSTM1. Approximately 50% of Caucasian populations lack measurable enzymatic activity for GSTM1, due to a gene deletion and corresponding absence of the GSTM1 enzyme. Most individuals with GSTM1 activity are heterozygote for the gene. Genotype predicts phenotype with 98% sensitivity and specificity.

Most studies of GSTM1 and lung cancer have reported a modest association between the null phenotype or genotype and elevated risk for lung cancer, with odds ratios in the range of 1.2–3.3, and some attaining statistical significance. A few studies have reported no association, but may be limited by their methods for subject ascertainment. These studies varied with respect to study populations (ethnicity, gender, smoking history, case histologies) and methods (for subject identification and in the lab), and this variation may have made it difficult to replicate findings of a modest association. However, several recent meta- and pooled analyses have confirmed a statistically significant weak or modest association between GSTM1 and lung cancer, with odds ratios for GSTM1 null status of

1.13–1.20, possibly stronger among Asians and for squamous and small cell cancers (Houlston 1999, Benhamou et al. 2002).

A methodologic concern is the use of hospital-based controls in studies of polymorphisms. If the enzymes under study are involved in diseases for which controls are hospitalized, high-risk genotypes may be overrepresented in a control series ascertained through a hospital or clinic, making the controls genotypically similar to the cases and biasing risk estimates toward null. It is plausible that GSTs (as well as other enzymes discussed in this chapter) are involved in multiple disease processes, particularly in smoking-related cancers, which occur with great frequency in hospital populations. This bias can be avoided by using population-based controls, or may be minimized by excluding from hospital-based controls those patients who are admitted to the hospital for smoking-related conditions (including conditions other than cancers).

Findings of a relationship between GSTM1 and lung cancer are supported by data from DNA and protein adduct studies, cytogenetics, urinary mutagenicity studies, and biochemical assays (Brockmoller et al. 1998). Higher levels of PAH-DNA adducts and more frequent p53 G:C to T:A transversions [mutations related to activated benzo(a)pyrene] have been found in lung tissue from GSTM1 null lung cancer patients and smokers, relative to those with functional GSTM1.

Efforts to determine whether the relationship between GSTM1 and lung cancer may differ according to histologic type of cancer, history of exposure to tobacco, subjects' ethnic background, or by function of other biotransformation enzymes, have also produced conflicting results (Houlston 1999), possibly due to sample sizes that are frequently inadequate to detect interactions. Interestingly, associations between GSTM1 and lung cancer are two- to seven-fold stronger in female smokers than in male smokers. Because the GSTM1 gene is on an autosomal chromosome, this gender effect cannot be explained by differing genotype frequencies between men and women. Instead, it appears that total GST activity is higher in lung tissue from women than in lung from men, suggesting that GST activity in women may be induced, more than in men, above the genetically determined level. Exposures that could explain this (because they differ in men and women, and are thought to be related to risk for lung cancer as well as to enzyme function) include tobacco use, steroid hormones, body characteristics (body weight, body mass index, waist-to-hip ratio), alcohol use, and diet (e.g., brassica vegetables and caffeine). Gender differences in GSTM1 function may explain, in part, why female smokers have a greater frequency of mutations in the p53 gene, and a higher average level of DNA adducts in lung tissue than do male smokers, even with lower exposure to tobacco carcinogens. Enzyme function differences may further explain, in part, reports that women may be at higher risk for lung cancer than are men, even after control for smoking habits and gender differences in baseline risk (Risch 1993, Zang 1996, Brownson 1992).

13.2.5.2. GSTT1. As with GSTM1, phenotypic absence of activity is due to a homozygous gene deletion. Frequency of the GSTT1 null genotype is 11–38% among Caucasians, and 11–64% among other ethnicities (Hengstler 1998). Smokers lacking a functional GSTT1 enzyme reportedly cannot metabolize monohalomethanes found in tobacco smoke and may have greater susceptibility to sister chromatid exchanges. Few studies of GSTT1 and lung cancer have been reported, and GSTT1 genotype has been only occasionally associated with lung cancer, or with particular histologic types, with most studies to date reporting no association.

13.2.5.3. GSTP1. Glutathione *S*-transferase P1 (GSTP1) is potentially important in determining risk for lung cancer because it is the most prevalent GST in lung tissue and has the highest activity toward the active benzo(a)pyrene diol epoxide metabolite of cigarette smoke, benzo(a)pyrene-7β,α-dihydrodiol-9α,10α-epoxide (BPDE). Two commonly expressed GSTP1 variants have differing specific activities and affinities for active substrates resulting from a single basepair difference (A → G at nucleotide 313) and an amino acid substitution (Ile → Val) at codon 105, a site close to the binding site for electrophilic substrates. Gene frequency for the val (or GSTP1b) variant has been reported as 0.11–0.34 among healthy control populations. The val variant has generally lower activity toward active polycyclic aromatic hydrocarbons (especially BPDE), and has been predicted to have lower detoxication potential and greater risk for cancer. Indeed, PAH adducts occur more frequently in GSTP1 val (GSTP1b) individuals than in those with GSTP1 iso (GSTP1a) (Haugen et al. 2000). However, the val variant has higher activity than does the isoleucine variant toward some other active compounds in tobacco smoke. If GSTP1 has an important role in both detoxification and activation, a measurable overall effect on risk may be small. There have been some reports of an excess of val variant homozygotes among lung cancer cases (16–27%) relative to controls (9–11%) (Ryberg 1997), especially among heavy smokers and among squamous cell cancer cases, although there have also been several reports of no association in varied ethnic study populations (Harris 1998).

13.2.5.4. GSTs and Dietary Isothiocyanates. Dietary isothiocyanates from cruciferous vegetables reduce risk for lung cancer through inhibition of CYP450-mediated activation and induction of detoxifying phase II enzymes, including GSTs that conjugate isothiocyanates for elimination. Isothiocyanate intake has been associated with strongest reductions in lung cancer risk among smokers, and among subjects genetically deficient in GSTT1, or deficient in both GSTM1 and GSTT1 (Spitz et al. 2000). Risk for lung cancer among current smokers with low isothiocyanate consumption and GST deficiencies were OR = 2.2 (95% confidence interval 1.2–4.1) for null GSTM1, OR = 3.2 (1.5–6.6) for null GSTT1, and OR = 5.4 (1.7–17.2) for subjects with both GSTM1 and GSTT1 null genotypes (Spitz et al. 2000). The strength of these associations suggests the need to consider diet in studies of inducible enzymes and risk for lung cancer. These data also show clearly that, while genotype measurements are easy and convenient markers for genetic susceptibility, they are limited because they may not reflect transcriptional and posttranslational events that can affect function and may bias risk estimates away from the true value. Two individuals with the same genotype for low enzymatic activity may, in fact, be operating at very different levels of induced activity. Smoking appears to affect expression of many of the biotransformation enzymes, as do alcohol consumption and diet (which tend to covary with smoking). Thus, it is possible that conflicting data on relationships between these enzymes and risk for lung cancer may result from serious confounding by these (and other less understood) exposures.

13.3. DNA REPAIR

Lung cancer risk is associated with elevated sensitivity to mutagens, measured as frequencies of bleomycin-induced DNA breaks, chromatid breaks induced by active benzo(a)pyrene, and BPDE adducts (Wei and Spitz 1997). Repair of DNA damage, before it is replicated, is critical to the prevention of genetic instability and cancer. Reduced DNA repair capability has been linked to elevated tissue levels of carcinogen–DNA adducts and high risk for lung cancer, with risk particularly high among young patients (<65 years of age at diagnosis) and among smokers.

There are at least four mechanisms for repair of specific types of DNA damage: base excision repair for small errors (e.g., nonbulky adducts or bases oxidized by ROS), nucleotide excision repair for larger lesions [such as bulky benzo(a)pyrene adducts], mismatch repair for base–base or insertion–deletion mismatches, and repair for double-strand breaks caused by exposures such as ionizing radiation. Because base and nucleotide excision processes are known to repair DNA damage resulting from tobacco use, their influence on risk for lung cancer has been studied. These repair processes involve polymorphic genes, although the functional significance of these variants has been established only in some cases.

13.3.1. Base Excision Repair

DNA damage caused by reactive oxygen species (ROS) from tobacco smoke has been linked to lung cancer. Through base excision repair, ROS-damaged bases are excised by base-specific DNA polymerases. Importantly, oxidized base 8-oxoguanine is commonly produced by ROS, and is removed by 8-oxoguanine DNA glycosylase (OGG1). Once damaged DNA is removed, the original base can be replaced by a variety of endonucleases, including X-ray cross-complementing group 1 (XRCC1). Both OGG1 and XRCC1 are genetically polymorphic and may determine, in part, risk for cancer from damage to DNA bases.

13.3.1.1. OGG1 Gene. ROS-induced modified base 8-oxoguanine (8-OHdG) is highly mutagenic, and occurs at high levels in leukocytes and lung tissue from smokers and lung cancer patients. A common S326C allele (C allele at codon 326 resulting in a Ser–Cys substitution) in OGG1 occurs with varying frequency of cys across ethnic groups (42% in Japanese, 22% in Caucasians) (Le Marchand 2002). The S326C allele may have reduced repair of 8-oxoG or reduced substrate specificity, but the functional importance of this polymorphism has been disputed. Nonetheless, the S326C polymorphism (Cys/Cys vs. Ser/Ser) has been associated with a two- to three-fold increased risk for lung cancer in hospital-based studies in Japan and Germany, and in a large multi-ethnic population-based study in Hawaii (Le Marchand 2002). In the German study, risk was elevated among Cys/Cys subjects (OR = 2.2) but reduced among Cys/Ser subjects (OR = 0.7), and, although the study was small and neither finding was statistically significant, these results show that categorization of heterozygote genotypes, in the absence of information on function, can affect results (Goode et al. 2002). In this situation, a useful approach is to present two odds ratios for each polymorphism (one for rare allele homozygotes vs. common allele homozygotes and one for rare allele heterozygotes vs. common allele homozygotes) so that patterns of dominance may be determined. A second, very rare, polymorphism in OGG1 (G-to-T transversion at position 18) was associated with lung adenocarcinoma in a Japanese population, although the functional significance of this polymorphism is unknown.

13.3.1.2. XRCC1 Gene. After excision of a damaged base, the X-ray cross-complementing group 1 (XRCC1) protein stimulates endonuclease action and provides a physical framework for restoration of the site. Although little is known of their functional importance, three XRCC1 polymorphisms (R194W, R399Q, and R280H) have been studied in epidemiology studies of lung cancer. For the R194W polymorphism, reduced risk of lung cancer was associated with the W allele (RW/WW vs. RR) with odds ratios of 0.4–0.7 among Chinese (Ratnasinghe et al. 2001) and African-American subjects, with possible interactions with tobacco and alcohol use. The direction of this effect on risk may provide some suggestion that the W variant has greater functional capability than does the wildtype. For R399Q, there have been inconsistent results. Regarding R280H, one small study has reported increased risk for lung cancer associated with either 1 or 2 H alleles (OR = 1.8, 1.0–3.4) (Ratnasinghe et al. 2001).

13.3.2. Nucleotide Excision Repair

Nucleotide excision repair is used for repair of tobacco-induced benzo(a)pyrene adducts (Wei and Spitz 1997). Repair involves a series of steps including the unwinding of DNA by the TFIIH (transcription factor IIH) complex with XPD (xeroderma pigmentosum complementation group D) in preparation for excision and repair of damage. XPD has several common genetic polymorphisms (including L751Q and D312N) that have been studied, but their functional significance and their effect on lung cancer risk is unclear. For XPD L751Q, two studies show increased risk associated with QQ (vs. LL) among nonsmokers only (Butkiewitcz et al. 2001), and two studies have reported no association and no smoking interaction. For XPD D312N, two studies have reported increased risk among NN genotypes, relative to DD, although only among nonsmokers, but these findings were not confirmed in another study. The effects of XPD may be apparent only in presence of lower levels of DNA damage than those caused by smoking (Goode et al. 2002).

13.4. COMBINED EFFECTS OF BIOTRANSFORMATION ENZYMES

Given the complexity of tobacco smoke metabolism, measurement of one isolated enzyme at a time may provide limited insight into the underlying biology of lung cancer. Unfortunately, few studies of combined effects or interactions of multiple polymorphisms have been reported because they require very large study populations (especially when rare alleles are considered). Most studies that have reported combined effects of more than one polymorphism have done so without completing the appropriate analytical methods, namely a formal test for interaction [for a discussion of testing for interaction, see Schlesselman (1982), pp. 63–66].

The most frequently considered enzyme combination is GSTM1/CYP1A1. Because procarcinogens in tobacco smoke are activated by CYP1A1 and detoxified by GSTM1, susceptibility may well depend on a balance in their activity. Persons with GSTM1 null/CYP1A1 *m1* or *m2* genotypes have been reported to have up to nine-fold risk for all lung cancer and for squamous cell cancers alone (Hayashi et al. 1992), relative to low-risk genotypes. It has also been reported that heavy smokers with GSTM1 null/CYP1A1 *m2/m2* have seven-fold higher risk for lung cancer than do heavy smokers with GSTM1 present/CYP1A1 *m2/m2*. With regard to the underlying mechanism, deletion of GSTM1 has been associated with strong inducibility of CYP1A1 transcription by dioxin. High CYP1A1 enzyme induction among GSTM1 deficient subjects may result from low levels of detoxification and the resulting higher intracellular concentrations of inducing compounds (Brockmoller 1998). In support of this, very high levels of benzo(a)pyrene adducts have been measured in lung tissue from smokers with homozygous CYP1A1 *m1*/GSTM1 null genotype (Alexandrov et al. 2002).

Genotype may influence location of cancer occurrence within the lung. Lung cancer patients with inducible CYP1A1 and GSTM1 present have been reported to have predominantly (88%) peripheral cancers, whereas patients with inducible CYP1A1 and no GSTM1 had both peripheral and bronchial cancers in equal proportions. Although many lung cancers are histologically heterogenous, some of the different histologies tend to occur in different areas of the lung—small cell cancers tend to occur in the bronchus, and adenocarcinomas tend to occur in the alveolar periphery. Genotype may be related to site and to histology, and further research efforts to examine histology-specific relationships are needed.

A few other combinations have been reported sporadically but are unconfirmed. A strong combined effect of homozygous susceptible CYP1A1 MspI (*m1*) or Ile–Val (*m2*) with GSTT1 homozygous null on risk for squamous cell cancer has been reported. Combined effects of GSTP1 and GSTM1 have been considered. One study reported lowest levels of DNA adducts in subjects with GSTM1 positive/GSTP1 ile/ile genotypes and highest levels in subjects with GSTM1 null/combined GSTP1 val/val and val/ile genotype combination. In another study of lung cancer cases, smokers with GSTM1 nonnull/GSTP1 ile genotype had significantly higher levels of DNA adducts, and the GSTM1 null/GSTP1val genotype was associated with lung cancer even after adjusting for adduct frequency, suggesting that adducts and genotypes may be independent predictors of risk [see Ryberg et al. (1997)].

13.5. SURVIVAL

Few studies have considered the role of polymorphisms in lung cancer survival. However, there are reasons to expect that GSTs, at least, may influence prognosis and survival:

1. Several chemotherapeutic drugs are detoxified by GSTs, and chemotherapy is increasingly used to treat lung cancer. Differential ability to metabolize these drugs could influence response to therapy and corresponding survival. In fact, GSTP1 has great affinity for several chemotheraputic drugs, and the GSTP1 polymorphism has been studied as a potential predictor of drug efficacy and success for cancers at sēveral anatomic sites.
2. High-risk GST variants are associated with p53 and K-*ras* mutations in lung cancers (Ryberg et al. 1997), which, in turn, may be related to tumor aggressiveness and survival.

More recent studies have found no effect of GSTM1 on 1-year survival, but GSTM1 homozygous null genotype has been associated with poor 3-year (42–44% of GSTM1 null vs. 48–55% GSTM1 positive patients) and 5-year survival (20% GSTM1 null vs. 29% GSTM1 positive patients), even after adjustment for chemotherapy, histology and stage at diagnosis, and among both non–small cell and small cell histologies (Sweeny et al. 2003). Neither GSTP1 nor GSTT1 polymorphisms were related to survival in this study, but power to detect a modest effect was low (Sweeny et al. 2003). If GSTM1 detoxifies chemotherapies, we might expect GSTM1 null patients to get full benefit from their chemotherapies and have improved survival, but the GSTM1 null genotype was found to be associated with poor survival. It is possible that toxicity was greater and treatment regimens reduced in GSTM1 null patients, or that these patients have more reactive carcinogens, DNA adducts, p53 or K-*ras* mutations, and correspondingly more aggressive tumors. The latter theory is consistent with evidence that high CYP1A1 inducibility may also shorten survival among lung cancer patients, perhaps by creating more unstable and aggressive cancers. CYP1A1 *Msp*I has been further associated with other indicators for prognosis [general condition and extent of primary tumor (T factor)], as has GSTM1 [general condition, extent of regional lymph node metastasis (N factor), and extent of distant metastasis (M factor)].

Reports that enzyme polymorphisms influence progression of lung cancer emphasize our need to study incident (newly diagnosed) cases rather than prevalent cases, and to achieve a high level of participation among eligible subjects. If patients with the GSTM1 null or high CYP1A1 inducibility do indeed have relatively short survival, we would expect to see a low prevalence of these genotypes among live patient populations, and a bias in analyses of GSTM1 and CYP1A1 as risk factors for lung cancer toward a null finding. This could account for some of the negative GSTM1 and CYP1A1 studies cited above. In the case of GSTM1, we have data to show that we can avoid this survival bias by studying incident cases, that is, only those patients ascertained within one year after diagnosis. In the absence of information on whether other genes influence survival, most accurate risk estimates for gene–environment interactions will be found among incident cases, ascertained as soon after diagnosis as possible.

Similarly, hospital-based cases could also bias study results because hospitalized patients may have cancers that differ in aggressiveness from cancers in nonhospitalized cases. Population-based studies, including all incident cases in a specified geographic area over a specific time period, will provide the best estimates of exposure–disease relationships. Cancer registries, with a mechanism for rapid reporting of new cases, are optimal for case identification. Where a registry is not available, cases can be defined as all incident cases seen at specified hospitals over a certain period.

13.6. CONCLUSIONS AND FUTURE RESEARCH

Since the 1980s collaborations between epidemiologists and laboratory scientists have contributed considerably to progress in understanding the molecular basis for differences in susceptibility to lung cancer among smokers. This progress has gone hand in hand with advances in laboratory methods and progress in understanding study design and analytical methods most suitable for gene–environment studies.

We have recognized that investigations and data interpretation must take into account the complexity of tobacco metabolism, in which multiple enzymes act at the same time on the same, or related, metabolic pathways. Indeed, function of one enzyme may influence functional level of another. Many of the enzymes within families (e.g., P450s) have overlapping substrate specificities and, along with other important pathways controlling neoplasia (e.g., DNA repair, immune response, and cell cycle control), are redundant and may compensate for deficiencies (e.g., both MPO and CYPs activate BaP). Also, complex interactions may change over time, and may even be in constant flux, depending on diet, exposures, and occurrence of diseases.

Despite these difficulties, the preponderance of evidence suggests that some enzymes are involved in determining risk for lung cancer among smokers. See Table 13.1 for a summary of the polymorphisms and the direction of their effect on lung cancer. Variant alleles for CYP1A1, CYP2E1, MPO, and GSTM1 polymorphisms each appear to have relationships, albeit modest, with risk for lung cancer that are in the direction predicted from our understanding of their functional significance. With further study, microsomal EH, NAT1, NAT2, OGG1, and XRCC1 may well be included on this list. Because we have compelling reasons, based on biology, to expect that GSTP1 variants may influence risk, GSTP1 also deserves further study. Studies of the combined effects of CYP1A1 and GSTM1 polymorphisms have been promising, and provide stimulus for further research on combinations of polymorphisms, as well as exposures (tobacco, alcohol, diet, ethnicity) that can influence enzyme function. Complementary work in the laboratory to determine exposures that do influence enzyme function can contribute importantly.

Interesting preliminary data suggest several additional areas for further research. Findings that relationships between CYP1A1 and GSTM1 and risk for lung cancer may be stronger in women than in men are particularly worthy of attention, given the current high rate of smoking and lung cancer among women. Confirmation and further study of these findings, as well as findings that risk associated with CYP1A1 variants varies by ethnicity, may result ultimately in identification of risk factors (in addition to tobacco) that influence enzyme function and that can be avoided by individuals through changes in lifestyle. Future research may also provide a better understanding of how different lung cancer histologies develop. For example, adenocarcinomas and squamous cell carcinomas may have different patterns of p53 mutation (Liu et al. 2001). If we find that particular polymorphism genotypes result in particular types of p53 mutation or other DNA damage, they may also influence histologic type of lung cancer.

Given the complex biology of tobacco metabolism, it is unsurprising that variation in function of an individual enzyme would have no more than a modest effect on risk, and that studies often have had some conflicting findings. Conflicting results can also result when studies differ with respect to choice of case and control populations, phenotyping or genotyping methods, lung cancer histologies, and study populations (ethnicity, exposure, diet, gender). While associations of these enzymes with lung cancer risk may be modest, their attributable risks may be quite large.

Table 13.1 Summary of Important Genetic Polymorphisms and Direction of their Effect on Risk for Lung Cancer

Polymorphism	Genotype/ Allele	Relative Function of Genotype/ Allele	Relative Relationship of Genotype/Allele to Risk for Lung Cancer	Main Ethnic Groups Studied	Selected References[a]
			Phase I		
CYP1A1 MspI	*m1*	Elevated activity	Elevated risk	Japanese, Caucasian, mixed, black Brazilian	1–3
CYP1A1 IleVal	*m2*	Elevated activity	Elevated risk	Japanese, Caucasian, mixed, black Brazilian	1–3
CYP2D6	Extensive metabolism	Elevated activity	Uncertain	Mixed European	4
CYP2E1 DraI	Intron 6	Uncertain	Reduced risk	Varied	5,6
CYP2E1 RsaI/PstI	c2 homozygous	Reduced activity	Reduced risk	Japanese, Mexican, Swedish, African-American	5,7,8
MPO	A/A	Reduced activity	Reduced risk	Caucasian, Japanese	9–11
NAT2	Rapid acetylation	Elevated activity	Elevated risk possible	German, Swedish	12,13
NAT1	Slow acetylation	Reduced activity	Limited data	Caucasian	14
mEH exon 3	His/His	Reduced activity	Reduced risk	Caucasian, Asian, African-American	9,15

			Phase II		
GSTM1	Null	Reduced activity	Elevated risk	Caucasian, Asian, mixed	16
GSTT1	Null	Reduced activity	Uncertain	Caucasian	17
GSTP1	val/val	Uncertain	Uncertain	Caucasian, Japanese	17–19
			DNA Repair		
OGG1 S326C	CC	Uncertain	Elevated risk	Japanese, German, Caucasian, Hawaiian	20,21
XRCC1 R194W	W	Uncertain	Reduced risk	Chinese, African-American, Caucasian	22,23
XRCC1 R399Q	QQ	Uncertain	Uncertain	Caucasian, Hispanic, Polish, African-American	23–25
XRCC1 R280H	HH	Uncertain	Limited data	Polish	25
XPD L751Q	Q	Uncertain	Uncertain	Caucasian, African-American	21,23,25

[a] *Key*:

1. Hayashi et al. (1992).
2. Garcia-Closas et al. (1997).
3. Sugimura et al. (1995).
4. Amos et al. (1992).
5. Le Marchand et al. (1998).
6. Guengerich and Shimada (1998).
7. Persson et al. (1993).
8. Wu et al. (1997).
9. Kiyohara et al. (2002).
10. Feyler et al. (2002).
11. Brockmoller et al. (2000).
12. Nyberg et al. (1998).
13. Cascorbi et al. (1996).
14. Bouchardy et al. (1998).
15. Lee et al. (2002).
16. Houlston (1999).
17. Nazar-Stewart et al. (2003).
18. Coles et al. (2000).
19. Harries et al. (1997).
20. Spitz et al. (2000).
21. Goode et al. (2002).
22. Ratnasinghe et al. (2001).
23. David-Beabes and London (2001).
24. Divine et al. (2001).
25. Butkiewicz et al. (2001).

In order to measure interactions that are part of a complex puzzle and modest in strength, epidemiology studies must be large, well designed, and conducted with great care. Population-based studies, with incident cases and controls enumerated from the general population, are optimal. Findings from study groups homogeneous with respect to histology, smoking history, ethnicity, gender, diet, and exposures that may influence function may not be generalizable but are most likely to produce estimates of relationships between enzymes and risk that are precise and not confounded by a variety of factors. Phenotype or data on potentially inducing exposures (e.g., tobacco and alcohol use, hormones, and diet) should be collected where enzymes are inducible and genotype measurements may not be indicative of enzyme function.

14

Gastrointestinal Cancers

Thomas L. Vaughan
University of Washington, Seattle, WA

14.1. INTRODUCTION

Cancers arise in the gastrointestinal (GI) tract more commonly than in any other organ system. Moreover, these tumors are the leading cause of cancer related death in the world (Parkin et al. 1999; Pisani et al. 1999). Much has been learned about the etiology of these diverse cancers over the past several decades of laboratory- and population-based research. These studies indicate that a person's likelihood of developing a particular cancer is determined largely by a complex interplay of risk factors involving environmental exposures, lifestyle factors, and inherited susceptibility. The challenge to the emerging field of ecogenetics is to understand in sufficient detail how these specific factors interact in causing (or preventing) these cancers, so that more effective prevention and early detection programs can be developed and applied toward persons at highest risk of the disease.

14.1.1. Incidence

While gastrointestinal (GI) tract cancers are uniformly important contributors to morbidity and mortality throughout the world, the relative contributions of cancers arising in the esophagus, stomach, and colon/rectum, the sites of origin discussed here, vary markedly in different populations (Table 14.1). For example, esophageal cancer occurs with extraordinarily high fre-

Gene-Environment Interactions: Fundamentals of Ecogenetics, edited by Lucio G. Costa and David L. Eaton

Table 14.1 Annual Incidence[a] of GI Tract Cancers by Site of Origin and Geographic Region, 1990

	Esophagus		Stomach		Colon/Rectum		% GI Cancers		All Cancers	
	Male	Female	Male	Female	Male	Female	Male	Female	Male	Female
East Asia	21.6	9.9	43.6	19.0	13.3	10.2	43.8	37.1	179.2	105.3
Japan	9.5	1.6	77.9	33.3	39.5	24.6	46.8	35.7	270.9	166.8
North America	5.2	1.4	8.4	4.0	44.3	32.8	15.7	13.8	369.9	277.5

[a] Per 100,000, age adjusted to world standard population.

Source: Parkin et al. (1999).

quency in parts of Central Asia, whereas in Japan, stomach cancer predominates. In contrast, among the U.S. population, GI tumors arise most frequently in the colon and rectum, which represents the second most common cancer in both men and women.

Cancers arising in the esophagus are intriguing to study, as they exhibit striking geographic and secular changes. Worldwide, it is estimated that more than 300,000 persons develop the disease every year, making it the eighth most common cancer (Parkin et al. 1999). The most frequently occurring type of esophageal tumor is squamous cell carcinoma, which is relatively common in many developing countries, but particularly so in north central China, central Asia, and southern Africa. The incidence of squamous cell carcinoma is relatively low in the United States (Figure 14.1), except among U.S. blacks, whose incidence rate is 6 times higher than that of whites. In the United States as well as most countries, the majority of squamous cell carcinomas occur among men. Overall, there is a gradual decreasing trend in incidence of these cancers over time.

Esophageal adenocarcinoma, once a relatively uncommon histologic type, has undergone a dramatic rise in incidence over the last several decades in western Europe, Australia, and New Zealand, as well as in the United States (Figure 14.1), where they now represent the most common histologic type (Bollschweiler et al. 2001). This cancer is substantially more common in whites and men than in blacks and women, although the time trends in incidence are similar. The underlying causes of the markedly different absolute rates of this cancer by gender and ethnicity, and of the rapid increases experienced by both genders and racial groups are somewhat unclear, although the epidemic of obesity in developed countries appears to be playing an important role (Chow et al. 1998).

Gastric cancer is the second leading cause of cancer-related mortality worldwide, accounting for over 600,000 deaths per year (Pisani et al. 1999). As with esophageal cancer, incidence rates vary widely. Japan has some of

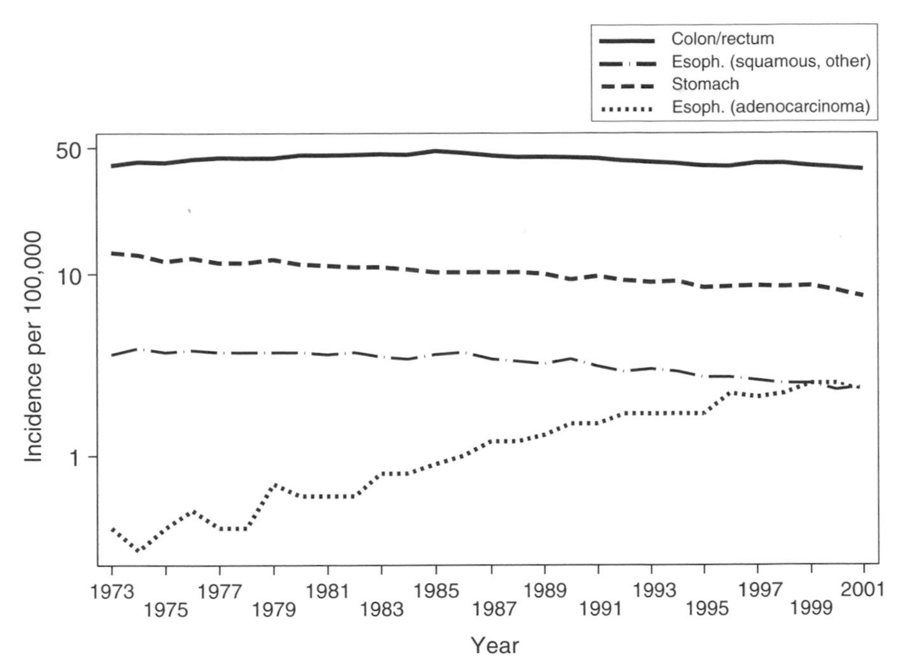

Figure 14.1. Incidence of GI cancers by site of origin, SEER 1973–2001 (SEER 2004).

the highest rates in the world, whereas low rates can be found among whites in the United States, the United Kingdom, New Zealand, and Australia (Kelley and Duggan 2003). In contrast to esophageal adenocarcinoma, the incidence of cancers arising in the stomach has undergone a slow but steady decrease for a number of decades (Figure 14.1). This decrease has been most apparent in low-risk countries, although decreases in Japan and other high-risk countries are also observed (Kelley and Duggan 2003). An exception to this has been an increase in incidence of cancers arising from the gastric cardia, which is adjacent to the esophagus. The vast majority of cancers arising in the stomach are classified as adenocarcinomas.

Cancers arising in the colon and rectum now account for approximately one out of every nine new cancer cases in the United States, and are the third leading cause of cancer-related mortality in the world (Pisani et al. 1999). Some of the highest incidence rates in the world occur in the United States (Table 14.1). Secular changes in colon cancer incidence have been relatively modest (Figure 14.1). As with gastric cancer, most cancers arising in the colon are adenocarcinomas.

14.1.2. Mortality

The prognosis of persons diagnosed with cancers arising in the GI tract also varies widely by site of origin (Ries et al. 2004). Esophageal cancers remain one of the most deadly neoplasms, with median survival less than 12 months, 5-year relative survival of 14.3%, and little evidence that more modern treatment techniques yield significant survival advantages. Stomach cancer is a disease that also confers a poor prognosis, although not as dismal as esophageal cancer: 5-year survival is 23.3%. On the other hand, 5-year survival for colon cancer is relatively good: 63.4%.

14.2. MAJOR RISK FACTORS AND MOLECULAR PATHWAYS

14.2.1. Esophagus

Cigarette smoking and alcohol consumption have long been known to be the most important risk factors for esophageal squamous cell carcinoma. Heavy smokers experience at least a sixfold increased risk of this cancer, which is similar to that observed among heavy alcohol users (Gammon et al. 1997). Together, tobacco and alcohol are estimated to account for approximately 85% of new cases occurring in the U.S. population.

Loss of function of the p53 gene, usually through mutation of one allele and loss of the other, can be detected in 50% or more of esophageal squamous cell carcinomas (Vaughan 2002). Allelic loss involving chromosome 9p is also relatively common in squamous cell carcinomas (Vaughan 2002). This loss likely targets the p16 gene, which is important in cell cycle regulation. There is only limited evidence for familial clustering of esophageal squamous cell carcinoma, with the strongest reports coming from high-risk areas of China. The rare tylosis syndrome also involves increased risk of esophageal squamous cell carcinoma, although the gene or genes involved have not been identified (Vaughan 2002).

In contrast to squamous cell carcinomas, tobacco use plays only a modest role in the etiology of esophageal adenocarcinomas, while alcohol intake has little effect on risk (Gammon et al. 1997). Most adenocarcinomas develop under conditions of chronic gastroesophageal reflux. Those with the most severe or frequent reflux are estimated to be at 7–16 times the risk of those without reflux (Farrow et al. 2000). Obesity also plays a key role in the development of these cancers (Chow et al. 1998). The mechanisms likely involve promotion of reflux, although other pathways, such as altered hormonal status, may also be important. A high percentage of fat in the diet also appears to increase risk, possibly by promoting reflux. As with many epithelial cancers, a low intake of fruits and vegetables appears to be associated with increased risk of esophageal cancer (Steinmetz and Potter 1996). However, the particular dietary components involved and the causal mechanisms are not yet clear. There is little evidence that genetic susceptibility plays a major role in esophageal adenocarcinoma, although case reports of familial clustering of esophageal adenocarcinoma or precursor abnormalities have been noted.

The vast majority of adenocarcinomas progress through a precancerous stage, termed *Barrett's esophagus*, in which the normal squamous epithelium is replaced by a specialized intestinal metaplastic epithelium. Persons with such metaplasia are at high risk of progressing to cancer, estimated at 0.5–1% per year. Molecular abnormalities that accumulate in the development of Barrett's esophagus and subsequent neoplastic progression include inactivation of the p16 gene, through mutation, methylation, or allelic loss; and inactivation of the p53 gene, also through mutation or allelic loss (Vaughan 2002).

14.2.2. Stomach

Until the mid-1990s the etiology of most cases of gastric cancer had remained a mystery. However, the discovery that the bacteria *H. pylori* plays a major role in the development of gastric cancer represents a major breakthrough, one which may eventually lead to effective prevention strategies in both developing and developed countries. In contrast to the esophagus, risk of gastric cancer appears to be little affected by cigarette smoking or alcohol intake. Specific factors in the diet, namely, high intake of nitrates and nitrites, as well as salted foods, have long been suspected of the increasing gastric cancer risk, as patterns of intake in different populations correlate to some degree with incidence rates. Familial clustering of gastric cancer has also been observed, with first-degree relatives experiencing a two- to threefold increase in risk. The degree to which these associations represent shared environment versus inherited genetic abnormalities is at present unknown.

In contrast to esophageal cancer and colon cancer (see discussion below), less is known about the molecular pathways involved in the development of gastric adenocarcinomas. There appears to be a substantial amount of heterogeneity in the combination of genetic and epigenetic abnormalities observed in neoplastic progression in the stomach, with *p53, APC, DCC, E-cadherin, ras*, and *p27* abnormalities observed in various frequencies and combinations (Chan et al. 1999).

14.2.3. Colon and Rectum

Certain dietary patterns, particularly those high in fat and energy intake, and low in consumption of fiber, fruits, and vegetables, appear to increase risk of colon cancer (Steinmetz and Potter 1996). Specific micronutrients, including selenium, folate, calcium, and vitamin D, have also been hypothesized to affect risk, but observational studies in humans have not produced consistent results. In addition to specific foods, methods of cooking have also been examined as possible risk factors (Lang et al. 1994). In particular, high-temperature methods of cooking meats, such as barbecuing and broiling, have been shown to produce relatively high levels of heterocyclic aromatic amines, which are carcinogenic in animal and in vitro experiments. Persons with inflammatory bowel disease, including ulcerative colitis and Crohn's disease, also experience markedly elevated risk of colon cancer.

Over the past several decades much has been learned about the molecular pathways involved in colon neoplasia (Neibergs et al. 2002). Loss of function, through mutation or other events, of the APC gene on chromosome 5q21 appears to be an early event in the formation of adenomatous polyps, which are precancerous lesions. A rare inherited predisposition to colon cancer, familial adenomatous polyps (FAP), involves germline mutations of the APC gene, with the resulting formation of hundreds or thousands of polyps in the colon as well as small intestines and stomach. Cumulative incidence of colon cancer among persons with this predisposition approaches 100% by the sixth decade of life. A more common inherited syndrome, hereditary nonpolyposis colorectal cancer syndrome (HNPCC), may account for up to 10% of colorectal cancers in the United States. The syndrome involves mutation of mismatch repair genes, and is often associated with increased risk of other cancers, such as ovarian and endometrial, in addition to the colon. In addition to these known inherited susceptibilities, family studies indicate that additional, unknown, pathways may be operating.

14.3. INTERACTIONS BETWEEN GENETIC AND ENVIRONMENTAL FACTORS

Well over 100 types of cancer have been observed to occur in humans, each with its own unique constellation of risk and protective factors. However, no matter how strong a particular risk factor might be, whether an environmental exposure such as tobacco or alcohol, or an inherited predisposition to cancer such as HNPCC, it is quite rare that the one factor completely determines the development of cancer. For example, the vast majority of regular cigarette smokers never develop lung cancer, and a sizable number of women who inherit a mutation in the BRCA1 gene do not develop breast cancer over their lifetimes.

Clearly multiple factors must therefore play a role in causing a normal cell to develop the myriad genetic abnormalities characteristic of the malignant phenotype. In some instances, risk factors may act through different biological pathways, and are therefore said to act independently of each other. In other instances, the effect of one risk factor may depend on the presence of another risk factor; that is, they may interact. The interacting risk factors might both be environmental in nature. For example, heavy smokers develop lung cancer at a rate approximately 10 times that of the general population, while nonsmokers who are occupationally exposed to asbestos have approximately an eightfold risk of cancer. However, persons exposed to asbestos who are also smokers experience an approximately 80-fold increased risk. In other instances environmental risk factors may interact with genetic factors in modifying disease risk. Finally, multiple gene products typically act in complex metabolic pathways, and they may interact with each other in determining disease risk (i.e., gene–gene interactions).

Given the wide variation in etiology of different cancers, it is important to consider each tissue or organ system in its own particular context. Exposures of particular relevance for the GI tract are, of course, those arising from the diet. The GI epithelium is constantly bathed in chemicals present in ingested foods and their metabolites. Also, in contrast to internal organs, much of the GI tract harbors microbial organisms, which can also play an important role in neoplasia. For each major cancer site of the GI tract, examples of etiologic studies examining the joint effects of environmental and genetic factors are given below.

14.3.1. Esophageal Cancer

Alcohol (ethanol) has long been known to be a major risk factor for esophageal squamous cell carcinoma. Studies throughout the world have confirmed an increased risk, ranging from seven- to tenfold, with heavy intake (Thomas 1995; Gammon et al. 1997). Although alcohol intake and cigarette smoking are usually correlated, studies carefully controlling for tobacco intake and those that have focused on nonsmokers have convincingly demonstrated the independent effects of alcohol. There is evidence that some types of alcoholic beverage (e.g., liquor, and to a lesser extent, beer) may be more strongly linked to increased risk than others (Gammon et al. 1997). For example, in some areas of the world, such as northern France, extremely high incidence rates have been observed. Epidemiologic studies have determined that intake of hot spirits ("Calvados"), a common beverage in the area, accounts for much of the clustering of cases (Launoy et al. 1997).

Interestingly, the mechanisms by which alcohol increases cancer risk are not known. Hypotheses include depressed immune function; local irritant effects; increased availability of solubilized cocarcinogens, including those found in tobacco smoke; and production of reactive intermediates.

An early intermediate in the metabolism of alcohol is acetaldehyde, which is a known carcinogen. This compound is detoxified via oxidation to acetic acid by the enzyme aldehyde dehydrogenase (ALDH2). A polymorphism in the *ALDH2* gene results in a protein with very little activity toward acetaldehyde; even those who are heterozygous for the variant allele metabolize acetaldehyde inefficiently, resulting in elevated concentrations of this carcinogenic molecule in the tissues and blood. This polymorphism is quite common in Japanese, but rare among Caucasians. Because an elevated concentration of acetaldehyde results in a variety of unpleasant symptoms, most people with the variant allele avoid regular alcohol use. Nevertheless, if the effects of acetaldehyde play a major role in the increased cancer risk associated with alcohol consumption, the association should be particularly strong among those with the variant protein. In fact, a number of studies have tested this hypothesis, one of which is discussed below.

Matsuo and coworkers studied 102 persons with esophageal squamous cell carcinoma and compared their alcohol intake history and *ALDH2* genotype with 241 controls in a hospital-based study in Japan (Matsuo et al. 2001). Heavy drinkers, defined as those with an intake of 75 mL or more of alcohol per day for 5 or more days per week, experienced a 13-fold increased risk of esophageal cancer after controlling for smoking history. However, this summary figure masks a large gene–environment interaction between *ALDH2* genotype and alcohol intake. Table 14.2 summarizes the

Table 14.2 Crude Odds Ratios[a] for Esophageal Squamous Cell Carcinoma According to Intake of Alcohol and *ALDH2* Genotype

	ALDH2 Activity	
Alcohol Intake	Normal	Low
Low	1.0	1.5
High	7.6	92.0

[a] Odds ratios calculated from published data (Matsuo et al., 2001).

independent and joint effects of these two risk factors, using crude odds ratios calculated from the published data. Among persons who were not heavy drinkers, there was little evidence of an association with the ALDH2 variant genotype (OR = 1.5). Conversely, among those with the more common, normally functioning, ALDH2 genotype, the odds ratio for heavy drinking was 7.6. However, heavy drinkers with the variant allele experienced a 92-fold increased risk of esophageal cancer, compared to those with neither risk factor. To view this in perspective, if these two risk factors (alcohol intake and ALDH activity) acted through different biological pathways, one would expect an additive relationship, resulting in a joint OR of approximately 8.1 [calculated as $(1.5 - 1) + (7.6 - 1) + 1$]. If both factors were roughly equally important in a shared pathway, one might expect them to interact in a multiplicative fashion, yielding a joint OR of approximately 11.4 (calculated as 1.5×7.6). The observed joint effect of 92.0 is described as a supramultiplicative interaction, implying that the variant, low-activity, ALDH2 genotype has an effect on esophageal cancer risk only among heavy drinkers. This makes biological sense in that there would be little reason to expect that the ALDH2 polymorphism would have any biological significance unless sufficient acetaldehyde were being produced. These results suggest that persons who drink significant quantities of alcohol and also have inherited the ALDH2 low-activity allele may be appropriate targets for primary and secondary (e.g., cancer surveillance) activities. They also suggest that a key mechanism underlying the carcinogenic effects of alcohol may be the production of the intermediate acetaldehyde.

14.3.2. Gastric Cancer

Cancers arising in the stomach develop through a process of neoplastic progression in which the normal mucosa becomes acutely inflamed (acute gastritis), and under conditions of continuing insult, develops features of chronic inflammation (chronic gastritis) (Correa 2003). A subset of those progress to atrophic gastritis, which is characterized by loss of glandular cells, and eventually to intestinal metaplasia. Persons with intestinal metaplasia are at particularly high risk of developing invasive gastric cancer.

Chronic infection with *H. pylori* clearly plays a role in the early stages of neoplastic progression in the stomach. There is strong evidence that infection with an *H. pylori* strain containing the gene encoding for the *cagA* pathogenicity island is particularly carcinogenic. However, a large proportion of the population worldwide have or have had *H. pylori* infection, many of them *cagA* positive, yet only a small fraction develop gastric cancer. Identification of a susceptible portion of the population would be extremely helpful in developing efficient and effective prevention and screening programs. Consequently an active area of research into possible cofactors in *H. pylori*–associated gastric cancer carcinogenesis relates to interindividual variations in immune response to agents or conditions promoting inflammation. In particular, a number of polymorphisms have been identified in genes regulating production of cytokines, many of which are instrumental in determining the type and extent of tissue response to chronic inflammatory stimuli (El-Omar et al. 2003; Macarthur et al. 2004).

Using data from a population-based case–control study of esophageal and gastric cancer, El-Omar et al. investigated the association between cancer risk and proinflammatory polymorphisms in *IL-1B, IL-4, IL-6, IL-10, TNF-A*, and *IL-1RN* (El-Omar et al. 2003). With regard to gastric cancers originating distal to the cardia, the site most likely to be related to *H. pylori*, four functional polymorphisms were observed to be related to cancer risk after controlling for the potential confounding effects of age, gender, and race: homozygosity for *IL-1RN**2 or *IL-10* (ATA), and carriage of *IL-1B*-511T or *TNF-A*-308. The effects on noncardia gastric cancer risk of having zero, one, two or three pro-inflammatory polymorphisms in these genes is described in Figure 14.2. The ORs are displayed (1) for all subjects, (2) those with positive serum antibodies to *H. pylori*, and (3) those who were also *cagA* positive. As can be seen, the slope of the graph relating

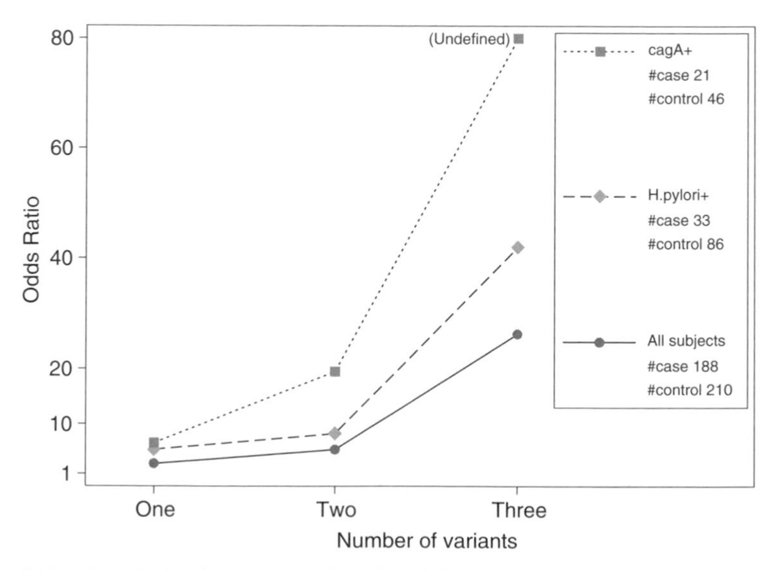

Figure 14.2. Association between number of proinflammatory cytokine polymorphisms and risk of gastric cancer according to *H. pylori* infection status (El-Omar et al. 2003).

number of cytokine polymorphisms to cancer risk is higher among those who are *H. pylori* positive and markedly higher among those who were also *cagA* positive. These results suggest some important conclusions: (1) cytokine polymorphisms may be significant markers of genetic susceptibility to gastric cancer and (2) they may be central to the pathogenicity of *H. pylori*. As such, they represent an interesting and unusual three-way interaction involving an external agent, genetic variation in that agent, and genetically determined variation in the host's immune response. If further research corroborates these findings, they may suggest a relatively simple screening test to identify those *H. pylori*–infected persons who are at particularly high risk of gastric cancer, and perhaps appropriate recipients of *H. pylori* eradication treatment and subsequent cancer surveillance.

14.3.3. Colon Cancer

The role of diet in the etiology of colon cancer has been a focus of epidemiologic research for a number of decades. A great deal of evidence has accumulated implicating a diet low in vegetables, and possibly fruit, as key risk factors in the development and progression of colon cancer. However, the specific nutrients whose relative deficit is responsible for increasing risk have been difficult to identify. There is accumulating evidence that folate, which is intimately involved in cell replication and DNA synthesis, may play an important role in colon cancer prevention. This nutrient, along with cofactors vitamin B_6 and vitamin B_{12}, interact in a complex series of reactions that provide substrates for the synthesis of DNA as well as its methylation. These reactions are modulated by a number of enzymes, many of which are polymorphic. There is a strong scientific basis, therefore, for hypothesizing that the effect of dietary intake of folate and related vitamins might differ depending on an individual's genotype.

One such enzyme in the folate metabolism pathway, for which substantial genetic heterogeneity has been noted across Caucasian populations, is methylenetetrahydrofolate reductase (*MTHFR*). A common variant of this gene results in substantially reduced enzymatic activity. Several studies have examined the joint effects of folate and related vitamins and *MTFHR* genotype in relation to risk of colon cancer or of key intermediate lesions in the progression to colon cancer. For example, Ulrich et al. examined this question in a large clinic-based study of colorectal adenomas (527 cases and 645 controls) (Ulrich et al. 1999). Folate intake was estimated from responses to a food frequency questionnaire. Salient results are seen in Figure 14.3. The protective associations between colorectal adenoma and intake of folate, vitamin B_6, and vitamin B_{12} was seen primarily among persons with the variant (low-activity) *MTHFR* genotype, adding additional

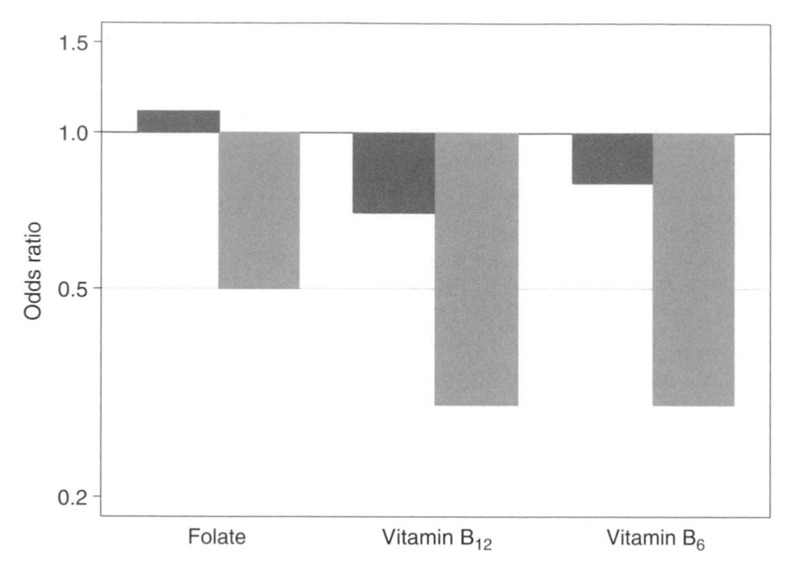

Figure 14.3. Association between vitamin intake (high vs. low tertile) and risk of adenomatous colon polyps, according to MTHFR genotype. Dark bar—normal activity (CC) genotype (247 cases, 297 controls); light bar—low activity (TT) genotype (48 cases, 69 controls) (Ulrich et al. 1999).

support to the hypothesized mechanism of action for folate, and implying that this subgroup of the population may receive the most benefit from vitamin supplementation.

14.4. CONCLUSIONS

Clearly there is much to learn about the interactions among environmental and genetic factors in the etiology of cancers of the GI tract. The studies discussed above are meant to serve as examples of the type of research needed to examine these questions, and to illustrate the complexity of such studies. Essentially, each hypothesis must be addressed individually, taking into account each cancer's unique constellation of risk factors, the mechanisms by which they are likely to act, and potentially relevant genes and gene products. A number of difficult challenges are faced in the study of ecogenetics:

1. Large sample sizes are typically required in order to precisely quantify joint effects of specific polymorphisms and related environmental exposures.

2. It is usually necessary to simultaneously evaluate the effects of multiple genes that are involved in a biologic pathway.
3. It is critically important to validate findings in multiple populations and ethnic groups.

A key challenge in the study of the ecogenetics of the GI system as well as other organ sites lies in identifying the important genes and related risk factors to study. Previously, the flow of information and hypotheses has been more or less unidirectional: from laboratory scientists, who were identifying new polymorphisms one at a time, testing each variant for possible functional significance, and then passing this lead on to population scientists who would examine its prevalence in specific populations and determine, taking into account other important risk factors, whether the variant played an important role in human carcinogenesis. This was a relatively slow and costly process, due not only to the expense of the discovery process but also to the costs of carrying out assays to identify variant alleles in large population studies.

As high-throughput genotyping methods have been developed, however, it has became increasingly efficient to examine a large number of polymorphisms at a time, even when their functional significance had not yet been determined. It is consequently now common for hypotheses to be generated in population studies, and passed along to the laboratory, where mechanistic studies can be undertaken to more fully explore the underlying biology. The more recent development of large-scale array-based studies of DNA and protein expression brings a powerful new set of tools to the study of gene–environment interactions. This set of technologies allows the simultaneous interrogation of tens of thousands of genes, and the profiling of large numbers of proteins in order to identify differences between persons with and without cancer, or between cancerous cells and normal cells (Weston and Hood 2004). A particular challenge inherent in such studies is the very large number of associations that are typically generated, and the consequent difficulty in interpreting them. Verification of hypotheses from such studies needs to take place in population-based studies as well laboratory-based experiments.

The exchange of information and expertise among investigators involved in (1) studying particular pathways in the laboratory or using animal models, (2) genomic and proteomic profiling, and (3) observational and experimental studies in humans is described in Figure 14.4. What once was a slowly developing field is now burgeoning into a rapidly growing collaboration among epidemiologists, molecular biologists, and biostatisticians, in which etiologic hypotheses can be cultivated from a number of sources and rapidly tested in complementary ways using high-throughput techniques.

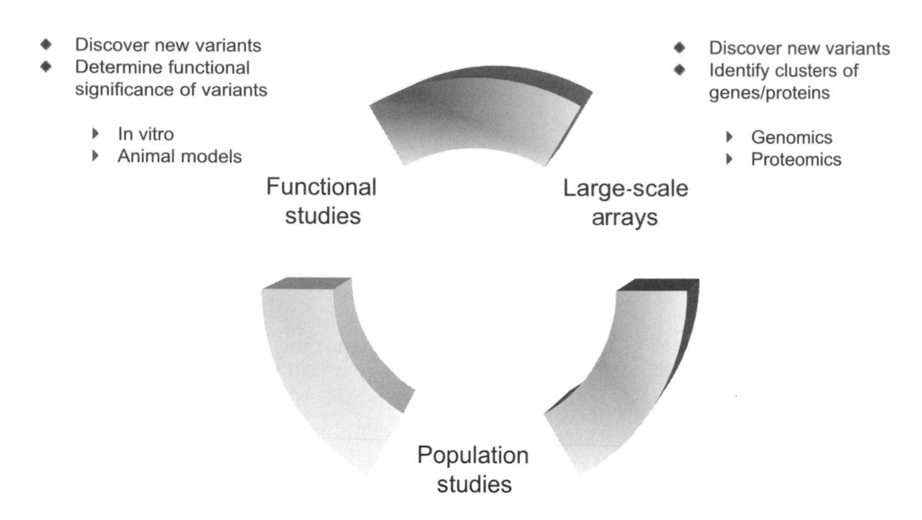

Figure 14.4. Model of information interchange among disciplines involved in identifying variant genes and proteins and their interactions with environmental and host factors in modifying cancer risk.

The ultimate goals of this interdisciplinary approach to ecogenetics are to more fully understand the mechanisms by which cancers occur, and to identify new and promising opportunities for prevention and cancer control in the population.

14.5. SUGGESTED READING

For further information on gastrointestinal cancer, the text by Khoury et al. (2004) and the journal articles by O'Byrne and Dalgleish (2001), Chan et al. (1999), Brennan et al. (2004), and Yokoyama et al. (2002) are recommended.

15

Neurodegenerative Diseases

Samir N. Kelada, Harvey Checkoway, and Lucio G. Costa
University of Washington, Seattle, WA

15.1. INTRODUCTION

Alzheimer's disease, Parkinson's disease, and amyotrophic lateral sclerosis (also known as *Lou Gehrig's disease*) are relatively common neurodegenerative diseases of late onset that are characterized by the progressive loss of specific groups of neurons, resulting in cognitive or motor function deficits. The burden of neurodegenerative diseases is of increasing concern to the medical and public health communities because the number of people who are reaching the ages at which these diseases typically occur is rising. Multiple sclerosis is also an important neurodegenerative disease of concern, but will not be discussed here because it is generally considered to be an autoimmune disease (probably with a viral etiology).

While mutations in specific genes causally related to the familial forms of Alzheimer's disease (AD), Parkinson's disease (PD), and amyotrophic lateral sclerosis (ALS) have been identified, the causes of the sporadic forms of these diseases remain largely unknown. Like many other complex diseases, they probably result from numerous complex interactions between genes and environment, but these, too, remain mostly unidentified. Interestingly, there is much evidence suggesting that these three diseases may share common pathological mechanisms, and hence there may also be overlap in their causal gene–disease associations and gene–environment interactions.

Gene-Environment Interactions: Fundamentals of Ecogenetics, edited by Lucio G. Costa and David L. Eaton

We begin this chapter by briefly discussing what is known about the relevant pathogenesis of each of these diseases before moving on to what is known about environmental risk factors, candidate genes identified from studies of the familial and sporadic forms of disease, and then gene–environment and gene–gene interactions. As we will explain, currently there is an overall paucity of consistent data suggesting important, reproducible gene–environment interactions, and thus this field of research remains ripe for study. We also discuss some of the logistical and methodological challenges to consider when interpreting the results of such studies.

15.2. ALZHEIMER'S DISEASE

Alzheimer's disease (AD) is the most common neurodegenerative disease, with a prevalence of roughly 30% among people 85 years and older. Incidence rates climb steeply from 0.5% per year for ages 65–75 to 6–8% per year for ages 85 and up. Disease onset is rare before the age of 50, except in cases of familial AD, representing ~5–10% of cases.

15.2.1. Pathogenesis

The primary clinical manifestation of AD is dementia, an accelerated loss of cognitive function beyond that due to normal aging. Alterations in mood and behavior often accompany the onset of dementia, followed in time by memory loss, disorientation, and aphasia. In AD, the hippocampus and cerebral cortex are severely affected. Pathologically, there are two hallmarks in affected tissues:

The first hallmark is the existence of senile or neuritic plaques in blood vessels and neurons of the hippocampus, which are composed primarily of the Aβ peptide aggregates (Figure 15.1). Aβ peptide monomers of varying size (amino acids) are formed from the cleavage of amyloid precursor protein (APP) by secretases. The 42-amino-acid Aβ peptide is of primary concern, as it is insoluble and forms aggregates of twisted β pleated sheets—the first step of the β-amyloid senile plaque formation. This is thought to be the first and central neurotoxic insult resulting in neuron loss. This toxicity may be exacerbated by an inflammatory response mediated by astrocytes and microglia.

The second pathological hallmark is the occurrence of neurofibrillary tangles (NFTs), filamentous bundles that accumulate in the cytoplasm of affected neurons (Figure 15.2). In this pathological cascade, the

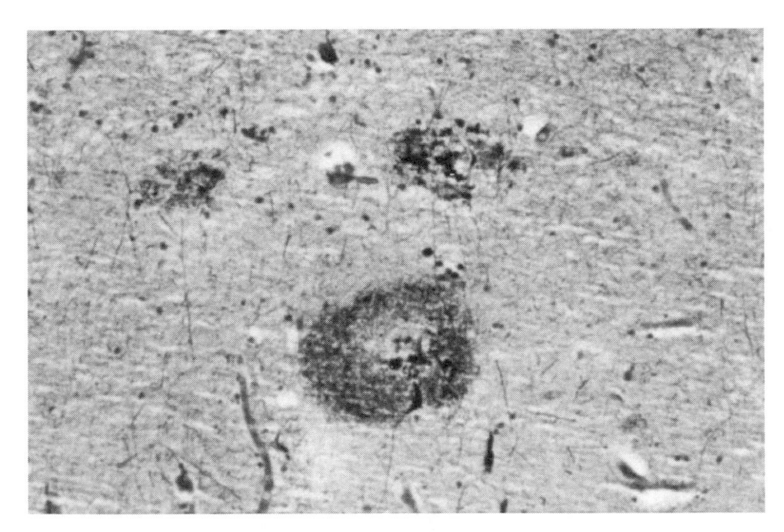

Figure 15.1. High-power micrographs show the first pathologic hallmark of AD, the neuritic plaque. In this silver stain the brown and black granular and fibrillar masses are abnormal accumulations of protein. The black globular material in the center is indicative of amyloid. (From Rodney Schmidt, M.D., Ph.D., *Systematic Pathology: An Interactive Tutorial*, with permission. Copyright University of Washington, 1998. http://eduserv.hscer.washington.edu/hubio546/.)

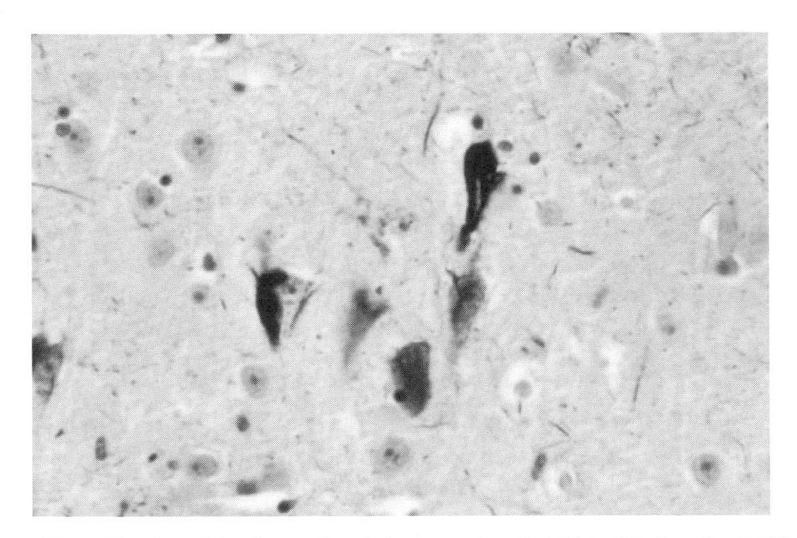

Figure 15.2. Silver precipitation stain of the second pathological hallmark of AD; the "flame-shaped" black inclusions within the neurons are neurofibrillary tangles from a brain of person with AD. (From Rodney Schmidt, M.D., Ph.D., *Systematic Pathology: An Interactive Tutorial*, with permission. Copyright University of Washington, 1998. http://eduserv.hscer.washington.edu/hubio546/.)

microtubule-associated protein Tau, which is normally involved in nutrient transport along axons, becomes hyperphosphorylated. Hyperphosphorylated Tau proteins bind to each other and subsequently form paired helical filaments, which are the primary components of NFTs.

15.2.2. Environmental Risk Factors

Aluminum (Al) has been studied intensely as a risk factor for AD. The initial hypothesis for a role of Al in AD originated in 1965, with the demonstration that injection of Al salts into the brains or cerebrospinal fluid of rabbits induced a progressive encephalopathy with a pathology reminiscent of NFTs. Subsequent evidence in support of a positive relationship between Al and AD consisted of pathological findings of higher Al concentrations in AD versus controls brains and ecologic and case–control studies linking aluminum in drinking water to AD. The issue overall has been highly controversial, and although there is some evidence indicating that Al can interact with Tau to form NFTs (Shin et al. 1995), more methodologically rigorous pathology studies have not shown elevations in brain Al concentrations in AD versus control brains (Bjertness et al. 1996). On the whole, the epidemiologic data are not convincing for an association of AD with aluminum, most notably because of inconsistent findings among studies and the lack of dose–response gradients (Muñoz 1998).

There is, however, increasing evidence that other metals such as copper, iron, and zinc, specifically their homeostatic disturbances that occur with aging, may be involved in AD. Copper appears to affect β-amyloid pathways in at least three ways: (1) in acidic conditions, copper promotes the formation of β-amyloid plaques; (2) in the presence of copper, β-amyloid produces hydrogen peroxide, a known generator of reactive oxygen species (ROS)—ROS have repeatedly been implicated in neuronal apoptosis in AD, as well as PD, which is an additional plausible mechanism of the neurotoxicity of β-amyloid; and (3) copper appears to lower the ability to clear β-amyloid deposits.

A report also showed that trace amounts of copper in drinking water induced β-amyloid aggregation in a rabbits fed high-cholesterol diets (Sparks and Schreurs 2003). The investigators noted senile plaque-like structures in the hippocampus and temporal lobe, and that the combination of copper and cholesterol significantly retarded the ability of rabbits to learn a difficult trace conditioning task. These observations suggest the possibility that copper in drinking water may be relevant to AD, but no epidemiologic evidence for such an association has yet been reported.

Iron participates in ROS production via the Fenton reaction. Some studies have also shown elevated levels of iron in AD versus control brains, as was noted for copper. Zinc, while not a transition metal capable of the same ROS production, is also present at higher levels in AD brains, implicating that it may be important as well (Finefrock et al. 2003). Overall, this evidence has prompted use of metal chelating agents in a clinical trial; initial results of this trial appear to be promising (Finefrock et al. 2003).

In 1995, Kukull et al. reported data showing that previous exposure to solvents is related to AD risk (Kukull et al. 1995). Although this finding has not been replicated consistently, solvents remain a plausible etiologic factor because many have neurotoxic properties, including CNS depression after short-term exposure. Exposure to electromagnetic fields (EMF) has also been studied in relation to AD, with some positive associations from cohort (Sobel et al. 1995) and case–control studies (Feychting et al. 2003), but the data are currently insufficient to make any conclusions.

Results from case–control studies have suggested that smoking is protective in AD. Cohort studies do not support this. Rather, results from these studies indicate that smoking is either unrelated to AD or results in modest risk increase (Kukull 2001).

15.2.3. Candidate Genes

15.2.3.1. Familial AD. Studies of the familial form of AD have identified four major loci: *APP*; *PS1*, which codes for presenilin 1; *PS2*, which codes for presenilin 2; and *ApoE*. *APP* mutations result in high levels of Aβ-42 peptide formation and subsequently amyloid plaques. Presenilins may play a role in controlling the activity of the secretase enzymes that cleave APP into peptide fragments. Apolipoprotein E (ApoE) is a very-low-density lipoprotein (VLDL) carrier that in part controls amyloid β accumulation or amyloid β peptide clearance (Morishima-Kawashima et al. 2000). In familial AD, presence of one (or two) variant allele(s) lowers the age of onset.

15.2.3.2. Sporadic AD. In the general population, three common *ApoE* alleles, ε2, ε3, and ε4, exist at varying frequencies. In addition to its role in amyloid β accumulation or clearance, studies of the protein products of the different *ApoE* alleles indicate that they differ with respect to ligand binding ability, catabolic rates, and in vivo plasma concentration, which probably explains their role in altered lipid and lipoprotein metabolism (Jarvik 1997). Results from meta-analysis show that *ApoE* genotype affects AD risk in a gene dose-dependent fashion (Farrer et al. 1997). Relative to ε3 homozygotes, carriers of one ε4 allele have a two- to threefold increase in disease risk, and ε4 homozygotes have about a 15-fold risk of AD. Additionally, the ε2 allele reduces risk of AD.

15.2.4. Gene–Environment Interactions with ApoE

On the basis of data showing different effects of *ApoE* genotype dependent on race/ethnicity, some have argued that in order for *ApoE* genotype to predispose to AD, some additional genetic or environmental risk factor must be present as well. Given that ApoE is a VLDL carrier, potential gene–environment interactions of *ApoE* with diet, specifically cholesterol, would seem to be a reasonable line of questioning. Evidence of such a gene–environment interaction would correspond well with that seen in cardiovascular disease, where *ApoE* is also a disease risk modifier dependent on lipid intake. Jarvik et al. (1995) reported that the relationship of *ApoE* genotype and AD was dependent on total cholesterol, as well as age and gender, indicating a gene–environment interaction. However, other investigators have not reported subsequent confirmation of this finding.

There are some epidemiologic reports of an association between head injury and AD. This may be consistent with the amyloid cascade theory of AD in that head trauma appears to increase expression of APP. Mayeux et al. (1995) have demonstrated that a synergistic interaction exists between head trauma and *ApoE*. More specifically, they reported a ten-fold increase in risk for the combination of head trauma and *ApoE* ε4, whereas *ApoE* ε4 alone gave only an increase of twofold risk. Head trauma without any *ApoE* ε4 alleles did not result in increased risk of AD.

With respect to smoking, results from a cohort study by Ott et al. (1998) show that the effect of smoking on AD risk may be dependent upon *ApoE* genotype as well. Current smoking was associated with a relative risk (RR) of 2.3 [95% confidence interval (CI): 1.3–4.1], and this effect was stronger among individuals without *ApoE* ε4 (RR = 4.6; 95% CI: 1.5–14.2). In contrast, there was no increase in risk associated with smoking among ε4 allele carriers (RR = 0.6; 95% CI: 0.1–4.8).

ApoE genotype has also been shown to influence the effectiveness of cholinesterase inhibition therapy, which is used to improve cholinergic tone in AD patients (e.g., with the drug tacrine) (Poirier et al. 1995). Continuing from the pharmacogenetic perspective, it would also be interesting to consider whether *ApoE* has a modulatory effect of metal chelation therapy currently on trial. Other pharmacogenetic studies have examined interactions with genes related to neurotransmission, neurotrophic factors, and inflammation, with mixed results (Maimone et al. 2001).

In summary, *ApoE* has been clearly and reproducibly linked to AD, and is related to both the sporadic and familial forms of AD. Genotype–phenotype correlation data provide a critical asset in the understanding of the role of *ApoE* in AD, including the plausibility of gene–environment interaction. Consequently many epidemiologic studies of AD now include *ApoE* as a primary risk or modifying factor. This will aid in confirming initial findings of the *ApoE*–environment interactions discussed here. Data regarding the role of environmental risk factors for AD, however, remain sparse. New evidence regarding a possible role of endogenous and perhaps also exogenous copper, iron, and zinc is accruing, and supports further exploration with respect to gene–environment interaction.

15.3. PARKINSON'S DISEASE

Parkinson's disease (PD) is a movement disorder characterized by the loss of dopamine-producing neurons in the substantia nigra region of the midbrain. It is the second most common neurodegenerative disease, with an incidence of 10–20 cases per 100,000 per year. There appears to be a modest gender difference in the incidence, with a 1.5–3-fold male preponderance, at least in Caucasians. Like AD, PD can manifest as both a rare, familial form and more common sporadic form of the disease. Familial PD onset is typically before age 50, whereas the sporadic form occurs predominantly at age 60 and older. Familial and sporadic PD have long been thought of as etiologically distinct disease entities, but newer evidence has challenged this notion considerably.

15.3.1. Pathogenesis

Intracytoplasmic inclusions known as *Lewy bodies* in dopaminergic neurons are the pathological hallmark of PD (Figure 15.3). Lewy bodies consist largely of proteins, especially α-synuclein and ubiquitin. Physiologic functions of Lewy bodies remain to be determined, although there is evidence that they may be either protective or neurotoxic, depending on the proposed mechanism of PD pathogenesis (Goedert 2001). Selective dopaminergic cell death results in diminished ability to control voluntary movement, and hence the cardinal symptoms for diagnosis include slowness of movement, rigidity, tremor at rest, and impaired balance or coordination.

PD can also be thought of as a disease of dopamine deficiency. Current therapeutic interventions hence include a drug for dopamine replacement

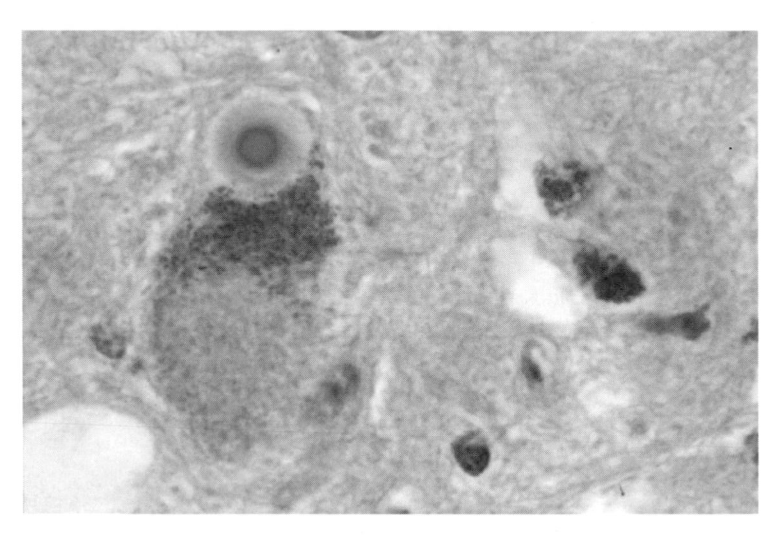

Figure 15.3. High-power micrograph from the substantia nigra of a patient with PD. Note the Lewy body in the upper left: a round hyaline inclusion with a pale surround. (From Rodney Schmidt, M.D., Ph.D., *Systematic Pathology: An Interactive Tutorial*, with permission. Copyright University of Washington, 1998. http://eduserv.hscer.washington.edu/hubio546/.)

(L-DOPA) and secondarily, drugs that inhibit the enzymes responsible for dopamine catabolism, specifically monoamine oxidase A and B (MAO-A and MAO-B). Inhibition of MAO also serves to limit the production of reactive oxygen species by MAO-mediated dopamine catabolisms, which are thought to contribute to the death of nigral neurons.

15.3.2. Environmental Risk Factors

Environmental factors have long been considered to be fundamentally important in PD. This idea is based primarily on two important findings in PD research: (1) twin studies have not identified a major role for genetics in sporadic PD (Tanner et al. 1999); and (2) a small cluster of cases of Parkinsonism was identified in which a contaminant of synthetic heroine, called MPTP, was shown to be causally related to the development of chronic Parkinsonism (Langston et al. 1983). These findings prompted many to look for other environmental risk factors for the disease. Pesticides, particularly the herbicide paraquat, based on structural similarity to MPTP, were intensely studied from this framework. There is some laboratory evidence suggesting that some pesticides can cause degeneration of the neurons in vitro or in animal models (Betarbet et al. 2000; Thiruchelvam et al. 2003). However, the epidemiologic evidence of such an association between pesticides and PD is not convincing, despite publication of an overall positive finding from meta-analysis (Priyadarshi et al. 2000), because specific categories of pesticide exposure, rather than a global category of exposure, have not shown strong associations and most studies have not employed rigorous exposure assessments.

Some pesticides (e.g., rotenone) are thought to cause neuron cell death by perturbing mitochondrial function, and consequent ROS production may also contribute to neurodegeneration in PD. Indeed, postmortem studies show increased indices of oxidative stress in PD brains, although the underlying cause has not been identified. Transition metals can also lead to the production of ROS through Fenton chemistry, and have thus been studied extensively in PD as well. For instance, Gorell et al. (1997) reported elevated risks of PD associated with occupational exposure to copper (OR = 2.5; 95% CI = 1.1, 5.9) and manganese (OR = 10.6; 95% CI = 1.1, 105.8). Additionally, higher risks were observed for long-term (≥20 years) exposure combinations of lead and copper (OR = 5.2; 95% CI = 1.6, 17.2), lead and iron (OR = 2.8; 95% CI = 1.07, 7.50), and iron and copper (OR = 3.7; 95% CI = 1.4, 9.7). Results from this study imply that chronic exposure to these metals is associated with PD, and that exposure combinations have a greater effect than does each metal alone. Powers et al. (2003) reported that iron and manganese in the diet can increase risk of PD as well. More specifically, comparing study subjects whose iron intake was in the highest quartile to those in the lowest quartile, they calculated an odds ratio of 1.7 (95% CI: 1.0–2.7) and also saw evidence of dose–response. Subjects with high dietary intakes of iron and manganese had roughly double the risk of PD of subjects with low intakes of each. Thus it appears that metals, whether from occupational exposure or diet, can increase the risk of PD, although more data, preferably from cohort studies, are necessary to solidify the evidence.

Interestingly, the most consistent evidence for a role of environment in PD comes from studies of smoking. Both case–control and cohort studies have consistently shown that smoking is associated with decreased risk of PD, approximately 40% less risk for ever smokers, 20% less for past smokers, and ~60% less for current smokers (Hernan et al. 2002b). Since cigarette smoking has been shown to decrease both MAO-A and MAO-B activity in vivo (Fowler et al. 1996), these findings have a plausible biologic basis, in that MAO inhibition would lead to less dopamine catabolism and less ROS production. Other work has also identified specific compounds in tobacco that can inhibit MAO at physiologically relevant concentrations (Khalil et al. 2000). There is also evidence that nicotine may have anti-apoptotic effects independent of MAO inhibition. Collectively, these findings have stimulated much new research on the role of tobacco smoking in PD, and have provided evidence that there may be several different mechanisms by which tobacco smoking alters PD risk, and hence one could assert that multiple gene–environment interactions in these pathways may exist.

15.3.3. Candidate Genes

15.3.3.1. Familial PD. Studies of different PD kindreds have identified several important genes (*PARK1–PARK10*), three of which have been extensively studied: *PARK1*, which codes for α-synuclein; *PARK2*, which codes for the protein known as parkin; and *PARK5*, which codes for ubiquitin carboxy-terminal hydrolase L1 (UCHL1). At least one of the rare familial mutations in *PARK1*, A53T, has altered functional characteristics that appear to favor fibrillization and accumulation of the α-synuclein protein into aggregates, a toxic gain of function. Collectively, studies on these three genes have shown that they are all involved in protein aggregation and proteosome dysfunction, prompting the theory that, like AD (in which Aβ-42 peptide aggregates), PD is also a disease of toxic protein aggregation (Trojanowski and Lee 2000). This theory has gained considerable weight and correlates well with the well-known pathological hallmark of PD, the Lewy body (Figure 15.3), which is an intracytoplasmic inclusion body observed in sporadic PD and some forms of familial PD, that is composed largely of aggregated α-synuclein (Goedert 2001).

15.3.3.2. Sporadic PD. Subsequent analysis of *PARK1* in population-based studies identified sequence variation in the 5′ region of the gene ("Rep1 polymorphism") that appears to modify protein expression (Chiba-Falek and Nussbaum 2001). In several case–control studies of PD, this polymorphism was associated with case–control status depending on *ApoE* genotype or other SNPs in *PARK1* (Kruger et al. 1999; Farrer et al. 2001).

Given the importance of dopamine in PD, many investigators have explored the role of polymorphisms in genes that influence dopamine disposition. Monoamine oxidase B, catechol *O*-methyl transferase (COMT), dopamine receptors, and the dopamine transporter have all been investigated in this framework. The gene-encoding monoamine oxidase B (*MAOB*), for instance, contains two intronic polymorphisms, for which conflicting evidence of association with PD exists. Results for other dopamine-related genes have been largely negative (Checkoway et al. 1998a; Tan et al. 2000).

15.3.4. Gene–Environment and Gene–Gene Interactions

Some studies have shown interactions of a *MAOB* intron 13 polymorphism with smoking (Checkoway et al. 1998b; Mellick et al. 1999), and others have shown gene–gene interactions with *COMT* (Wu et al. 2001); however, there are other reports showing no such association (Hernan et al. 2002a). A further complicating feature for interpretation is that the identified polymorphisms are in intronic regions, and no commonly occurring polymorphisms have been identified in coding or promoter regions (Costa-Mallen et al. 2004). Lacking a corresponding phenotype limits the ability of investigators to infer how such gene–gene and gene–environment interactions impact disease risk.

The occasional positive findings of a role of pesticides, metals, or other environmental risk factors prompted many to consider how polymorphisms in genes involved in environmental response could modify the effect of these exposures. Initial positive findings with *CYP2D6* led some to propose a theory of poor metabolic status and PD (Christensen et al. 1998). This hypothesis posits that individuals with polymorphisms in biotransformation enzyme genes may have a diminished ability to metabolize toxic xenobiotics, which could induce neurodegeneration of the substantia nigra. Many other cytochromes P450, N-acetyltransferases, and glutathione *S*-tranferases, as well as paraoxonase 1 (PON1), have been studied in relation to PD (Tan et al. 2000). Yet convincing evidence for gene–environment or gene–gene interactions with low penetrance polymorphisms in these genes is lacking.

There is evidence that transition metals (as well as some pesticides) can induce the fibrillization of α-synuclein (Uversky et al. 2001a, 2001b). Iron-induced aggregation of α-synuclein may be dependent on A53T genotype (Ostrerova-Golts et al. 2000). However, no data regarding gene–environment interaction between low-penetrance α-synuclein polymorphisms and any potential environmental risk factor have been gathered in the context of an epidemiologic study, but this does appear to be a promising avenue of future research.

In summary, α-synuclein (*PARK1*) appears to play a role in both familial and sporadic PD. Smoking is the environmental factor with the most consistent evidence in PD; new studies could further elucidate the potential roles of metals and pesticides—and studies of dopamine related genes have yielded interesting but inconsistent results.

15.4. AMYOTROPHIC LATERAL SCLEROSIS

Amyotrophic lateral sclerosis (ALS) is a disease of the motor neurons of the anterior horns of the spinal cord and motor neurons in the cerebral cortex. As with AD and PD, there are both familial and sporadic forms of ALS. The annual incidence of sporadic ALS is about one to two cases per 100,000; ALS appears to be slightly more common in men, and onset usually occurs in the middle to later years of life.

15.4.1. Pathogenesis

As a result of motor neuron cell death, muscle atrophy occurs, resulting in diminished muscle strength and bulk as well as fasciculations and hyper-reflexia. Effects on respiratory muscles can lead to pulmonary infection, and eventually amyotrophy leads to paralysis and death. In similarity to AD and PD, ALS also has characteristic cellular pathology. Intracytoplasmic inclusion bodies are found in affected tissues from ALS patients, and these are composed of neurofilaments and spheroids containing aggregated copper and zinc superoxide dismutase (CuZnSOD), an enzyme that catalyzes the conversion of superoxide (a ROS) to hydrogen peroxide.

15.4.2. Environmental Risk Factors

Most studies to date have not found a major role for environmental agents in the development of ALS (Nelson 1995; Mitchell 2000). Smoking appears to increase risk to a modest extent, as reported by Nelson et al. (2000) and Kamel et al. (1999), both case–control studies. Exposure to agricultural chemicals may be related to the development of ALS. Surveys of the prevalence of ALS among agricultural workers compared to the general population initially suggested an association, and subsequently a case–control study showed supporting evidence (McGuire et al. 1997), specifically among males.

Prompted by case reports of severe electrical shock and subsequent motor symptoms, interest in the role of electrical injury in ALS has increased over the last several years (as of 2005). Case–control studies provide some evidence for a role of electrical trauma, and a cohort study conducted in Sweden showed a modest effect of electrical and electronics work on ALS (OR = 1.4; 95% CI: 1.1–1.9) (Feychting et al. 2003). There have also been some reports of associations between exposure to heavy metals and ALS, particularly lead (Armon et al. 1991; Kamel et al. 2002). For example, Kamel and colleagues reported that occupational exposure to lead was associated with ALS (OR = 1.8; 95% CI = 1.0–3.1) among a population of cases and controls in New England (Kamel et al. 2002).

15.4.3. Candidate Genes

Evidence suggests that ~15–20% of familial ALS cases have mutations in the CuZnSOD gene (SOD1), and perhaps 2% of sporadic ALS cases do as well. Thus, two theories of CuZnSOD involvement in neurodegeneration have been advanced: either (1) mutant or wildtype CuZnSOD participates in deleterious redox reactions that affect motor neuron viability or (2) mutant or wildtype CuZnSOD becomes misfolded and oligomerizes into aggregates, forming inclusion bodies that are cytotoxic at some stage of their formation.

15.4.4. Gene–Environment Interaction

With respect to the role of lead in ALS, one case–control study examined interactions with two candidate genes, *ALAD* and *VDR*. *ALAD* codes for δ-aminolevulinic acid dehydratase, a heme synthesis enzyme strongly inhibited by lead; *VDR* encodes the vitamin D receptor, which may mediate lead's interfering effects on calcium homeostasis. No evidence of interaction was found (Kamel et al. 2003). Thus, gene–environment interaction in ALS, if it exists, remains to be uncovered.

15.5. COMMON THEMES

By now, it is probably clear that AD, PD, and ALS share some common characteristics. According to Trojanowski and Lee (2000), these diseases likely share a common pathogenic sequence involving interactions between normal and abnormal proteins, assembly into filaments, and subsequent accumulation to form toxic proteinaceous deposits. Interactions of the critical disease proteins, Aβ and Tau in AD, α-synuclein in PD, and CuZnSOD in ALS, with environmental or endogenous agents (e.g., neurotransmitters) may also explain the selectivity of neuronal loss. Thus, identifying which interactions and which specific forms of protein conformations are most toxic is critical, and has clear therapeutic implications (Volles and Lansbury 2003).

There may also be overlap in the role of protein aggregation between these diseases. Consider, as an example, the *Tau* gene (chromosome 17q21.1), which codes for the microtubule-associated protein that is critically involved in the formation of paired helical filaments that proceed to form the NFTs observed in AD. Some studies have shown that *Tau* SNPs are also important in certain types of PD cases (Martin et al. 2001; Farrer et al. 2002). Mutations in *Tau* are also involved in frontotemporal dementia linked (linked to the region of chromosome 17 where *Tau* is), a rare autosomal dominant disorder also characterized by abnormal protein aggregation. Perhaps, then, one can use evidence gathered about one disease to generate hypotheses for the other two. Additionally, if the pathological processes are similar across AD, PD, and ALS, then preventive strategies that would apply to all might be revealed through further study of these processes. Besides abnormal protein–protein interactions, oxidative stress and inflammation are two other pathological processes that are common to these diseases, either as initiators or perhaps more likely in response to initial cellular injury. Gene–disease and gene–environment interactions may therefore be similar with respect to these etiologic pathways as well.

Moreover, studies of the rare, familial forms of these diseases have given tremendous new insights into disease etiology (Table 15.1). These influence not only how one thinks of the etiology of the familial form but also the more common sporadic form. The same *ApoE* polymorphism that affects age of onset in familial AD cases is also associated with AD risk in sporadic cases, implying that the etiology of the two forms may have common

Table 15.1 Genes Implicated in Both the Familial and Sporadic Forms of AD, PD, and ALS

Disease	Gene	Familial Phenotype	Sporadic Phenotype	Reference(s)
AD	*ApoE*	Lower age of onset with ε4 allele(s)	Increased disease risk and lower age of onset with ε4 allele(s)	(Farrer et al. 1997)
PD	*PARK1*	Autosomal dominant disease	Possible Rep1 polymorphism effect	(Kruger et al. 1999; Farrer et al. 2001)
	PARK2	Autosomal recessive juvenile disease	Unclear, possible role for promoter polymorphisms	(West et al. 2002)
	PARK5	Autosomal dominant disease	Lower disease risk among 18Y vs 18S carriers	(Maraganore et al. 1999)
ALS	*SOD1*	Young onset ALS	Unclear; remains to be determined	(Andersen 2001; Alexander et al. 2002)

features. α-Synuclein in PD is perhaps a subtler example of this. Mutations in the α-synuclein gene, *PARK1*, were originally identified in two kindreds with Mendelian PD inheritance and α-synuclein was soon shown to be a major component of Lewy bodies. Not long after, studies showed associations of a common variant in *PARK1*, the Rep1 polymorphism, with sporadic PD. It now appears that levels of α-synuclein expression are important in PD. This initial hypothesis derived from results of familial PD studies. The role of mutations in *SOD1* in familial ALS has also lead to new hypotheses regarding the role of other *SOD1* mutations in sporadic ALS, although the relationship remains to be fully determined. In the near future, then, candidate genes for gene–environment interaction in the common sporadic forms diseases studies will likely derive from findings gathered in the study of respective familial forms.

15.6. ISSUES TO CONSIDER

In total, there is not much in the way of convincing evidence of gene–environment interaction with the typical environmental risk factors studied in these diseases. Given that most believe these diseases result from complex interactions between genes and environment, why have investigators not been able to identify and quantify them? Perhaps these interactions are exceptionally hard to identify, or perhaps there have been shortcomings of studies conducted to date.

There are several important issues that can significantly affect study results. First, for many neurodegenerative diseases, diagnosis is not straightforward until the condition is advanced, thus complicating the estimate of the time of onset and associating onset with exposure. There can also be study-to-study differences in diagnostic criteria employed. Second, the lack of population-based registries in most places, compared to cancer registries, which are readily accessible, makes neurodegenerative disease cases difficult to ascertain (especially new onset). Subsequently, obtaining accurate information from study subjects is often difficult, and subject to potential recall bias, and can be even harder because of dementia. Reliance on proxy respondents represents an obvious problem as well (see Chapter 4).

The next issue to consider is potentially the most difficult. It is possible, perhaps even likely, that a neurodegenerative disease manifests in one person because of some genetic cause, whereas another case may be due to a gene–environment interaction involving a completely different gene and pathobiological pathway. This is referred to as *etiologic heterogeneity*. An investigator who designs a study to assess the effect of a polymorphism in one gene may be addressing disease causation in only a fraction of the total cases. Thus evidence of the hypothesized effect would be diluted among a pool of cases for whom potentially many different causes exist. The issue of etiologic heterogeneity again points to the complexity of disease and has convinced some to employ different methods that might be able to tease apart the relative contributions of certain factors in some cases and different factors in other cases. In one example, a team of PD investigators decided to test whether gender modified an interaction between *MAOB* and smoking, since there is a gender difference in the incidence of disease. Indeed, they found that the interaction of smoking and *MAOB* exists only among men (Kelada et al. 2002). For women, then, other factors, perhaps hormones, must be playing a role in the causation of disease. Alternatively, other investigators have used different analytical approaches to identify gene–environment interactions that underlie etiologic heterogeneity. For example, Maraganore et al. used a recursive partitioning method and found that *UCHL1* polymorphisms interact with the Rep1 polymorphism in *PARK1* to modulate PD risk among women but not among men (Maraganore et al. 2003). Given the lack of solid mechanistically based hypotheses and exposure assessments, these types of approaches can be used to generate new and interesting hypotheses to test.

Gene or locus heterogeneity can also be a complicating factor. A particular polymorphism may be measured in a study, but study subjects may have other polymorphisms that affect gene function and can therefore also be important in disease causation. There are multiple sequence variations in *SOD1* and *PARK2*, for instance, and thus investigators need to assess the presence and genotype–phenotype relationships of each mutation in ALS and PD cases, respectively. Not doing so could lead to a bias in the estimate of effect of gene–disease association and/or gene–environment interaction for any single variant measured.

From the study design perspective, several important issues need to be considered as well. First, exposure assessment is of paramount importance in studies of gene–environment interaction. Typically, efforts aim to characterize type, duration, intensity, and timing of exposure. Exposure misclassification is a major concern, since it can bias the estimate of the effect of exposure and the joint genotype–environment effect (Rothman et al. 1999). Obviously this is more easily done in the pharmacogenetic setting.

The selection of appropriate candidate genes is another critical methodological step. Consider, as an example, that since the late 1990s, many PD investigators have taken the approach that general biotransformation genes or genes involved in mediating oxidative stress may be important (because exposure and oxidative stress, respectively, appear to be important in the disease). To conduct them without an exposure assessment to a putative biotransformation enzyme substrate or agent that induces oxidative stress, which was routinely done in PD (and many other diseases for that matter) constitutes a limited approach from the ecogenetics perspective. While such studies can be viewed as hypothesis generating (especially considering probable "endogenous" exposures), more enlightening studies use exposure data to drive the selection of candidate genes. (Another benefit is that this type of study design allows for analysis of the independent and joint effects of a genetic polymorphism and environmental risk factor.) Additionally, focusing on polymorphisms in selected genes for which intermediate phenotype data exist is advantageous in that it allows the investigator to make better a priori hypotheses about which variants might result in altered disease risk. To that end, there is a critical need to generate genotype–phenotype data to aid in the development of better epidemiologic studies.

15.7. CONCLUSIONS

The study of gene–environment interactions in neurodegenerative diseases has provided interesting but inconsistent findings. Future research efforts employing more robust study strategies and designs may offer new insights into the potential role of low-penetrance genetic polymorphisms and interactions with exposures. Research aimed in this direction would benefit greatly from the establishment of cooperative research centers, to ascertain greater numbers of cases and controls, as well as AD, PD, and ALS disease registries. Additionally, thorough exposure assessments must become a part of the methodology of gene–environment interaction studies; collaboration with exposure assessment experts is one solution to this problem.

Future studies should aim to identify environmental and genetic factors that promote or decrease putative protein–protein interactions, which could influence the development of all three diseases. There has been an impetus to develop therapeutic strategies that slow down the development or toxicity of filamentous protein aggregates. In this way, pharmacogenetic studies assessing the role of polymorphisms in related genes should shed light on gene–environment interaction as was well. Additionally, as cell death in these diseases appears to be at least in part apoptotic, consideration of the role of sequence variation in genes related to be apoptosis could prove to be a fruitful line of questioning as well.

16

Cardiovascular Disease

Melissa A. Austin and Stephen M. Schwartz
University of Washington, Seattle, WA

16.1. INTRODUCTION

Cardiovascular disease (CVD) is the leading cause of illness and death in the United States. It has been estimated that there are 62 million people with CVD in this country, and approximately 946,000 deaths were attributable to CVD in 2000, accounting for 39% of all deaths (National Heart Lung and Blood Institute 2002). Coronary heart disease (CHD) is the most common form of CVD, including myocardial infarction (MI) or heart attack. In 2003, approximately 650,000 Americans had their first heart attack, and another 450,000 experienced a recurrent CHD event (American Heart Association 2002). The cumulative risk of CHD by age 70 is 35% in males and 24% in females. Thus, CVD is a major public health concern in the United States. At least one study has determined that CHD itself is influenced by genetic factors among both men and women at younger ages, although the effects may decrease at older ages (Marenberg et al. 1994).

CVD is manifest as a variety of clinical conditions, including CHD (which itself includes myocardial infarction (MI), unstable angina, and other clinical manifestations), cerebrovascular disease (stroke), venous thrombosis, and peripheral vascular disease. The pathophysiological processes underlying these clinical events are also varied, and include lipid metabolism, inflammatory responses, oxidative stress, endothelial function, platelet function, thrombosis and hemostasis, homocysteine metabolism, insulin sensitivity, and hypertension (Scheuner 2003), as reflected by the

Gene-Environment Interactions: Fundamentals of Ecogenetics, edited by Lucio G. Costa and David L. Eaton

many intermediate phenotypes that can often be used as biomarkers for disease risk. Thus, not only is CVD is common, but its etiology is extremely complex.

Further, each process underlying CVD risk is regulated by structural proteins, enzymes, receptors, and ligands, all of which are encoded by genes. Data combined from 14 randomized clinical trials (Khot et al. 2003) and from three large observational studies (Greenland et al. 2003) have demonstrated that the conventional CVD risk factors—high cholesterol, high blood pressure, and cigarette smoking—explains the occurrence of CHD in 80–90% of patients. However, most individuals who have these risk factors do not develop CHD. Although environmental and behavioral risk factors for CVD, including dietary fat, smoking, alcohol consumption, and lower exercise levels, are well established (Haskell 1986; Doll et al. 1994; Savolainen and Kesaniemi 1995; Ginsberg et al. 1998), even these conventional risk factors may be genetically influenced (Austin et al. 1987; Heller et al. 1993). Thus, most of CVD is likely the result of environmental exposures acting on a background of inherited susceptibility (Marenberg et al. 1994).

The contribution of specific inherited susceptibility factors to CVD risk has been delineated for only a few, relatively rare settings. For example, familial hypercholesterolemia and familial defective Apo B-100 are caused by mutations in the low-density lipoprotein receptor (LDLR) and the apolipoprotein B-100 gene, respectively, and are associated with substantially increased risk of CHD (Austin et al. 2004). Similarly, other single-gene defects are known to cause hypertension and arrhythmias (Nabel 2003). These genetic disorders, however, explain only a small proportion of CVD in the general population.

Sing and colleagues have argued that understanding CVD etiology requires a framework that organizes an immense amount of information about extremely complex phenotypes (Sing et al. 2003). They note that CVD, like other complex diseases, is a consequence of many intermediate biochemical and physiological systems that interact nonlinearly, that are population-specific, and that change over time. Finally, they emphasize that equal weight must be placed on understanding the environmental risk factors and how they interact with genetic variants to influence CVD risk. In addition to increasing our understanding of disease etiology, examining associations between combinations of environmental risk factors and genetic variants will potentially enhance the predictive value of risk factor information, and may inform development of preventive strategies and effective therapeutic measures.

In this chapter, we focus on examples to illustrate the importance of gene–environment interactions in elucidating risk for CVD. We examine several traditional epidemiologic research study designs, including both observational and experimental studies, and demonstrate how these approaches can be used. We conclude with an overview and examples of several special topics, including pharmacogenomics, the methodological importance of distinguishing additive and multiplicative interaction effects on disease risk, and the need to incorporate emerging information on characteristics of the human genome into these studies.

16.2. RESEARCH APPROACHES TO UNDERSTANDING GENE–ENVIRONMENT INTERACTIONS IN CVD

16.2.1. Case–Control Studies

The challenges of mapping causal genes for common, complex diseases such as CVD, are now well recognized (Page et al. 2003). Many investigators are using large-scale, case–control studies to identify polymorphisms that contribute to disease risk. In perhaps one of the largest such studies to date, Yamada and colleagues measured 179,402 genotypes for 112 polymorphisms in 71 candidate genes based on their potential role in atherosclerosis, hypertension, diabetes mellitus, or hyperlipidemia (Yamada et al. 2002).

In a two-stage design, they first examined the association of these variants with myocardial infarction among a subset of 909 study subjects in a "screening study." Using a p value of 0.1, they then selected 19 polymorphisms in men and 18 in women for the large-scale study in the remaining 4152 study subjects. The results, based on multivariate logistic regression analysis adjusting for age, body mass index, smoking status, and other risk factors, are summarized in Table 16.1 under the assumption of a dominant effect of the variant allele. (The assumption of a dominant effect implies that the risk of disease is increased among carriers of the mutant allele, whether heterozygotes or homozygotes.)

Among men, only the C1019T variant in the connexin 37 gene, which is involved in communication between vascular endothelial cells, was associated with increased risk of MI. Among women, two variants, one in the plasminogen activator inhibitor type 1 gene and one in the stromelysin-1 gene (also known as *matrix metalloproteinase-3*), were associated with risk. Although this study likely minimized potential population stratification (confounding due to population substructure) since it was based on a homogeneous ethnic group (Thomas and Witte 2002; Wacholder et al. 2002), it is somewhat surprising that so few variants were found to be significant risk factors, and that such striking sex differences were seen. Importantly, however, this study did not consider gene–environment interactions that

Table 16.1 Association of Polymorphisms and Myocardial Infarction in Japanese Subjects Based on the "Large-Scale" Study of 4,152 Subjects

	Gene	Polymorphism	Odds Ratios (95% Confidence Interval)
Men ($N = 2858$)	Connexin 37	C1019T (Pro391Ser)	1.4 (1.1, 1.6), $p < 0.001$ for CT plus TT genotypes with CC genotype
Women ($N = 1294$)	Plasminogen activator inhibitor type 1	4G-668/5G	1.6 (1.2, 2.1), $p < 0.001$ for 4G/5G plus 5G/5G genotypes with 4G/4G genotype
	Stromelysin-1	5A-1171/6A	4.7 (2.0, 12.2), $p < 0.001$ for 5A/6A plus 6A/6A with 5A/5A genotype

Source: Adapted from Yamada et al. (2002).

could mask important associations, and perhaps even explain the sex-specific relationships.

Perhaps one of the most important environmental risk factors that could interact with genetic susceptibility for CVD is cigarette smoking. Smoking approximately doubles the risk of CHD (Doll et al. 1994), probably through several different mechanisms, including direct damage of the vascular endothelium, perturbing lipid metabolism, and inducing inflammatory response (Talmud and Humphries 2002). Identifying whether there are subsets of cigarette smokers who are at increased CVD risk due to genetic susceptibility is of considerable interest.

The potential importance of smoking–gene interactions was clearly demonstrated in a case–control study of MI and factor V Leiden (resistance to activated protein C) among young women (Rosendaal et al. 1997). Factor V Leiden (FV 506Q) is the most common, hereditary procoagulation disorder, with a prevalence of heterozygous carriers of approximately 3–5% in European-descent populations. In this population-based study conducted in three counties in the state of Washington, 84 incident MI cases among women less than age 45, and 388 age-matched controls identified by random digit dialing, were genotyped. This study population was especially well suited to examine smoking–gene interactions since 74% of the cases were current, regular smokers at the time of their MI. Among all women, the odds ratio (OR) for factor V Leiden carriers was 2.4 [95% confidence

Table 16.2 Association of Cigarette Smoking and Factor V Leiden on Risk of Myocardial Infarction in Young Women

	Odds Ratio (95% Confidence Interval)[a] Factor V Genotype	
Current Smoker	Wildtype	Leiden
No	1.0	1.1 (0.1, 8.5)
Yes	9.0 (5.1, 15.7)	32.0 (7.7, 133)

[a] Odds ratios relative to reference category of nonsmokers with factor V wildtype genotype.

Source: Adapted from Rosendaal et al. (1997).

interval (CI) = 1.0, 5.9]. However, when the sample was stratified by smoking status, major differences in risk were seen (Table 16.2). Smoking alone was associated with a ninefold increased risk of MI, whereas factor V Leiden genotype alone was not related to increased risk. However, the combined effect of smoking and factor V Leiden genotype resulted in highly significant odds ratio of 32.0, much greater than would be expected on the basis of the multiplicative combined effects of smoking and factor V Leiden ($1.1 * 9.0 = 9.9$; see Section 16.3.2, below). Thus, with respect to MI risk in young women, smoking appears to be a "prerequisite for the risk-increasing effect for factor V Leiden" (Rosendaal et al. 1997; Tanis et al. 2003).

16.2.2. Prospective Studies

Unlike case–control studies, prospective studies can measure the absolute incidence of CVD, and ensure that the occurrence of risk factors (including the measurement of biomarkers) precede onset of disease. Germline genetic polymorphisms, of course, do not change with time, so the primary advantage of prospective studies over case–control studies derives primarily from their ability to (1) include individuals who develop disease endpoints regardless of whether the disease is fatal and (2) reduce the influence of measurement error (particularly differential measurement error in cases and noncases) of nongenetic characteristics (e.g., lifestyle characteristics).

Many studies have investigated the role of apolipoprotein structural genes for their effects on plasma lipid risk factors and their relationship to risk of CVD (Wilson et al. 1997). One of the best-documented associations is that 5–8% of interindividual variation in total cholesterol levels are attributable to the ε2, ε3, and ε4 alleles of the APOE gene that are deter-

Table 16.3 Mean Baseline Blood Cholesterol Values (mmol/L) among Men by APOE Genotype and Smoking Status

	APOE Genotype			
Smoking Status	ε3/ε3	ε2/ε2 or ε2/ε3	ε4/ε4 or ε4/ε3	*p* Value[a]
Nonsmokers	5.80	5.37	5.91	<0.0001
Smokers	5.77	5.42	5.83	<0.005

[a] Based on analysis of variance with adjustment for age and medical practice.

Source: Adapted from Humphries et al. (2001).

mined by polymorphisms at amino acids 112 and 158 of this gene (Sing and Davignon 1985; Boerwinkle and Utermann 1988; Kaprio et al. 1991). The APOE ε4 allele in particular has been associated with an increased risk of CHD in multiple studies (Eichner et al. 2002).

In a study of 2052 men ages 50–61 years free of CHD at enrollment in the Northwick Park Heart Study, potential interactions between baseline smoking status and APOE genotypes were examined in relation to CHD at follow-up (fatal and nonfatal coronary heart disease event and nonfatal MI) (Humphries et al. 2001). To ensure sufficient sample sizes in each genotype group, the ε2/ε2 and ε2/ε3 genotypes were combined to constitute APOE ε2 carriers, ε3/ε4 and ε4/ε4 genotypes were combined to constitute APOE ε4 carriers, and the ε2/ε4 subjects were excluded from the analyses. As shown in Table 16.3, this study confirmed many previous findings of significant differences in plasma cholesterol among the genotype groups, with increased levels among the ε4 carrier group (Eichner et al. 2002). This relationship was independent of smoking status.

However, during 18,836 person-years of follow-up there was an interaction between baseline smoking status and APOE genotype in relation to incident CHD. The strongest effect was seen among smokers, with those who carried APOE ε4 having nearly a threefold increased risk, even after adjustment for a variety of other risk factors including lipid levels (Table 16.4). In contrast, there was no increased risk of CHD among exsmokers who were APOE ε4 carriers. These findings illustrate the necessity for considering genetic, environmental, and behavioral factors in understanding the etiology of CVD.

16.2.3. Experimental Studies

In the preceding observational study designs, investigators do not control whether an individual is exposed to the factors under study. Experimental study designs, by virtue of random assignment of exposures, are particularly

Table 16.4 Risk of Coronary Artery Disease among Men by APOE Genotype and Smoking Status

	Adjusted Hazard Ratio (95% Confidence Interval)[a]		
Smoking Status	ε3/ε3	ε2/ε2 or ε2/ ε3	ε4/ε4 or ε4/ε3
Exsmokers	1.5 (0.9, 2.4)	0.5 (0.1, 1.9)	0.7 (0.4, 1.6)
Smokers	1.5 (0.9, 2.5)	0.9 (0.3, 2.4)	2.8 (1.6, 4.9)

[a] Hazard ratios relative to reference category of never smokers with all Apo E genotypes combined; adjusted for medical practice, age, systolic blood pressure, and plasma levels of cholesterol, triglyceride, and fibrinogen.

Source: Adapted from Humphries et al. (2001).

robust to sources of systematic error. Although genetic factors cannot be assigned, individuals can be randomized to be exposed to nongenetic factors hypothesized to interact with genetic factors. Such designs are particularly valuable not only because the possibility of confounding is greatly reduced, but also because the exposure to the nongenetic factor can be extremely well characterized. Further, in some settings, subjects can serve as their own controls, adding further robustness to the conclusions regarding the genetic factors.

For example, Friedlander et al. (2000) sought to determine whether the well-documented changes in lipid levels following increasing intake of saturated fat and cholesterol varies according to carriership of alleles in several genes known to be involved in cholesterol metabolism (APOB, CETP, and LPL) (Friedlander et al. 2000). They conducted a crossover design in which 214 individuals alternately were administered, for 4 week periods, either high-saturated-fat/high-cholesterol diets or low-saturated-fat/low-cholesterol diets that were prepared by the researchers. The order of the diets was randomly assigned for each person, and between the two diets the subjects followed their usual diets. Serum total cholesterol, high-density lipoprotein (HDL) cholesterol, low-density lipoprotein cholesterol, and triglyceride were measured multiple times during the trial so that changes attributable to the diets could be assessed. The authors reported that, in response to the high-saturated-fat/high-cholesterol diet, carriership of certain APOB alleles was significantly associated with increasing serum total cholesterol and LDL cholesterol, and carriership of certain LPL alleles was significantly associated with increasing triglyceride and reduced HDL cholesterol (sometimes called the "good" type of cholesterol). Carriers of other alleles in these genes, however, did not experience adverse changes in serum lipoproteins or triglyceride. In addition to helping to clarify mechanisms through which intermediate CVD phenotypes are affected by environmental factors, experimental studies such as that of Friedlander et al. (2000) have the potential to provide particularly novel evidence as to the value of recommending individualized diet, exercise, and other lifestyle choices based on genomic information.

16.3. SPECIAL TOPICS IN CARDIOVASCULAR ECOGENETICS

16.3.1. Pharmacogenomics

Drugs are used to manage an increasingly large proportion of CVD risk factor levels—hypertension, hypercholesterolemia, and insulin resistance—in CVD patients as well as adults in the general population. It has been known for decades that interindividual variation exists in the efficacy of pharmaceutical agents, and a substantial component of such variation is due to genetic factors. In some instances the variation produces unwanted side effects, whereas in others, therapeutic effects are enhanced or are blunted. Identifying specific genetic variation that causes individuals to differ in response to the same therapy holds the potential for greater individualization of medical care, for example, by altering doses or choosing different classes of medications on the basis of genetic information, to ensure that harm is reduced and benefits are maximized (Mukherjee and Topol 2002). Pharmacogenomics, the study of how a human genome influences that person's response to drugs, is thus a subdiscipline of ecogenetics in which the "environmental" exposures are pharmaceutical agents.

For example, intraindividual variation in response to antihypertensive medications is of tremendous research and clinical interest by both academic and pharmaceutical groups. Blood pressure homeostasis is influenced strongly by components of the renin–angiotensin system (Sealey and Laragh 1990). Angiotensinogen (ATG), a polypeptide produced by the liver under influence of a variety of endogenous and exogenous stimuli, is cleaved by renin to produce angiotensin I, which in turn is activated through cleavage in the kidney by angiotensin-converting enzyme (ACE) to angiotensin II. Angiotensin II causes vasoconstriction, which increases blood pressure, and treatment with ACE inhibitors causes blood pressure reduction and reduced risks of CVD among hypertensives. ATG contains a Met235Thr polymorphism, and homozygosity for the Thr allele is associated with an increased risk of hypertension (primarily in Caucasian populations) (Sealey and Laragh 1990) but greater efficacy of ACE inhibitors in reducing blood pressure (Hunt et al. 1998). Evidence that this polymorphism is associated with a reduced risk of CVD among hypertensive individuals treated with ACE inhibitors specifically was reported in a case–control study of incident nonfatal stroke ($n = 118$) and MI ($n = 208$), and approximately 700 age- and sex-matched controls (Bis et al. 2003). Patients receiving ACE inhibitors who carried two copies of the AGT235Thr allele had a reduced risk of stroke (OR = 0.37, 95% CI = 0.14, 0.99), whereas carriers of the AGT235Met allele were not at reduced risk (OR = 1.4, 95% CI 0.9, 2.4), although this analysis was not adjusted for smoking. The authors suggested that no similar interaction was found with MI because hypertension may be a more dominant risk factor for stroke than CHD. This study provides an example of the types of data that will eventually be required to determine whether clinical prescribing practices will need to be altered for individuals carrying genotypes that influence the actions and metabolism of particular antihypertensive medications. Assessment of gene–drug interactions in the context of experimental study designs (e.g., as part of clinical trials of drug efficacy in preventing clinical outcomes) will be particularly important for making progress in this field.

16.3.2. Additive and Multiplicative Models of Interaction

At least six qualitative patterns of gene–environment interactions have been described (Khoury et al. 1993). From an epidemiologic viewpoint, however, two factors are said to be involved in a statistical interaction [also termed "effect measure modification" (Rothman 2002)] if the combination of the two are associated with a greater (or smaller) risk compared to what would be expected under a particular statistical model of independent associations. The most common model (or "scale") against which interactions are assessed are additive effects and multiplicative effects. Interactions on the additive scale ("departures from additivity") would assess how the absolute excess risk of disease due to an environmental exposure varies according to genotype, whereas interactions on a multiplicative scale ("departures from multiplicativity") would assess the same relationships with respect to the relative risk of disease.

For the simple situation where both environmental and genetic factors are dichotomous, assessments on both scales can be made using the same data from either case–control or cohort studies. As shown in Table 16.5,

Table 16.5 Relative Risk Data Layout for Evaluating Gene–Environment Interactions

	Environmental Exposure	
Genetic Factor	No	Yes
No	1	RR_{E+}
Yes	RR_{G+}	RR_{G+E+}

relative risks (RR) are estimated for study subjects with only the genetic variant of interest only (RR_{G+}), only the environmental exposure (RR_{E+}), or both (RR_{G+E+}), each relative to persons who lack both factors.

In evaluating additive interactions, the question of interest is whether the risk for those subjects with both the genetic variant and the environmental exposure is higher than the sum of the risk for the environmental exposure and the genetic variant alone. Using the nomenclature in Table 16.5, it can be shown that additive interaction is present when $RR_{G+E+} > RR_{G+} + RR_{E+} - 1$ (Rothman 2002) and multiplicative interaction is present when the RR for individuals with both the genetic variant and the environmental exposure is greater than the product of the RR for the genetic variant alone and the environmental exposure alone (i.e., when $RR_{G+E+} > RR_{G+} \times RR_{E+}$). Although it is technically easier to evaluate departures from multiplicativity because they can be assessed in multivariate logistic regression models (which by default assume a multiplicative relationship between genetic and environmental factors), assessment of departures from additive relationships is generally viewed as necessary to determine whether there are excess cases of a disease that occur specifically as a result of the interaction (Rothman 2002).

The importance of considering the underlying model against which gene–environment interactions are tested (i.e., additive vs. multiplicative) is illustrated by a study of the factor V Leiden variant, use of oral contraceptives (OC), and risk of deep vein thrombosis (DVT) (Vandenbroucke et al. 1994) (Table 16.6). Factor V Leiden carriership is associated with an

Table 16.6 Association between Factor V Leiden, Oral Contraceptive Use (OC) Use, and Deep-Vein Thrombosis (DVT)

Factor V Leiden	OC Use	RR[a] (95% CI)	RR[b] (95% CI)	Absolute Risk Difference[c]
Negative	No	1.0	1.0	
				2.2
	Yes	3.7 (2.2, 6.1)	3.7	
Positive	No	1.0	4.0	
				22.8
	Yes	5.0 (0.8, 31.8)	19.8	

[a] Relative risks (RR) for OC use (vs. nonuse) estimated separately for women with and without factor V Leiden.

[b] RR for individuals who both used OCs and carried factor V Leiden, used OCs and did not carry factor V Leiden, and did not use OCs but carried factor V Leiden, each relative to individuals who neither used OCs nor carried factor V Leiden.

[c] Risk of DVT per 10,000 women-years in OC users minus risk of DVT in OC nonusers, separately for individuals with and without factor V Leiden.

Source: Adapted from Vandenbroucke et al. (1994).

approximately four- to fivefold increased risk of DVT in both OC users and OC nonusers (see RR, footnote *a* in Table 16.6). Because RR_{G+E+} (19.8) is similar to $RR_{G+} \times RR_{E+}$ ($3.7 \times 4.0 = 14.8$), multiplicative interaction is not present. However, because the absolute incidence of DVT is higher among OC users compared to OC nonusers, the same relative increase in risk of DVT associated with OC use in factor V Leiden carriers and noncarriers translates into a larger absolute risk difference due to OC use among factor V Leiden carriers; that is, RR_{G+E+} (19.8) is almost 3 times larger than $RR_{G+} + RR_{E+} - 1$ (6.7). In short, assessing the interaction between OC use and factor V Leiden by comparing RRs across subgroups (which corresponds to looking for departures from predicted multiplicative effects) fails to identify an important interaction between these characteristics; excess DVT cases occur among women who are both OC users and factor V Leiden carriers compared to what would be expected according to the sum of the individual effects of these factors. Such information could potentially be used, for example, in counseling factor V Leiden carriers as to the risks and benefits of choosing OCs for birth control as opposed to other available methods.

16.3.3. Linkage Disequilibrium and Haplotype Blocks in the Genome

A growing body of evidence demonstrates that a block-like structure of linkage disequilibrium (LD) exists in the human genome (Gabriel et al. 2002; Wall and Pritchard 2003). In other words, there are DNA sequences in which there is strong LD, separated by hot spots of recombination (Goldstein 2001), and there are a limited number of haplotypes within each region of LD. This has been convincingly demonstrated in a study of the sequence of the entire length of chromosome 21 (Patil et al. 2001). These studies suggest that such regions, or "haplotype blocks," may extend for 10–100 kb in the genome, facilitating investigation of causal disease variants using genetic association approaches within these linked regions (Weiss and Clark 2002). Further, it appears that the boundaries of haplotype blocks are highly correlated across populations (Gabriel et al. 2002). The goal of the international HapMap project is to develop a haplotype map of the entire human genome to characterize these patterns of human DNA sequence variation (The International HapMap Consortium 2003). It will be a key resource of researchers for finding genes that affect health, disease, and response to drugs and environmental factors.

The APOE gene is located in just such a genomic region of linkage disequilibrium that includes the APOC1, APOC4, and APOC2 genes, and is characterized by strong LD that is likely to contain haplotype blocks. Thus, it is difficult to determine which portion of this complex may be functional. For example, one study suggested that the variation in the APOE gene may influence triglyceride levels (Lussier-Cacan et al. 2002). However, another study of five variants in the closely linked APOC4 gene has also reported an association with triglyceride levels, at least in women (Kamboh et al. 2000). Thus, more extensive studies of the genomic characteristics of this region in different populations, as well as functional genomic investigation in human and nonhuman systems, will be needed to determine the specific nucleotide variation that is associated with variation in intermediate and clinical CVD phenotypes (Page et al. 2003).

It is also now clear that the extent of genomic variation, even within an individual candidate gene, is much more extensive and diverse than previously recognized. This is illustrated by the assumption for many years that functional variation in the APOE gene was fully characterized by the ε2, ε3, and ε4 alleles. More recently, a survey of 72 individuals from two different ethnic groups identified 21 single nucleotide polymorphisms (SNPs) in the APOE gene alone, 14 of which had not been previously reported (Nickerson et al. 2000). Thus, epidemiologic association studies must develop a strategy for not only selecting SNPs that may be potentially involved in disease susceptibility (Calafell 2003) but also for determining whether they interact with environmental factors. Several such methods are under development (Li et al. 2003b; Stram et al. 2003a, 2003b; Carlson et al. 2004). These and new approaches will be increasingly useful in understanding genetic susceptibility to common diseases.

16.4. SUMMARY

CVD is the leading cause of illness and death in the United States. It is manifest as a variety of clinical conditions, including coronary artery disease, stroke, and peripheral vascular disease. The pathophysiological processes underlying these clinical events are also varied, and each process is regulated by structural proteins, enzymes, receptors, and ligands, all of which are encoded by genes. Thus, not only is CVD common, but its etiology is extremely complex. This chapter focused on examples to illustrate the importance of gene–environment interactions in elucidating risk for CVD, including traditional observational and experimental epidemiologic research study designs, and has demonstrated how these approaches can be used. Additional special topics are considered, including pharmacogenomics, additive versus multiplicative interaction effects on disease risk, and the need to incorporate emerging information on characteristics of the human genome into these studies.

ACKNOWLEDGMENTS

This work was supported by NIH grant P30ES07033 from NIEHS.

17

Type 2 Diabetes

Karen L. Edwards
University of Washington, Seattle, WA

17.1. INTRODUCTION

Glucose homeostasis depends on the balance between synthesis and catabolism of many products, which may be determined in part by genes. Diseases such as diabetes mellitus result from imbalances within the complex physiologic system that may arise through defects at the genetic level, through environmental insults (e.g., exposure to arsenic or lifestyle factors), or interactions between the two. The purpose of this chapter is to review evidence for underlying genetic influences in type 2 diabetes and provide examples illustrating gene–environment (G×E) interactions in type 2 diabetes.

17.2. BACKGROUND

Diabetes is one of the most costly and burdensome chronic diseases of our time (American Diabetes Association 2002). Diabetic complications are a significant cause of morbidity and mortality. For example, diabetes accounts for the majority of amputations and is the leading cause of blindness in U.S. adults. Diabetics are also at significantly increased risk for coronary heart disease, peripheral vascular disease, and stroke. Estimated economic costs of diabetes in the United States in 2002 were $132 billion.

Gene-Environment Interactions: Fundamentals of Ecogenetics, edited by Lucio G. Costa and David L. Eaton

Diabetes mellitus refers to a diverse group of metabolic diseases characterized by chronic elevation of plasma glucose:

Type 1 diabetes, previously referred to as "juvenile-onset" or "insulin-dependent" diabetes mellitus (IDDM), accounts for less than 10% of all cases of diabetes mellitus, and is characterized by failure of the pancreatic β-cells due to autoimmune destruction.

Type 2 diabetes, previously referred to as "adult-onset" or "non-insulin-dependent diabetes mellitus" (NIDDM), accounts for 90–95% of all diabetes. The metabolic hallmarks of type 2 diabetes are insulin resistance, impaired pancreatic β-cell function, and increased hepatic glucose production. In addition, people with type 2 diabetes are usually overweight.

Maturity-onset diabetes of the young (MODY) includes several rare monogenic forms of type 2 diabetes that account for 1–5% of type 2 diabetes cases. MODY is genetically heterogeneous, characterized by an autosomal dominant mode of inheritance, an early age of onset (usually <25 years of age), and abnormal pancreatic β-cell function (but not insulin resistance). In contrast to people with type 2 diabetes, MODY subjects are typically lean.

Hyperglycemia (elevated plasma glucose levels), which characterizes both major types of diabetes, results from an absolute deficiency of insulin in type 1 diabetes and a relative lack of insulin in type 2 diabetes. Although symptoms associated with hyperglycemia can be problematic, it is the chronic complications, including accelerated macrovascular disease (particularly ischemic heart disease, stroke, and gangrene) and microvascular disease (particularly retinopathy, neuropathy, and nephropathy) that reduce life expectancy and confer the enormous public health burden associated with diabetes.

17.3. EPIDEMIOLOGY

While the prevalence of type 2 diabetes is increasing in epidemic proportions in the United States and throughout many regions of the world, the global variation in the prevalence of type 2 diabetes is marked (Table 17.1). The geographic pattern of variation suggests that the prevalence is lowest in rural areas of developing countries, is generally intermediate in developed countries, and is highest in certain ethnic groups who have adopted Western lifestyle patterns (King et al. 1998).

Table 17.1 Geographic/Economic Prevalence Pattern of Diabetes for 2000 and Projections for 2005

	2000		2005		
Country or Economic Region	Population Size	Percent with Diabetes	Projected Population Size	Projected Percent with Diabetes	Percent change
Latin America and the Caribbean	305,400	6.0	486,097	8.1	2.1
India	577,814	4.0	857,337	6.0	2.0
Middle Eastern Crescent	335,553	6.5	653,472	8.2	1.7
Former socialist economies of Europe	249,867	7.2	256,185	8.8	1.6
United States	196,407	7.6	244,959	8.9	1.3
Established market economies	633,575	5.8	697,518	7.1	1.3
China	859,086	2.2	1,116,209	3.4	1.2
Other Asia and islands	453,613	3.2	739,466	4.3	1.1
Sub-Saharan Africa	304,644	1.1	665,976	1.3	0.2

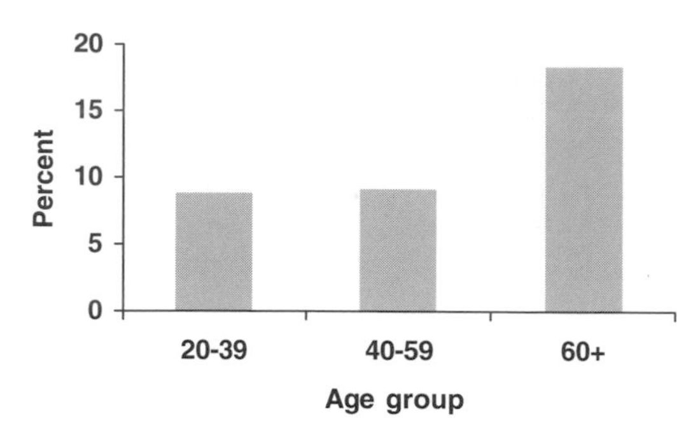

Figure 17.1. Total prevalence of diabetes in people aged 20 years or older, by age group—United States, 2002. [Data from Centers for Disease Control and Prevention (2004) (see Bibliography for full source).]

In the United States the overall prevalence of type 2 diabetes has increased by 33% since the mid-1990s (Centers for Disease Control and Prevention 1997; Mokdad et al. 2000), and in 2002, 18.2 million people, or 6.3% of the U.S. population, had diabetes (American Diabetes Association 2002). While the aging of the U.S. population accounts for part of this increase, most attribute the current epidemic to the increasing prevalence of obesity and physical inactivity in this country (Mokdad et al. 2000).

The prevalence of type 2 diabetes is known to increase with age. For example, 8.7% of all people aged 20 years or over have diabetes. The prevalence increases to 18.3% among those 60 years and older (Figure 17.1).

The prevalence of diabetes also varies by race and ethnicity, with higher rates among African-Americans (11.4%), Hispanic/Latin Americans (8.2%), Native Americans (8.2% among Alaska Natives and 27.8% among certain Native American tribes from the Southwest), and some Asian-American and Pacific Islander groups (Figure 17.2). Prevalence data for diabetes among Asian-Americans and Native Hawaiians or other Pacific Islanders are limited, but indicate that some groups are at increased risk compared to non-Hispanic whites. For example, Native Hawaiians, Japanese, and Filipino residents of Hawaii aged 20 years or older were approximately 2 times as likely to have diagnosed diabetes as were white residents of Hawaii of similar age (Centers for Disease Control and Prevention 2004).

Other important and well-established risk factors for type 2 diabetes include physical inactivity, dietary fat intake, obesity (particularly central or visceral adiposity), family history of diabetes, and gestational diabetes mellitus. Further, randomized trials have demonstrated that changing lifestyle

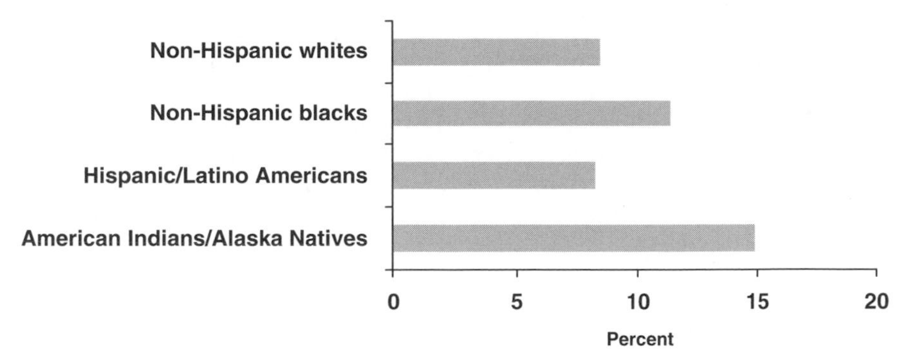

Figure 17.2. Age-adjusted prevalence of diabetes in people aged 20 years and older by race/ethnicity. [Data from Centers for Disease Control and Prevention (2004) (see Bibliography for full source).]

factors (weight loss, improved diet, and increased physical activity) results in a decreased risk of diabetes (American Diabetes Association 2002). There is also limited evidence suggesting a role for environmental contaminants and occupational exposures in disease development.

17.4. ENVIRONMENTAL EXPOSURES

The notion that environmental contaminants could increase the risk of type 2 diabetes is a relatively new concept (Kostraba et al. 1992; Rahman et al. 1998; Longnecker and Michalek 2000). However, a handful of studies have evaluated the association between exposure to arsenic and 2,3,7,8-tetrachlorodibenzo-*p*-dioxin (TCDD) and risk of type 2 diabetes. These studies have been reviewed in Longnecker and Daniels (2001) and are described briefly below.

17.4.1. Arsenic

The association between exposure to arsenic and development of type 2 diabetes has been evaluated by studying populations with high exposure to arsenic, including studies of selected industrial groups. Two of three published studies suggest a positive association between occupational exposure to arsenic and risk of type 2 diabetes, however, none of the associations from these studies were statistically significant (Rahman and Axelson 1995; Rahman et al. 1996; Bartoli et al. 1998). In other studies of populations with high exposure to arsenic, investigators have not reported associations with type 2 diabetes (Tsuda et al. 1990; Cuzick et al. 1992; Hopenhayn-Rich et al. 1998; Smith et al. 1998). These studies were all limited in size, and it is not known whether the investigators evaluated associations with type 2 diabetes. Several occupational mortality studies evaluating exposure to arsenic have also been published and reviewed by Longnecker and Daniels (2001) (McMichael et al. 1974; Andjelkovic et al. 1976; Blair et al. 1979; Morgan et al. 1980; Katz and Jowett 1981; Wong et al. 1984; Schwartz 1988; Marsh et al. 1991; Wingren et al. 1991; Park and Mirer 1996; Weiland et al. 1996; Wong et al. 1996; Longnecker and Daniels 2001). In general, these studies have several limitations, including the use of death certificates to identify diabetes, lack of detail in assessing exposure to arsenic, and limited numbers of cases.

The relationship between arsenic exposure and risk of type 2 diabetes has also been evaluated by measuring arsenic exposure in drinking water. While these tend to be larger than the occupational studies, they also tend to be cross-sectional or ecologic studies, rather than cohort or case–control. In general, these studies are consistent with an increased risk of type 2 diabetes among those exposed to the highest levels of arsenic compared to those with less exposure (Rahman et al. 1998; Tsai et al. 1999; Tseng et al. 2000).

Research examining the relationship between arsenic exposure and the risk of type 2 diabetes is limited in a number of ways: a small number of studies, limited sample sizes, the use of population level data rather than individual level data, and a failure to evaluate dose–response effects. In addition, no study has reported potential interactions between arsenic exposure and genetic susceptibility. Currently, the evidence linking exposure to arsenic and risk of type 2 diabetes is suggestive, but too limited at this point to draw any firm conclusions.

17.4.2. Tetrachlorodibenzo-*p*-dioxin (TCDD)

Exposure to TCDD and risk of diabetes, including mortality, has been evaluated in a limited number of studies (Zober et al. 1994; Henriksen et al. 1997; Longnecker and Michalek 2000). Additional studies have evaluated exposure to TCDD and plasma glucose or insulin levels (Pazderova-Vejlupkova et al. 1981; Suskind and Hertzberg 1984; Ott et al. 1994; Henriksen et al. 1997; Calvert et al. 1999; Cranmer et al. 2000; Longnecker and Michalek 2000). While some of these studies are consistent with an increase in risk for those with the highest level of exposure, taken together the evidence for an association between exposure to TCDD and risk of type 2 diabetes is not strong. However, Enan et al. (1996) have shown that TCDD decreases cellular glucose uptake, providing evidence for a biologically plausible diabetogenic effect of TCDD.

17.5. EVIDENCE FOR GENETIC INFLUENCES

Considerable efforts have been directed toward identifying susceptibility genes for type 2 diabetes. As a result, several lines of evidence suggest that type 2 diabetes and its hallmark features of insulin resistance and impaired insulin secretion include a strong genetic component. However, relatively little is known about specific genetic factors that increase susceptibility to type 2 diabetes. Identifying susceptibility genes for type 2 diabetes has proved difficult because of genetic heterogeneity and the context dependence of disease expression, which is strongly related to environmental factors such as diet and lifestyle. Yet, despite these difficulties, progress in identifying specific genes involved in type 2 diabetes has been made. In addition, major breakthroughs in understanding rare monogenic forms of type 2 diabetes have also been made. For example, mutations in pancreatic β-cell genes have been linked to specific MODY subtypes. By increasing our understanding of the genes and pathways that are crucial for normal β-cell function, these findings have also advanced our understanding of the more common form(s) of type 2 diabetes.

17.5.1. MODY

Major breakthroughs in understanding genetic susceptibility to type 2 diabetes first occurred by identifying a subgroup of diabetics with earlier age of onset and evaluating evidence for linkage in their families. To date, mutations in at least six genes, including glucokinase, insulin promoter factor (IPF)1, hepatocyte nuclear factor (HNF)-1alpha, HNF-1beta, HNF-4alpha, and NEURO-DI/BETA-2, have been identified and account for the majority of MODY cases (Table 17.2). Currently, only approximately 11% of MODY (referred to as MODY X) cases have no known genetic cause.

17.5.2. Pharmacogenomic Effects in MODY3

While there appears to be little evidence for G×E interactions in the development of MODY, there is evidence for pharmacogenomic effects with a specific sensitivity to sulfonylureas (Pearson et al. 2000, 2003). In a small case series ($n = 3$), Pearson et al. (2000) observed that MODY patients with HNF-1alpha mutations (MODY3) were unusually sensitive to the effects of sulfonylureas, and that the introduction of sulfonylureas was associated with risk of hypoglycemia. As a follow-up to the case series, Pearson and

Table 17.2 Genetic Mutations Responsible for MODY Cases

	Mutated Gene	Function	Known Variant(s)	Percent MODY Cases
MODY 1	HNF-4alpha	β-Cell transcription factor	12 mutations	3
MODY 2	Glucokinase	Hexokinase enzyme that catalyzes phosphorylation of glucose	>130 mutations	14
MODY 3	HNF-1alpha	β-Cell transcription factor	>100 mutations	69
MODY 4	IPF-1	β-Cell transcription factor	Pro63fsdelC	<1
MODY 5	HNF-1beta	β-Cell transcription factor	6 mutations	3
MODY 6	NEURO-DI/BETA2	Helix–loop–helix transcriptional factor, regulates insulin gene expression	206+C	<1
MODY X	Unknown gene(s)	—	—	11

colleagues conducted a randomized crossover trial of glicazide (a sulfonylurea) and metformin in patients with HNF-1alpha MODY and type 2 diabetes. This study demonstrated that MODY patients with HNF-1alpha mutations had a 5.2-fold greater response to gliclazide than to metformin, whereas patients with type 2 diabetes did not respond differently to the two medications. Metformin does not appear to be as effective at lowering plasma glucose levels in patients with HNF-1alpha mutations (Pearson et al. 2000). Although results of these studies suggest that genetic information may be used to improve the management of specific groups of patients, larger studies are needed to confirm these associations.

17.6. GENETICS OF TYPE 2 DIABETES

Evidence from a variety of studies, including familial aggregation, twin, linkage, and genetic association studies indicate that genetic factors make a major contribution to the development of type 2 diabetes. Researchers have used a variety of phenotypes in the search for susceptibility genes for type 2 diabetes, including central or visceral obesity, insulin resistance, and type 2 diabetes itself. As a result of these efforts, numerous candidate genes or chromosomal locations have been implicated; however, few of these initial reports have been replicated in independent samples. Thus, despite this seemingly abundant evidence, the genetic basis of type 2 diabetes is still not well understood. Evidence for genetic susceptibility to type 2 diabetes (but not the many intermediate and related phenotypes) is reviewed below, beginning with twin studies and concluding with the most promising results from linkage and association studies.

17.6.1. Evidence from Twin Studies

Twin studies have been used extensively to demonstrate evidence for both genetic and environmental influences on numerous diseases and traits, including diabetes. Some of the earliest evidence for genetic susceptibility to type 2 diabetes has come from studies of twins. Monozygotic (MZ) twins are genetically identical and should theoretically be concordant for a disease with a genetic basis. In contrast, dizygotic (DZ) twins share 50% of their genes on average, and are as genetically alike as nontwin siblings. In classical twin studies, similarity for a discrete trait (known as *concordance*) is compared in MZ and DZ twins and is used to evaluate the potential degree of genetic control over the trait. Classical twin studies indicate that concordance for type 2 diabetes is in the range of 50–92% among MZ twins (Beck-Nielsen et al. 2003), and most studies indicate that the concordance among DZ twins is significantly lower than in MZ twins. The greater concordance among MZ compared to DZ twins is consistent with genetic influences. However, the incomplete concordance among MZ twins, suggests a role for environmental factors as well.

Twin studies can also be used to quantify the proportion of the total variance in a quantitative trait attributable to possible genetic influences and is estimated by comparing the intraclass correlations in MZ and DZ twins. The resulting measure is called the *heritability*. Heritability estimates range from 0 to 1 (or within 0–100%), with higher values suggesting greater genetic influences. Heritability estimates for fasting and postprandial plasma glucose and insulin levels are generally less than 50% (Eisenbarth and Rewers 1995; Rewers and Hamman 1995; Beck-Nielsen et al. 2003) and are consistent with modest genetic influences.

17.6.2. Linkage and Association

Early attempts to identify susceptibility genes for type 2 diabetes relied on two general approaches: the candidate gene approach and the positional cloning or genomewide scanning approach. The candidate gene approach relies on identifying genes whose dysfunction might reasonably be expected to cause the disease or phenotype of interest. In contrast, the positional cloning approach relies on identifying chromosomal regions through linkage, followed by finer genetic mapping, and then physical mapping and sequence analysis of the region harboring a possible susceptibility gene. More recent approaches are relying on linkage disequilibrium mapping in unrelated subjects and in families.

Genomewide scans for type 2 diabetes have identified several regions containing possible susceptibility genes. Briefly, these scans have been performed in a variety of ethnic and racial groups, including Mexican-Americans (Hanis et al. 1996; Duggirala et al. 1999), American Pima Indians (Hanson et al. 1998), Canadian Oji-Cree (Hegele et al. 1999), Ashkenazim (Permutt et al. 2001), Scandinavians (Parker et al. 2001), the Swedish-speaking population of Bothnia, Finland (Mahtani et al. 1996), Utahans of northern European descent (Elbein et al. 1999), and Chinese Han (Luo et al. 2001) (see Table 17.3). These studies have identified possible susceptibility loci for type 2 diabetes on over a dozen different chromosomes or chromosomal regions. The potentially large number of susceptibility genes suggests that type 2 diabetes is genetically heterogeneous and that underlying mechanisms might differ between study samples.

Several positional candidate genes have been identified as a result of the genomewide linkage studies of type 2 diabetes, including PON2 on chro-

Table 17.3 Possible Susceptibility Genes and Linked Regions Identified by Genomewide Linkage Scans in Various Ethnic and Racial Groups

Study Population	Linked Region	Putative Susceptibility Gene(s)[a]
Mexican-Americans[b,c]	2q37.3	CAPN10
Pima American Indians[d]	1q, 7q, 11q	LMNA, PON2
Canadian Oji-Cree[e]	6q, 8p, 16q, 22q	
Ashkenazim[f]	4q, 8q, 14q, 20q	HNF-4α
Scandinavians[g]	18p11	
Bothnia, Finland[h]	12q24.3	HNF-1α
Utahans of north European descent[i]	1q21–q23	LMNA, APOA2, PKLR, LMX1
Chinese Han[j]	9p21, 9p13–q21, 20q13.2	X25 (frataxin)

[a] Susceptibility genes listed include only those genes identified by the individual study authors.
[b] Hanis et al. (1996).
[c] Duggirala et al. (1999).
[d] Hanson et al. (1998).
[e] Hegele et al. (1999).
[f] Permutt et al. (2001).
[g] Parker et al. (2001).
[h] Mahtani et al. (1996).
[i] Elbein et al. (1999).
[j] Luo et al. (2001).

mosome 7q21.3, LMNA on chromosome 1q21, CAPN10 on chromosome 2q37.3, and the HNF-1alpha gene on chromosome 12q. Additional candidates for intermediate phenotypes such as insulin resistance and for rare syndromes including diabetes or insulin resistance have also been identified.

17.6.3. NIDDM1/CAPN10

CAPN10 encodes a ubiquitously expressed member of the calpain-like cysteine protease family, calpain-10. Calpain-10 is expressed in many tissues including those involved in the pathogenesis of type 2 diabetes (pancreatic islets, muscle and adipose tissue). Although the family of nonlysosomal cysteine proteases catalyze the endoproteolytic cleavage of specific substrates involved in a number of cellular functions, including intracellular signaling (Sreenan et al. 2001), the specific functions of calpain-10 are not yet understood.

CAPN10 was identified as a promising positional candidate gene on the basis of the initial mapping of the NIDDM1 gene on chromosome 2q37.3 in Mexican-Americans (Hanis et al. 1996). CAPN10 was proposed to be the NIDDM1 gene after it was shown that a G>A polymorphism within intron 3 was associated with risk of type 2 diabetes in Mexican-Americans and in a sample of Bothnian Finns (Horikawa et al. 2000). Haplotypes consisting of two additional single-nucleotide polymorphisms (SNPs) in CAPN10 were also associated with increased risk of type 2 diabetes in these samples (Horikawa et al. 2000). However, this locus was not linked with type 2 diabetes in Pima Indians, but was associated with measures of insulin resistance (Baier et al. 2000). Although associations between CAPN10 and type 2 diabetes have not been demonstrated in every population (Hegele et al. 2001), the identification of CAPN10 as a putative diabetes gene is a major achievement in understanding genetic susceptibility to type 2 diabetes, and provides insights into new biochemical pathways regulating blood glucose levels.

17.6.4. Hepatocyte Nuclear Factor 4a (HNF-4α)

Mutations in the HNF-4alpha gene on chromosome 20q12–q13 have been found in families with MODY, and although evidence for a type 2 diabetes locus at chromosome 20q12–q13 had been found in several Caucasian and Asian populations, MODY genes were thought to play, at most, a minor role in the common form of type 2 diabetes. However, the simultaneous publication of results from two independent samples provides increasing evidence for HNF-4alpha as a type 2 diabetes susceptibility gene.

The Finland–United States Investigation of NIDDM Genetics (FUSION) study consists of 737 Finnish affected sibling pair families (Vale et al. 1998; Ghosh et al. 2000; Silander et al. 2004b). Fine mapping results in a case–control sample from the FUSION study indicate associations between markers (SNPs) in the downstream region of the primary β-cell promoter of HNF-4alpha and type 2 diabetes (Silander et al. 2004a). This marker was also associated with several additional diabetes related traits. Nine additional markers (SNPs) in this region were also associated with type 2 diabetes (Silander et al. 2004a). In an independent sample of Ashkenazi Jewish subjects, Love-Gregory and coworkers also report associations between type 2 diabetes and SNPs around a 78-kbp region of the HNF-4alpha gene. Together, these studies suggest that variants located within or near the HNF-4alpha gene increase susceptibility to type 2 diabetes.

17.7. GENE–ENVIRONMENT INTERACTIONS

The dramatic rise in the prevalence of type 2 diabetes likely reflects interactions between genetic and environmental factors, particularly lifestyle factors. In complex diseases such as type 2 diabetes, it is likely that only those with high-risk genetic profiles that are exposed to a high-risk environment (e.g., lack of physical activity and/or excessive caloric intake) will develop the disease. Understanding gene–environment interactions is one of the important challenges faced by current researchers and public health agencies because while our genetic makeup is not modifiable, many environmental risk factors are.

Gene–environment interaction (G×E) implies that in combination, the risk associated with the genotype and the environmental factor under study is more than the additive effects of each factor independently (Talmud and Stephens 2004). In other words, interaction suggests that environmental factors modify the molecular function of the gene or product. A biologist might frame this as environmental dependence of gene expression, or as a particular genetic response to an environmental factor (Guo 2000). Thus, a statistically different effect of an environmental factor in two groups defined by a genetic factor would be consistent with G×E. Despite the belief that G×E interactions are important in the development of type 2 diabetes, direct evidence for interactions between functional gene polymorphisms and environmental factors are lacking. However, the following are examples of likely interactions between genetic and environmental factors.

17.7.1. Thrifty Gene Hypothesis

The thrifty gene hypothesis was proposed by J. V. Neel in 1962 to explain the emergence of type 2 diabetes in populations transitioning from vigorous activity and subsistence nutrition to inactivity, over nutrition and consequent obesity (Neel 1962). The thrifty gene hypothesis speculates that a genetic predisposition to obesity and diabetes would be advantageous in an evolutionary sense in times of food scarcity by promoting the efficient retention of energy stores in the form of adipose tissue. However, this efficiency and ability to store fat would become disadvantageous in times of relative food abundance and low energy expenditure (Neel 1962; Swinburn 1996). Further, populations that had survived periods of marked food scarcity in their history would become enriched for these thrifty genes and, thus susceptible to developing diabetes (Wendorf 1989; Houghton 1991). Although not a direct evaluation of G×E, this hypothesis provides one of the best examples for the importance of gene-environment interactions in type 2 diabetes.

The thrifty gene hypothesis may explain the increases in type 2 diabetes among many aboriginal populations, including Native American and Canadian populations and populations undergoing Westernization. For example, the Oji-Cree population is a geographically and genetically isolated community in the subarctic boreal forest of northern Ontario that is accessible only by air during most of the year. As recently as the mid-1950s, type 2 diabetes was virtually unknown in these people. Since the development of the reservation, the lifestyle of the Oji-Cree has changed from nomadic hunting-gathering to a sedentary lifestyle. In addition, the traditional diet of low-fat, nutrient-dense foods has been replaced by high-energy processed foods. The incidence of type 2 diabetes among the Oji-Cree of the Sandy Lake Reserve is now approximately 40% (Busch and Hegele 2001) and is 6 times higher than the general Canadian population (Young et al. 1990; Harris et al. 1997). It is likely that the genetic basis of the thrifty genotype is due to polymorphisms at multiple sites, rather than a single abnormality.

17.7.2. Family History

Because an individual's family history information reflects both genetic and environmental factors, it, too, may serve as a unique measure of G×E interaction. Many studies have evaluated the risk of type 2 diabetes associated with a positive family history of diabetes, and a review paper describes these findings (Harrison et al. 2003). Briefly, most studies report a twofold to sixfold increased risk of type 2 diabetes associated with a positive family history (first-degree relative affected) compared to a negative family history and that the risk is greater when both parents are affected. Associations with family history are consistently demonstrated in different ethnic groups and regardless of study design. Further, the risk associated with family history appears to be independent of other known risk factors including age, body mass index, glucose status, and smoking. Some studies suggest that having a mother with diabetes confers greater risk than if the father has diabetes, and could be consistent with mitochondrial DNA inheritance. However, the relative importance of maternal versus paternal diabetes is unclear and differences in risk may reflect bias (Khoury and Flanders 1996; Harrison et al. 2003).

While most investigators recognize the importance of evaluating the influence of genetic susceptibility on phenotypic response to environmental risk factors, there is surprisingly little direct statistical evidence for G×E interactions in the literature. This may be due to the inherent difficulties in evaluating G×E interactions, including inadequate statistical power of many genetic epidemiologic studies.

17.7.3. Alternative Designs for Evaluating Gene–Environment Interactions

Statistical issues relating to G×E evaluation can be complex, and the sample size required to have adequate statistical power can be so large that traditional case–control studies are sometimes not feasible unless the interaction effect is very large, and the genetic and environmental factors are both relatively common (Goldstein et al. 1997). As a result, several alternative epidemiologic designs have been proposed to assess G×E, including the case-only design and the case–parent design (Piegorsch et al. 1994; Khoury and Flanders 1996). Other familial methods of linkage analysis have been developed to account for G×E, although they have rarely been applied to real data.

17.7.3.1. Case-Only Design. The case-only or case–case design uses only subjects with the disease of interest. Because this design does not require control subjects, the required sample size is reduced by at least half that required in case–control studies. Cases are distributed in a 2 × 2 matrix according to their genetic and environmental exposure status, and the odds ratio (OR) for the interaction is calculated in the usual way. Importantly, the main effects of the gene and the environmental factor cannot be estimated, but a measure of association between the two factors is easily computed as a cross-product ratio, which is approximately equal to the OR for G×E interactions computed from case–control data (Schmidt and Schaid 1999). This approach assumes independence between the genetic and environmental factors and under this assumption offers better precision for estimating interactions compared to the case–control approach. However, the case-only design measures interaction only as a departure from multiplicative effects and cannot detect G×E under departures from additivity.

17.7.3.2. Case-Parent Design. Cases and their parents are frequently used to evaluate evidence for genetic association and linkage. In particular, the transmission disequilibrium test (TDT) proposed by Spielman et al. (1993) is a popular method to assess genetic association and linkage. The TDT may be used to evaluate G×E and avoids false positive detection of G×E due to potentially confounding factors such as race and ethnicity (population stratification) (Umbach and Weinberg 2000). Briefly, because the TDT evaluates the probability that a parent with an Aa genotype transmits the A allele to an affected child, G×E is suggested when the observed transmission probability differs between exposed and unexposed cases. Simulation studies have shown that the case–parent design can sometimes be more powerful than the case–control design in detecting G×E, particularly when the disease susceptibility allele is rare and the environmental factor has a large effect in the absence of the susceptible genotype (Schaid 1999). However, the validity of the case–parent design to assess G×E also requires independence of the genotype and exposure in the general population. Although the TDT is an appealing approach, it may be difficult to conduct in late-onset diseases, such as type 2 diabetes as parents of the cases may be deceased.

17.8. SUMMARY

In summary, the evidence clearly indicates underlying genetic susceptibility in type 2 diabetes. It has also been clearly established that nongenetic factors are involved in type 2 diabetes and that preventing or delaying the onset of type 2 diabetes is possible in high-risk groups through simple lifestyle modifications such as a healthy diet or increased physical activity. Although it is widely believed that environmental factors interact with genetic factors in increasing susceptibility to type 2 diabetes, there is little direct statistical evidence to support this hypothesis.

Developing and evaluating strategies to identify at-risk individuals who may benefit from targeted interventions are important and challenging public health responsibilities. However, the question of whether intervention would be more effective in people with a particular genetic makeup has not yet been addressed. Genetic epidemiologic studies should be designed with these goals in mind. Understanding G×E is one of the important challenges faced by current researchers and public health agencies because, whereas our genetic makeup is not modifiable, many environmental risk factors are.

17.9. SUGGESTED READING

Three excellent literature sources on type 2 diabetes are Busch and Hegele (2001), Longnecker and Daniels (2001), and the American Diabetes Association (2002).

ACKNOWLEDGMENTS

This project was supported by a cooperative agreement (U36/CCU300430-23) from the Centers for Disease Control and Prevention (CDC) through the Association of Schools of Public Health. The contents of this article are the responsibility of the author and do not necessarily represent the official views of CDC or ASPH.

18

Infectious Disease Ecogenetics

David R. Sherman
University of Washington, Seattle, WA

> ... to say our lives have been completely changed just because we have been infected with a microbe is not a satisfying answer. We want to know, why me? Or if our sister or brother is sick, we want to know, why not me?
>
> —*Jeanette Farrell, Invisible Enemies* (*1998, p. 9*)

18.1. INTRODUCTION

Because of its seemingly random nature and potentially devastating impact, infectious disease has been deeply disturbing to people since time immemorial. For millennia, even the most basic aspects of why one person became ill while another remained well were shrouded in magic and mystery. The modern view of infectious disease has been traced to the Italian physician Girolamo Fracastoro, who proposed in 1546 that some diseases might be spread by invisible *seminaria contagium*, seeds of contagion that could be transmitted from person to person by touch or through the air. This idea was appealing because it helped explain the epidemiology of infectious epidemics, but it was, of course, impossible to verify at the time. Then only a century or so later, Antony van Leeuwenhoek invented the microscope and rendered microorganisms visible to the human eye for the first time. Even with this revolutionary new technology, it still took another 200 years before the forefathers of modern microbiology such as Robert Koch and Louis Pasteur began to define the etiological agents responsible

Gene-Environment Interactions: Fundamentals of Ecogenetics, edited by Lucio G. Costa and David L. Eaton

for major infectious illnesses. This was a huge advance—malarious fevers could be explained by the *Plasmodium* coursing through a patient's veins, and patients coughing up blood could be shown the *Mycobacterium* in their sputum. Still, faced with the ubiquity of microbes, explaining why a particular person is sick but not her husband, child, or neighbor has been all but impossible to address until very recently.

Of all subdisciplines in the emerging field of ecogenetics, host susceptibility to infectious disease is one of the most challenging. In addition to the usual complexities of polygenic traits, host genetic polymorphism and environmental variability, pathogens carry their own genomes and are continually evolving and adapting to conditions within their hosts. Despite this added layer of complexity, recent advances have been significant. As described below, most big successes in the study of host susceptibility to infectious disease have come via the candidate gene approach, in which research first identifies a host gene with a key role in the infectious process, and then population studies seek to demonstrate pertinent variation at that locus. However, the field is rapidly progressing with new technology to map the position of loci and manipulate genes within mammalian chromosomes and to determine the sequence of large swaths of DNA from multiple donors with great precision and speed. It seems likely that whole genome scans will play a bigger part in determining host susceptibility loci in the future, even without a candidate gene to drive the process. We seem poised for another major advance in our understanding of susceptibility to infectious disease.

This chapter describes the progress made in delineating susceptibility to two prominent human pathogens: the causative agents of malaria and AIDS. This chapter highlights general approaches and principles that will apply to other infectious diseases, as well as some challenges and opportunities for the near future.

18.2. MODELING SUSCEPTIBILITY TO INFECTIOUS DISEASE

Aspects of the multifaceted nature of susceptibility to infection can be modeled along a continuum of disease pathogenesis (McNicholl et al. 2000; Qureshi et al. 1999) (Figure 18.1). A particular host gene (e.g., Figure 18.1b) may play a key role in whether exposure to a pathogen leads to infection, but then play no role in the determining the subsequent course of that infection. Another host gene (Figure 18.1c) may affect disease severity without altering rates of infection. Still another host gene (Figure 18.1a) could play a role at all stages of the infectious process.

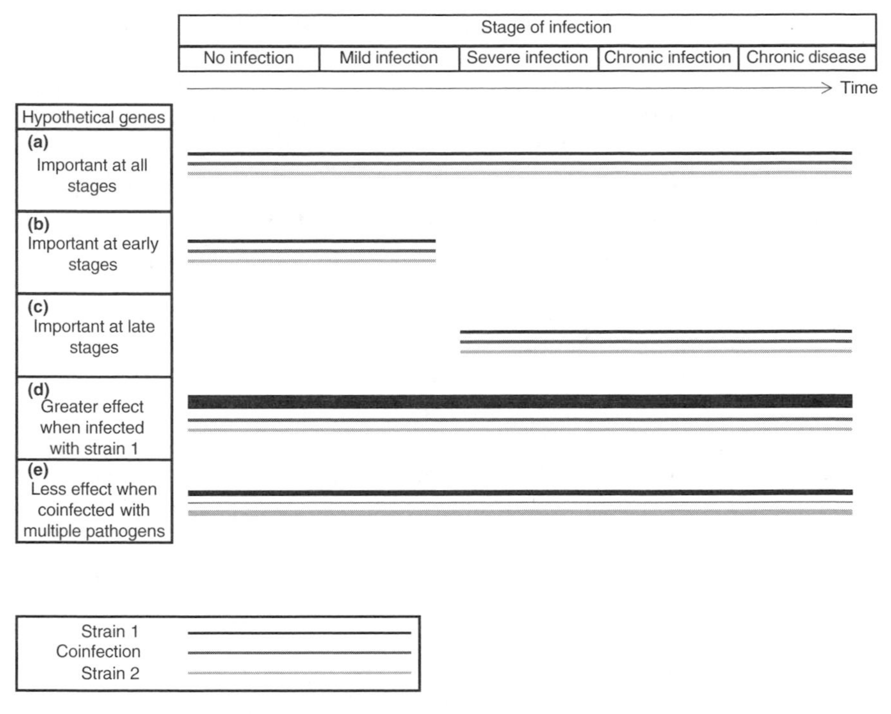

Figure 18.1. Mapping the interaction of host genes and infection outcomes. Thicker bars represent an increased role of gene at the indicated stages of infection. [Adapted from McNicholl et al. (2000).]

This model can only hint at the genetic complexities that underlie the host–pathogen dynamic. For example, some host genes will only exert their effects following infection with particular strains of a given pathogen, or in certain host genetic backgrounds. Similarly, the role of particular host genes may change in individuals coinfected with multiple pathogens. Also, virtually every aspect of the susceptibility to infectious disease is likely to be influenced by multiple genetic loci. Individual alleles that strongly affect outcome or transmission of widespread deadly diseases would experience intense selective pressure. Depending on the nature of the effect, such alleles will either dominate or disappear from a population, so individual mutations with dramatic phenotypes should be rare.

Because two genomes are at work, susceptibility to infectious disease can result in rapid coevolution of both host and pathogen. Since rapidly killing one's host can reduce a pathogen's opportunities to spread, it is often assumed that newly introduced pathogens are more likely to display heightened virulence that attenuates as host and pathogen adapt. Numerous examples have been cited. One especially well-studied case involves the semi-accidental introduction of myxoma virus to Australia in an effort to contain the European rabbit, which had become a serious agricultural pest (Hayes and Richardson 2001). In the first year that the virus circulated in Australia, >99% of infected rabbits died. However, by the second year, only ~90% of infected rabbits died, and eventually mortality was reduced to ~25%. This adaptation involved demonstrated changes in both virus and rabbit genomes (Best and Kerr 2000). Other examples that suggest coadaptation to commensalism are less clear. For instance, scarlet fever caused by group A streptococci was one of the deadliest childhood diseases of the nineteenth century, causing widespread epidemics throughout Asia, Europe, and the Unites States. Rates began to decline sharply around 1880. Today scarlet fever is rare, and generally produces much milder disease when it occurs (Quinn 1989). However, while rates and morbidity from scarlet fever have unquestionably declined, pathogenic variants of the group A streptococci with new virulence properties and profiles have begun to circulate since 1980 (Krause 2002).

Clearly, the evolution of disease susceptibility is not simply a one-way march from virulence to coexistence. Evolution selects for the genes and traits that are passed on most successfully. Depending on the course of disease and the mode of transmission, a pathogen may succeed by reproducing to high titers, even if the host is killed (Ewald 1993). Malaria is spread by an insect vector, the Anopheles mosquito, which can transmit very effectively even when a person is too sick to get out of bed. With malaria, high parasite titers in the blood are crucial for effective transmission, increasing the odds that each mosquito bite will also transmit the pathogen. In this case selection pressure would seem to favor traits that result in increased virulence. Similarly, HIV and *M. tuberculosis* (MTB), while different in so many respects, both usually take years to kill their hosts. Clearly, ample time for transmission and high pathogen titers can promote pathogen success even though millions of persons are killed each year.

The study of host–pathogen interactions relies heavily on animal models. The advantage of animal model systems to investigate infectious disease is obvious. Aspects of both pathogen and host can be manipulated under carefully controlled conditions, and conclusions can be tested in repeat experiments. Questions that cannot be addressed in humans for practical or ethical reasons become accessible. Even those questions that can be addressed in human studies are almost always easier to investigate with an animal model. However, researchers eager to exploit the advantages of animal models often overlook the assumptions that underlie and sometimes threaten to undermine their models (Druilhe et al. 2002). Infectious dis-

Table 18.1 Global Incidence of the World's Deadliest Infectious Diseases[a]

Pathogen	People Infected	Active Disease	Deaths Annually
HIV	40 million (2003)	—	3 million (2003)
Tuberculosis	1.86 billion (1999)	8.8 million (2002)	2 million (2003)
Malaria	300–500 million (2002)	—	1.5 million (2002)

[a] Year during which statistics were collected is shown in parentheses.

eases generally act differently in different hosts. This problem is exacerbated when human pathogens are studied in lab animals that are not their natural hosts. In addition, to speed up the experimental process, researchers often infect by an artificial route with an unnaturally high infectious dose. Such experiments run the risk of defining important susceptibility determinants in mice that may have little relevance to the disease in people. As a result, in studying the ecogenetics of susceptibility to infectious disease, there will always be a place for studies with the "human model" (Casanova and Abel 2002, 2004) as well.

Given all these confounding issues, the difficulty in identifying loci that confer susceptibility or resistance to infectious disease in humans becomes clear. Not surprisingly, the biggest infectious killers that exert the strongest selection pressure—malaria, HIV, and tuberculosis (TB) (Table 18.1)—have each been studied extensively. Several host genes have been implicated in the response to malaria and HIV, while in contrast, susceptibility to TB is still poorly understood. Hope is strong that continued study of host response to infectious agents will contribute to lessening the burden from these and other global scourges.

18.3. MALARIA

18.3.1. Epidemiology, Life History, and Pathology

Malaria is probably the first infectious illness for which host genetics were shown to play a role in susceptibility and resistance. Malaria is a disease of life-threatening fevers and chills whose distribution generally coincides with the world's hottest (and poorest) regions. For centuries, malaria was associated with the foul gasses (*mal aria*) of marshes and swamps. Then in the late nineteenth century it was determined that malaria is caused not by bad air but by protozoan parasites of the genus *Plasmodium* transmitted by the mosquitoes that infest tropical regions. *Plasmodium vivax* and *P. falciparum* are the most serious health threats, although *P. ovale* and *P. malar-*

iae also cause disease in humans. Worldwide, there are probably more than 300 million cases of malaria and at least 1 million deaths each year. Falciparum malaria is by far the most lethal and, because it occurs predominantly in Africa, roughly 90% of malaria deaths occur there. Most deaths are in young children and pregnant women. In highly endemic areas, children are infected several times each year and carry parasite in their blood almost continually. Worldwide, human activities often exacerbate the malaria pandemic: generation of drug-resistant parasites and insecticide-resistant mosquitoes, as well as war, climate change, population expansion, and migration.

As typified by *P. falciparum*, the lifecycle of the malaria parasite is remarkably complex, with an array of morphologically distinct forms. When an infected mosquito takes a blood meal, sporozoites from the insect salivary gland enter the bloodstream and migrate rapidly into the liver. Within hepatocytes, *Plasmodium* differentiates into schizonts and then merozoites, which are released back into the bloodstream. Merozoites infect red blood cells, where they undergo further growth and differentiation (into trophozoites and schizonts, and then back to merozoites) before lysing the infected red cell. Released merozoites can then infect new red blood cells, completing what is called the *erythrocytic cycle*. Occasionally a merozoite differentiates into a gametocyte that, when ingested by a mosquito, can initiate a sexual reproductive cycle involving yet other morphological forms.

Given the magnitude of malaria's impact throughout the world, one might predict that any host genetic trait affecting susceptibility to this disease would experience strong selective pressure. These traits could impact on any of the varied stages of *Plasmodium* development in humans, but the strongest selective pressures seem to be exerted during the erythrocytic cycle. Devoid of nuclei and many organelles and filled with the protein hemoglobin, erythrocytes are highly adapted to their role as oxygen transporters. *Plasmodium* in turn is highly adapted to thrive in these specialized cells. Exploiting hemoglobin as their primary food source, a single merozoite can produce 12–24 merozoite progeny that lyse their host cell and go on to infect new erythrocytes. This cycle of growth and lysis causes rapid, dramatic fluctuations in parasite numbers in the blood and is responsible for the periodic fevers and chills that characterize malaria. With *P. falciparum*, this process takes only 48 h and the parasites can rapidly grow to prodigious levels. Parasite numbers in the blood can exceed $1 \times 10^9\,mL^{-1}$, at which point parasitemias (the percentage of infected red blood cells) can exceed 30% or more. Under these circumstances, the parasites will consume 175 g of hemoglobin in a single day. Such overwhelming infections are nearly always fatal. One cause of death is catastrophic anemia. With *P. falciparum*, a second cause may be even more important. Infected erythrocytes can stick to capillary walls and thus avoid the filtering action of the spleen. In high numbers these sequestered cells can obstruct blood flow to the brain or other vital organs (see Figure 18.2).

18.3.2. Haldane and Hemoglobinopathies

The study of host susceptibility/resistance to malaria had an unlikely beginning. In the 1940s, human geneticists were grappling with the challenge posed by common hemoglobin disorders, including sickle cell anemia and the α- and β-thalassemias. The hemoglobinopathies, which inherit in an autosomal recessive fashion, may be the world's most common genetic illness (see also Chapter 2, on historical perspectives of ecogenetics). In homozygous form these traits are deleterious (some are lethal); however, heterozygous carriers make up at least 5% of all people and can reach above 90% in certain populations. The very high frequency of carriers could not be explained and was extremely troubling to geneticists of the day [reviewed in Clegg and Weatherall (1999)]. Then, in 1948 at the 8th International Congress of Genetics in Stockholm, J. B. S. Haldane proposed a startling hypothesis: "I believe that the possibility that the heterozygote is fitter than normal must seriously be considered. . . . It is at least conceivable that they are . . . more resistant to attacks by the sporozoa which cause malaria, a disease prevalent in Italy, Sicily and Greece, where the gene is frequent" (Haldane 1949). Even in the molecular age, biology is generally an empiric science, with explanation following observation. Predictions like Haldane's that prove correct are the remarkable exception.

The first evidence to support Haldane's hypothesis came from studies of sickle cell anemia. Hemoglobin is a tetramer, with the adult form (HbA) consisting of two α and two β chains. Sickle cell hemoglobin (HbS) has a single nucleotide change in the β-chain gene that mutates the glutamic acid at position 6 to a valine. Deoxygenated HbS can aggregate in ways that physically distort (or sickle) the erythrocytes. Sickled cells can themselves aggregate, blocking capillaries and producing extensive tissue damage. Because about half of their hemoglobin is normal, heterozygotes usually suffer sickle cell disease only under extremely hypoxic conditions, such as at high altitude. However, in the absence of significant, frequent medical intervention, persons homozygous for the sickle cell mutation nearly always die at an early age (Friedman and Trager 1981).

To achieve a high frequency in a population, an allele so costly when homozygous must confer a major advantage in the heterozygous state. Numerous epidemiologic studies over decades bear out the forecast that persons carrying one copy of the HbS mutation are significantly protected against severe malaria. Rates of disease are not affected, but infected indi-

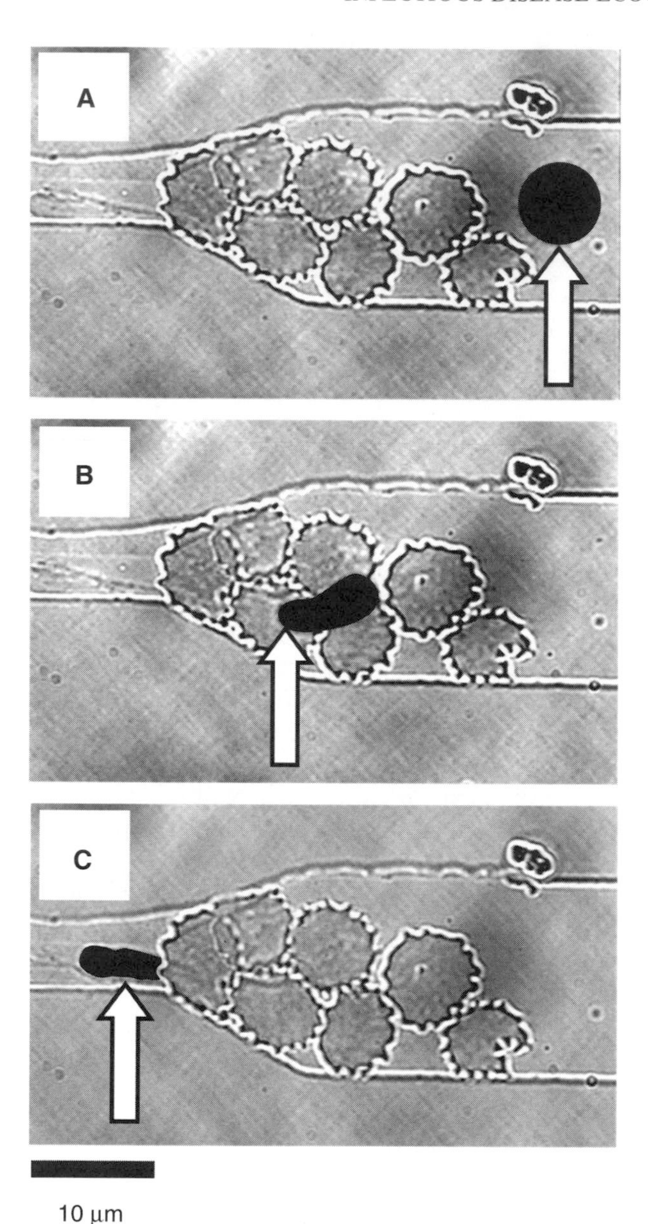

Figure 18.2. A model for capillary obstruction by *P. falciparum*–infected erythrocytes. Erythrocytes were added to an artificial channel with an engineered constriction of 6 μm and a series of video images were captured. Infected erythrocytes cannot pass through the constriction, forming a blockage at the entrance. An uninfected erythrocytes (red, with arrow) (A) flows freely in the main channel, (B) passes through the blockage formed by infected cells, and (C) emerges beyond the blockage and passes easily through the constriction. [Adapted from Shelby et al. (2003).]

viduals are much less likely to be hospitalized or to die. The greatest protection is seen in those most at risk—children under 5. The biochemical basis for this protection is still not clear. It seems to correlate with reduced growth during the erythrocytic stage, although HbS erythrocytes can support *Plasmodium* growth in culture as long as they remain oxygenated and do not sickle. While less well studied, other hemoglobin variants, HbC and HbE, are also known to confer protection against malaria.

The other class of hemoglobinopathy, the thalassemias, results from defective synthesis of either the α or β globin chain, which in turn causes abnormal maturation and premature loss of the affected red blood cells. Though Haldane's original hypothesis linked malaria with β-thalassemia, decades passed before this association was convincingly demonstrated. It is now clear that the situation with thalassemias is far more complex than with the sickle cell mutation. Literally dozens of different mutations can produce thalassemias of varying severity, which greatly complicates the analysis. In addition, while their protective effects can be significant, the thalassemias probably don't confer as much protection against malaria as the HbS heterozygote. Finally, the thalassemias seem to afford some protection against some other infectious diseases as well. Confounding variables proved less problematic among islanders of the South Pacific, where a series of population studies begun in the 1980s finally vindicated Haldane (Clegg and Weatherall 1999). As with the sickle cell mutation, thalassemics are protected against the worst effects of malaria but not against malaria infection, and the mechanism of protection is unknown. In fact, one study showed that thalassemics are even more likely than normal to be infected with malaria, especially as small children. Some researchers have suggested that thalassemia may coincide with a more robust immune response to malaria, perhaps as a result of heightened early exposure to *Plasmodium* antigens (Clegg and Weatherall 1999; Fortin et al. 2002). If correct, this explanation has profound implications for malaria vaccine development.

18.3.3. Erythrocyte Invasion Polymorphisms

Since the erythrocytic cycle is central to malaria pathogenesis, host factors that affect invasion of erythrocytes are likely to contribute to malaria susceptibility. In the 1970s, studies by Miller and colleagues at NIH demonstrated that the Duffy antigen on red cell membranes is the only receptor used by the simian malaria *P. knowlesi* and its close relative *P. vivax* for attachment and invasion of erythrocytes (Miller et al. 1994). This finding led to an explanation for the curious absence of *P. vivax* and corresponding predominance of *P. falciparum* among malaria cases in Africa. The Duffy antigen functions as a chemokine receptor (DARC, also called Fy) on the red cell surface. A mutation at a transcription factor binding site in the *DARC* gene promoter reduces *DARC* expression, which produces a Duffy blood group negative phenotype and blocks invasion by *P. vivax*. This mutation is very prevalent among Africans, resulting in failure of *P. vivax* to colonize this continent. Interestingly, this resistance may actually be harmful to Africans. Probably many factors contribute to the stifling grip of deadly falciparum malaria on the continent, but the lack of competition or immunostimulation by *P. vivax* may well play a significant role.

Ovalocytosis is a condition marked by oval erythrocytes that may or may not also exhibit other morphological anomalies. Southeast Asian ovalocytosis (SAO) is caused by a 27-bp (basepair) deletion that removes amino acids 400–408 from the abundant erythrocyte protein known either as band 3 or as anion exchange protein 1 (AE-1). SAO inherits in dominant fashion, is associated with heightened red cell rigidity, and confers measurable protection against malaria. SAO erythrocytes are resistant to *Plasmodium* invasion in culture, and in field studies SAO is associated with reduced parasite burdens in the blood and fewer deaths (Fortin et al. 2002).

A second erythrocyte polymorphism associated with ovalocytosis may bestow resistance to falciparum malaria on people of Papua New Guinea (Zimmerman et al. 2003). The Gerbich-negative (Ge^-) blood group phenotype caused by deletion of exon 3 of the glycophorin C gene is remarkably common in parts of Papua New Guinea where malaria is endemic. Glycophorin C is an integral membrane glycoprotein with an important role in maintaining the structural integrity of erythrocytes. In addition, evidence indicates that glycophorin C also acts as a receptor for binding and subsequent invasion of red cells by *P. falciparum*. The merozoite surface protein EBA140 binds to normal but not mutant glycophorin C, and antibodies to EBA140 reduce *P. falciparum* invasion of red cells in culture. These data coupled with the strong colocalization of the Ge^- phenotype with malaria in the region argue that this polymorphism must offer at least some protection against malaria. Thus far field studies in the region have failed to provide evidence of protection. Of course, field studies did not always link the thalassemias or SAO ovalocytosis with malaria resistance either, but those connections are no longer in doubt. There may be many confounding factors obscuring an association between the Ge^- phenotype and malaria protection, not the least of which is the fact that *P. falciparum* has at least three other independent receptor–ligand interactions that mediate erythrocyte invasion.

18.3.4. Malaria Severity Polymorphisms

Some known human polymorphisms are associated with increased rather than decreased risk of severe malaria. Tumor necrosis factor alpha (TNF-

α) is a potent cytokine with powerful effects on a variety of host immune processes. The TNF-α promoter region has three independent single-nucleotide changes that each correlate with heightened severity of malaria (Fortin et al. 2002; Kwiatkowski 2000; McNicholl et al. 2000). Persons homozygous for the polymorphism at position -308 are about sevenfold more likely to die of cerebral malaria. Various studies have also linked the *TNF-308A* allele with other infectious illnesses, including leprosy, trachoma, and leishmaniasis. This polymorphism appears to increase the basal level of TNF gene expression. Paradoxically, a second promoter polymorphism, *TNF-238A*, is associated with severe falciparum anemia in children and lower levels of circulating TNF. A third allele, *TNF-376A*, alters recruitment of a transcription factor that modulates TNF gene expression and is associated with a fourfold greater risk of cerebral malaria. TNF and malaria severity are plausibly linked since TNF levels help determine the extent and duration of malarial fevers, but the precise mechanism by which TNF promoter mutations exacerbate malaria is unclear. Maintenance of these alleles in malarious regions argues that altered TNF expression may confer some as-yet-undetermined selective advantage.

18.3.5. Malaria Susceptibility and Animal Models

There are no convenient animal models in which to study human malaria because the parasites responsible infect only humans and a few other primates. However, four rodent malaria species, *P. berghei*, *P. chabaudi*, *P. yoelli*, and *P. vinckei*, have been exploited to study the host response to malaria in a laboratory setting. In mice, the severity of disease caused by these pathogens varies dramatically with the host genetic background. Susceptible mouse strains are characterized by high parasitemias, with >50% of red blood cells parasitized at the peak of infection, and by extremely high mortality. In resistant strains, parasitemias and mortality are both substantially lower. When susceptible and resistant mouse strains are mated, the offspring are generally more resistant than either parent, indicating that resistance to malaria is genetically dominant and that multiple host loci must be involved. Genetic linkage studies are beginning to map those loci. At least six different loci on six different mouse chromosomes have been strongly linked with malaria resistance. The implicated chromosomal regions contain numerous interesting candidate genes, encoding chemokines and cytokines, receptors and proteins involved in erythrocyte structure and iron storage (Fortin et al. 2002). Still, identifying the relevant genes, defining how they influence malaria resistance in mice, and determining whether the same effects occur in humans remain as significant future challenges.

18.4. HIV/AIDS

18.4.1. Epidemiology, Life History, and Pathology

From its presumed origins in an obscure African jungle, acquired immunodeficiency syndrome (AIDS) has spread globally with astonishing and terrifying speed. The first cases of the disease now known as AIDS were reported in 1981, but within 20 years it had disseminated to all regions of the earth. AIDS now claims about 3 million lives each year, more than any other single infectious agent (Table 18.1), and the pandemic continues to worsen. The World Health Organization estimates that roughly 14,000 persons were newly infected each day in 2003. By the end of that year, about 40 million were infected worldwide. Virtually all of those people will die of AIDS unless they receive lifelong therapy that is both complex and expensive to administer.

AIDS is caused by the human immunodeficiency virus HIV. AIDS spreads when body fluids such as blood or semen from an HIV+ individual enter someone who is uninfected. Specific high-risk behaviors are necessary to fuel the AIDS pandemic, especially unprotected sex, intravenous drug use, and, in countries with poor surveillance, transfusions with tainted blood. In addition, perinatal transmission is very widespread. More than 15% of all AIDS deaths occur in children under 15 years old.

HIV infects cells of the immune system that express the surface protein CD4, primarily T lymphocytes, but also macrophages and monocytes. These cells normally form the heart of host defense against infectious disease. On entry, HIV subverts the host cellular machinery for its own reproduction until the cell is killed. HIV has an RNA genome of about 10,000 nucleotides, which is copied into DNA by the viral enzyme reverse transcriptase. This process is very efficient but not especially precise. After an initial ramp-up, billions of new virus particles are released each day. On average, each particle contains a mutated nucleotide somewhere in its genome.

Despite this massive immunological onslaught, it typically takes about a decade before T-cell levels decline substantially and the immune system falters, signaling the onset of AIDS disease. Once T-cell levels drop sufficiently, the person becomes susceptible to a variety of other infectious agents and eventually succumbs. Certain infections that plague AIDS patients but are rarely encountered in immunocompetent individuals are considered AIDS-defining illnesses. These include *Pneumocystis carinii* pneumonia (PCP), oral and systemic candidiasis, cytomegalovirus, lymphoma, and Kaposi's sarcoma. Globally, tuberculosis (TB) is much too prevalent to be considered an AIDS-defining illness, but the synergy between TB and HIV is especially sinister. TB is responsible for almost

10% of AIDS-related deaths, far more than any other single infectious agent.

18.4.2. Host Response to HIV

The search for variation in the host response to AIDS began very soon after HIV was identified as the causative agent (O'Brien and Dean 1997). We might expect such variation to be evident, since HIV has exerted a powerful selective pressure on the human population in an extremely brief evolutionary timeframe. Still, identifying host loci that confer resistance to a particular disease can be an enormous challenge even when the effect is robust, as the study of susceptibility to malaria makes clear. The progress made in identifying AIDS resistance genes underscores the resourcefulness and perseverance of the scientists working in this field.

For years, anecdotal and indirect evidence suggested that not everyone is equally susceptible to AIDS. There were occasional reports of people who engaged in repeated high-risk behaviors such as frequent sex with HIV+ individuals yet remained uninfected. Also, HIV contamination was not eliminated from the U.S. blood supply until 1985. Before then, hemophiliacs, who depend on concentrated blood products from many donors, were extremely vulnerable and more than 10,000 became infected. However, at least 10% of hemophiliacs who received tainted blood remained HIV-negative. In other studies, the speed with which HIV-positive persons progress to AIDS was also shown to vary substantially. While most HIV-positive people exhibited signs of AIDS within 10 years, roughly 1% of individuals remained disease-free for 15 years or more. Until recently (as of 2005), however, the molecular basis for these variations could not be explained.

Clues to understanding variations in host response to AIDS first emerged from studies of HIV pathogenesis. Soluble factors secreted by CD8 T cells were known to inhibit invasion of new cells by HIV. In the mid-1990s, these factors were identified as particular chemokines, soluble peptides with amino terminal cysteines that mediate immune response by stimulating leucocyte motility. At the same time, it was clear that the CD4 ligand on the host cell surface acts as a receptor for HIV, but CD4 alone is not sufficient for viral uptake. An intensive search for the coreceptor paid off with the 1996 discoveries of two different proteins, CCR5 and CXCR4, each of which are coreceptor for particular common variants of HIV. Both proteins normally function as chemokine receptors. These discoveries immediately suggested that variations in certain chemokines and/or their receptors might alter various aspects of the host response to AIDS (O'Brien and Dean 1997).

18.4.3. CCR5

The first variations in the gene encoding CCR5 were identified just months after the role of this protein as HIV coreceptor was reported. A deletion of 32 bp from within the CCR5 coding region produces a stop codon and a severely truncated protein that fails to reach the cell surface. Homozygotes for the CCR5-Δ32 allele have no surface-exposed CCR5, and heterozygotes have only 20–30% of wildtype levels. Since 90–95% of primary HIV isolates use CCR5 as their major coreceptor, this mutation has profound impact on AIDS susceptibility (O'Brien and Moore 2000). CCR5-Δ32 homozygotes are almost completely resistant to HIV infection. Heterozygotes are readily infected with HIV, but they show reduced viral loads and a 2–3-year delay in the onset of symptoms associated with AIDS. In addition, the spectrum of AIDS-related illnesses is altered in these individuals. CCR5-Δ32 heterozygotes with AIDS are only half as likely to develop non-Hodgkin's B-cell lymphoma as are those with two copies of the wildtype allele.

The CCR5-Δ32 allele shows a curious distribution in human populations. It is quite common among Caucasians (allele frequency 5–15%), but vanishingly rare among East Asians and Africans. Population genetics studies suggest that the mutation first appeared roughly 1000–5000 years ago and that its frequency in Caucasians rose substantially in the last 1000 years. This pattern and frequency argue that the mutation was adaptive well before the appearance of HIV, perhaps by conferring resistance to some other widespread and deadly infectious disease such as smallpox, tuberculosis, or bubonic plague. Whatever the agent, this selection must be powerful since the allele frequency has increased rapidly even though CCR5-Δ32 also has negative consequences, namely, an increased risk of hypertension (Mettimano et al. 2003).

Screening for other CCR5 mutations revealed a surprising degree of polymorphism in both its coding region and promoter. Over 80% of the coding region mutations produce a protein with altered amino acid sequence, suggesting that selection pressure may be acting to maintain diversity at this locus. None of the coding region polymorphisms have been linked with altered host response to HIV. However, at least one promoter polymorphism has been associated with a 2–3-year faster progression to AIDS. The CCR5P1 promoter mutation has been proposed to increase CCR5 expression leading to enhanced cell entry by HIV, but evidence supporting that idea is lacking.

18.4.4. Other Chemokines and Coreceptors

It is now clear that at least a dozen different chemokine receptors can also serve to facilitate HIV entry into cells. While CCR5 is the major corecep-

Table 18.2 Genes Implicated in Resistance or Susceptibility to HIV

Gene	Allele	European	African	East Asian	
CCR5	Wildtype	0.86–0.96	1.0	1.0	Susceptible
	Δ32	0.044–0.14	0.0	0.0	Resistant
CCR2	Wildtype	0.9	0.77	0.75	Susceptible
	64I	0.1	0.23	0.25	Resistant
SDF1	Wildtype	0.79	0.98	0.74	Susceptible
	3′A	0.21	0.02	0.26	Resistant
CCR5P	P1	0.56		0.44	Susceptible
	P2	0.09		0.23	
	P3	0.14		0.15	
	P4	0.35			

Source: O'Brien and Moore (2000).

tor in the early stages of HIV infection, late-stage viruses more frequently use the receptor CXCR4. Intensive analysis of CXCR4 genes in human populations has revealed few polymorphisms, none of any apparent consequence. However, variation was revealed in the gene for the only known ligand of CXCR4, the chemokine SDF-1. The mutant SDF-1 allele frequency is over 0.2 among East Asians and European Caucasians, but only 0.02 among Africans (Table 18.2). In homozygous form, this mutation is strongly associated with a delay in the onset of AIDS symptoms. Similarly, mutation in the CCR2 coreceptor also delays AIDS onset, although the effect is less pronounced. Interestingly, the protective effects of the SDF-1 and CCR2 mutations are complementary, both with one another and with the CCR5-Δ32 heterozygote.

18.5. CONCLUSIONS AND FUTURE PROSPECTS

Malaria and AIDS, so different at first glance, share several features that make them fertile ground for the study of host susceptibility. Both diseases exert very strong selective pressure on the human population. Because of their powerful impact on human health, both are subjects of intense research efforts, with good candidate host genes that have been identified. Also quite important, with both illnesses it is relatively easy to determine who in a population has been exposed, who has been infected, and who is suffering acute disease. This is not always the case. Consider tuberculosis, a well-studied disease with similar impact on human health (Table 18.1), for which much less is known about variation in host susceptibility. It is commonly reported that one-third of all people globally harbor TB bacteria, but that figure is extremely uncertain because the test on which it is based cannot distinguish exposure from infection (Cosma et al. 2003). To define host susceptibility loci, it is obviously essential to determine reliably who does and does not have the disease.

Plasmodium and HIV also show some relevant similarity in terms of pathogenesis. To cause disease and spread, both agents must replicate to exceptionally high titers very quickly. Any change in the rate at which new host cells are infected is likely to alter the disease outcome. In addition, *Plasmodium* and HIV both depend on a few specific host proteins for their rapid expansion. Again, tuberculosis demonstrates how each infectious disease can take a different course. The bacteria that cause tuberculosis replicate quite slowly and pathogen burden correlates only loosely with disease symptoms. Also, while TB bacteria reside within host macrophages, it appears that their uptake does not depend critically on any one host receptor (Cosma et al. 2003).

The majority of characterized host susceptibility/resistance loci are known or thought to act directly on pathogen uptake or metabolism. Surprisingly few effects are clearly mediated through a difference in host immune response. However, classes of immune response genes that are very likely to affect infectious disease outcomes have been described. For example, the major histocompatibility complex (MHC) of humans is an extremely polymorphic chromosomal region containing more than 125 genes, many of which are crucially important for acquired immunity. Differences in MHC genotype have been associated with resistance to malaria (Hill 2001), AIDS (Carrington and O'Brien 2003), and other diseases. Similarly, the toll-like receptor (TLR) family of proteins recognizes specific conserved pathogen features to initiate innate immune signaling (Janssens and Beyaert 2003). TLR gene polymorphisms may well underlie some variation in susceptibility to infectious agents.

While malaria and AIDS offer particularly rich examples of infectious disease ecogenetics, serious efforts are underway to understand host variation in susceptibility to virtually every major pathogen, and progress has been significant. As it accumulates, this information has the potential to revolutionize medical care. With each pathogen, it may eventually be possible to personalize treatment based on the sum of a patient's susceptibility and resistance loci for that particular agent. Additional efforts and resources could be directed specifically to those who are most vulnerable. And it may finally become possible to draw a satisfying answer the age-old question about why your sibling, parent, neighbor became ill, but not you.

18.6. SUGGESTED READING

For further reading on infectious disease ecogenetics, the papers by Clegg and Weatherall (1999), Fortin et al. (2002), Hill (2001), McNicholl et al. (2000), O'Brien and Moore (2000), and Qureshi et al. (1999) are recommended.

19

Genetic Variation, Diet, and Disease Susceptibility

Johanna W. Lampe and John D. Potter
Fred Hutchinson Cancer Research Center, Seattle, WA

19.1. INTRODUCTION

Food is a mixture of compounds that are essential for life and important for long-term health. Considered as a whole, diet encompasses the most diverse and complex assortment of compounds to which humans are exposed. Genetic differences in taste preference dictate food choices. Genetic differences in immune response determine food tolerance and thus, ultimately, food choice. The interaction between dietary constituents and genetic differences in the regulation of their absorption, transport, metabolism, and effects on target tissues has a potent impact on human health and disease.

Food constituents that have been identified as necessary for life are called *nutrients*. They provide energy and building materials for the substances that are essential for growth and survival (Table 19.1). Absence or inadequate intake of any one of these can be fatal or can severely reduce lifespan. A wide range of other compounds present in foods contribute to overall health both positively and adversely. Some of these are bioactive plant compounds, or "phytochemicals," which have a variety of effects and can play a role in prevention of chronic disease. In contrast, several classes of foodborne chemicals that are carcinogenic or toxic are associated with

Gene-Environment Interactions: Fundamentals of Ecogenetics, edited by Lucio G. Costa and David L. Eaton

Table 19.1 Nutrients in Human Nutrition

Class of Nutrient	Constituents
Carbohydrate	Poly-, di-, and monosaccharides
Protein	Essential amino acids: arginine, leucine, isoleucine, valine, tryptophan, phenylalanine, methionine, threonine, lysine, histidine
	Conditionally essential amino acids: proline, serine, tyrosine, cysteine, taurine, glycine
	Nonessential amino acids: glutamate, alanine, aspartate, glutamine
Lipids	Fatty acids: saturated and mono- and polyunsaturated
	Essential fatty acids: linoleic and α-linolenic
Vitamins	Fat-soluble: retinol (vitamin A), vitamin D, vitamin K, vitamin E
	Water-soluble: ascorbic acid (vitamin C), biotin, thiamin (vitamin B_1), riboflavin (vitamin B_2), niacin (vitamin B_3), pyridoxine (vitamin B_6), folate, cobalamin (vitamin B_{12}), pantothenic acid
Minerals	Calcium, phosphorus, magnesium, sodium, chloride, potassium, sulfur, iron, zinc, copper, iodine, manganese, fluoride, molybdenum, cobalt, selenium, chromium, tin, nickel, vanadium, silicon
Water	
Dietary fiber	Nonstarch polysaccharides, lignin, resistant starch

increased risk of cancer and other chronic diseases. How all these compounds are metabolized in the body influences their potency; genetic differences in the associated metabolic pathways can alter risk of disease.

The rare and severe genetic defects that cause monogenic diseases (e.g., pheylketonuria) are frequently associated with pathways related to handling of a specific nutrient (e.g., phenylalanine). By comparison, the genetic factors that modulate individual susceptibility to multifactorial diseases (e.g., cancer, diabetes, and cardiovascular diseases) are common, but functionally different forms (polymorphisms) of genes. These generally have a modest effect at an individual level but, because of their high frequency in the population, can be associated with high attributable risk. Environmental factors can reveal or facilitate the phenotypic expression of such susceptibility genes. As outlined in previous chapters, most of the susceptibility genes for common diseases do not have a primary etiologic role in predisposition to disease, but act as response modifiers to environmental factors, such as diet. Our understanding of the extent to which subtle genetic differences influence optimal nutrition, nutrient requirements, and susceptibility to disease is still rudimentary.

The objective of this chapter is to outline the key concepts related to diet and nutrition and their interactions with genetic backgrounds and to illustrate these with canonical examples. Often, the ultimate results of these interactions are differences in disease risk. However, in the case of multifactorial diseases, the interactions are complex, involving multiple genes and several exposures. In some cases, for pathways that are still poorly understood, genetics and genetic variants have helped elucidate which compounds are key players in pathways. In other cases, animal models, including transgenics, have improved our understanding of gene–diet interactions. This chapter briefly reviews genetic determinants of nutrient-dependent phenotypes and then discusses the influence that genetic differences have on food choice and metabolism and impact of nutrients, carcinogens, and phytochemicals.

19.2. MOLECULAR DETERMINANTS OF NUTRIENT-DEPENDENT PHENOTYPES

Genetic variation in components of nutrient digestion, absorption, distribution, transformation, storage, and excretion affect nutritional health and individual nutrient requirements. However, the gap between genotype and a biochemical phenotype is spanned by multiple steps, each with additional levels of genetic variation (Figure 19.1). Further, nonmutational epigenetic modifications to DNA that can be propagated faithfully with cell division (notably methylation) are also susceptible to dietary factors.

19.2.1. Genotype and Phenotype

Most human DNA sequence variation is attributable to single-nucleotide polymorphisms (SNPs), with the rest attributable to base insertions or deletions, repeat length polymorphisms, and rearrangements. Despite the high prevalence of these variants, ultimately, differences in phenotype, not merely differences in genes, determine the impact of nutrition on health. Many of the identified SNPs never translate to altered protein function because the triplet genetic code is degenerate and the amino acid sequence remains the same (i.e., they are "synonymous" SNPs), or because the base change produces an amino acid change that has subtle or no effects or is not at a site that is crucial to protein function. Many SNPs also occur with great frequency outside coding regions. Although these have no effect on

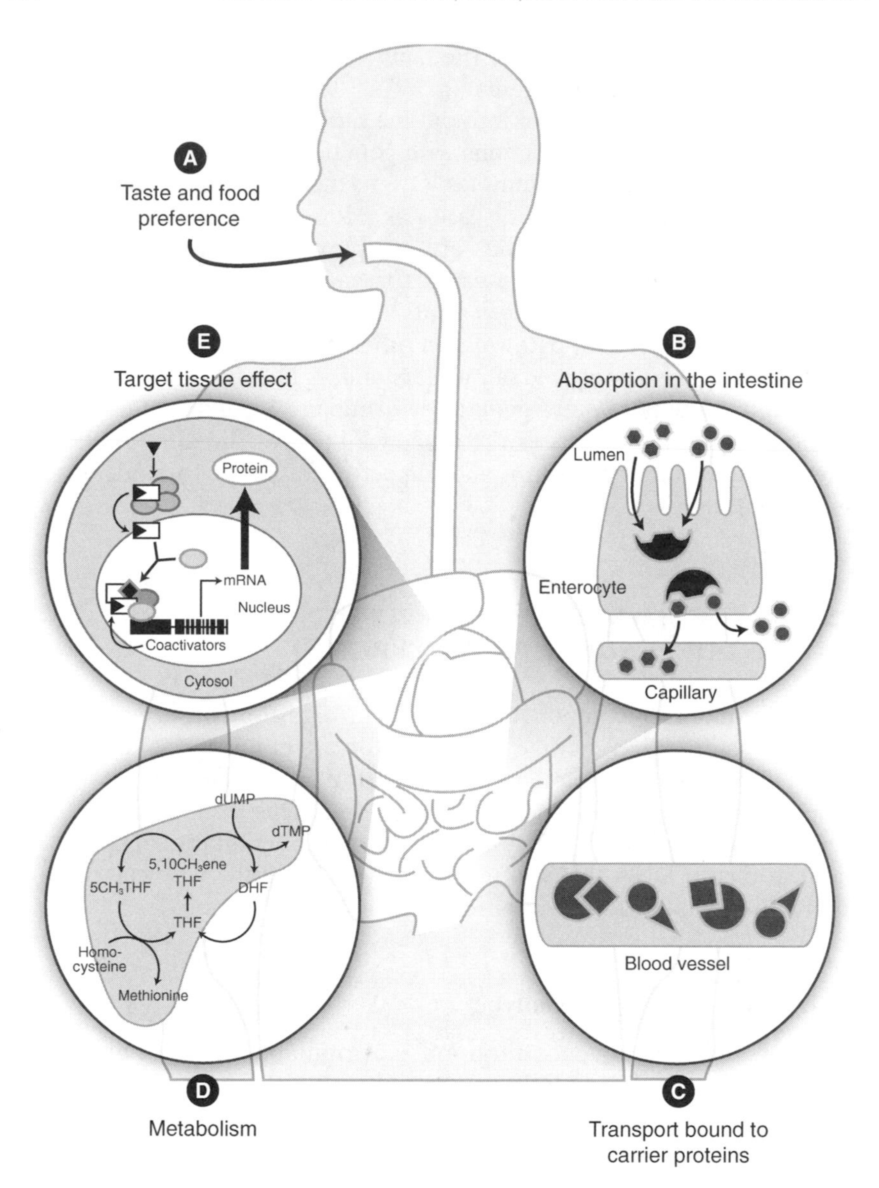

Figure 19.1. Polymorphisms in genes that regulate (A) taste, (B) nutrient absorption, (C) transport, (D) metabolism, and (E) uptake and effects in target tissues can contribute to interindividual response to diet and disease risk (for details, see text).

protein sequence, they can affect gene expression if they occur in regulatory regions. See, for example, Section 19.6.1.

Enzyme-catalyzed metabolic reactions are a key component in the processing of nutrients. Typically, enzymes involved in nutrient metabolism have high substrate specificity; the ability of enzymes to distinguish, with high fidelity, between two very similar macromolecules is one approach to effectively separating flow through anabolic and catabolic pathways. Because most pathways carry out a single specialized metabolic function, regulation in these pathways can be achieved without adversely affecting the flux of other metabolites. Efficient and economical regulation is often achieved through controlling key rate-determining enzymes that catalyze one of the committed steps in the pathway. A committed step is one whose reaction product is unique to that pathway (i.e., the only enzyme for which that compound serves as substrate is the one that catalyzes the next step in the pathway). Complete blockage of any one step in an enzyme-catalyzed pathway will shut down flux through the pathway as a whole; however, either slowing or accelerating the rate of flow is often desirable. As a result, functional polymorphisms in these key (committed-step) enzymes often have the greatest impact on a system and, sometimes, have devastating consequences. For example, SNPs that result in amino acid changes (i.e., nonsynonymous SNPs) in the active site of pheylalanine hydroxylase, the enzyme that catalyzes the conversion of the amino acid phenylalanine to tyrosine, result in severe phenotypes; see Section 19.5.1.

By comparison, biotransformation enzymes are more promiscuous and typically catalyze the metabolism of a wide variety of endogenous and xenobiotic substrates. Furthermore, multiple isozymes (i.e., enzymes with highly similar structure and function, but usually derived from different genes), often expressed in the same tissue, may have significant substrate overlap, albeit with differing affinities, providing redundancy in the system. Consequently, alteration in enzyme function might affect clearance of not only an endogenous substrate but also a carcinogen or a potential chemoprotective agent. For example, glutathione *S*-transferases (GST) are important in the conjugation of several classes of carcinogens as well as in the clearance of isothiocyanates (probable protective agents) found in cruciferous vegetables. Thus, a polymorphism that alters GST isozyme activity may alter risks in unpredictable ways.

19.2.2. Developmental Epigenetic Determinants of Adult Phenotype

Fetal or neonatal nutrient exposure can also hardwire gene expression patterns that affect adult metabolism and disease risk. Since the 1920s, it has been recognized that nutrient exposures during critical periods of development can influence birth outcome and, in the 1980s, the "fetal origins hypothesis" became a driving postulate directing research to understand how in utero exposures influence risk for chronic degenerative diseases that appear during adulthood (Barker 1998). Although much of the focus has been on the importance of overall maternal nutrition status and birth weight (Harding 2001), there are several studies, primarily in animal models, that suggest that in utero or early-life exposures to several different dietary components affect adult phenotypes (Waterland and Jirtle 2003; Singhal et al. 2004).

The biologic mechanisms underpinning adaptations to prenatal and early postnatal nutrition have been largely unknown. However, it has been established relatively recently that some of these differences in gene expression patterns are determined by epigenetic events (i.e., nonmutational DNA changes) that can be faithfully propagated during cell proliferation. Cytosine methylation at crucial periods during development has surfaced as one of the likely epigenetic mechanisms. Cytosine methylation within CpG dinucleotides of DNA acts in concert with other chromatin modifications to maintain heritably specific genomic regions in a transcriptionally silent state. Genomic patterns of CpG methylation are reprogrammed in the early embryo and maintained thereafter.

Most regions of the adult mammalian genome exhibit little interindividual variation in degree of tissue-specific methylation. Specific transposable element insertion sites are an exception; in these sites, CpG methylation can be determined probabilistically, causing cellular epigenetic mosaicism and phenotypic differences between cells within the same tissue. Transposable elements (including retrotransposons and DNA transposons) are parasitic nucleotide sequences that are scattered throughout the genome; in humans, they constitute over 35% of the genome and are found within approximately 4% of genes. Most transposable elements in the mammalian genome are normally silenced by CpG methylation. However, the epigenetic state of a subset of transposable elements is metastable (i.e., marginally stable) and can affect regions encompassing neighboring genes. The epigenetic metastability of such regions renders them susceptible to influences during early development. Because diet-derived methyl donors and cofactors are necessary for synthesis of *S*-adenosylmethionine, a substrate required for CpG methylation, early nutrition can influence adult phenotype via DNA methylation. Waterland and Jirtle (2003) showed that in yellow agouti (A^{vy}) mice, modest supplementation of a pregnant *a/a* mouse's nutritionally adequate diet with extra folic acid, vitamin B_{12}, choline, and betaine permanently affected the A^{vy}/a offspring's DNA methylation at epigenetically susceptible loci (Figure 19.2). This transient supplemental dietary exposure in utero also resulted in an associated

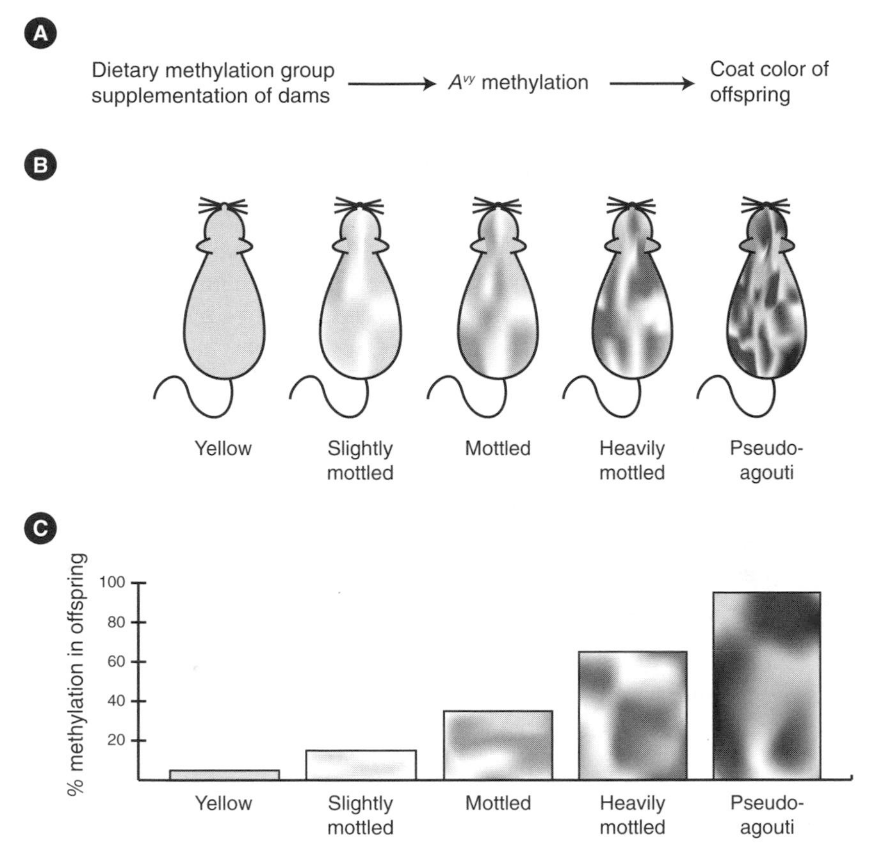

Figure 19.2. In utero exposure to supplemental dietary methyl sources results in epigenetic effects and distinct phenotypes that persist into adulthood. (A,B) Methyl group supplementation in the dams' diet results in increased A^{vy} CpG methylation in the offspring and subsequent phenotypic differences, including increased mottling and darkening of the coat color. (C) Mean percent methylation of seven CpG sites in tissues from the five coat color classes; the proportion of offspring in methyl-supplemented dams also was shifted toward the pseudoagouti phenotype. [Adapted from Waterland and Jirtle (2003).]

phenotype—heavily mottled or pseudoagouti coat color—that persisted into adulthood. Methylation patterns of the A^{vy} transposable elements differed between offspring exposed to the regular diet in utero and those methyl supplemented in utero. Methylation of the transposable element explained the effect of dietary supplementation on coat color—ectopic *agouti* expression. In these animals, epigenetic variability results in not only variation in coat color but also adiposity, glucose tolerance, and tumor susceptibility in the adult mice. These findings have important implications for human nutrition in that they demonstrate the crucial role that early nutritional exposures have on DNA methylation and gene expression and ultimately on phenotype and disease risk.

Similarly, exposure to hormonally active dietary compounds during critical periods of development can permanently alter predisposition for biochemical insult. For example, in rats, early postnatal exposure to the soy isoflavone genistein enhances mammary gland differentiation, decreases cell proliferation, and reduces susceptibility to mammary cancer in adult animals (Lamartiniere et al. 2002). Although the mechanism has not been elucidated, this early exposure to genistein appears to determine how the mammary tissue will respond, in the adult, to hormone and growth factor stimuli; exposure to genistein in adult life was more protective when the female mammary gland had already been exposed to genistein prepubertally. These results are consistent with epidemiologic data that suggest that soy intake as a child, compared to soy intake as an adult, is associated with lower breast cancer risk (Shu et al. 2001). It is likely that other dietary exposures during early development have a similarly important impact on establishing epigenetic gene regulation; however, these have yet to be identified.

19.3. GENETIC DIFFERENCES IN FOOD PREFERENCE: SENSITIVITY TO BITTER TASTE

Factors that influence food intake will affect dietary exposure and nutritional status. Although social and economic factors have a strong influence on food choices, physiologic factors, such as taste perception, also determine what a person puts in their mouth. This is probably most evident in relation to bitter-tasting foods, to which humans have developed a strong aversion (Drewnowski and Gomez-Carneros 2000). A wide range of structurally unrelated compounds, such as peptides, amino acids, ureas, thioureas, terpenoids, phenols, and polyphenols, give rise to a uniform bitter taste suggesting that there are a large number of distinct bitter taste receptors (T2Rs)—maybe as many as 60. T2Rs, receptors involved in bitter-taste detection, are one of several families of G-protein-coupled receptors that are selectively expressed in subsets of taste receptor cells. In contrast, various members of the T1R family combine to function as sweet-taste or L-amino acid receptors associated with the umami taste (Nelson et al. 2002a). *Umami*, translated from Japanese to a number of not entirely consistent English equivalents, including "savory," "essence," "pungent," "deliciousness," and "meaty," is the taste triggered by constituents of some amino acids, such as glutamates or aspartates, especially in the flavor-enhancing substance monosodium glutamate (MSG).

Phenylthiocarbamide (PTC) and 6-*n*-propylthiouracil (PROP), two compounds that taste bitter to some individuals but are tasteless to others, have been used in genetic linkage studies. People can be distributed into three PROP-tasting categories: nontasters, medium tasters, and supertasters. Given the distribution of these phenotypes (25% nontasters, 50% medium tasters, and 25% supertasters), there is some conjecture that genetically, nontasters may have two recessive alleles, medium tasters have one recessive and one dominant allele, and supertasters have two dominant alleles. The ability to taste PROP has been linked with a locus at 5p15; however, the relevant polymorphisms have not been identified.

Sensitivity to bitter taste appears to affect food preference. Higher sensitivity to the bitter taste of PROP has been shown to be associated with lower acceptance of particular foods. Compared to nontasters of PROP, women identified as medium and supertasters, had lower acceptance scores for grapefruit juice, green tea, brussels sprouts, and soybean curd (tofu) (Drewnowski et al. 2001). Bitter-tasting foods are frequently disliked, and bitter taste is a reason for low acceptance of cruciferous and leafy green vegetables. Thus, genetic differences in bitter-taste perception may affect taste preferences and food choices, and, ultimately, disease risk. Further, it is likely that polymorphisms in other genes for the other families of receptors may influence taste preferences for sweet and umami flavors; however, these have not received the same attention.

19.4. GENETIC DIFFERENCES IN IMMUNE RESPONSE TO DIETARY ANTIGENS

The immune response is a complicated process by which the immune system generates both antibodies and sensitized lymphyocytes that recognize specific antigens. A wide spectrum of immune-mediated conditions is inherited as complex polygenic traits resulting from the interaction of multiple gene loci and environment (e.g., diet). Genetic differences in immune response determine food tolerance and therefore have consequences for food choice.

19.4.1. Atopy Genes and Specific Food Proteins: Food Allergy

Food allergy is characterized by, and dependent on, the induction of an immune response (of sufficient vigor, and of an appropriate quality) to one or more proteins (Kimber and Dearman 2002). It involves sensitization via the generation of a specific IgE antibody response. The antibody is distributed systemically, associated with tissue mast cells. In a sensitized individual, subsequent exposure to the inducing allergen will provoke a reaction. Most food allergies are associated with a relatively limited number of food sources, including cow's milk, eggs, tree nuts, peanuts, wheat, fish, shellfish, and soy. The common symptoms of food allergy include nausea and vomiting, abdominal pain, distension, flatulence, and diarrhea. Other organ systems can also be involved, such as the skin (acute urticaria and angioedema, and atopic dermatitis) and respiratory tract (allergic rhinitis and asthma). Severe systemic (anaphylactic) reactions can also occur.

Although a variety of factors, such as timing, duration, and extent of exposure, are key determinants, genetic predisposition also plays a key role. An atopic phenotype (i.e., predisposition to mount an IgE response) in one or both parents substantially increases the chances of atopy in their offspring; however, inheritance from the mother carries a higher risk, and the particular proteins against which allergic responses will be mounted do not appear to be programmed genetically. The initiation and maintenance of IgE antibody responses are tightly regulated, and are dependent on the reciprocal activities of subpopulations of T lymphocytes. Given the complexity of the system, genetic variation in several components of the system probably contribute to an atopic phenotype. For example, STAT6 (signal transducer and activator of transcription 6) is a key transcription factor involved in both interleukin-4 (IL-4) and IL-13-mediated biological responses. A combination of dinucleotide repeat polymorphisms (13/15-GT) of the STAT6 exon 1 gene and the G2964A variant are strongly associated with allergic diseases, including food-related anaphylaxis among Japanese.

19.4.2. The HLA Complex and Gluten: Celiac Disease

Celiac disease, also known as *celiac sprue* and *gluten-sensitive enteropathy*, is a multifactorial disorder of the small intestine. The clinical manifestation ranges from asymptomatic or mild to severe malabsorption resulting from inflammatory injury to the mucosa of the small intestine after the ingestion of wheat gluten or related rye and barley proteins. The specific protein subunits that cause the damage are gliadins (the ethanol-soluble component of wheat gluten) and the phylogenetically related rye secalins and barley hordeins.

Unlike the majority of food allergic reactions, in which IgE-dependent mechanisms are implicated, celiac disease is IgA- and IgG-mediated and involves an autoimmune component. There is a strong genetic association with human leukocyte antigen (HLA) class II antigens; up to 95% of celiac disease patients are HLA-DQ2 carriers, and most of the remainder are HLA-DQ8 carriers. However, given that 20–30% of healthy controls are also carriers of HLA-DQ2, clearly additional factors determine development of the disease. The enzymatic activity of tissue transglutaminase (tTG) in small intestinal mucosa is a key factor in the pathogenesis of celiac disease, and tTG has been identified as an autoantigen. tTG, which catalyzes the covalent and irreversible crosslinking of a protein with a glutamine residue, has a preference for gliadins as glutamine donor substrates; tTG-catalyzed deamidation of these gluten peptides reveals neoepitopes with enhanced HLA-DQ2/DQ8 binding and T-cell stimulatory capacity. Cytokines released from the activated gliadin-specific T cells apparently cause the mucosal damage typical of celiac disease. In addition, antibodies have been detected that are directed against tTG-catalyzed crosslinks between gliadin and tTG. The recognition of these additional neoepitopes may favor intermolecular epitope spreading from gliadin to tTG and contribute to the autoantibody production (Dieterich et al. 2003).

Celiac disease can be treated with strict adherence to a gluten-free diet, omitting wheat, rye, barley, and possibly oats. Poor compliance with the lifelong diet and undiagnosed disease are associated with numerous and diverse complications, including increased mortality. With the advent of improved diagnostic tests, such as serologic screening for gliadin antibodies and autoantibodies against tTG, the disorder is now known to be relatively common, affecting 1 in 100–200 persons in both Europe and North America.

19.5. GENETIC DIFFERENCES IN AMINO ACID AND PROTEIN METABOLISM

Most high-penetrance, low-prevalence inborn errors of metabolism, particularly in amino acid metabolism, are associated with severe clinical illness soon after birth. Mental retardation and severe neurologic involvement may result quickly if dietary intake of the relevant substrates is not carefully regulated. Nutritional therapy for amino acid disorders most frequently consists of substrate restriction, limiting one or more essential amino acids to the minimum requirement, and product supplementation, augmenting intake of the deficient product of the enzymatic pathway, while providing adequate energy and nutrients to promote normal growth and development. Selected amino acid and protein metabolism disorders that require dietary adjustments are outlined in Table 19.2.

Table 19.2 Selected Amino Acid and Protein Metabolism Disorders that Require Dietary Adjustments

Gene Product	Gene	Disorder	Incidence in Live Newborns —General U.S. Population[a]	Dietary Adjustments
		Amino Acid Disorders		
Phenylalanine hydroxylase	PAH	Phenylketonuria	1:10,000–1:25,000	Low phenylalanine, supplemented tyrosine
Fumarylacetoacetate hydrolase	FAH	Tyrosinemia, type 1	1:120,000 (Sweden)	Low phenylalanine, tyrosine
Branched-chain ketoacid dehydrogenase complex	BCKDHA, BCKDHB, DBT, and DLD	Maple syrup urine disease	1:205,000–1:400,000	Low leucine, isoleucine, valine
Isovaleryl–CoA dehydrogenase	IVD	Isovalaric acidemia	1:50,000	Low leucine
Cystathionine-β-synthase	CBS	Homocystinuria	1:50,000–1:150,000	Low methionine, supplement cystine
		Organic Acid Disorders		
Methylmalonyl–CoA mutase	MUT	Methylmalonic aciduria	1:50,000	Low protein, isoleucine, methionine, threonine, valine
Propionyl–CoA carboxylase	PCCA or PCCB	Propionic acidemia	1:50,000	Low protein, isoleucine, methionine, threonine, valine, long-chain unsaturated fatty acids
		Urea Cycle Disorders		
Ornithine transcarbamylase	OTC	Ornithine transcarbamylase deficiency	1:14,000[b]	Low protein, supplement essential amino acids, increased energy
Arginase	ARG1	Arginase deficiency	1:350,000[b]	Low protein, supplement essential amino acids, increased energy

[a] Source unless otherwise indicated: Committee on Genetics (1996).
[b] Gropman and Batshaw (2004).

19.5.1. Phenylalanine Hydroxylase and Phenylalanine: Hyperphenylalaninemia and Phenylketonuria

Phenylketonurias, a group of inherited disorders of phenylalanine metabolism caused by impaired phenylalanine hydroxlyase (PAH) activity, are probably the most familiar of the gene–diet interactions in amino acid metabolism. PAH is the rate-determining enzyme in phenylalanine catabolism and protein and neurotransmitter biosynthesis. Over 300 mutations that result in a deficient enzymatic activity have been identified in the gene encoding PAH and lead to hyperphenylalaninemia (HPA) and phenylketouria (PKU). Deficient PAH enzymatic activity is the most common cause of HPA, with 99% of mutant alleles mapping to the *PAH* gene; the remainder map to genes coding for enzymes involved in homeostasis of tetrahydrobiopterin, a PAH co-factor (Jennings et al. 2000).

PAH deficiency is highly heterogeneous, and, depending on the specific mutations, there is a wide range in ability to metabolize phenylalanine to tyrosine. In a systematic analysis of the structural basis of 120 mutations with a classified biochemical phenotype and/or in vitro expression data, Jennings et al. (2000) determined that mutations that resulted in PAH protein with truncations and large deletions, fusion proteins, and active-site mutations typically resulted in severe phenotypes, namely, "classic PKU."

In "classic PKU," progressive, severe mental retardation can be prevented by early restriction of the PAH substrate, phenylalanine, and supplementation of the product, tyrosine. Because phenylalanine is essential, as well as toxic at pathophysiologic levels, nutritional therapy involves careful titration of the amino acids in this pathway in order to balance the risks of deficiency and toxicity. This is especially important in infants and young children; however, studies also suggest that phenylalanine restriction for life may help maintain normal cognitive function.

19.5.2. Branched-Chain Ketoacid Dehydrogenase and Leucine, Isoleucine, and Valine: Branched-Chain Ketoaciduria

Branched-chain ketoaciduria, or maple syrup urine disease (MSUD), results from a defect in decarboxylation of the branched-chain amino acids (BCAA) leucine, isoleucine, and valine. It can be caused by a mutation in one of least four genes: *BCKDHA*, *BCKDHB*, *DBT*, and *DLD*. These genes encode the catalytic components of the branched-chain α-ketoacid dehydrogenase complex (BCKD), which catalyzes the catabolism of BCAA. There are five major clinical subtypes of MSUD: the "classic" neonatal severe form, an "intermediate" form, an "intermittent" form, a "thiamine-responsive" form, and a "subunit E3-deficient with lactic acidosis" form. All of these subtypes can be caused by mutations in any of the 4 genes mentioned above, except for the E3-deficient form. In classic MSUD, which is the most common form of the disorder, infants appear normal at birth but by 4 or 5 days of age demonstrate poor feeding, vomiting, lethargy, and periodic hypertonia. A characteristic sweet, malty odor from perspiration and urine becomes evident by 7 days of age. Failure to treat this condition leads to acidosis, neurologic deterioration, seizures, and coma, and eventually death. Long-term nutritional therapy requires very careful titration of BCAA content of the diet and routine monitoring of blood BCAA concentrations (especially leucine and alloisoleucine), growth, and general nutritional adequacy.

19.6. GENETIC DIFFERENCES IN CARBOHYDRATE METABOLISM

Dietary carbohydrates are a major source of energy for humans, and glucose is an essential source of energy for certain tissues. Carbohydrates are consumed as polysaccharides (i.e., starches) and mono- and disaccharides (i.e., sugars). Impaired function of the enzymes responsible for carbohydrate metabolism and transport has varying levels of impact on human health. Typically, intolerances of dietary saccharides are seldom life-threatening because the offending sugar can be removed from the diet and symptoms avoided; however, severe disturbances in glucose–galactose malabsorption and galactose metabolism can be fatal if not diagnosed early.

19.6.1. Lactase and Lactose: Adult Lactase Persistence

In mammals, lactase activity is necessary for obtaining full nutritional benefit from milk because it is needed for digestion of lactose, the major carbohydrate in milk. Thus, in baby mammals, lactase or lactase phlorizin hydrolase (LPH), which hydrolyzes the disaccharide, lactose, to the monosaccharides, glucose and galactose, is essential. Developmental expression of lactase in small-intestinal enterocytes (the only place where it is expressed) is tightly controlled, with protein levels remaining low in fetal life, increasing around birth, and declining after weaning.

Lactase expression is highly variable in adult humans. This is the consequence of an unusual genetically determined regulatory polymorphism with large differences in allele frequency in human populations. Lactase nonpersistence is caused by decreased expression of *LPH* around the age of 5–10 years. In most adults, lactase is absent; however, its persistence tends to be the most frequent phenotype in northern Europeans and pastoral nomadic tribes where fresh milk forms a significant part of the adult diet.

For example, the prevalence of lactase persistence among Scandinavians and western Europeans is 92–98%, whereas among African Bantus, Japanese, and Thais, it is less than 10%. Lactase-nonpersistent, or lactose-intolerant, individuals can usually consume only limited amounts of fresh milk without experiencing flatulence and diarrhea.

The lactase persistence/nonpersistence phenotype appears to be controlled by the ability of a developmentally regulated DNA binding protein to act with a *cis* element upstream from the LPH gene (–13910) (Swallow et al. 2001; Troelsen et al. 2003) (Figure 19.3). During infancy, LPH expres-

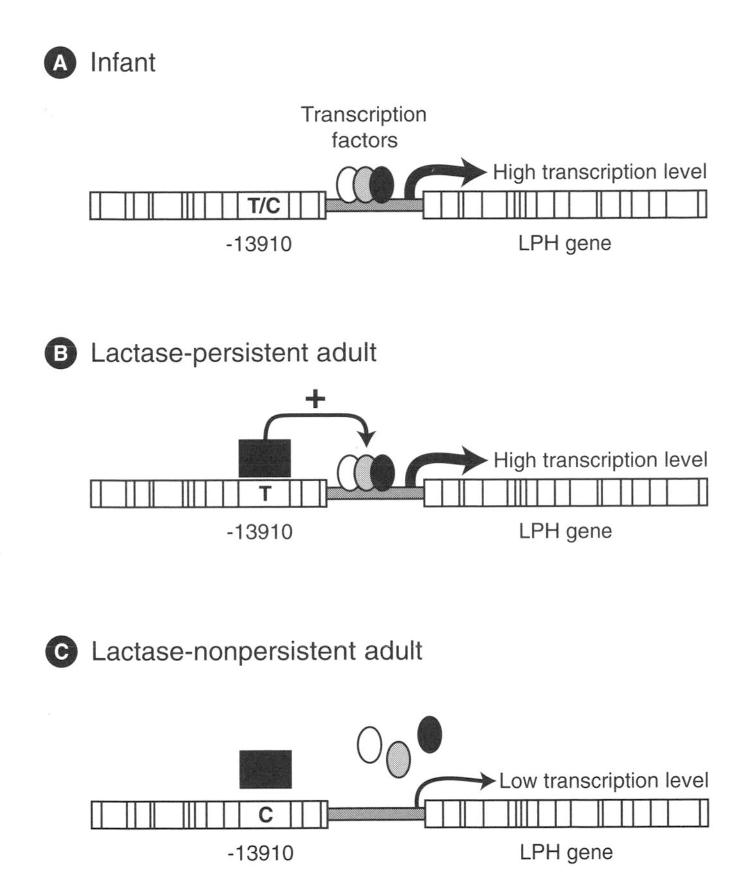

Figure 19.3. (A) In infants, lactase (LPH) expression is high because of sufficient levels of relevant transcription factors. (B,C) Adult lactase persistence involves an interaction between the LPH gene promoter and the polymorphic region at position –13910 upstream from the start codon of the gene; enhancer activity in –13910T variants (B) is stronger than in –13910C variants (C), allowing for more effective recruitment of transcription factors, and subsequent LPH expression. [Adapted from Troelsen et al. (2003).]

sion is high because levels of relevant transcription factors are high. During childhood, as transcription factor expression decreases or the transcription factors are coopted by other genes necessary for digestion of a more varied diet (e.g., sucrase–isomaltase with high-starch diets), enhancer activity in the upstream region is necessary in order to maintain LPH expression. The –13910T variant, found in lactase-persistent individuals, has approximately 4 times more enhancer activity than does the –13910C variant found in lactase-nonpersistent individuals; this ensures an active *LPH* gene into adulthood. Thus, in this example, the gene–diet interaction results from a polymorphism that controls developmental regulation of enzyme expression. Further, evolutionarily selection for lactose persistence has been driven by dietary exposure to lactose in populations for which milk is a significant part of the diet. Adult lactase persistence has become more prevalent since the introduction of dairy culture in approximately 10,000–12,000 B.P.

19.6.2. Sucrase–Isomaltase and Sucrose: Sucrose Maldigestion

The enzyme sucrase–isomaltase (SI) is an integral protein in the small-intestinal brush border that cleaves the disaccharides sucrose, maltose, and isomaltose into their respective monosaccharides. Because hydrolysis of these to monosaccharides is required for absorption, SI deficiency results in osmotic diarrhea, abdominal pain and cramping when the dissacharides are ingested. In the case of SI, the prevalent mutations do not affect the catalytic activity of the enzyme (in vitro enzymatic activity is normal) but cause defects in posttranslational processing that affect intracellular trafficking of the enzyme. These mutations can result in protein malfolding and mistargeting, such that the protein is not routed appropriately to the cell membrane.

Sucrose maldigestion is common, but varies globally; it is estimated to affect approximately 0.2% of Caucasians, but 10% of Inuit in Greenland. Severity of the condition depends on dietary sucrose intake and possibly is becoming more apparent with the increased intake of diets containing high amounts of sucrose (Swallow et al. 2001). Interestingly, the high prevalence of the enzyme deficiency became evident in Greenland when adoption of the practice of feeding excessive amounts of sucrose to babies led to clinically manifest failure to thrive.

19.6.3. Galactose-1-Phosphate Uridyl Transferase and Galactose: Galactosemia

Classic galactosemia is caused by mutations in the galactose-1-phosphate uridyl transferase (GALT) gene. Disturbance in the conversion of galactose to glucose due to GALT deficiency results in an accumulation of galactose and/or galactose-1-phosphate. If transferase activity is absent, illness occurs within 2 weeks of birth, and if diagnosis and treatment are delayed, mental retardation can result. Galactosemia is treated with lifelong galactose restriction, primarily through strict avoidance of milk and milk products, other lactose-containing foods, and fruits and vegetables (e.g., papaya, watermelon, bell pepper, tomato) that are significant sources of galactose.

The incidence of GALT deficiency varies in different populations, ranging from 1 in 30,000–40,000 in European countries to as few as one in a million in Japan. The disorder exhibits considerable allelic and phenotypic heterogeneity, with over 150 recorded base changes and ranges in enzymatic activity and disease severity (Tyfield et al. 1999). Most base changes (>60%) are missense mutations scattered the length of the gene, and a few are nonsense or splice-site mutations (14%) or small deletions or insertions (10%). Even the phenotypes associated with the missense mutations are heterogeneous. Some (e.g., Q188R) severely affect catalytic activity, where as others (e.g., N314D) are associated with intact activity but altered protein abundance.

19.7. GENETIC DIFFERENCES IN LIPID METABOLISM

Lipids or fats, primarily triacylglycerols (triglycerides), are a major energy source in the human diet throughout life. Dietary fats absorbed into the intestine are packaged into large triacylglycerol-rich chylomicrons for delivery to sites of lipid metabolism or storage. During their transit to the liver, the particles interact with lipoprotein lipase (LPL) and undergo partial lipolysis to form chylomicron remnants. The remnants pick up apolipoprotein E (apoE) and cholesteryl ester from high-density lipoproteins (HDLs) and are taken up by the liver by a process mediated by the interaction of apoE with hepatic receptors. Typically, this is a rapid process; however, there is great interindividual variation in postprandial exogenous lipoprotein metabolism.

Further endogenous metabolism is required to handle the lipoproteins. In the liver, hepatocytes synthesize and secrete triacylglycerol-rich very-low-density lipoprotein (VLDL), which can be converted to intermediate-density and low-density lipoprotein (LDL) through lipolysis, similar to the mechanism involved in chylomicron handling. The excess protein components from the LDL may be taken up by HDL and recycled by the liver. Polymorphisms in the exogenous and endogenous lipoprotein pathways influence the efficiency of transport and handling of lipids and alter risk for cardiovascular disease (Ordovas 2001). More than 250 SNPs have been identified in about fifteen genes encoding for key proteins in the pathways

Table 19.3 Polymorphic Proteins in Exogenous and Endogenous Lipoprotein Pathways

Exogenous Lipoprotein Metabolism
Apolipoprotein B
Apolipoprotein A-IV
Apolipoprotein E
Apolipoprotein C-III
Lipoprotein lipase
Microsomal triacylglycerol transfer protein
Intestinal fatty-acid-binding protein
Endogenous Lipoprotein Metabolism
LDL receptor
3-Hydroxy-3-methylglutaryl–coenzyme A reductase
Reverse Cholesterol Transport
Apoprotein A-I
Cholesteryl ester transfer protein
Hepatic lipase
Cholesterol 7 α-hydroxylase
Scavenger receptor B type I
ATP-binding cassette I

involved in lipid and lipoprotein metabolism (Table 19.3). These result in a range of phenotypes, some with consequences for disease, but the interactions with diet are not always well established.

19.7.1. Intestinal Fatty-Acid-Binding Protein and Lipoprotein Metabolism

Intestinal fatty-acid-binding protein (FABP2) is a cytosolic intracellular protein that binds nonesterified fatty acids and delivers them to the membrane. It plays important roles in several steps of fat absorption and transport, including uptake and trafficking of saturated and unsaturated long-chain fatty acids, targeting of free fatty acids toward different metabolic pathways, protecting the cytosol from cytotoxic effects of free fatty acids, and modulating enzymatic activity involved in lipid metabolism. A common SNP at exon 2, codon 54, with a G-to-A mutation results in a substitution of alanine (wildtype) to threonine (variant). This polymorphism is very common, with a 54Thr allele frequency of approximately 30% in most populations. The functional changes resulting from the 54Thr variant are higher fatty-acid-binding and transport and greater triacylglycerol secretion from intestinal cells. Associated clinical traits include higher body mass index, higher insulinemia and insulin resistance, higher circulating fasting LDL cholesterol, apoB, and triacylglycerol concentrations. In addition, postprandial insulin and triacylglycerol responses are also influenced by genotype, although results are not consistent across all studies. Interestingly, in a study of Keewatin Inuit, whose diets are high in omega-3 fatty acids from fish, the 54Thr allele was associated with lower 2-h glucose concentrations, suggesting that the type of fatty acid consumed may interact with functional differences in the gene products to produce phenotypic differences (Ordovas 2001). Diets high in soluble fibers (guar gum, pectins, etc.) also are more effective at lowering LDL cholesterol and apoB in individuals with the 54Thr variant.

19.7.2. Apolipoprotein E and Lipoprotein Metabolism

Apolipoprotein E is involved in uptake of triacylglycerol-rich lipoprotein (intermediate-density lipoprotein and chylomicron remnants) by the liver and peripheral tissues through interaction with apoB, apoE, and remnant receptors. Genetic variation at the apoE locus results in three common alleles, E4, E3, and E2, with frequencies in Caucasian populations of approximately 15%, 77%, and 8%, respectively. These produce three homozygous (*E2/2*, *E3/3*, and *E4/4*) and three heterozygous (*E2/3*, *E2/4*, and *E3/4*) genotypes. The sequence changes in variants E4 and E2 result in alterations in two charged amino acids, affecting plasma clearance of the proteins and related cholesterol-rich lipoproteins. Population studies have shown that plasma cholesterol, LDL cholesterol, and apoB levels are highest in individuals with the apoE4 allele, intermediate in those with the apoE3 allele, and lowest in those with the apoE2 allele, whereas postprandial triacylglycerol levels are highest in individuals carrying the apoE2 allele and lowest in those with the apoE4 allele. Diet further influences manifestation of the phenotypes; however, the effects have been discordant. This possibly reflects differences in diets and study populations across studies, in that age, physical activity levels, body weight, and fasting lipid levels all affect response. In general, individuals with the E4 allele appear to be most responsive to changes in dietary fat and cholesterol but less responsive to dietary fiber supplementation than are individuals carrying the E2 allele (Masson et al. 2003). Among men and women with coronary heart disease, a high sucrose intake has been shown to be associated with high plasma triacylglycerol concentration only in patients with the E2 allele. These results suggest that gene–diet interactions in relation to lipoprotein metabolism are as multifactorial as the diseases with which they are associated and also suggests that dietary management of cardiovascular disease risk may need to be tailored to genotype.

19.8. GENETIC DIFFERENCES IN VITAMIN METABOLISM

Vitamins are non-energy-providing, organic compounds that are essential for specific metabolic reactions and that cannot be synthesized in sufficient quantities from simple metabolites. Vitamins are usually classified into two groups on the basis of solubility (i.e, water-soluble and fat-soluble), which to a certain extent determines their stability, occurrence in foodstuffs, distribution in body fluids, and tissue storage capacity (Table 19.1). Many vitamins act as coenzymes or as parts of enzymes responsible for promoting essential chemical reactions. SNPs in the various enzymes, carrier proteins, and receptors involved in vitamin metabolism affect nutrient requirements.

19.8.1. Methylenetetrahydrofolate Reductase and Folate: Hyperhomocysteinemia

Tetrahydrofolate is a carrier for single-carbon formyl, hydroxymethyl, or methyl groups. It plays an important role in the synthesis of the purines, guanine and adenine, and of the pyrimidine, thymine, compounds central to the synthesis of DNA and RNA. It also participates in amino acid interconversion (e.g., serine and glycine), oxidation (e.g., glycine), and methylation (e.g., homocysteine to methionine) and in methylation of DNA. Folate deficiency results in poor growth, megaloblastic anemia and other blood disorders, glossitis, and gastrointestinal tract disturbances. Folate deficiency also has been associated with increased risk of neural tube defects, vascular diseases, several cancers, impaired cognitive function, and psychoneurological conditions such as Alzheimer's disease and affective disorders.

The enzyme methylenetetrahydrofolate reductase (MTHFR) catalyzes the irreversible conversion of 5,10-methylenetetrahydrofolate to 5-methyltetrahydrofolate and toward methionine synthesis at the expense of DNA synthesis. Two MTHFR polymorphisms (C667T and A1298C) are associated with reduced enzymatic activity and impaired conversion of homocysteine to methionine. The homozygous C667T variant (TT; 5–15% of most populations) is associated with altered distribution of reduced folates, aberrant DNA methylation, and lower plasma folate and elevated plasma homocysteine concentrations. Individuals who are homozygous for A1298C do not have lower plasma folate or increased homocysteine; however, individuals who are heterozygous for both polymorphisms (approximately 15% of the Caucasian population) are phenotypically similar to individuals with the TT genotype with respect to reduced

MTHFR activity, and lower plasma folate and higher homocysteine. The TT genotype and concomitantly elevated circulating homocysteine concentrations are associated with increased risk of cardiovascular disease. In contrast, the TT genotype in conjunction with a sufficient folate intake is associated with lower risk of several cancers and precancers (e.g., adenomatous polyps). The reduced MTHFR activity probably allows for more substrate channeling through the purine and pyrimidine synthesis pathways, improving DNA repair capacity, and maintaining fidelity of DNA synthesis. Thus, as genotype and folate intakes jointly vary, risks of several developmental outcomes and chronic diseases change in predicatable, but sometimes contradictory, ways.

19.8.2. Vitamin D_3 Receptor and 1,25-α Dihydroxycholecalciferol: Calcium Absorption

(*Note*: This topic is also discussed in Chapter 11, on receptors and ion channels.) Vitamin D receptors (VDR) are intracellular polypeptides that specifically bind 1,25-α dihydroxycholecalciferol, the active form of vitamin D, and interact with target cell nuclei to produce a range of biologic effects. The VDR protein belongs to the superfamily of *trans*-acting transcriptional regulatory factors, including steroid and thyroid hormone receptors, and is most closely related to the thyroid hormone receptors. VDR and vitamin D exert effects in almost every tissue in the body. Targets for vitamin D signaling include the central nervous system, skin and hair follicles, the immune system, and endocrine glands. Studies in mice deficient in VDR have shown that VDR is critical for growth, bone formation, and female reproduction in the postweaning stage. At the cellular level, vitamin D signaling affects proliferation, differentiation, and apoptosis of both normal and transformed cells. Thus, VDR polymorphisms are recognized as important in relation to susceptibility toward a wide range of diseases, including osteoporosis and cancer.

Polymorphisms in the VDR gene have been associated with bone turnover and bone density in humans (Eisman 2001). In the gut, VDR and vitamin D regulate calcium absorption. Several VDR polymorphisms predict differences in gut calcium absorption and long-term bone density response to calcium intake and active vitamin D analog treatment. For example, "bb" individuals (homozygous for the VDR BSM-1 restriction site) and "BB" individuals (homozygous wildtype, without the restriction site) had similar levels of calcium absorption on high calcium intakes (1500 mg/day), but "BB" individuals did not experience increase in their gut calcium absorption when exposed to lower dietary calcium intakes (<300 mg/day) (Dawson-Hughes et al. 1995). Urinary excretion of calcium has also been shown to be higher in "bb" genotypes probably because of increased intestinal absorption (Ongphiphadhanakul et al. 1997). Nonetheless, despite these effects on calcium handling, the resulting phenotypes do not consistently include lower bone mineral densities, suggesting that additional control mechanisms help maintain bone mass. Relationships between VDR and bone mass are complicated further by the evidence that VDR polymorphisms are also associated positively with body weight, one of the strongest positive predictors of bone density. Some of the relationships between VDR genotypes, bone size, and bone mass are proposed to be mediated through insulin or insulin-like growth factor pathways. Calcium homeostasis phenotypes are affected by the several polymorphisms in VDR, coupled with differences in exposure to calcium and vitamin D, as well as polymorphisms in other genes associated with the pathway (e.g., estrogen, calcitonin, and parathyroid receptors).

19.9. GENETIC DIFFERENCES IN MINERAL HANDLING

Collectively, minerals represent about 4–5% of body weight. About half of this weight is calcium, and another quarter is phosphorus. The five other macrominerals (magnesium, sodium, chloride, potassium, and sulfur, which are required in amounts of ≥100 mg/day) and the 14 microminerals (iron, zinc, copper, iodide, manganese, fluoride, molybdenum, cobalt, selenium, chromium, tin, nickel, vanadium, and silicon) make up the remaining 25%.

Macrominerals have essential roles as dissolved ions in body fluids and as structural constituents of body tissues. The balance of ions in body fluids regulates enzymatic activity, maintains acid–base balance and osmotic pressure, facilitates membrane transfer of essential compounds, and maintains nerve and muscular irritability. The microminerals function primarily as cofactors in enzyme systems by (1) direct participation in catalysis, (2) combination with substrate to form a complex on which the enzyme acts, (3) formation of metalloenzymes, (4) combination with reaction end products, or (5) maintenance of quaternary protein structure. Frequently, the phenotypes that are manifest as a result of polymorphisms are alterations to the enzyme systems for which the relevant mineral is a cofactor. For example, in Menke's disease, a copper malabsorption disease, many of the phenotypic features are due to interference with crosslinking of collagen and elastin, for which copper is a key component. Polymorphisms in proteins involved in absorption, transport, metabolism, and excretion of the different minerals influence the processes in which they are involved and influence dietary requirements.

19.9.1. HFE and Iron: Hemochromatosis

Globally, iron deficiency and the ensuing debilitating anemia have long been a primary public health concern. However, with the advent of iron fortification of foods and increased use of dietary supplements, excessive iron intake among individuals susceptible to iron overload has also become a clinical concern (Heath and Fairweather-Tait 2003). Iron homeostasis is controlled by changes in absorptive efficiency from the gut, and there are no general mechanisms for excretion or elimination of absorbed iron. Women of childbearing age can lose iron through menstruation, pregnancy, or lactation, whereas men and postmenopausal women, lacking these routes of elimination, are more susceptible to iron overload. Prolonged, high intakes of iron are unlikely to cause iron overload in the general population because iron absorption is so tightly regulated, although control of absorption of heme iron (i.e., organic iron bound to the porphyrin ring of hemoglobin and myoglobin in animal-derived foods) may not be as strict as that for nonheme iron (i.e., inorganic iron from plant sources).

The precise mechanism by which iron is absorbed and how this absorption is regulated is unknown; however, several different proteins that are involved either in the iron transport process itself or its regulation have been identified. These include HFE, a class 1 HLA molecule involved in hereditary hemochromatosis, the divalent metal transporter (DMT-1), hephaestin, the transferrin receptor, and mobilferrin. On the basis of genetic data, these proteins clearly play a role in iron trafficking, but how they fit into the process of getting a highly reactive metal safely through the enterocyte is not understood. HFE binds to transferrin receptor, DMT-1 appears to be involved in intracellular iron transport, and hephaestin is a ceruloplasmin-like protein that probably serves to oxidize iron at the luminal side of the cell membrane so that it may be bound by plasma transferrin (Beutler et al. 2001).

Most cases of iron overload found in populations of northern European ancestry are attributed to mutations in the *HFE* gene. The protein product complexes with β_2-microglobin and transferrin receptor and somehow regulates iron trafficking in epithelial cells in the small intestine. Mutations in *HFE* impair HFE function and set up a low-iron state in the epithelial cell indistinguishable from that found in anemia. The hypothesis is that this stimulates iron absorption and eventually leads to iron overload. Two *HFE* mutations are common in patients with hemochromatosis: c.845G $\rightarrow$ A (C282Y) and c.187C $\rightarrow$ G (H63D). Most patients are homozygous for the 845A mutation, a cysteine-to-tyrosine substitution at position 282 in the HFE preprotein that appears to result in a protein product that prevents binding with β_2-microglobin. The structural effect of 187G is not known, but the 845A and 187G mutations are in complete linkage disequilibrium, and a large proportion of hemochromatosis patients have the combined 845A/187G genotype (Beutler et al. 2001). A number of other functional mutations that result in milder forms of the condition also have been found in the *HFE* coding sequence and, more recently, other types of hemochromatosis have been identified: juvenile or type 2 hemochromatosis, not related to *HFE* gene mutations; and type 3 hemochromatosis, which is associated with mutations of the transferrin receptor 2 gene but has a phenotype indistinguishable from that of *HFE* hemochromatosis. Additionally, African iron overload appears to be due to another yet-to-be-identified non-*HFE* iron-loading gene.

Hemochromatosis results in a gradual, progressive accumulation of iron in body tissues. Patients with hereditary hemochromatosis are at higher risk for death from liver cancer, cardiomyopathy, liver cirrhosis, and diabetes. If the condition is diagnosed early, treatment by regular phlebotomy to remove excess iron is very effective and results in normal lifespan. Dietary interventions that decrease both intake and bioavailability of dietary iron can reduce the number of routine phlebotomies required to manage the condition. Individuals with iron overload are also counseled to lower their intakes of heme iron (i.e., meat, fish and poultry) compared to nonheme iron (plant sources) and to avoid alcohol and high levels of vitamin C, which both enhance iron absorption. Thus, the form as well as amount of iron ingested are important in relation to this gene–diet interaction.

19.9.2. Cu^{2+}-Transporting ATPase and Copper: Menkes Disease

Menkes disease is an X-linked recessive disorder that results in copper malabsorption and abnormal intracellular copper transport and is characterized by early retardation in growth, and focal cerebral and cerebellar degeneration. Severe neurologic impairment begins within a month or two of birth and progresses rapidly to decerebration.

Gross deletions in the gene for Cu^{2+}-transporting ATPase alpha polypeptide result in reduced or altered mRNA expression. The protein is normally localized in the *trans*-Golgi network (TGN) of cells. When cells are exposed to excessive copper, the enzyme is relocalized to the plasma membrane, where it functions in copper efflux. With Menkes, this efflux system is impaired and a marked accumulation of copper occurs in the intestinal mucosa, although serum copper and ceruloplasmin levels are very low. Biochemical features of Menkes disease reflect the copper dependence of many enzyme systems. Connective tissue and arterial abnormalities occur because lysine-derived crosslinks in elastin and collagen are impaired; copper-dependent amine oxidase is responsible for the initial modification of lysine. Defective formation of disulfide bonds in keratin leads to characteristic kinky hair. Parenteral administration of copper, bypassing the defective absorption system, can result in transient improvement; however, the classic form of Menkes disease usually ends in death in early childhood.

19.10. GENETIC DIFFERENCES IN METABOLISM OF DIETARY CARCINOGENS

Exposure to dietary carcinogens depends on food choices and options and preferences in food storage and food preparation methods. Typically, carcinogens are metabolized by xenobiotic biotransformation enzymes to facilitate excretion. In the biotransformation process, reactive intermediates with enhanced carcinogenicity are often formed during the initial metabolic steps. Other biotransformation pathways usually eliminate the reactive intermediates, although not with 100% efficiency. Thus, the potency of a specific compound is often determined by the relative rates of activation and detoxification.

Genetic variability in biotransformation enzymatic activity or expression further influences the metabolic fate of a carcinogen. Depending on the effect of a polymorphism on gene expression or on enzyme function (i.e., increased or decreased activity toward a particular substrate), certain variants could result in higher or lower risk for mutagenesis in the presence of the dietary carcinogen. Numerous epidemiologic studies have examined the impact of gene–carcinogen interactions in relation to cancer risk; these are addressed in the specific disease-oriented chapters (Chapter 13–18). In contrast, few studies have measured the direct impact of the variants on response to exposure in humans.

19.10.1. Polymorphisms in Biotransformation Enzymes and Effects of Cooked Meat

Epidemiologic studies have reported an association between high intake of meat and risk of several cancers, including colon and pancreas. Cooking meat generates many compounds that are mutagenic in bacterial assays, and some which are carcinogenic in animal models. The heterocyclic amines (HA) 2-amino-3,8-dimethylimidazo[4,5-*f*]quinoxaline (MeIQx), 2-amino-3,4,8-trimethylimidazo[4,5-*f*]quinoxaline (DiMeIQx), and 2-amino-1-methyl-6-phenylimidazo[4,5-*b*]pyridine (PhIP), are formed when food, particularly red meat, is cooked at high temperatures (e.g., fried, charbroiled, or roasted). Thus, exposure to these depends on dietary choices and cooking methods. They are readily absorbed and bioavailable and are extensively, but variably, metabolized (Gooderham et al. 2001). Metabolic activation of HA involves primarily CYP1-mediated *N*-hydroxylation followed by esterification by conjugating enzymes.

Results of several small-scale controlled dietary studies in humans have shown that biotransformation enzyme polymorphisms or associated drug-metabolizing phenotypes influence clearance of some of these mutagens. For example, when individuals were exposed to HA in a diet containing ground beef cooked at high temperature, higher CYP1A2 activity (determined by caffeine metabolite phenotyping) was associated with lower levels of total unconjugated MeIQx in urine (Sinha et al. 1995). In another study, controlled 2-day exposure to a "high meat" diet (200 g smoked sausages, two slices of fried meat, and four hamburgers daily) resulted in increased levels of DNA strand breaks in exfoliated colorectal mucosal cells. This effect was particularly pronounced in individuals who had an *N*-acetyltransferase 2 (NAT2) genotype associated with rapid acetylation and who were null for the GSTM1 genotype (Kiss et al. 2000). These studies demonstrate that genotypic differences in biotransformation enzymes influence metabolism of and exposure to the reactive DNA-damaging intermediates formed during metabolism of dietary precursors in cooked meats.

19.10.2. Polymorphisms in Biotransformation Enzymes and Aflatoxin: AFB_1–Albumin Adduct Formation

Mycotoxins, secondary metabolites produced by fungi and molds, are found in cereal grains, forages, corn, rice, peanuts, and peanut oil. Of the common dietary mycotoxins, aflatoxins, and aflatoxin B_1 (AFB_1) in particular, are strong carcinogens; AFB_1 is the most potent hepatocarcinogen that has been identified. In humans, AFB_1 is activated to a reactive epoxide intermediate by CYP1A2 and subsequently is hydrolyzed spontaneously or by epoxide hydrolase, or is conjugated with glutathione by GST. Reduced formation and increased clearance of the reactive epoxide is key to preventing formation of N-7 guanine adducts in DNA, a first step in the carcinogenic pathway. The interaction between polymorphisms in the key metabolizing enzymes and aflatoxin exposure influences disease risk; however, the relationship is not straightforward, and results of various studies are inconsistent (McGlynn et al. 2003). Just as AFB_1 forms DNA adducts, it also forms protein adducts, and AFB_1–albumin adducts in blood can be used as a marker of AFB_1 exposure and metabolism. Depending on the population studied, researchers have reported positive, inverse, and no associations between GSTM1 null genotype and AFB_1–albumin adduct levels. Similarly, polymorphisms in EPHX1, which encodes for microsomal epoxide hydrolase, are not consistently associated with AFB_1–albumin adducts. These results point to the complexity of the interrelationships and the likely effect of combined variation in enzymatic activity. For example, being null for GSTM1 may not be a risk factor if epoxide hydrolase activity is intact or if CYP1A2 activity is reduced.

19.11. GENETIC DIFFERENCES AND PHYTOCHEMICALS

A wide range of other organic compounds make up a substantial part of the human diet. In humans, many plant compounds—phytochemicals—are potent effectors of biologic processes and therefore have the capacity to influence disease risk via several complementary and overlapping mechanisms. Mechanisms include antioxidant activity, alteration of biotransformation enzymatic activity, antibacterial and antiviral effects, alteration of immune function, reduction of inflammation, modulation of steroid hormone concentrations and hormone metabolism, arrest in cell cycle progression, and stimulation of apoptosis. In theory, as in the case of nutrients, genetic variation in pathways affecting absorption, metabolism, and distribution of phytochemicals is likely to influence exposure at the tissue level. Similarly, genetic variation in the pathways with which these compounds interact can alter response. However, beyond a few well-recognized conditions (see, e.g., Section 19.11.1), very little is known about the effects of genetic variation on these gene–phytochemical interactions.

19.11.1. Glucose-6-phosphate Dehydrogenase and Vicine and Covicine: Favism

Glucose-6-phosphate dehydrogenase (G6PD) is the first enzyme in the pentose phosphate pathway and is the rate-determining step in the oxidative portion of the pathway. In red blood cells, its main role is to produce the NADPH necessary to protect cells against free radicals and peroxides. G6PD deficiency affects over 400 million people worldwide, primarily populations where malaria is prevalent. Individuals whose red blood cells are G6PD-deficient are more resistant to effects of malaria, consistent with genetic selection for this protection against malaria. Hundreds of G6PD variants have been described, and there are marked variations in prevalence rates. Typically G6PD deficiencies are not extreme; however, clinical manifestations occur when this lowered reducing capacity is coupled with an exogenous oxidative stress, such as exposure to an oxidative agent, diabetic ketoacidosis, or viral infection. Ironically, certain antimalarials (e.g., primaquine), and sulfonamides and sulfones—drugs that have strong oxidative capacity—also elicit adverse reactions in G6PD-deficient individuals (Beutler 1996).

One well-recognized dietary oxidant source is fava beans (*Vicia faba*), a common human food in the Mediterranean regions of Europe and in the Middle East. The glycosides, vicine and covicine, present in fava beans are hydrolyzed into the pyrimidine aglycones, divicine and isouramil, in the intestinal tract. Without normal G6PD activity, these strongly oxidative metabolites absorbed into the bloodstream attack the membranes of red blood cells. Among susceptible individuals, eating fava beans or even inhaling pollen from the flowers can cause favism. The symptoms of favism include headaches, dizziness, and nausea, which progress to vomiting, abdominal pain and fever, and finally acute hemolytic anemia and renal failure, which can be fatal. Given the high prevalence of G6PD deficiency in Mediterranean populations, it is not surprising that the Greek philosopher and mathematician Pythagoras warned his students "Eat no beans."

19.11.2. Metabolism of Phytochemicals

Phytochemicals are metabolized by biotransformation enzymes in a manner similar to that for other xenobiotics. Many classes of phytochemicals are rapidly conjugated with glutathione, glucuronide, and sulfate moieties and excreted in urine and bile. Thus, polymorphisms in biotransformation enzymes, such as the glutathione *S*-transferases, UDP–glucuronosyltransferases, and sulfotransferases, have the capacity to affect phytochemical metabolism in the same fashion as they do carcinogens. However, the difference between carcinogen and phytochemical metabolism is that, whereas polymorphic enzymes that reduce the circulating half-lives of carcinogens are thought to lower cancer risk, when these same (or other) polymorphisms reduce the half-lives of phytochemicals that may be protective, they may in fact increase risk.

The primary route of in vivo metabolism of isothiocyanates (ITC) is by the mercapturic acid pathway, a major pathway for elimination of many xenobiotics (Conaway et al. 1999). Thiol conjugates of ITC are formed by conjugation with glutathione, catalyzed by GST. Subsequent, stepwise cleavage of glutamine and glycine yields L-cysteine-ITC, which are acetylated to produce *N*-acetyl-L-cysteine ITC conjugates (mercapturic acids); these are excreted in urine. Thus, GSTs play an important role in disposition of ITC in humans. Benzyl ITC, phenethyl ITC, allyl ITC, and sulforaphane—common ITC in cruciferous vegetables—are all catalyzed by the four major human GSTs: GSTA1-1, GSTP1-1, GSTM1-1, and GSTM2-2; however, reaction velocities of isozymes can differ by as much as 700-fold, and there is wide variation in the extent to which ITC are disposed (Zhang et al. 1995).

Cruciferous vegetables have been shown to protect against several cancers, particularly via interaction with GST polymorphisms (see also Chapter 7). Polymorphisms in the GSTM1 and GSTT1 genes that result in complete lack of GSTM1-1 and GSTT1-1 protein, respectively, appear to confer greater protection from cruciferous vegetables in individuals with these genotypes. For example, in one study, individuals with the highest quartile of broccoli intake had the lowest risk for colorectal adenomas compared to individuals who reportedly never ate broccoli; this inverse association was observed only in those with the *GSTM1*-null genotype (Lin et al. 1998). Similarly, in studies of colon cancer and lung cancer, risk was altered by cruciferous vegetable intake only in particular subgroups defined by age, smoking status, and *GSTM1* genotype (Slattery et al. 2000; Spitz et al. 2000). Results of one population-based study of ITC excretion have shown that urinary ITC was higher among *GSTT1*-positive, relative to *GSTT1*-null, individuals, but that *GSTM1* and *P1* genotypes had no effect (Seow et al. 1998). These data support the in vitro evidence that both GSTM1 and T1 metabolize ITC and that the combination of cruciferous vegetables and the *GST* genotypes may modify cancer risk.

Dietary flavonoids are present in a wide variety of fruits and vegetables and other plant foods, and daily intakes are estimated at 23–1000 mg/day. This is a structurally diverse class of compounds. It includes the flavones and flavonols (apigenin, chrysin, galangin, luteolin, quercetin, etc.), flavanes (catechin, hesperetin, naringenin, etc.), and isoflavonoids (genistein, daidzein, etc.). The selectivity of glucuronosyl conjugation of the flavonoids is dependent on both the structure of a particular flavonoid and the UGT isozyme involved in its conjugation (see also Chapter 7). For example, UGT1A1, UGT1A8, and UGT1A9 have been shown to be especially active in conjugating luteolin and quercetin, whereas UGT1A4 and UGT1A10 and UGT2B7 and UGT2B15 in the UGT2B family are less efficient (Boersma et al. 2002). In contrast, the isoflavone genistein appears to be more effectively conjugated by the UGT2B family. Although the effects of the polymorphisms in UGT on flavonoid clearance have not been examined, studies showing that polymorphisms affect glucuronidation of several drugs suggest that it is likely that similar effects may be seen for the dietary flavonoids.

19.12. SUMMARY

Polymorphic genes that are involved in the absorption, transport, metabolism, and excretion of nutrients can affect nutrient requirements. Moreover, subtle genetic differences influence an individual's optimal nutrition, their nutritional requirements, and susceptibility to disease. The most effective intervention or prevention of chronic diseases will occur when targeted changes in environmental factors, including diet, are matched to an individual's specific genetic susceptibility. Improved understanding of gene–diet interactions will facilitate satisfaction of the nutritional needs of individuals. However, in the case of some aspects of metabolism (e.g., lipids), the often wide rift between genotype and phenotype, as well as the wide range in responses to a dietary exposure, suggest that there is still much to be learned before we could effectively tailor dietary interventions to individual genotypes in order to reduce disease risk.

19.13. SUGGESTED READING

The online publication by McKusick et al. (no date), the text by Coulston et al. (2001), and the paper by Friso and Choi (2002) are excellent sources of information on the possible links between genetic variation, diet, and disease susceptibility.

20

Genetic Determinants of Addiction to Alcohol, Tobacco, and Drugs of Abuse

Andrew J. Saxon
Veterans Affairs Puget Sound Health Care System, Seattle, WA

20.1. INTRODUCTION

As in other areas of medicine, more recent years have witnessed rapidly emerging evidence for the importance of genetic explanatory models in the etiology of addictive disorders. Numerous genetic epidemiologic studies demonstrate conclusively that heritability plays a major role in the development of these disorders. The challenge ahead lies in determining the specific genes and gene products that exert this impact.

This chapter addresses this area of intensive scientific scrutiny by first describing the epidemiological work that has verified genetic contributions to addictive disorders. It will then summarize biological aspects of addiction essential to understanding potential mechanisms of genetic contributions. Next will come sections devoted to a synopsis of current knowledge concerning potential genetic markers for several of the major addictive substances including alcohol, tobacco, opioids, and cocaine. Finally, a summary section will seek to integrate the current findings in the context of their meaning for genomics and in the clinical management of patients.

A few general points, which underscore the complexity of the topic of the genetics of addiction, deserve mention at the outset:

Gene-Environment Interactions: Fundamentals of Ecogenetics, edited by Lucio G. Costa and David L. Eaton

1. Defining phenotypes in addiction, as in most other psychiatric disorders, poses a challenge (Merikangas and Risch 2003). Contrary to the situation with many medical disorders such as cancer, diabetes, or cardiac disease for which a tissue, blood, or radiological examination confirms the diagnosis and phenotype, in the case of brain disorders such as addiction, no specific, objective test currently exists that definitively confirms the phenotype. Since the mid-1990s a set of seven physiologic and behavioral criteria as set forth in the *Diagnostic and Statistical Manual for Mental Disorders* (DSM-IV) has served as the most widely accepted approach to defining a diagnosis and/or phenotype of substance dependence. The presence of a minimum of three criteria is required for the diagnosis. A less serious diagnosis (or possibly less severe form of the phenotype) of substance use disorder is substance abuse, which is determined by the presence of a minimum of one problematic behavior that is related to the use of the substance. However, individuals with substance abuse frequently go on to develop substance dependence. Generally, the presence of a diagnosis can be established with relative certainty by administration of a structured clinical interview. The precise type of structured interview may vary across studies, and older studies frequently applied slightly different diagnostic criteria to define the phenotype.

2. Addiction quite obviously represents a disease state that cannot be explained purely on the basis of genetics. At the very least, environmental exposure to the addictive substance (and possibly other environmental insults) must occur to create this disease state. Thus, the disease of addiction necessarily develops as a consequence of gene–environment interactions (Merikangas and Avenevoli 2000).

3. Data accrued thus far and reviewed below indicate that genetic propensity to addiction, as with other psychiatric and complex disorders, most likely arises from the combined influence of several genes rather than from a single gene (Insel and Collins 2003). Thus, genetic models of addiction, while evolving, remain from far complete.

Dividing this chapter into sections that consider the genetics of each addictive substance makes organizational sense and may have some theoretic underpinnings, but in the real world most individuals with addiction problems suffer from dependence on multiple substances. It will become readily apparent that substantial overlap exists for potential genetic causes of addiction to the various substances.

Finally, considerable previous work has looked at the genetic aspects of addiction in laboratory animals, but most of the information presented here will concern humans. Animal studies will be noted only to the extent that they inform efforts at grappling with the conundrums posed by human addiction genetics.

20.2. GENETIC EPIDEMIOLOGY OF ADDICTION

A number of epidemiologic approaches have elucidated the genetics of addiction and reached similar conclusions. These methods include family studies, adoption studies, and twin studies. A few pertinent studies have also examined subjective and biological responses to a test dose of alcohol in individuals who do not have an alcohol use disorder in an effort to determine the heritability of the direct pharmacological effects of alcohol.

Family studies recruit probands with the phenotype of a specified substance use disorder. An attempt then ensues to determine the phenotypes of blood relatives. The rates of the disorder of interest in the blood relatives of affected probands may be compared to (1) rates of the disorder in blood relatives of a specifically recruited control sample whose members do not have the disorder or (2) known rates of the disorder in the general population. If blood relatives of the affected probands have elevated rates of the disorder, this suggests that either a genetic factor, a familial-specific environmental factor, or both contribute etiologically to the disorder. Overall family studies demonstrate that relatives of affected probands have 2–8 times the risk of substance dependence as relatives of nonaffected probands or of members of the general population, and the family studies also convey some suggestions that familial inheritance of addictive disorders may be in part substance-specific (Merikangas and Avenevoli 2000). An obvious limitation to the family studies is that they cannot sort out conclusively the degree to which familial transmission springs from genes versus environment.

This limitation can be addressed in part by *adoption studies*, which generally involve longitudinal follow-up of adoptees whose biological parents either had or did not have a substance use disorder. Using adoptees as subjects, this methodology attempts to eliminate the confound of the environmental effects of a child growing up in a household with an addicted parent. If the adoptees with a biological parent positive for the disorder develop the disorder at a higher rate than do adoptees with a biological parent negative for the disorder, researchers can reasonably conclude that heritable traits contribute to development of the disorder. In regard to alcohol dependence, research suggests that adoptees who have a biologic parent with alcohol dependence develop alcohol dependence at 2–3 times the rate of adoptees without such a parent (Sigvardsson et al. 1996). One study that examined drug abuse in adoptees (242 male, 201 female) found that the adoptees with a biologic first-degree relative with an alcohol problem had 4 times the rates of drug abuse compared to adoptees without an alcoholic first degree relative (Cadoret et al. 1986).

Twin studies provide further support for the high degree of heritability of substance use disorders. In these types of studies a comparison typically occurs between monozygotic twins who possess identical genotypes and dizygotic twins who share genes to the same degree as do any other sibling pair. Thus, the extent to which concordance rates or correlations for a disorder in monozygotic twins exceed the rates in dizygotic twins affords an estimate of the heritability of the disorder. Twin studies on alcohol dependence suggest that the overall genetic influence on the likelihood of developing alcohol dependence ranges from 48% to 64%. (True et al. 1999). In regard to smoking, a metaanalysis of 10 twin studies on persistent smoking found mean overall genetic influences on the likelihood of becoming a persistent smoker of 59% for men and 46% for women (Li et al. 2003a). Twin studies looking at risk for dependence on different classes of illicit substances also show high levels of heritability that is nonspecific (Kendler et al. 2003). In other words, the same genetic factors probably account for liability to dependence on any of these substances.

Alcohol challenge studies have investigated the heritability of the degree of response to test doses of alcohol in subjects who use alcohol but do not have alcohol abuse or dependence. Young men with and without first-degree relatives with alcohol dependence were evaluated after ingesting test doses of alcohol. Subjects who had affected first-degree relatives exhibited less subjective effect, less body sway, and decreased cortisol levels compared to those without affected relatives (Schuckit et al. 1996). Lower subjective response to alcohol as well as a positive family history served as a predictor of the future development of alcohol dependence (Schuckit and Smith 1996). Comparisons of subjective responses to alcohol between monozygotic and dizygotic twins indicate also that the subjective response to alcohol has a significant genetic component (Viken et al. 2003). In these studies the level of alcohol measured in the blood or breath did not account for the differences in alcohol effects experienced between groups. Thus, some as-yet-unknown biologic, at least in part genetically determined, responses of the body and brain must explain these differences.

20.3. BIOLOGIC ASPECTS OF ADDICTION

Since epidemiologic studies demonstrate that a substantial proportion of risk for substance use disorders is heritable, genes must be affecting some biologic characteristics of the organism to produce this risk. These characteristics would likely involve either the way the body handles or metabolizes a specific substance (pharmacokinetics) or the way the body responds to the substance (pharmacodynamics). These events could occur in the central nervous system or on the periphery. Metabolic events commonly occur in both places. The important response events that lead to addiction will occur mainly in the brain. A discussion of the peripheral biologic events that could be genetically determined and could predispose to addiction will precede a synopsis of the neurobiology of addiction that is also susceptible to genetic influence.

The development of substance dependence necessarily requires frequent ingestion of a foreign substance. Many individuals ingest substances that can cause dependence, yet never progress to excessive use or dependence on that substance; alcohol is the best example. Clearly, the body's capacity to metabolize the substance in question would have an influence on the likelihood of developing excessive use or dependence. If the body metabolizes the substance very quickly, the individual might either have no interest in the substance as it produces very little effect or, conversely, may use large quantities of the substance intending to build on the small effect experienced. At the opposite end of the spectrum, if the body metabolizes the substance slowly, small quantities ingested will exert large effects. In this case the individual might either experience a preponderance of toxic effects and dislike the substance or might appreciate the positive effects that it produces and continue to use it in order to repeat that experience.

The cytochrome P450 system, an enzyme system quite subject to genetic polymorphisms that affect its function, has a key role in the metabolism of nicotine and some opioids such as codeine, oxycodone, buprenorphine, and methadone (Howard et al. 2002). Alcohol can also be metabolized via this system, although its major metabolic pathway, susceptible as well to polymorphisms with functional significance, is through the alcohol dehydrogenase and aldehyde dehydrogenase systems (Howard et al. 2002). The opioids heroin and morphine are metabolized via enzyme systems distinct from the P450 system (Brzezinski et al. 1997; Coffman et al. 1997; Salmon et al. 1999), some of which also display potentially significant polymorphisms. The relevance of the genetics of each of these enzyme systems to the phenotype of substance dependence will be discussed in more detail below in each of the sections relevant to each specific substance of dependence.

It is estimated that more than half of the human genome is expressed in the central nervous system, but scientific inquiry about genes and the brain has thus far been directed at less than 1% of the genome (Insel and Collins 2003) so this effort remains in its infancy. In regard to substance dependence, most genetic investigation has quite obviously targeted the brain areas known to be involved in the dependence process, which may vary somewhat by substance, although there is substantial overlap.

The primary neurobiologic system that appears to mediate substance dependence is the brain reward pathway (Koob and Le Moal 1997). This system relies primarily on the neurotransmitter, dopamine (Koob and Le Moal 1997). The cell bodies that contain the dopamine reside in an area in the midbrain called the *ventral tegmental* area and project anteriorly to an area called the *nucleus accumbens* and also to prefrontal cortex (Koob and Le Moal 1997). Dopamine release in these areas results in reward or reinforcement, and current theory suggests that all substances of dependence directly or indirectly cause dopamine release in this system (Cami and Farre 2003). In normal physiology brain reward pathways serve an important evolutionary function. These pathways are stimulated when the organism performs an activity conducive to survival including eating when hungry, ingesting fluids when thirsty, sex, nest building, and care of young. When substances of dependence are used repeatedly and stimulate this pathway by directly or indirectly increasing the effects of dopamine, the organism mistakenly behaves as if use of the substance will enhance survival and continues to use it (Koob and Le Moal 2001). There are at least five different subtypes of dopamine receptors to which dopamine can bind so that its signal is transmitted to the receiving neuron after it has been released into the synapse (Emilien et al. 1999). Each of these receptors is a distinct protein (although the amino acid sequences are quite similar) coded for by distinct genes (Emilien et al. 1999). Neurons that release dopamine into the synapse also have a dopamine transporter, another distinct protein coded for by a distinct gene, which can take dopamine back out of the synapse and terminate the transmission (Cami and Farre 2003). Cocaine and amphetamine increase dopamine activity primarily by blocking the dopamine transporter (Cami and Farre 2003). Since the dopamine system exhibits such critical involvement in substance dependence, considerable interest has developed in the possible relationship of polymorphisms in the genes coding for the dopamine receptors and transporter to substance dependence. Indeed, experimental evidence suggests that variability in the length or sequence of the dopamine transporter gene may affect its expression in brain (Miller and Madras 2002).

However, several other neurotransmitter systems, each with its own set of specific proteins, also impinge on and help regulate the dopamine system. Each of these systems could have involvement in the genetics of substance dependence. Important among these are brain inhibitory and excitatory systems. The widespread inhibitory system in the brain is based on the neurotransmitter α-aminobutyric acid (GABA) (Simeone et al. 2003). GABA can bind to two different types of GABA receptors. Each of these receptors is extremely complex and is assembled from a total of 19 possible protein subunits, each of which is coded for by a separate gene. As with dopamine, there are a variety of GABA transporters, all distinct proteins, that can remove GABA from the synapse (Gadea and Lopez-Colome 2001). When GABA binds to a receptor, it tends to hyperpolarize or inhibit the neuron that contains the receptor (Simeone et al. 2003). Thus, in most instances, GABA binding would inhibit dopamine release.

Generally opposed to the GABA system is the major excitatory system based on the neurotransmitter glutamate or chemically similar analogs (Yuzaki 2003). Three different families of glutamate receptors exist (Meldrum et al. 1999). As with the GABA receptors, a great deal of complexity characterizes the glutamate receptors. At least 17 different genes encode the various proteins that constitute different subunits of the three receptor subtypes (Goff and Coyle 2001). When glutamate or its analogs bind to their receptors, they tend to depolarize or excite the neuron that contains the receptor (Goff and Coyle 2001). Various types of glutamate transporters that remove glutamate from the synapse and are coded for by specific genes also exist (Meldrum et al. 1999). Although the interactions between glutamate receptors and dopamine are complex, in general glutamate binding would promote dopamine release.

Another critically important brain system is the endogenous opioid system, which directly mediates the effects of endogenous opioid neuropeptides and of exogenously administered opioids such as heroin, morphine, methadone, buprenorphine, and codeine. There are three subtypes of opioid receptors (Cami and Farre 2003) and several different types of endogenous opioid peptides (Bodnar and Hadjimarkou 2002) that serve as neurotransmitters and bind to these receptors. A different gene codes for each subtype of opioid receptor and for specific proteins from which endogenous opioid peptides are cleaved. Abused opioids bind primarily to the μ-opioid receptor (Cami and Farre 2003). When an exogenously administered opioid or endogenous opioid peptide binds to an opioid receptor, it, similarly to GABA, tends to hyperpolarize or inhibit the neuron that contains the receptor (Johnson and North 1992). Since GABA-releasing neurons inhibit dopamine release, when opioid receptors on these GABA neurons are activated by binding of an exogenous or endogenous opioid, the GABA inhibition of dopamine neurons is reduced. Thus opioids acting via opioid receptors and GABA neurons indirectly induce dopamine release in brain reward pathways (Johnson and North 1992). Opioids also, quite obviously, suppress pain, an action that may contribute in some cases to their potential to cause dependence (Inturrisi 2002).

Just as the brain contains an endogenous opioid system and opioid receptors, it contains an endogenous cannabinoid system and cannabinoid receptors that serve as the mechanism through which marijuana exerts its effects (Fride 2002). There are two types of cannabinoid receptors; CB1, expressed in the central and peripheral nervous systems; and CB2, expressed predominantly in the immune system (Lutz 2002). Three endogenous cannabinoids that act as the neurotransmitters for this system and are ethanol amide molecules, not proteins, have been discovered thus far (Fride 2002). A separate gene codes for each type of cannabinoid receptor (Onaivi et al. 2002). Since CB1 resides in many brain areas (Fride 2002), the effects of stimulating this receptor are complex and not fully understood. However, the cannabinoid system is believed to modulate the effects of several other neurotransmitters including GABA, glutamate, opioids, and monoamines such as dopamine and serotonin (see discussion below) (Fride 2002). Marijuana causes dopamine release in brain reward pathways via an as-yet-undetermined mechanism (Mechoulam and Parker 2003). Like opioids, cannabinoids and marijuana can modulate pain (Fride 2002). The effects of alcohol (Mechoulam and Parker 2003) and other substances of dependence (Solinas et al. 2003) may be mediated in part through indirect activation of cannabinoid receptors.

Nicotine affects the brain through direct activation of nicotinic cholinergic receptors (Leonard and Bertrand 2001). Nicotinic receptors consist of structures of five related but distinct protein subunits (Le Novere et al. 2002). To make matters more complex, there are at least eight different nicotinic receptor subunits expressed in human brain (Leonard and Bertrand 2001), so a variety of combinations of subunit types may form receptors, each of which may have somewhat different functions (Picciotto 2003). A distinct gene codes for each of these eight subunit proteins (Leonard and Bertrand 2001). Acetyl choline is the normal neurotransmitter for this system. Activation of various subtypes of nicotinic receptors modulates the actions of several other neurotransmitter system, including GABA and serotonin and also causes release of dopamine in brain reward pathways (Picciotto and Corrigall 2002).

Serotonin (5-hydroxytryptamine), a monoamine neurotransmitter, has effects on mood, sleep, appetite, impulse control, and cognition (Kroeze et al. 2002). The serotonin-containing neurons in the brain exist in a midbrain area called the *raphe nuclei* and send fiber tracts to numerous other areas in the central nervous system (Azmitia and Whitaker-Azmitia 1991). There are at least 15 subtypes of serotonin receptors that are related proteins all coded by different genes (Kroeze et al. 2002). As in other neurotransmitter systems, neurons that release serotonin into the synapse also have a serotonin transporter, again a distinct protein coded by a distinct gene, which can remove serotonin from the synapse and terminate the transmission (Inoue et al. 2002). This gene has functional polymorphisms (Lesch et al. 1996). Cocaine and amphetamine also block the serotonin transporter (Cami and Farre 2003) and, undoubtedly exert some of their dependence-inducing effects via this mechanism. Stimulation of a specific subtype of serotonin receptor, known as the 5-HT_3 receptor, causes rapid release of dopamine in brain reward pathways (Johnson and Ait-Daoud 2000).

20.4. GENETIC DETERMINANTS OF ADDICTION

Tables 20.1, 20.2, and 20.3 summarize information currently available regarding specific gene products that have potential influence on the development of alcohol dependence, nicotine dependence, and other types of substance dependence, respectively. This information is elaborated in more detail in the text below.

20.4.1. Genetic Determinants of Alcohol Dependence

Awareness of the basic biology of substance dependence will facilitate a discussion concerning current knowledge of the genetics of alcohol dependence. As noted above, alcohol is metabolized primarily by the alcohol dehydrogenase (ADH) and aldehyde dehydrogenase (ALDH) systems (Howard et al. 2002) enzymes, which are both expressed in brain. Alcohol is first converted to acetaldehyde by ADH then acetaldehyde is converted to acetate by ALDH (Agarwal 2001). A large body of evidence firmly supports the conclusion that polymorphisms in the genes that code for these enzymes are associated with the heritability of alcohol dependence (Dick and Foroud 2003). Among the three ADH isoenzymes that catalyze the conversion of alcohol and that are all closely linked on chromosome 4q22, the ADHIB gene exhibits three different polymorphic alleles (Dick and Foroud 2003). The capacity of these different enzymes coded by these differing alleles to catalyze the conversions varies by more than 30-fold (Dick and Foroud 2003). The form of the enzyme that causes the conversion most rapidly occurs primarily in Asian populations (Dick and Foroud 2003), and its presence decreases the risk for alcohol dependence by a factor of 3 (Whitfield 1997), probably because it causes a prompt buildup of acetaldehyde that exerts unpleasant aversive effects including flushing (Dick and Foroud 2003). A second form of the ADH1B enzyme found largely in populations of African descent or in some Native Americans (Dick and Foroud 2003) may also provide some protective effect against the development of alcohol dependence (Wall et al. 2003).

Table 20.1 Summary of Studies of Potential Genetic Determinants of Alcohol Disorders

Gene Product	Alcohol Disorder	Association Found	References
Alcohol dehydrogenase	Dependence	Yes	Wall et al. (2003) Whitfield (1997)
Aldehyde dehydrogenase	Dependence	Yes	Dick and Foroud (2003)
CYP450 2E1	Dependence	Yes	Howard et al. (2003)
Dopamine D_1 receptor	Dependence	No	Sander et al. (1995)
Dopamine D_2 receptor	Dependence	Yes (males or severe only)	Connor et al. (2002) Limosin et al. (2002)
	Dependence	No	Dick and Foroud (2003)
Dopamine D_3 receptor	Alcohol withdrawal delirium	Yes	Sander et al. (1995)
	Dependence	No	Gorwood et al. (2001)
Dopamine transporter	Dependence	Yes	Ueno et al. (1999)
	Dependence	Trend	Dobashi et al. (1997)
	Alcohol withdrawal	Yes	Sander et al. (1997)
	Dependence	No	Franke et al. (1999b)
γ-Aminobutyric acid subunits	Dependence	Yes	Long et al. (1998) Noble et al. (1998) Parsian and Zhang (1999) Song et al. (2003)
	Dependence	No	Dick and Foroud (2003)
Glutamate receptor subunit NR2B	Dependence	No	Schumann et al. (2003)
μ-Opioid receptor	Dependence	Yes	Schinka et al. (2002) Town et al. (1999)
	Dependence	No	Franke et al. (2001)
	Alcohol and opioids combined	Yes	Luo et al. (2003)
δ-Opioid receptor	Dependence	Possible	Mayer and Hollt (2001)
	Dependence	No	Franke et al. (1999a)
Serotonin receptor subtypes	Dependence	No	Cigler et al. (2001) Gorwood et al. (2002) Kranzler et al. (2002b) Parsian and Cloninger (2001) Sinha et al. (2003)
Serotonin transporter	Dependence	No	Kranzler et al. (2002a)

Table 20.2 Summary of Studies of Potential Genetic Determinants of Nicotine Dependence

Gene Product	Association Found	References
CYP450 2A6	Yes	London et al. (1999)
		Minematsu et al. (2003)
		Tyndale and Sellers (2001)
CYP450 2D6	Yes	Boustead et al. (1997)
	No	Cholerton et al. (1996)
Dopamine D_1 receptor	Yes	Comings et al. (1997a)
Dopamine D_2 receptor	Yes	Caporaso et al. (1997)
		Comings et al. (1996)
		Hamajima et al. (2002)
		Lee (2003)
		Wu et al. (2000)
		Yoshida et al. (2001)
	No	Bierut et al. (2000)
		Singleton et al. (1998)
Dopamine D_4 receptor	Yes	Shields et al. (1998)
		Hutchison et al. (2002)[a]
Dopamine transporter	Yes	Jorm et al. (2000)
	No	Jorm et al. (2000)
		Vandenbergh et al. (2002)
Monoamine oxidase	Yes	Ito et al. (2003)
		McKinney et al. (2000)
	No	Johnstone et al. (2002)
Nicotinic receptor β_2 subunit	No	Lueders et al. (2002)
		Silverman et al. (2000)
Serotonin transporter	Yes	Ishikawa et al. (1999)

[a] Nicotine craving only.

The form of ALDH principally involved in the conversion of acetaldehyde to acetate, ALDH2, coded for by a gene on chromosome 12, also exhibits polymorphisms (Dick and Foroud 2003). In this case the form of the enzyme that causes slower metabolism of acetaldehyde leads to acetaldehyde buildup with flushing and other toxic reactions (Dick and Foroud 2003). This form occurs almost exclusively in Asian populations (Dick and Foroud 2003). Individuals with this enzyme have one-tenth the risk of developing alcohol dependence compared to those with other forms (Dick and Foroud 2003). Other forms of ALDH have not been thoroughly investigated but may also bear some relationship to risk for alcohol dependence (Agarwal 2001).

Table 20.3 Summary of Studies of Potential Genetic Determinants of Other Substance Dependence

Gene Product	Substance Use Disorder	Association Found	References
CYP450 2D6	Oral opioids	Yes	Tyndale et al. (1997)
Dopamine D_2 receptor	Heroin	Yes	Lawford et al. (2000)
	Stimulants	Yes	Persico et al. (1996)
	Cocaine	Yes	Comings et al. (1999b)
	Cocaine	No	Gelernter et al. (1999b)
Dopamine D_3 receptor	Opioids	Yes	Duaux et al. (1998)
	Cocaine	Yes	Comings et al. (1999b)
	Cocaine	No	Freimer et al. (1996)
Dopamine transporter	Opioids	Yes	Galeeva et al. (2002)
Dopamine β-hydroxylase	Cocaine paranoia	Yes	Cubells et al. (2000)
μ-Opioid receptor	Alcohol and opioids combined	Yes	Luo et al. (2003)
	Opioids	No	Franke et al. (2001) Gelernter et al. (1999a)
	Cocaine	No	Luo et al. (2003)
δ-Opioid receptor	Opioids	Yes	Mayer et al. (1997)
	Opioids	No	Franke et al. (1999a)
Proenkephalin	Opioids	Yes	Comings et al. (1999a)
Prodynorphin	Cocaine (Protective)	Yes	Chen et al. (2002)
Cannabinoid CB1 receptor	Heroin	No	Li et al. (2000a)
	Cocaine	Yes	Comings et al. (1997b)
	Cocaine	No	Covault et al. (2001)
	Cannabis Abuse	No	Hoehe et al. (2000b)
Serotonin receptor subtypes	Cocaine	No	Cigler et al. (2001)
Serotonin transporter	Cocaine	No	Patkar et al. (2002)

A more minor, although still clinically meaningful, route of alcohol metabolism occurs through the cytochrome P450 2E1 enzyme (Lieber 1994). The activity level of CYP2E1 is increased by exposure to alcohol, a characteristic not shared by other alcohol-metabolizing enzymes, and so may contribute to the tolerance observed in many regular consumers of alcohol (Lieber 1994). Wide variation exists between individuals in

CYP2E1 activity (Agarwal 2001), and some polymorphisms discovered in the mid-1990s can explain at least part of this variation (Lucas et al. 1995). Two allelic variants can affect the capacity of the enzyme to have its activity induced. One of these (c2) appears to confer less capacity for induction, whereas the one-dimensional (1D) variant allows greater capacity for induction (Howard et al. 2002). In either case the biological and behavioral clinical events relating to alcohol tolerance could be altered. More recent work demonstrates that both polymorphisms show some relationship to the development of alcohol dependence or increased use of alcohol in various populations (Howard et al. 2003). Although nicotine itself is apparently not metabolized by 2E1 (see discussion below), tobacco does appear to induce 2E1 (Howard et al. 2002). Thus, smoking behavior itself could alter the pharmacokinetics of alcohol and the risk for development of alcohol dependence. Genetic linkage studies have found putative markers associated with both smoking and alcohol dependence near the gene coding for CYP2E1 (Bergen et al. 1999; Tiwari et al. 1999).

When considering other genes that could contribute to alcohol dependence risk, attention naturally turns first to genes that code for dopamine receptors or dopamine transporters because of dopamine's key role in substance mediated reinforcement. Several studies have indicated that a polymorphism (Taq1 A) in the gene that codes for the dopamine D_2 receptor occurred more frequently in alcohol-dependent individuals compared to controls, but subsequent studies failed to confirm this finding (Dick and Foroud 2003). Inadequate screening or matching of control populations could account for some of the discrepancy among studies (Dick and Foroud 2003). However, four family studies also failed to find any association (Dick and Foroud 2003). The most recent findings suggest that this association is limited to males with alcohol dependence (Limosin et al. 2002) or to individuals with severe forms of alcohol dependence (Connor et al. 2002). Thus, any potential role for the dopamine D_2 receptor gene in alcohol dependence remains unresolved at present. Investigations of the relationship of other subtypes of dopamine receptor genes to alcohol dependence have found no evidence for any such relationship (Gorwood et al. 2001; Sander et al. 1995) except for a weak association with alcohol withdrawal delirium (Sander et al. 1995). The gene coding for the dopamine transporter exhibits polymorphisms (Vandenbergh et al. 2000). Only a handful of studies have examined the association between dopamine transporter polymorphisms and alcohol dependence. One found only a trend toward association between a specific allele and alcohol dependence in a Japanese population (Dobashi et al. 1997), one found a significant association in a Japanese populations (Ueno et al. 1999), one uncovered a relationship between a specific dopamine transporter gene and alcohol dependence manifesting severe withdrawal symptoms (Sander et al. 1997), but another found no association (Franke et al. 1999b). The role of the dopamine transporter gene in alcohol dependence will require further investigation.

Since alcohol wields some of its effects including impaired coordination and sedation through the inhibitory, GABA, system, a search for a genetic role of GABA in the heritability of alcohol dependence has ensued (Davies 2003). A study of linkage to alcohol dependence in an American-Indian population implicated the area on chromosome 4p that includes the gene coding for one of the GABA receptor subunits (Long et al. 1998), findings subsequently supported by a case–control association study finding a higher rate of a specific allele of this gene in individuals with alcohol dependence (Parsian and Zhang 1999) and by a family association study (Song et al. 2003). Similarly, an association between alcohol dependence and a gene coding for another GABA subunit on chromosome 15 has been detected in a case–control study (Noble et al. 1998) and via paternal transmission only in a family study (Song et al. 2003). A number of studies have examined the relationship between alcohol dependence and several GABA subunits the genes for which exist on chromosome 5 with some supportive evidence for an association but to date conflicting and unsubstantiated results (Dick and Foroud 2003).

Alcohol acts in part through the endogenous opioid system (Gianoulakis 2001), and medications that block the μ-opioid receptor show efficacy in the treatment of alcohol dependence (Johnson and Ait-Daoud 2000). The μ-opioid receptor gene manifests numerous polymorphisms (Mayer and Hollt 2001). There is some evidence for a relationship between one of these polymorphisms (+118A) and alcohol dependence (Town et al. 1999). A subsequent study could not replicate this finding (Franke et al. 2001), although later work suggests that this polymorphism is not unique to alcohol but is associated with all types of substance dependence (Schinka et al. 2002). A study that looked at eight separate polymorphisms of the μ-opioid receptor gene did find an association between a distinct polymorphism (-2044A) and individuals who manifested both alcohol and opioid dependence among European Americans only (Luo et al. 2003). One polymorphism in the δ-opioid receptor gene (T921C) was found to have no association with alcohol dependence in general (Franke et al. 1999a), but some association with subgroups of alcohol-dependent individuals could not be ruled out (Mayer and Hollt 2001). Clearly, with the number of functional polymorphisms present in the endogenous opioid system, future work may show that genetic variations in this system contribute to the development of alcohol dependence.

Considerable evidence indicates that serotonin function has an impact on alcohol consumption and development of dependence (LeMarquand et al. 1994). A few studies have now examined the relationship between some of the serotonin receptor subtypes ($5HT_{1B}$, $5HT_{2A}$, $5HT_{2C}$) and alcohol dependence without finding any robust associations even when looking at subtypes of alcohol dependence (Cigler et al. 2001; Gorwood et al. 2002; Kranzler et al. 2002b; Parsian and Cloninger 2001; Sinha et al. 2003). Numerous other studies investigated the relationship between the genetics of the serotonin transporter and alcohol dependence. The results have been contradictory (Dick and Foroud 2003), but the most recently and carefully performed study indicates no association (Kranzler et al. 2002a). A definitive answer about the relationship of serotonin genetics with alcohol dependence will have to await evaluation of serotonin receptor subtypes that have not yet been studied in this regard.

In addition, knowledge of the neuropharmacology of alcohol (Mechoulam and Parker 2003; Schumann et al. 2003) suggests that the genes coding for glutamate or cannabinoid receptors would likely have an influence on development of alcohol dependence, but, aside from a single, negative study of the NR2B subunit of the *N*-methyl-*d*-aspartate subtype of glutamate receptors (Schumann et al. 2003), extensive human studies of these genes in alcohol dependence have yet to be conducted.

20.4.2. Genetic Determinants of Nicotine/Tobacco Dependence

Nicotine is, of course, the constituent of tobacco that confers its addictive properties (Picciotto and Corrigall 2002). Nicotine is metabolized primarily by the cytochrome P450 enzyme 2A6 (Messina et al. 1997) with a potential minor level of activity of 2D6 (Batra et al. 2003). CYP2A6 demonstrates considerable interindividual variation in activity on the basis of functional genetic polymorphisms (Howard et al. 2002). The bulk of the evidence to date suggests that this genetic variation in enzymatic activity has an impact on likelihood of smoking. Individuals with versions of this enzyme that exhibits low activity (poor metabolizers), leading to higher nicotine plasma and tissue levels for a given amount of ingestion, have greater likelihood of never smoking or quitting (Tyndale and Sellers 2001), although one study showed only a weak trend for this effect (London et al. 1999). In contrast, individuals with versions of the enzyme that have high activity (rapid metabolizers), who have lower plasma levels following a given dose of nicotine, are more likely to be smokers (Tyndale and Sellers 2001) or, at least heavy smokers (Minematsu et al. 2003).

Similarly, CYP2D6 is highly polymorphic and expressed in brain (Howard et al. 2002). One study found that individuals with high 2D6 activity tend to be smokers or heavy smokers (Saarikoski et al. 2000), while another study found no association between 2D6 and smoking (Cholerton et al. 1996), and a third study found that activity levels of this enzyme had no relationship with likelihood of smoking but did influence degree of smoking behavior (Boustead et al. 1997).

In consideration of nicotine's ability to release dopamine in brain reward pathways, attention has turned to the genetics of dopamine-related proteins in relationship to tobacco dependence. The same polymorphism (Taq1 A) in the gene that codes for the dopamine D_2 receptor evaluated for alcohol dependence has also been explored for smoking. As with the studies on alcohol, results for smoking appear promising, although somewhat mixed. An initial study indicated that Caucasian smokers were nearly twice as likely as nonsmokers to carry the A1 allele (Comings et al. 1996), a finding supported in a separate study (Caporaso et al. 1997). A subsequent study in a British Caucasian population (Singleton et al. 1998) failed to confirm these results, and a family-based study of smokers found no relationship between familial transmission of the A1 allele and smoking status (Bierut et al. 2000). Investigation of the presence of this allele in African-Americans and Mexican-Americans revealed an association between the A1 allele and smoking only for Mexican-Americans (Wu et al. 2000). Among a sample of Koreans, the A1 allele positively associated with smoking among males but actually had an inverse association with smoking among females (Lee 2003). To add somewhat to the uncertainty in this area, two studies of Japanese smokers found somewhat opposite results to those seen in other populations; individuals homozygous for the A2 allele were more likely to be smokers in one study (Yoshida et al. 2001), but this effect occurred only for males in the second study (Hamajima et al. 2002). One study to date has found an association between polymorphisms in the dopamine D_1 receptor gene and smoking status (Comings et al. 1997a). Some evidence also implicates the dopamine D_4 receptor gene in tobacco dependence. This gene demonstrates a tandem repeat polymorphism in which the seven repeat allele decreases affinity of dopamine for the receptor (Shields et al. 1998). African-Americans but not Caucasians who had six to eight repeats versus two to five repeats were more likely to be smokers (Shields et al. 1998). Furthermore, smokers who have the seven-repeat or longer allele had greater craving and more attention to smoking cues than did smokers with shorter repeat alleles (Hutchison et al. 2002).

The potential role of polymorphisms in the dopamine transporter gene has also been examined in relationship to smoking. Two large studies of racially and ethnically mixed populations concluded that individuals with the SLC6A3–9 allele had less likelihood of smoking and better ability to quit if they did smoke (Jorm et al. 2000). The combination of the SLC6A3–9 allele and the D_2 A2 allele apparently provided even greater protection against smoking (Jorm et al. 2000). Unfortunately, two subsequent studies, one in a purely Caucasian sample, could not confirm these findings (Jorm et al. 2000; Vandenbergh et al. 2002), leaving uncertainty regarding the relationship of the dopamine transporter gene to smoking.

A number of enzymes are involved in the synthesis and metabolism of dopamine itself (Batra et al. 2003). Some effort to relate the genetics of these enzymes to smoking behavior has occurred. Monoamine oxidase and dopamine β-hydroxylase both metabolize dopamine. Two studies in Britain regarding the relationship of polymorphisms in the genes coding for the monoamine oxidase enzymes and smoking reached contradictory conclusions with the smaller and earlier work finding an association (McKinney et al. 2000) and the larger, later study finding none (Johnstone et al. 2002a). However, a study in Japanese that did separate analyses for male and female subjects determined that more serious nicotine dependence occurred for males who smoked and had the monoamine oxidase A four-repeat allele and that this allele demonstrated a positive association with smoking among females (Ito et al. 2003).

Given the complexity of nicotinic receptors, research regarding the possible association of polymorphisms in human genes coding for nicotinic receptor subunits has been underway since only 2000. A potential candidate, the nicotinic β-2 subunit gene, appears, by virtue of gene knockout studies in mice, to have a role in nicotine reinforcement and addiction (Shoaib et al. 2002) and demonstrates polymorphisms in humans (Silverman et al. 2000). However, in the two human studies conducted thus far, polymorphisms in the β-2 subunit gene fail to associate with smoking behavior (Lueders et al. 2002; Silverman et al. 2000).

Some evidence implicates a polymorphism in the serotonin transporter gene in smoking. One study found an independent association with the L allele for this gene in a Japanese population (Ishikawa et al. 1999), while two other studies found in mixed populations that the combination of the S allele and the personality trait of neuroticism predicted smoking behavior (Arinami et al. 2000; Jorm et al. 2000).

No research has yet assessed the potential influence of GABA, glutamate, opioid, or cannabinoid receptor genes on tobacco use.

20.4.3. Genetic Determinants of Opioid Dependence

As mentioned above, different opioid drugs have different metabolic routes. Heroin (diacetyl morphine) is first transformed to monoacetyl morphine by acetylcholinesterase, butyrylcholinesterase (Salmon et al. 1999), and carboxylesterase (Brzezinski et al. 1997). Monoacetyl morphine is metabolized to morphine by human erythrocyte acetylcholinesterase (Salmon et al. 1999). Morphine undergoes transformation to one active and one possibly toxic glucuronide metabolite via the enzyme uridine diphosphate glucuronosyltransferase 2B7 (Coffman et al. 1997) and other glucuronosyltransferases (Stone et al. 2003), some of which are expressed in brain (King et al. 1999a). The semisynthetic opioids oxymorphone and hydromorphone are also glucuronidated by these enzyme systems (Armstrong and Cozza 2003). Although no studies have been done regarding the genetics of these enzyme systems in relationship to dependence on heroin or morphine, some of them exhibit functional polymorphisms (Darvesh et al. 2003) which have the capacity to alter morphine pharmacokinetics in some individuals (Sawyer et al. 2003). Thus, they would make good candidate genes to evaluate for opioid dependence.

The opioids codeine, oxycodone, and hydrocodone serve as prodrugs and undergo transformation, respectively, to the active drugs morphine, oxymorphone, and hydromorphone via CYP 2D6 (Tyndale et al. 1997). When CYP 2D6 polymorphisms were examined among individuals with oral opioid dependence, none of these individuals had genotypes consistent with being poor metabolizers (Messina et al. 1997). This finding suggests that having a poor metabolizing genotype provides protection against dependence on these specific opioids.

The semisynthetic opioids methadone and buprenorphine are metabolized by a different cytochrome P450 enzyme system, CYP3A4 (Kobayashi et al. 1998). CYP 3A4 also demonstrates numerous polymorphisms (Lamba et al. 2002), some of which may have functional significance (Hesselink et al. 2003). No studies thus far examined the relationship of these polymorphisms to dependence on methadone or buprenorphine.

Since opioids also indirectly facilitate the release of dopamine in brain reward pathways, a few studies have considered the relationship of dopamine receptor genes to opioid dependence. One study assessed the Taq1 A polymorphism in the gene that codes for the dopamine D_2 receptor in regard to opioid dependence and found not only an association but also a relationship between homozygosity for the A1 allele and amount of heroin used (Lawford et al. 2000). Two studies suggest an association between the dopamine D_4 exon III seven-repeat allele and opioid dependence, particularly in combination with the personality trait of sensation seeking (Kotler et al. 1997; Li et al. 1997). Also in a study with a Chinese sample the dopamine D_3 receptor Bal 1 polymorphism showed an association with opioid dependence and sensation seeking (Duaux et al. 1998). Variable nucleotide tandem repeat polymorphisms of both the dopamine and serotonin transporters were examined in a sample of male Russians and Tatars, and homozygosity for specific polymorphisms of both of these genes demonstrated an association with opioid dependence (Galeeva et al. 2002).

When considering the genetics of opioid dependence, attention has naturally turned to the opioid receptor genes. The gene that codes for the μ-opioid receptor has numerous polymorphisms. The most common and most studied is the A118G polymorphism (Mayer and Hollt 2001). One study found that the receptors created by this polymorphism (a single amino acid alteration) had much higher binding affinity for β-endorphin, an important endogenously occurring ligand for this receptor (Bond et al. 1998). Several other studies have looked at the relationship of this and other μ-opioid receptor gene polymorphisms and found either no association (Franke et al. 2001; Gelernter et al. 1999a) or have found that these polymorphisms represent a general risk factor for substance dependence not necessarily specific to opioid dependence (Hoehe et al. 2000a; Schinka et al. 2002). One, as noted above, concluded that a single polymorphism (-2044A) in the gene represented a risk factor for combined opioid and alcohol dependence (Luo et al. 2003). A polymorphism in the δ-opioid receptor gene had an association with opioid dependence in one study of a German, Caucasian sample (Mayer et al. 1997) but not in another (Franke et al. 1999a).

One study considered a polymorphism in the gene coding for proenkephalin, the precursor peptide to the enkephalins, important endogenous opioid peptides. In this work the CA(*n*) repeat polymorphism did demonstrate a positive association with opioid dependence (Comings et al. 1999a).

A single study that looked specifically at a polymorphism in the cannabinoid receptor (CB1) gene among a Chinese population found no association with heroin dependence (Li et al. 2000a).

The genetics of GABA, glutamate, nicotinic, and serotonin receptors await exploration in regard to human opioid dependence.

20.4.4. Genetic Determinants of Cocaine Dependence

Cocaine has one major and two minor metabolic pathways (Warner and Norman 2000). The major route involves transformation to benzoylecognine mediated by hepatic microsomal carboxylesterase (Warner and Norman 2000). It can also be metabolized by the plasma enzyme butyrylcholinesterase to ecognine methyl ester (Warner and Norman 2000) and to norcocaine by CYP3A (Ladona et al. 2000). It is certainly of interest that some of the enzymes involved in opioid metabolism are also involved in cocaine metabolism, and, as described above, these enzymes have apparent functional polymorphisms. However, the possible genetic contribution of polymorphisms among these enzyme systems to the development of cocaine dependence has not yet been assessed.

Cocaine directly blocks dopamine and serotonin transporters, thereby preventing the reuptake of these neurotransmitters back into the releasing neuron (Cami and Farre 2003). Therefore, potential genetic contributions of dopamine and serotonin systems to cocaine-dependence stir considerable interest. An initial study suggested that, as with other forms of substance use, the Taq1 A and B polymorphisms in the gene that codes for the dopamine D_2 receptor showed an association with heavy stimulant use (Persico et al. 1996). Two subsequent studies looked specifically at these polymorphisms in cocaine-dependent individuals. One confirmed the findings in a purely Caucasian sample (Comings et al. 1999b), while the other, using a racially stratified sample, failed to demonstrate any association (Gelernter et al. 1999b). The former study also delineated an additive association of a polymorphism in the D_3 dopamine receptor gene and the polymorphism in the D_2 receptor gene with cocaine dependence (Comings et al. 1999b) despite an earlier assertion that polymorphisms in the D_3 gene did not affect the likelihood of cocaine dependence (Freimer et al. 1996). A study of a polymorphism in the dopamine-metabolizing enzyme dopamine β-hydroxylase indicates an association of specific alleles with both low dopamine β-hydroxylase plasma activity and cocaine-induced paranoia in cocaine users (Cubells et al. 2000).

The two studies that have examined serotonin related genes have found no association of polymorphisms either in the 5 HT_{1B} gene (Cigler et al. 2001) or in the serotonin transporter gene (Patkar et al. 2002) with cocaine dependence.

An attempt to find an association between cocaine dependence and polymorphisms in the μ-opioid receptor gene failed to do so (Luo et al. 2003). A single study looked at the gene coding for prodynorphin, the precursor to another opioid peptide, and found that an alleleic variation that results in enhanced transcription of the gene may protect against cocaine dependence (Chen et al. 2002).

Two studies have investigated the relationship of the CB1 receptor gene to cocaine dependence. One found an association with CB1 polymorphisms (Comings et al. 1997b), but the other did not (Covault et al. 2001).

Obviously, numerous genes potentially related to the neurobiology of cocaine dependence have not yet been studied.

20.4.5. Genetic Determinants of Cannabis Use and Abuse

The single study that examined the genetics of cannabis use naturally focused on the CB1 cannabinoid receptor. It found no differences in molecular structure of the CB1 receptor gene in a sample of individuals who had an adverse response to cannabis with the development of psychotic symptoms compared to a sample of individuals with heavy cannabis abuse (Hoehe et al. 2000b).

20.5. SUMMARY AND CONCLUSIONS

The foregoing compilation of findings makes it quite apparent that, apart from a few exceptions involving the metabolism of alcohol and nicotine, the search for specific genes that definitively contribute to the risk for substance dependence remains elusive. Initially positive results seem to be contradicted more often than confirmed. Nevertheless, review of the research performed to date in this field offers intriguing hints. Some evidence at least supports the notion that the genetics of the dopamine, GABA, endogenous opioid, endogenous cannabinoid, and serotonin systems could be related to the development of various forms of substance dependence. The potential explanations for the numerous contradictions in prior research also point the way toward design of future studies that might obtain more conclusive outcomes.

Future investigations must take account of the fact that, in many instances, the contribution of a single gene product to the risk for substance dependence is small. Therefore, more definitive studies will require relatively large sample sizes. In addition, findings can vary considerably among different racial and ethnic groups and between genders. Future research must either confine itself to a single ethnic group and gender or perform separate analyses for these different subsets of subjects. Since uncertainty still exists regarding whether genetic risk for substance dependence is substance-specific (Luo et al. 2003) or generalized (Kendler et al. 2003), extensive, careful phenotyping must examine all subjects for the full range of substance dependence diagnoses and exclude these diagnoses from controls. Finally, the high probability that substance dependence has a polygenic etiology, along with the tantalizing evidence from a few studies that concomitant polymorphisms in two distinct candidate genes may have an additive impact (Comings et al. 1999b; Galeeva et al. 2002), supports the strategy of testing subjects for polymorphisms in a multitude of candidate genes. Obviously, including all these design elements demands a lot of time, a huge organizational commitment, and infusion of resources, but ongoing examples of this kind of endeavor indicate its feasibility (Reich et al. 1999).

The idea of making such an enormous investment immediately raises the question of what, other than pure scientific knowledge, might evolve from this effort. Although the idea of altering an individual's genome to change the phenotype seems remote at this juncture, the era of changing the phenotype through targeted pharmacologic manipulation is already upon us. In fact, the very first medication used to treat alcohol dependence, disulfiram (Antabuse), available for more than 50 years, does just that. Disulfiram blocks the enzyme ALDH, rendering the individual taking it similar in phenotype to someone with the form of the enzyme that causes slower metabolism of acetaldehyde (Veverka et al. 1997). In this circumstance the acetaldehyde buildup that occurs with ingestion of alcohol can render alcohol aversive by causing flushing and other toxic reactions. Of course, disulfiram was discovered by serendipity before its mechanism of action was even known. Using it as an example, however, it is not hard to imagine developing and targeting pharmacologic treatments specific to an underlying genetic etiology of substance dependence. Certainly, if we knew that individuals with nicotine, opioid, or cocaine dependence were rapid metabolizers of those substances, we could conceive of using medications to block the relevant metabolic enzymes, perhaps causing toxic effects of the abused substances and making them aversive instead of reinforcing. If we knew the specific genetic underpinnings of an individual's substance use disorder, we could also potentially determine in advance which individuals might respond well to medications we already have, such as opioid antagonists for alcohol dependence or bupropion for nicotine dependence, which have efficacy in only some patients. If we knew that particular neuronal receptors or transporters were abnormal in certain individuals, we could develop pharmacotherapeutic agents targeted to rectify the precise abnormality and perhaps change responses to substances of dependence in that fashion.

With the knowledge about addiction genetics already accrued, these clinical scenarios seem reasonably close at hand. We know with certainty that genetics contribute a large proportion of the risk for substance dependence. We have launched a pursuit of numerous candidate genes. If we invest the resources, we have the technical capacity and scientific acumen ultimately to identify the combinations of genes that underlie the development of substance dependence. Armed with that information, we will have much improved capacity to treat, and even perhaps prevent, this common and often lethal disorder.

20.6. SUGGESTED READING

For further information on genetic determinants in substance addiction and abuse, the following journal articles and recommended:

Batra et al. (2003)
Cami and Farre (2003)
Dick and Foroud (2003)
Howard et al. (2002)
Kendler et al. (2003)
Mayer and Hollt (2001)
Merikangas and Avenevoli (2000)
Reich et al. (1999)

Part IV

21

Overview of Section IV

Lucio G. Costa and David L. Eaton
University of Washington, Seattle, WA

This final section address the so-called ELSI (ethical, legal, and social implications) issues related to ecogenetics, and the potential applications of ecogenetics to risk assessment of environmental hazards. The medical and societal implications of genetic testing have, of course, been around for decades. Genetic councilors and others with expertise in the field of medical genetics are generally well versed in how to explain to their patients the implications of a positive, or negative, genetic test for familial "disease genes," such as those responsible for cystic fibrosis, Huntington's disease, and sickle cell anemia. Individuals or families harboring such mutations in their genome are provided with advice that is generally based on a solid scientific understanding of both the consequences of the disease gene, and the probabilities that such consequences will be realized in an individual. The confidence comes from the great strength of association between the genotype (the specific genetic variant) and the phenotype (the expression of the disease); in other words, the gene penetrance is very high (e.g., 90–100%). However, the frequency of such alleles in the population is very low (e.g., 1 in 10,000–100,000), and thus only small proportions of the population are actually affected. In other words, individual risk associated with the genetic variant is high, but the population attributable risk is very low.

In contrast, for the area of ecogenetics, the penetrance of an "environmental susceptibility" gene is generally quite low (e.g., increased risk less than two-fold over background), but the allele frequencies in the popula-

Gene-Environment Interactions: Fundamentals of Ecogenetics, edited by Lucio G. Costa and David L. Eaton

tion are often quite high (10–50%), giving rise to a small individual risk, but substantial population attributable risk. Furthermore, because the strength of association for a low-penetrant allele is often weak (usually determined through population-based epidemiology studies rather than controlled clinical trials), there is generally a great deal of uncertainty as to the overall significance of the variant allele to an individual. Thus, the ELSI issues in ecogenetics are quite different from those associated with rare, high-penetrance genetic variants.

In Chapter 22 the ethical issues related to ecogenetics are explored in more detail. Three key principles of bioethics—autonomy, beneficence, and justice—are explored in the context of genetic susceptibility to environmental exposures. Of particular interest is the relatively new concept of environmental justice, which focuses on socioeconomically vulnerable populations who may be disproportionately exposed to potentially toxic chemicals because of their socioeconomic status. What ethical considerations should be factored into the utilization of genetic tests that might identify genetically sensitive individuals—or ethnically identifiable subpopulations—already disproportionately burdened because of their socioeconomic status? This chapter provides a case study approach to consideration of the ethical issues associated with genetic testing for common, but low-penetrance, allelic variants, using asthma in children as the example. This example also raises many interesting pharmacogenetics issues, since new drugs developed for asthma may work in only a subset of patients (which is likely true of almost any drug).

Although ethical considerations are clearly important elements in how society thinks about the applications of genetic and genomic technologies, the social and psychological aspects of identification of common, low-penetrant environmental susceptibility genes has received relatively little attention. In Chapter 23 the authors describe a framework for considering social and psychological issues in general, and then discuss how issues related to ecogenetics can fit within this overall framework. The chapter examines environmental and genetic risk from different hierarchical perspectives, including individual, family, sociocultural, organizational, residential, and political levels. Although there is a general dearth of psychosocial research in this area, it is likely that the various levels interact to ultimately affect risk information and communication to individuals and communities, which in turn will impact behaviors and thus, potentially, health outcomes.

Chapter 24 explores several legal issues associated with our increasing ability to identify genetically susceptibility to environmentally induced diseases. The chapter discusses several legal issues pertaining to both research and direct use of genetic susceptibility testing in environmental and occupational health. A particularly important legal issue is the use—and potential abuse—of genetic susceptibility testing in the workplace. Numerous environmental and occupational regulations invoke the concept that "safe" levels of exposure should protect "the most sensitive" individuals in the population. The identification of putative genetic susceptibility genotypes raises many interesting issues pertinent to employment discrimination, employer liability, workers compensation, and medical coverage, to name only a few. Examples of previous case law where genetic susceptibility issues have come before the courts and the implications of genetic testing for susceptibility to such laws as the Americans with Disabilities Act (ADA), and even in "toxic tort" law, are discussed.

Finally, the last chapter of this textbook focuses on how genetic susceptibility, and the tools of toxicogenomics, might impact the field of risk assessment of environmental hazards. Many state and federal laws utilize the process of quantitative risk assessment to make reasoned, scientifically based judgments about the level of harm or disease that might result from exposure at a specified level to a toxic substance in the workplace or general environment. Often, risk assessments rely on experimental animal data, making projections of risk measured in one species (rats or mice, usually) to the target species of interest—humans. Generally, a safety, or uncertainty, factor is used to adjust for species differences. A second uncertainty factor is also applied to compensate for innate biological variability between individuals in the population. Typically, a factor of 10 is used as the uncertainty factor for each of these (cross-species, interindividual variability). But is a factor of 10 enough—or too much—to allow for variability in the entire human population? Will the new tools of toxicogenomics, including the study of genetic susceptibility and gene–environment interactions in human populations, allow us to improve the accuracy of risk assessments? This chapter explores how ecogenetics research might be able to improve the area of risk assessment for environmental hazards, as well as the many hurdles and limitations that will be encountered to do this.

22

Ethical Issues in Ecogenetics

Kelly Fryer-Edwards
University of Washington, Seattle, WA

Lindsay A. Hampson
National Institutes of Health, Bethesda, MD

Christopher R. Carlsten and Wylie Burke
University of Washington, Seattle, WA

22.1. INTRODUCTION

Ecogenetics brings together the two scientific disciplines of environmental health and genetics. Current knowledge of the ethical issues in these distinct disciplines provides a starting point for understanding the ethical issues of ecogenetics. Historically, environmental health questions have raised our awareness about justice and fairness, while ethical debates around emerging genetics applications have focused on personal impact of scientific knowledge. In this chapter, we bring these two discourses together and argue that ethical issues in ecogenetics are particularly concerned with vulnerable individuals and populations. Questions of justice lead to questions of responsibility concerning exposure to environmental hazards and genetic contributors to risk. We develop a case study of asthma to illustrate the range of possible applications for ecogenetics in the future and the range of ethical issues that will need to be considered when moving forward with research and policy in this area.

Gene-Environment Interactions: Fundamentals of Ecogenetics, edited by Lucio G. Costa and David L. Eaton

22.2. CORE ETHICAL CONCEPTS AND ISSUES

As with any medical or scientific endeavor, questions of value and ethics arise in ecogenetics. Scientific developments create new technical opportunities, raising questions about the wisdom of doing what has become possible. For any new technology, we can ask questions such as:

Who is benefiting? At the expense of whom?
Do the risks outweigh the benefits?
What are appropriate levels of evidence for taking action?
Whose responsibility is it to take action?
In the face of uncertainty, how should we respond?

Questions of ethics rarely involve simple choices between right and wrong. Rather, questions of ethics are queries about values and priorities, requiring justifications for doing what is best, among many possible solutions.

Traditional bioethics has centered around three principles: autonomy, beneficence, and justice (Beauchamp and Childress 2001). A version of these principles has guided medical ethics since the Hippocratic Oath, with a primary focus on the individual relationship between patient and physician. Public health and research activities shift the focus of our concern beyond this individual relationship, but the core principles remain the same. For example, the National Commission for the Protection of Human Subjects of Biomedical and Behavioral adopted the same three principles in the late 1970s to guide decisionmaking in human subjects research, using a broader interpretation of the principles to include questions of population and overall benefits of the proposed study (National Commission 1979).

Justice has been a central ethical concern for environmental health. Environmental justice is a concept that relates to differential exposure to environmental hazards among different groups, including groups defined by race or socioeconomic status. First defined by the Environmental Protection Agency (EPA) in the early 1990s, environmental justice is (Institute of Medicine 1999):

> the fair treatment and meaningful involvement of all people regardless of race, ethnicity, income, national origin or educational level with respect to the development, implementation, and enforcement of environmental laws, regulations, and policies. Fair treatment means that no population, due to policy or economic disempowerment, is forced to bear a disproportionate burden of the negative human health or environmental impacts of pollution or other

> environmental consequences resulting from industrial, municipal, and commercial operations or the execution of federal, state, local and tribal programs and policies.

Areas of concern for environmental health, including workplace exposures, unsafe living conditions, and toxic waste disposal, less commonly involve exposures or hazards for members of higher-class communities. Rather, environmental justice concerns typically focus on socioeconomically vulnerable populations. How to remedy disproportionate distributions of the risks and benefits of research or practice is a primary function of ethics work within environmental health.

Growing genetic knowledge, on the other hand, has generated a different set of ethical issues focusing on how genetic information may adversely affect individuals in the form of stigma, job or insurance discrimination, or invasions of privacy (Clayton 2003). With genetics, the focus shifts to those who might be considered genetically vulnerable. Respect for persons becomes a central concern in this area of research and practice, incorporating a careful weighing of whether the risks of genetics outweigh the hoped-for benefits.

22.3. ETHICAL ISSUES IN ECOGENETICS

When environmental health and genetics are brought together into ecogenetics, concerns about justice and the personal impact of research and practice become equally significant. Possible applications of ecogenetics are presented in Table 22.1. Applications such as presymptomatic testing for susceptibility to workplace or other environmental exposures invoke the need for respect for persons and concerns about stigma and discrimination seen in other areas of genetics. An important difference, however, is that in ecogenetics, genetic susceptibility is a mediator of environmental risk, leading to the possibility of differentiating between individuals who may be harmed by an exposure and those who are presumably at lower risk of harm. In this potential application of genetic information, one must be careful to ask when it is appropriate to remove an individual from harm's way and when the harm must be alleviated for all. For example, chronic beryllium disease (CBD) occurs as a result of exposure to beryllium, an essential component in ceramics and other industries. Recent work has documented an increased risk for CBD among individuals exposed to beryllium who carry specific variants of the HLA gene. Testing might reduce the rate of CBD significantly. However, the predictive value of the genetic information is limited, enhancing the risk of adverse social consequences

Table 22.1 Examples of Ethical, Legal, and Social Issues Raised by Ecogenetics Practices

Ecogenetic example	Type of Policy	Key Setting/ Group(s) Affected	ELSI Issues
Use of large databanks to study gene–environment interactions	Policies governing recruitment of research subjects and data protection	General public, affected families, researchers	Avoidance of coercion; inclusion of minorities; adequate protection of privacy; adequate subjects accrual for valid results
Genetic testing to guide decisions about diet and vitamin intake	Policies governing direct to consumer marketing of genetic tests	General public, commercial developers	Adequate consumer protection; fair market opportunities
Differential management of environmental health exposures or drugs based on genetic susceptibility	Clinical practice guidelines	Healthcare professionals, patient advocacy groups	Minimize adverse effects of genetic labeling and discrimination; assure cost-effectiveness of new technology
Differential management of workplace exposure based on genetic susceptibility	Policies for workplace exposure to toxic chemicals	Workplace employees and employers, regulatory agencies	Adequate protection of worker privacy and employment opportunities; potential job discrimination; appropriate workplace protection
Genetic classification of toxic responses or information about genetic susceptibilities as factors in environmental policies	Clean air and water standards; hazardous substances regulations	General public, manufacturers, industry	Adequate protection of the public; avoidance of unnecessary and burdensome restrictions

of testing. The prevalence of genotypes that confer increased risk is 30–35%. These genotypes have a positive predictive value of only about 15% and a negative predictive value that is very high, but not 100%. Thus, genetic testing could make a third of the workforce ineligible for jobs involving beryllium exposure, despite the fact that most would not develop CBD, and could shift all CBD to the "low risk" genotype population. Further complicating matters is current evidence of different genotypes that confer risk for sensitivity to CBD, which may result in a preclinical stage rather than developing the full-blown disease. With these complications in mind, is genetic testing a reasonable strategy for reducing CBD burden? The answer is likely to depend on contextual features, including the availability of other jobs for people with CBD-associated genotypes and the feasibility of measures to further reduce beryllium exposure within the workplace.

Workplace testing is not the only place where questions of appropriate, or fair, responses arise. For example, if a newborn is detected to have an increased susceptibility of developing type 1 diabetes based on a genetic screening test, a healthcare plan that provided greater access to care could be justified in order to follow the development of the predicted condition. Indeed, if no improvement in care or monitoring occurred, the testing would be difficult to justify. However, other children without a positive genetic test could also benefit from improved access to healthcare, particularly children of vulnerable populations for whom access to healthcare may be limited. The question from an ethics perspective is whether we have discharged our duty to provide quality healthcare by identifying children who demonstrate genetic predispositions to developing disease and following them closely. The answer to this question clearly depends on which genetic tests we have access to, how accurate they are at predicting disease onset, and how important specific interventions are in improving health outcome. Another key issue is which populations would be tested, given recent data connecting lower health status to lower socioeconomic status and less access to health care (Institute of Medicine 2003). We will pursue these questions more specifically within the case of asthma.

22.4. CASE STUDY: ASTHMA

Asthma is a chronic lung condition characterized by airway inflammation, hyperreactivity, and reversible airway obstruction. The disease is found disproportionately among children and minorities, and prevalence has increased significantly since the early 1980s. There is strong evidence for both genetic and environmental contributors to the development of asthma.

Table 22.2 Potential Future Applications of Asthma Genomics

Application	Health Implications	Current Research
Pharmacogenomics	Genetic testing to determine safest and most effective medications for individual asthmatic patients	Identification of association between polymorphisms in the β_2-adrenergic receptor and response to β_2-adrenergic agents
Workplace testing	Genetic testing to identify individuals with increased susceptibility to workplace exposure	Association between polymorphisms in *HLA DQB1* gene and isocyanate-induced asthma
Early identification of asthma risk	Newborn screening or early-childhood testing to identify increased asthma risk	Linkage and candidate gene studies identifying multiple loci associated with asthma risk

Genomics research has identified numerous genes and gene loci associated with asthma; further studies of genes, protein functions, and biological pathways associated with asthma are likely to yield new information about disease biology and innovative therapeutic and preventive approaches.

Although much remains to be learned about the genetic contributors to asthma, current research points to a range of potential applications of genetic information in addressing the public health burden of asthma (see Table 22.2). The earliest clinical applications of this research effort will likely be in pharmacogenomics (University of Washington Center for Genomics and Public Health 2004). Genomic strategies will aid in the identification of new drug targets, and may lead to drugs designed for use in specific subsets of asthmatic patients defined by genotype (Palmer et al. 2002). In addition, pharmacogenomics research may produce genetic tests designed to predict drug responses and adverse side effects.

In the long term, genomics research may also produce genetic tests that aid in disease classification and prognosis, or identify unaffected children who are at increased risk of developing asthma. One possible application of the latter capability would be testing of newborns or young children to identify those at increased risk for developing asthma for the purpose of implementing early prevention efforts. Another application, with different policy implications, would be the use of genetic testing to predict workplace asthma risk. For example, gene variants in *HLA DQB1* appear to be associated with susceptibility to isocyanate-induced asthma and variants in other genes may also contribute to the development of this work-related asthma syndrome (Mapp et al. 2000, 2002, Wikman et al. 2002; Piirila et al. 2001).

The value of these genetic testing opportunities will depend on the interventions available for individuals who test positive. In the case of pharmacogenomics, targeted drugs with greater efficacy may be available—but these drugs may be expensive. As a result, access may be difficult for minorities and those with low income—the very groups that bear the highest risk of asthma. For other testing opportunities, the value of identifying people with a genetic susceptibility will depend in large part on the environmental interventions that can be used to improve health outcome. If a susceptible child's risk can be reduced by specific efforts to reduce antigen exposure in the home, testing may provide a tangible benefit—but if the changes require resources unavailable to low-income families, testing could only serve to increase health disparities. Some public health programs exist that target healthy homes for families who need them, and arguably genetic testing can help identify families who would be most likely to benefit from such a resource (Public Health—Seattle & King County, 2004). However, achieving these benefits requires resources be used to support such innovative public health programs.

22.5. BASIC ETHICS TOOLS FOR ISSUES RAISED BY ASTHMA RESEARCH AND PRACTICE

In Section 22.1, we referred primarily to principles that highlight ethical issues. There are three basic strategies to guide one's thinking in this regard (see Box 22.1). In the traditional rule-based strategy, core principles can provide a useful guide when one is trying to determine the best course of action. However, there is room for considerable disagreement about how any one of the principles might be interpreted in practice. What it means to be of benefit in any given circumstance, for example, is often widely debated. Alternative strategies, one based on consequences and the other on core virtues, broaden the range of considerations used in addressing a specific question. The first of these alternatives attends to potential consequences of the policy or practice. Rather than relying on a set of rules, this approach requires that one project forward and play out the implications of the alternatives under consideration. This approach can also be challenging as it is often difficult to predict a full range of consequences for an action. Different groups will also differ in their assessment of the value of different consequences.

Box 22.1 Traditional Approaches to Ethical Analysis.

Principle- or rule-based. *An action is right if it follows fundamental moral rules.* The reasoning process here involves identifying the appropriate moral rule for the situation. Some describe this approach as "top–down" in that it begins by identifying previously agreed-on rules to inform the decision. Rules and principles may come from multiple sources, including one's profession, society, or religion, or an institution. Rules or principles, even from within the same system, may come into conflict at any one time. Justifying why one principle might be privileged over another in a given situation constitutes much of the work of ethics in this model.

Consequence-based. *An action is right if good consequences outweigh bad consequences.* The reasoning process here involves identifying specific anticipated, as well as unintended, outcomes of various options. Deciding which consequences to consider, and how to "weigh" them against one another, is the challenge of this approach. Who determines what counts as a "good" or "bad" outcome is also an important issue.

Virtue-based. *An action is right if it enacts a core purpose.* The reasoning process in this approach involves identifying what role the decision-maker will take in the situation. (Is it one of citizen? Policymaker? Employer? Scientist? Physician? Mother?) From there, one must decide what the core values are for that position (e.g., what the ideal researcher would do). These core values should capture the core purpose (e.g., as a mother, my core purpose is to protect my children from harm). The challenge of this approach is negotiating competing interpretations of core purpose and resolving conflicts between roles.

Rather than looking back to principles or forward to consequences, the virtue-based approach requires that one look inward and ask the following question: Do I have core beliefs or commitments that will guide me in this case? Core commitments are most likely to come from one's professional identity or personal roles in society. For example, a physician can have certain core commitments, as can a mother, simply because of the role she has adopted and the responsibilities she believes go along with that role.

To illustrate the three approaches to ethical analysis, we can take the following broad question: Should we work toward prevention of asthma? As a chronic disease affecting many people, the goal of asthma prevention is prima facie a desirable goal. Appeal to basic principles, such as beneficence and justice, should be sufficient to convince us that reducing the disease burden of asthma would be of benefit to people. Because low-income groups are disproportionately affected by the disease, working to ameliorate it would work toward restoring the disadvantages conferred by economic disparities. Certainly, the consequences following from eliminating disease burden would be desirable. Then, finally, from a virtue-based perspective, we would consider what our vision and values are for living in the world. All other things being equal, we could argue that many of us would choose to live in a world with reduced disease burden.

On one hand, this ethical analysis seems straightforward. However, when one turns to the next tier of questions, related to the actions one might take to achieve the goal, the simple conclusions we have drawn may be challenged. For example, we might agree with the general statement "Reducing disease burden is a good," but this statement does not indicate who has the responsibility to achieve (and finance) that good. While we can easily agree that something should be done about asthma, there is little consensus about who has the responsibility to act. For example, if we decide that people relying on public assistance for housing have a right to clean, allergen-free homes (e.g., see scenarios listed in Box 22.2), who has the

Box 22.2 Scenarios to Consider [Note to Editors: this box could be used upfront with the purpose of stimulating interest in readers].

Scenario 1. One hundred families will undergo screening to determine who will have access to the 40 new clean homes available. A public health worker proposes using genetic testing to screen children in the families to identify those with increased susceptibility to asthma. The new allergen-free homes would then be distributed among those families who have higher genetic risk for asthma. Families in this community are concerned about the proposal, as they feel that this is just another form of discrimination to tolerate. They ask: Don't we all have a right to a clean home?

Scenario 2. An employer is concerned about her increasing health insurance premiums. She considers using genetic testing to identify workers who have increased susceptibility to developing asthma so that she can move them to positions where they will have reduced exposure. Initially, there is great enthusiasm for the program, as workers, too, are supportive of protecting their health. However, tension within the company develops when, subsequent to initial job modifications, workers hear of an updated genetic screening test that is rumored to be superior than the first.

responsibility to ensure that those homes are provided? The government, through tax dollars and public health systems, is a key player, but landlords and employers are also players. For someone to claim a right to something, there must be another party with a corresponding duty to ensure that this right is achieved. While our public health and public safety systems are built on values that would support a responsibility to provide clean housing, there are practical limitations to the ability of those systems to enact those values. Resources would be required, which would implicate taxpayers and would ask citizens to consider the degree of responsibility they have to the health and well-being of fellow citizens. Many might conclude that the responsibility is discharged by programs such as Medicaid, and does not encompass broad-scale efforts to improve the quality of low-income housing.

Questions of responsibility and justice run throughout public health and healthcare ethics. The question before us in this chapter is whether our responsibilities or considerations change with the addition of genetic information. The answer to this question looks different depending on which perspective one takes. To examine this further, let us first consider the range of stakeholders who would be involved in considering genetic testing to prevent asthma (University of Washington Center for Genomics and Public Health 2004). These include patients with asthma and their families, who have a primary concern for affordable, effective healthcare. Genetic testing or other clinical applications of genetics will be of interest to the extent that they improve healthcare; but they will also raise the question of additional risks, such as stigma, worry, or discrimination resulting from genetic labeling. Issues of stigma may be particularly important from the perspective of the community. For example, genetic research could lead to the impression that a particular community is prone to asthma or to simplistic assumptions about genetics as the cause of disparities in asthma prevalence.

Further complicating matters, even if responsibility can be determined, it remains unclear what the best course of action should be in any given circumstance. Genetic interventions are only one set of possibilities. A potential outcome of focusing on genetic risks alone could be reduced attention to remediation of environmental contributors to asthma risk. If we do decide that innovative therapies are the best route to remediation of asthma risk, a new set of questions arise:

Who do we test for genetic risk?

What evidence is required to justify the use of new therapies?

How can access be ensured for low-income patients?

What healthcare provider education will be needed to ensure the appropriate use of new therapies and testing options, and who will provide it?

Public health professionals will need to address both research needs and population needs for effective care. As ecogenetic research reveals the gene–environment interactions that contribute to asthma, public health is likely to have an important role in determining whether genetic susceptibility should be a factor in environmental remediation efforts—for example, should families in which children have a genetic susceptibility to asthma receive preference in the allocation of allergen-free public housing?

Researchers and public health officials have different but related concerns, specifically, what genetic research can best address the public health implications of asthma, what funding mechanisms and research procedures will ensure that needed research is done, and how the results of research can be used to improve healthcare. One important source of new asthma treatments, particularly in pharmacogenomics, will be commercial developers. Pharmaceutical and other biotechnology companies are likely to play an essential role in the application of genomic information to improved asthma care, but have financial interests that potentially conflict with the interests of patients, healthcare providers, and communities.

22.6. DISCUSSION

How one reconciles the different interests and values within these different stakeholder groups is an important question. A key issue from an ethics perspective is simply to name and acknowledge that there are multiple stakeholders at the table. Too often, key stakeholders are not part of the virtual or actual conversations that lead up to policy or practice decisions. Leaving people out of the decisionmaking process is seldom intentional. Simply the fact that public health workers function in different contexts from primary care providers, who work in different contexts from policymakers, which is different again from citizens, can lead to key players missing from the table during policymaking discussions. Particular values and interests held by each group are also often unspoken. The policy debate can center around practical issues, such as funding, without articulating underlying assumptions or the ultimate goals of the decision at hand. Naming these values helps clarify what is at stake in the different decisions. Points of contrast as well as overlap can be found. For example, in the case of asthma, presumably all stakeholders, including the commercial interests, care about reducing disease burden for people with asthma. However, the conflicts emerge at the point of secondary interests where other values, such as concern about stigma or commercial interests, may compete with an interest in reducing disease burden.

We can play out some of the ethical implications of ecogenetics applications suggested above to learn more about the scope of issues that should be explored when working within this new area. For example, as we have discussed, pharmacogenomics research offers great promise for improving asthma therapies, but raises questions about allocation of healthcare resources and adverse labeling of patients. If new drugs require genetic testing prior to use to determine which patients should receive the drug, this process will add to the initial cost of care (although the cost may be compensated by reduced use of ineffective drugs). This practice strategy will require development of new practice guidelines and health provider education. Perhaps more importantly, genetic profiles that predict drug response may also provide other predictive information unrelated to asthma, such as information about other disease risks or susceptibility to occupational exposures (Maier et al. 2003). Practice guidelines will need to address the obligations of healthcare providers to provide ancillary information, and the potential risks to patients of unsought predictive information.

Commercial incentives are an important factor in pharmacogenomics generally, with a potential for both positive and negative effects on patient care. Commercial investment is critical to drug research and development, but the high cost of research and development contributes to high prices for new drugs. Commercial incentives (or the lack thereof) may also limit some pharmacogenomic opportunities. Potentially promising drugs might not be pursued if the market for them is perceived to be too small or nonremunerative. In addition, some important research findings will be proprietary and might not be publicly disclosed for market reasons. For example, a company might choose not to disclose data on genotypes predicting nonresponse to medications it manufactures, because such data might lead to tests that reduce market share.

These issues suggest that careful consideration should be given to the process by which clinical practice guidelines are developed in relation to new asthma drugs, with particular attention to the standards for use of genetic profiling to determine drug regimens. If new drugs are very expensive, access to these drugs by the medically underserved is a potential concern. Expensive drugs that are recommended for use in a particular clinically defined subset of asthma patients, or require prior genotype testing, will represent a challenge for publicly funded healthcare programs. Careful consideration will be needed to construct drug formularies that ensure appropriate access to such treatments in the context of cost-effectiveness. Efforts to address this problem will be aided by public health efforts to ensure adequate outcome data comparing new and established therapeutic strategies.

It may also be important to invite collaborative discussion among representatives of commercial, public health, and academic research sectors in considering guidelines for disclosure of information of high public interest—such as data concerning genotypes that predict nonresponse to commonly used asthma drugs—that has been gained in drug trials.

In addition to knowing who the stakeholders are and being explicit about the values that they hold, other questions must be considered. For example, genetic testing can be an exciting and promising area of research. As such, it grabs the attention of funders, the media, patients, and scientists. We all are interested in where these new developments will take us. However, we risk becoming so enamored with the latest emerging trends in genetics that we leave unexplored other, less technological or innovative, solutions. In the case of asthma prevention, for example, a thoughtful cost–benefit analysis would need to be done before we could argue for the technological solution over the environmental health solutions, such as providing clean living conditions for families in low-income housing. We should look carefully at the evidence so that we can be clear about what is gained by pursuing and including genetic testing in the public health equation.

The question of responsibility returns when one considers other solutions to asthma prevention. The current benefit of genetic testing strategies is that they are currently being paid for as part of research protocols. A less innovative solution, such as providing clean housing, would not necessarily be grant-funded. A new host of stakeholders come to the table when the question shifts to who has the responsibility to provide clean housing for low income families. The government and public health systems arguably have a central mission to promote health and prevent disease among the public. From a virtue-based perspective, one could argue that these public agencies do have a core duty to work to ameliorate health risks. However, from a practical side, one might argue that there are inadequate public funds to cover new housing project developments. Funds would need to be shifted from other public health efforts to pay for this one, with potentially adverse consequences for groups that would receive less funding as a result. These are systemic issues that require complex solutions.

As discussed above, it is particularly vulnerable groups that are affected most adversely by asthma. We might return to core ethical principles to outline how a principle of justice can guide our decisions, with attention to balancing the benefits of research and the burdens of disease. When thinking about population-based health issues, the concept of justice emerges as a primary ethical consideration. *Justice*, broadly speaking, refers to the principles or rules that guide fair distribution of goods and guards against inequitable distribution of harms. However, neither theorists nor policymakers are in agreement about how concepts of "fairness" and "equitable distribution" should be understood. Philosopher John Rawls outlined a process for achieving procedural justice (Rawls 1971). According to Rawls, if we were to define policies and make decisions as if we were behind a "veil of ignorance," the resulting system would be fair and just. Rawls' veil of ignorance presumes that the decisionmakers behind the veil do not know which position in the new system they will be assigned and that they are eligible to be in any position. Thus, if a group were deciding about fair allocation and provision of clean public housing, the decisionmakers would have to presume that they could end up as tenants of the resulting system, as taxpayers, as landlords, or as policymakers. The presumption behind this model of decisionmaking is that the resulting system or decision will maximize the position of the least well-off in the system, as the decisionmaker would have an equal chance of becoming that person.

Of course, Rawls' veil of ignorance does not represent an actual state of affairs. Critics have argued that it is not possible, nor necessarily desirable, for people to act as if they are not positioned in a certain way (Young et al. 1990). In practice, most decisionmakers do not have an equal chance of ending up in the least well-off position in a system. An alternative theory of justice requires that one start with an examination of the culture or institution within which the decision is being made. If the culture or institution is already affected by systems of domination and oppression, it will be difficult to enact a single decision that could be considered "just." Such an approach argues that we take the particular circumstances of each affected party into account in our decisionmaking, not from a privileged perspective of impartiality, but from a situated perspective of embeddedness (Young et al. 1990). Indeed, many population-based initiatives now begin with an attempt to bring all the stakeholders to the table. It is not easy, however, to have a conversation once at the table, when many competing interests and imbalances of power are present (Habermas 1987). Open discussions between employers, workers, public health officials, and citizen taxpayers, are challenging, particularly if you are asking participants to imagine other possible perspectives and not simply speak from an unbending stakeholder position.

22.7. CONCLUSIONS

In this chapter we have presented both traditional frameworks for ethics, key principles in bioethics, and an expanded analysis offered by considering stakeholder interests. The approaches demonstrated here should be seen as tools to be used in different circumstances. No one tool will be effective in all situations; becoming familiar with a range of tools can strengthen one's ability to respond appropriately. Often just learning which questions to ask can help open important areas for exploration and consideration.

As the case study of asthma highlights, different concepts are emphasized in most ecogenetics cases from traditional bioethics. Because of the concern with populations, be they children or workers, there is a need to understand concepts such as responsibility and vulnerability. Issues in genetics and the environment, particularly when linked, have profound implications for communities, and we will need to think more carefully in the future about what it means to be responsible for one another.

23

Social and Psychological Aspects of Ecogenetics

Deborah Bowen
University of Washington, Seattle, WA
Fred Hutchinson Cancer Research Center, Seattle, WA

Shirley Beresford
University of Washington, Seattle, WA

Brenda Diergaarde
Fred Hutchinson Cancer Research Center, Seattle, WA

23.1. INTRODUCTION

The field of ecogenetics is new, and presents exciting opportunities for improving the health outcomes of the public (Omenn 2001). Given the complexities of the social and psychological issues raised by the Human Genome Project (Collins and Mansoura 2001), and the psychosocial issues connected with environmental exposure (Sharp 2003), we must consider the emerging issues in ecogenetics from all human angles, including social and psychological. This chapter outlines a framework for considering social and psychological issues in general, and subsequently discusses possible issues related to ecogenetics in each of the framework areas. To illustrate the complexity of the problems, a current and relevant example from the literature is given for each area. Finally, we provide some thoughts for future scholarly activity in the applications of ecogenetics to social and psychological settings.

Gene-Environment Interactions: Fundamentals of Ecogenetics, edited by Lucio G. Costa and David L. Eaton

Key to understanding of the psychological and social issues in ecogenetics is recognition of the potential for teasing apart some of the influences on and reactions to environmental and genetic risk. Phrased another way, what are the levels of our social system, and how do environmental and genetic risk factors interact with these levels? Social and cultural forces act at a collective level to shape individual biology, risk behaviors, environmental exposures, and access to resources that promote health and quality of life. One way of looking at this is that neighborhood and community-level variables set the stage by providing the surroundings in which an individual lives, works, plays, worships, and exists. Health risks, both environmental and genetic, can by viewed through different elements of the surroundings and will be shaped accordingly. Therefore, organization of this chapter will be determined by examining environmental and genetic risk factors through the multiple lenses of the world around us.

Table 23.1 shows the various lenses through which the impact of genetics and environment on health can be viewed. This framework was developed as part of the University of Washington School of Public Health, Social and Behavioral Sciences Program strategic plan. It was adapted from the models of McLeroy (1988) and others (Patrick et al. 1988; Welton et al. 1997). In this depiction, the distal levels of influence (e.g., the political level) are placed farther away from the individual, but are able to influence the other, more proximal levels. Intermediate levels shown are the sociocultural and residential levels that have influence on the behavioral and biological mechanisms leading to health outcomes. This model will form the structure of this chapter and of this analysis. For each level, we will define the level and analyze its relevance relative to the effects of ecogenetics. We will then discuss an example of how environmental and genetic risk factors might be viewed through the prism of that particular level. At all times, we discuss "response" to risk in social, psychological, and behavioral ways. We will not discuss biological response in this chapter. We will close with conclusions and directions for future research in this area.

23.2. LEVELS OF ANALYSIS

23.2.1. Individual

23.2.1.1. Description and Analysis. At the simplest and most straightforward level of analysis, the psychological and social issues surrounding environmental and genetic differences impact individuals. Individuals have

Table 23.1 Biopsychosocial Determinants of Health

A. Political

- Federal, state, and local policy and law
- Political/geographic environment (public safety, food safety, physical environment protection, worksite environmental protection, and safety)
- Agriculture and food production
- Education system
- Economy
- Public health system
- Health system
- Law enforcement system
- Institutionalized biases (e.g., racism, agism)

B. Residential

- Residential environment
- Local community health service access
- Local public health prevention/health promotion
- Community economic development and employment access
- Community social service access
- School/educational services access

C. Organizational

- Healthcare benefits
- Occupational issues
- Work and school issues
- Social interaction
- Work or school related stress

D. Sociocultural

- Cultural risk environment
- Cultural SES/educational environment
- Social capital
- Cultural context/language
- Community social networks
- Religious affiliations

E. Familial

- Family coping/cohesion/support
- Shared genetic risk
- Shared exposure
- Family norms and guidance
- Family roles

F. Individual

- Demographic information (e.g., age)
- Individual values/preferences/emotions
- Social/economic resources
- Stress and reactivity
- Health behavior levels

Sources: Adapted from McLeroy (1988), Patrick et al. (1988), Welton et al. (1997).

many layers of psychological variability that will influence the ways in which environmental and genetic risk will work to influence health, probably the most heavily studied area presented here. How people respond to information regarding their personal risk is an important influence on the ways that risk information helps in making choices.

Studies have found differences in the ways in which people react to genetic risk information based on their age, sex, attitudes, and social and economic personal resources (Petersen et al. 1999). Other variables that cause individual variability in behavioral response to environmental and genetic risk include individual levels of stress and levels of health behaviors. High levels of chronic stress have been associated with poorer levels of health promoting behaviors, such as dietary and exercise behaviors (Cartwright et al. 2003; Wardle et al. 2000), or higher levels of adverse health behaviors, such as cigarette smoking. An entire issue of the *Journal of the National Cancer Institute* (Klausner 1999) summarized the literature on risk perceptions and risk communications, and concluded that we are at a very early stage of research into this issue, without much knowledge about how people react to and use risk information to make health decisions. Individual reactions to risk information underlie all areas of ecogenetics (see Chapter 25, on risk communication and assessment) and, therefore, more research on how individuals process and understand risk-related information, how best to communicate risk information, how that risk information influences health strategies and behaviors, and how these, subsequently, affect health outcomes and costs is needed (Collins et al. 2003; Miller et al. 2004).

23.2.1.2. Example. One hope for the use of information about genetic and environmental risk factors is that it will provide people with guidance on health choices and health behavior (Collins 1999). This as-yet-unproven strategy has generated early research on the effects of genetic and environmental risk feedback. Feedback of genetic information to increase risk perceptions and thereby reduce exposure to risk, by motivating smoking cessation, has been the topic of much attention (Johnstone et al. 2002; Marcy et al. 2002; McClure 2001). In these discussions smokers receive feedback about their genetic susceptibility to lung cancer from the results of genetic tests. In theory, smokers who are at high or enhanced risk for lung cancer will become more concerned about their personal risk for lung cancer and will try to reduce this risk by reducing environmental exposure to carcinogens through quitting smoking. More recent empirical evidence, however, suggests that this does not happen (McBride et al. 2002). Smokers attending an inner-city community health clinic were randomized to receive either usual care or biomarker feedback. Individuals in the biomarker feedback arm were given the opportunity to have a blood test for genotyping the glutathione *S*-transferase gene (*GSTM1*) (see Chapter 20 for discussion of GSTs, smoking, and cancer risk). While short-term cessation was higher for the biomarker feedback group at the 6-month follow-up compared to the usual care group, at 12 months postvisit there were no significant differences in cessation rates and sustained cessation between the two arms. This is particularly a problem given the high interest reported by smokers in receiving the results of genetic testing for susceptibility to tobacco-related cancers (Ostroff et al. 1999). We need to reconsider the use of biological feedback as sole motivator of health behavior change, and test new models of motivation that incorporate both personal variables and the larger social context (Bowen et al. in press).

23.2.2. Familial Level

23.2.2.1. Description and Analysis. Nuclear families are often defined as the immediate spouse or partners and their children; however, in genetics "family" is usually defined as a group of people sharing genetic material. Legally, "family" is defined as two or more persons who are related by blood, marriage, or adoption. The members of families often have legal roles, such as parent or child, caregiver, or next-of-kin. Family members can have social roles, too, such as "the person we all consult for health information." Whatever the definition of family, the relevance of family for ecogenetics is likely to be important.

The family exhibits a powerful influence on our choices and options, and certainly has a role in issues surrounding genetic and environmental risk. First, families share genetic material and genetic risk is transmitted through family members in the form of germline mutations. Environmental exposure differences can possibly alter these inherited risks, potentially making the family contributions to risk more (or less) prominent. In addition, families set health-related norms and values that the family members must follow, or they receive social or other reinforcement (Roberts and McElreath 1992). Between parent and child, responsibilities often include formal and informal guidance in health information and health choices (Meischke and Johnson 1995), including advice about genetic and environmental risk factors. The norms and values transmitted between parents and children certainly influence choices made about later healthcare. For example, children's sun-protective behaviors are often under the control of parents or other adults, and studies have found that parents do not always use the best protective strategies for their children (Lewis and Hammond 1996). As we learn more about genetic risk factors for melanoma, the lack of sun protection negotiated by parents for their children could become a complicated issue, especially since the most common barriers to good sun protection for children described are lack of perceived need and inconvenience (Buller et al. 1995; Diffey et al. 1996; Kakourou et al. 1995; Olson et al. 1997; Zinman et al. 1995). Information about increased genetic risk may increase motivation for parents to overcome the inconvenience factor and increase sun-protective behaviors. These elements of the familial world will need to be accounted for in any consideration of genetic and environmental risk factors.

23.2.2.2. Example. One example of a familial influence on environmental exposure and genetic risk is the issue of take-home pesticide exposure among agricultural workers and their family members. Agricultural workers who work in the fields where products are grown, or who work in plants where the food products are processed and preserved, are routinely exposed to pesticides as part of their occupation. Safety laws and policies are in place to prevent toxic short-term exposure to these pesticides in the form of monitoring devices, protective clothing, and other methods of reducing exposure. However, the issue of take-home exposure has received little attention. "Take-home exposure" is when an agricultural worker transports contaminants from the workplace to a vehicle or home, and individuals in these settings are subsequently exposed to and absorb these contaminants. This phenomenon has been documented with organophosphorus pesticides in agricultural workers and their children (Curl et al. 2002). Emerging genetic differences in individuals' biological reactions to pesticides are currently receiving research attention [e.g., see Costa et al. (2003a, 2003b) and Mackness et al. (2003)], and policy is under consideration to deal with these differences. For example, genetically sensitive individuals may need more frequent testing, or may not be allowed to work in certain areas of field management. This, however, may result in genetically based employment discrimination (see Chapter 24). Moreover, these types of restrictions do not protect children at home who may have different genetic susceptibilities than their parents, who have the primary exposures to the pesticides and who may bring home pesticide residues that produce measurable and harmful levels to their children. Current policy and law regarding the protection of workers from pesticide exposure does not take into account familial exposure through the take-home pathway. This issue needs both research and policy attention, as it illustrates the role of the family in the assessment of and protection from ecogenetic risk.

23.2.3. Sociocultural Level

23.2.3.1. Description and Analysis. This level, often called "community," is very diverse and involves many of the social ties that do not rely on geography or genetic connection but bind us by some other part of our identity. Sociocultural connections include ethnic or minority culture, language spoken and read, and religious affiliations.

Culture can have powerful effects on health, health outcomes, and behavioral reactions to personal risk. Differences among cultural perceptions and beliefs not only affect a person's trust in healthcare providers (Boulware et al. 2003) but also influence access to appropriate health care and social services (Saha et al. 2003) and availability of social networks and contacts. The density of social organizations and the degree of neighborhood cohesion are components of "social capital," which has great influence on the choices and sense of support and empowerment we have as individuals and as members of cultural groups (Lochner et al. 1999).

23.2.3.2. Example. One of the complicating factors in health disparities research and public health practice is that there are little data to address the issue of why health disparities exist (Baquet et al. 2002). We know very little, for example, about the causes of differential reaction to airborne pollutant exposure, leading to asthma (Brown et al. 2003). One hypothesis is that the genetic factors that influence reactions to airborne pollutants differ by ethnic group, and that these are the differences in genetic makeup that lead to differences in asthma rates between children of different ethnic groups. Another hypothesis is that exposure to airborne pollutants differs by ethnic group, and that this causes the differences in asthma rates. Whether it is genetics or environmental exposure, or a combination of both, has large implications for policy to prevent this chronic disease.

Since the mid-1970s multiple policies and laws have been implemented to improve the cleanliness of the environment, reducing exposures to pollutants in air, water, and soil. However, there still remain considerable amounts of these pollutants, and there are some data to indicate that they are particularly concentrated in settings with large and disproportionate numbers of low-income and ethnic minority individuals. This issue has prompted organization of the environmental justice and environmental equity movements, with the purpose of advocating for a more fair and equal distribution of toxic waste and other pollutants. Epidemiologic research that documents examples of environmental injustice is poorly developed (Bowen 2002; Rene et al. 2000); thus, few conclusions can definitively be drawn using the currently existing data. However, there are data to suggest that income and race play a role in the distribution of environmentally hazardous sites (Faber and Krieg 2002), and these data have lead to communities and scientists organizing to reduce bias and cope with the exposures in specific communities (Bonham and Nathan 2002; Corburn 2002; Faber and Krieg 2002). The potential genetic differences in reactions to exposure, together with the potential placement of toxins in neighborhoods of low-income and ethnic minority families, call for solutions that reduce the environmental and genetic burden of disease risk.

23.2.4. Organizational Level

23.2.4.1. Description and Analysis. People organize themselves into groups, or organizations, and these organizations create meaning and mission for many of us (Smedley et al. 2000). Organizations can be characterized by their structures and their functions. Organizations can be big or small, proactive or nascent, structured or unstructured, formal or informal, and effective in reaching goals or very ineffective and disorganized and thus unable to meet goals. Types of organizations include workplaces, schools and daycare centers, religious organizations, clubs, special-interest groups, support groups, play and study groups, and any other regular gatherings of people that have some mission or purpose.

One of the ways that environmental and genetic risk can be viewed through the lens of the organization is that organizations have policies, both formal and informal, which are made to guide behavior and choices within the organization. Workplaces and employment settings are probably the best example of how consideration of the organization is key to understanding ecogenetic risk and risk management. The issue of occupational exposure communication and management has long been a subject of social research (see Chapter 24). Issues considered here include how people react to workplace exposure to toxic substances, how the information is processed and used to make decisions, and what the long-term effects of exposure are to quality of life and stress, in additions to levels of harm and chronic disease.

23.2.4.2. Example. Exposure to beryllium in the workplace provides a complicated example of the social implications of genetic information in occupational settings. Even when the workplace meets or exceeds federal standards for acceptable levels of beryllium exposure, some workers may develop chronic beryllium disease (CBD), a scarring lung disease. CBD is costly for the worker, the employers, and society. Individual susceptibility to CBD is at least in part genetically mediated, and testing for genetic predisposition to CBD could enable workers at risk to avoid jobs involving exposure or could allow employers to identify and remove those workers from exposed workplaces. In these circumstances workplace genetic testing could potentially benefit highly susceptible workers by reducing their risk of CBD. It could also potentially shield employers from the costs associated with worker illness and the costs of implementing workplace changes to further reduce—or remove—beryllium exposure for all workers.

Worker interest in limiting the use of predictive genetic screening tests in the occupational setting makes sense, in that discrimination could ensue from using testing to determine the availability of jobs or positions. Historical experience with workplace genetic testing demonstrates the potential for applications of genetic testing to discriminate, stigmatize and unnecessarily exclude workers (OTA 1983; Omenn 1982; Schill 2000). The occurrence of these social harms relates in part to the accuracy with which a given test identifies the genetic trait of interest, the extent to which the presence of that trait is associated with an adverse health outcome when the carrier is exposed to the workplace hazard, and the availability of interventions to reduce risk or treat the health outcomes in workers who carry the genetic trait. These properties of genetic tests, designated as analytic validity, clinical validity, and clinical utility (Holtzman and Watson 1997; Secretary's Advisory Committee on Genetic Testing 2000), influence the ability of a given test to accurately label at risk workers and meaningfully predict their degree of risk for an adverse health outcome (Newill et al. 1986; OTA 1983, 1990; Van Damme et al. 1995). Because experience with the application and outcomes of genetic susceptibility testing in populations is limited, the predictive value of these tests, and therefore the ability to use them accurately to assess a worker's risk, is largely untested.

23.2.5. Residential Level

23.2.5.1. Description and Analysis. At the residential level we consider the local neighborhood influences and issues that impact health and health-care outcomes. These include the quality and quantity of health and social services in one's neighborhood, local health promotion or health-impairing messages, education and school services provided, and other community resources. Underlying wealth and economic development issues also are important in defining the residential or neighborhood level.

Why might these local resources affect how environmental and genetic risk factors play a role in health? Local resources are often related to health service options and health behavior choices, made by individuals but driven by one's immediate surroundings. Choices about healthy activities and diet are, in part, choices made with consideration of the local environmental options (Humpel et al. 2002; Owen et al. 2000). Personal risk information regarding high susceptibility to specific dietary factors will not be helpful if availability of healthier choices is low or nonexistent. Individuals living near the nuclear power plant at Three Mile Island (Pennsylvania, USA) found that they could not sell their houses after the plant had released potentially toxic gases and experienced increased levels of stress unless they reduced the amount of danger that they perceived from the gas release (Baum et al. 1983). Imagine that medical providers advised them to leave the area because of a high genetic susceptibility to the gases that they and their children had been exposed to! This relatively unresearched area may be important in future examples of environmental risk factors whose effects differ by genetics.

23.2.5.2. Example. Food that has been genetically engineered or modified to have certain properties, such as palatability, high shelf life, or resistance to herbicides, is currently the topic of much debate and discussion in the popular literature (Beachy et al. 2002; Wendo 2003). The public's acceptability of and preference for or against genetically modified foods has been the topic of several survey efforts. For example, measurements of opinions and attitudes toward genetically modified foods have been conducted in Finland (Backstrom et al. 2003), the United States (Brown and Ping 2003), Sweden (Magnusson and Koivisto Hursti 2002), the European Union (Pardo et al. 2002), and the United Kingdom (Kirk et al. 2002), among many other places. The results of these surveys generally indicate that most people do not trust the safety of genetically modified foods. However, the data also show variation by country or nation, and also by region of the country and by socioeconomic status, suggesting that occupation, income, and other local variants may play a role in shaping individuals' attitudes toward the acceptability of genetically engineered food products. In fact, there is evidence that many social variables play a role in acceptance of these foods (Finucane 2002), including media coverage that varies by media outlet (Frewer et al. 2002). These studies tell us that ideas about personal risk from both genetic and environmental sources are sensitive to local and regional influences.

23.2.6. Political Level

23.2.6.1. Description and Analysis. The political level of the world in which we live houses the laws and policies that govern interactions among nations, states, and local municipalities, as well as the people who live within them. These laws and policies in turn shape agriculture and food production; environmental protection legislation; educational, health, and public health systems; the economy; and law enforcement and military systems. Inherent in most systems at the political level are degrees of bias, including discrimination and disparity in resource allocation and incorporation into decisionmaking structure and process.

23.2.6.2. Example. Low folate status is a risk factor for the development of neural tube defects (Kirke et al. 1993; Molloy et al. 1985). Randomized trials evaluating of supplementary folic acid intake in the periconceptual period demonstrated its efficacy in reducing the risk of a neural tube defect affected pregnancy (MRC 1991, CDC 1992). As a result of these studies, and consultations between scientists, policymakers and the general public (Feinleib et al. 2001), folate fortification of grain products has been a standard public health intervention in the United States since 1998 to ensure that women consume adequate amounts of folate before and during pregnancy. For folate fortification to be efficacious, it needs to be at a sufficiently high level to have the optimum effect for most women, but low enough that it causes no other morbidity (Beresford 1994), not only in the target group (women of childbearing age) but also in the general population, where both old and young are exposed to folate-fortified foods [e.g., early concerns focused on masking of symptoms of vitamin B_{12} deficiency (Butterworth and Tamura 1989; Shojania 1980)]. Folate status is also influenced by genetic variants in the folate metabolic pathway (van der Put et al. 1997), and for women with a specific polymorphism in the methylenetetrahydrofolate reductase gene (MTHFR 677C $\rightarrow$ T) fortification may likely not be enough to provide adequate folate status (Ashfield-Watt et al. 2002). Women with this specific polymorphism may actually be rendered a disservice under such a public health policy, as they may believe that their dietary intake of folate is adequate through fortification. In fact, they may still have a low folate status because they exhibit this genetic variant. Therefore, although fortification is a simple low-cost measure designed to increase folic acid intake among all women similarly, it may deliver inadequate amounts of folate, and in effect give false assurance, to different subgroups of women, depending on their genetic makeup. National differences in policy, such as leaving the responsibility for obtaining adequate folate to women and their physicians versus providing a passive avenue of increased folate through food fortification, might lead to different groups of women being at risk.

23.3. CONCLUSIONS

The issues discussed here illustrate the complexity of social and psychological issues surrounding genetic and environmental risk. It is likely that these levels interact to affect health outcomes and health behaviors, the ulti-

Table 23.2 Puget Sound Partners for Global Health Proposed 2005 Timeline

Date	Task	Date Completed
January 24, 2005	Distribute request for pilot projects Notification to scientists	
March 15, 2005	Electronic letters of intent due	
March 31, 2005	Electronic applications due	
April 1–29	Committee review and determination of awardees	
May 2, 2005	Notification of awards	

mate outcomes of risk information. There is an incredible scarcity of research on the effects of exposure to ecogenetic risk information. All the topics presented here lend themselves to research, and this type of research can often be added to existing molecular studies with little additional cost. Most of the existing data are qualitative (Marteau and Lerman 2001) and cross-sectional survey data (Durfy et al. 1999). Very little data exist in longitudinal, multilevel, or randomized designs, and these types of studies must be done to assist in understanding ecogenetic risk, and communicating it effectively, in order to improve the health of the public.

An appropriate research agenda, then, might consist of cross-sectional and longitudinal observational studies that ask questions of the experiences, motivations, and intentions, as well as behaviors, of a group undergoing provision of ecogenetic information of different types. This information could include such situations as environmental exposure to arsenic, combined with differing levels of personal vulnerability, or the encouragement of extra protections, beyond the usual regulated and enforced protections, for agricultural workers and their families exposed to pesticide residue. The examples presented here are (see also the Puget Sound example presented in Table 23.2 and Boxes 23.1) simply ones that we have come across in the current infancy of this field, and many more will emerge as molecular discoveries keep pace with data on important exposures and disease risk. Once mechanisms of human understanding and behavior change in these contexts are identified, they can be applied to intervention design to improve people's reactions to ecogenetic information. For example, if we identify the ways in which information on personal vulnerability to arsenic should be combined with bulletins on community exposure, then we can improve our system of public health alerts, and we can also apply these strategies to situations other than arsenic exposure. These interventions will be designed and tested in randomized trials to identify public health aids to improving the public's health and safety using ecogenetics.

Box 23.1 Puget Sound Partners for Global Health 2005 Request for Pilot Projects

The Puget Sound Partners for Global Health announces the availability of awards for pilot projects in global health. We define "global health" as diseases and/or health conditions that have both a larger burden in the developing world and have received inequitable attention. Funding is available for up to $50,000, which includes a maximum of $7,000 (14%) of indirect expenses. Preference will be given to new research directions, investigators who are not now working on global health issues, and/or new research collaborations. Please submit a short application (3–4 pages, 12-point font, 1-inch margins) containing

1. A statement of the problem being investigated
2. The research proposed
3. The significance of the study
4. A brief budget.

In addition, applicants are requested to include the following:

5. An abbreviated CV
6. A one-sentence description of the proposal
7. An explanation of how this proposal satisfies the requirement of bringing a new investigator, a new collaboration, or a new idea to global health research, and
8. An explanation of what you expect to accomplish during the course of the pilot project that is reasonable for the time and effort.

To be eligible, a PI must be a faculty member at a nonprofit research institution in the Seattle area.

Please send a brief email of intent to "name" at "email" by March 15, 2005. If respondent is a PI, include institutional affiliation(s). If respondent is representing an institution, include the number of applications anticipated and from which PI(s) if known.

Applications must be submitted electronically by March 31, 2005. Indicate in your email whether your proposal includes human and/or animal research.

Notification of awards will be May 2, 2005.

24

Legal Issues

Kate Battuello and Anna Mastroianni
University of Washington, Seattle, WA

24.1. INTRODUCTION

The prospect of using genomic technology to identify individuals with increased susceptibility to environmentally induced diseases, and thereby potentially minimize their risk of adverse health outcomes, raises a number of legal issues, many of which have yet to be addressed directly by the courts, legislative initiatives, or regulatory bodies. This chapter introduces the legal issues regarding the research and use of genetic susceptibility testing to promote environmental and occupational health. It then more fully discusses the law as it pertains to one particularly controversial application of environmental genomics—genetic testing in the workplace. It ends with an overview of the issues surrounding the use of genetic susceptibility information, and information regarding genetic biomarkers for potentially toxic environmental exposures, in toxic tort litigation involving claims for damages allegedly caused by environmentally induced diseases.

24.2. GENETIC SUSCEPTIBILITY TESTING: THE ROLE FOR LAW AND SCIENCE

One of the more compelling public health applications of ecogenetics involves genetic testing to identify individual susceptibility to disease or other adverse clinical reactions following exposure to environmental

Gene-Environment Interactions: Fundamentals of Ecogenetics, edited by Lucio G. Costa and David L. Eaton

hazards. Obvious examples include pharmacogenetic testing to predict drug response (Rothstein and Epps 2001) and workplace testing of workers to predict risk for occupational induced injury or disease (Miller 2000). The goal of such testing is to mitigate individual risk through targeted interventions that either limit exposure or effectively treat the adverse response.

Predictive genetic testing offers the potential for improvements in environmental health and safety. However, a dynamic tension exists between the potential for health benefit and the risks associated with premature applications of genetic tests with uncertain or limited predictive value for disease onset. Historical experience with workplace genetic testing demonstrates that premature use of genomic advances can have a stigmatizing and discriminatory impact on individuals who are labeled as genetically susceptible (Draper 1991). For example, workplace applications of genetic testing for sickle cell trait, glucose-6-dehydrogenase deficiency, and α_1-antitrypsin deficiency—genetic variations that, at the time of application, had a poorly defined impact on risk for anemia, hemolytic crisis, emphysema and other health problems associated with exposure to workplace chemicals—unnecessarily excluded individuals from the workplace (OTA 1983). We anticipate that the law will increasingly be used as a tool to define socially acceptable practices in the development and application of environmental genomics, particularly genetics susceptibility testing. Statutory enforcement, regulatory proscription, litigation, and other legal approaches at the state, federal, and local levels together provide a means of deterring unacceptable practices and compensating those who are harmed by them. However, before the legal system can effectively assist efforts to realize the true promise of this complex scientific area, lawyers and others in the legal system face a daunting challenge of understanding the scientific basis for the risk of harm. With respect to susceptibility testing, two critical scientific considerations bear on the predictive value of a positive test result:

1. The tests predict genetic risk for multifactorial disease—that is, disease triggered by an interaction between one or more genetic variations and one or more environmental factors. By way of analogy, the individual's genetic makeup may be thought of as the gun and the environment as the trigger: a gun presents a certain risk of harm, but harm is realized only if the trigger is pulled (Olden and Wilson 2000).
2. The predictive value of a positive test result is dependent on a series of established epidemiologic parameters, notably: the population frequency of the genetic marker, the cumulative incidence of the disease outcome of interest, and the strength of the association between the

marker and the disease (Van Damme et al. 1995). Because our experience with the application and outcomes of genetic susceptibility testing in populations is limited, our ability to define these parameters—and therefore the predictive value of these tests—is largely untested.

In short, the current state of scientific evidence surrounding predictive genetic tests militates in favor of a cautious approach to clinical, occupational, and forensic applications of ongoing developments in environmental genomics. Our discussion of relevant law will illustrate how the courts and legislatures are attempting to balance benefits against harms, as well as reveal the gaps in the current legal framework that oblige thoughtful, multidisciplinary policy guidance.

24.3. GENETIC SUSCEPTIBILITY TESTING: RELEVANT LEGAL ISSUES

Research and use of genetic susceptibility testing in the context of environmental exposures raises many legal issues. The law can act prospectively and proscriptively to set standards and prevent misuse of technology. This can be accomplished through federal or state laws, regulations, and guidelines administered through various federal and state bodies (such as the National Institute for Occupational Safety and Health, or NIOSH, the federal agency responsible for conducting research and making recommendations for the prevention of work-related injury and illness, and the Equal Employment Opportunity Commission, the lead federal enforcement agency in the area of workplace discrimination) or through the courts, which decide cases that may have a deterrent effect on future practices. The law can also act retrospectively, to redress injuries, through administrative or statutory enforcement or review through application to the judicial system. In this section we discuss several legal issues that have arisen or are expected to arise in the research and commercial application of genetic susceptibility testing.

24.3.1. Issues Pertaining to Research

Programs such as the Environmental Genome Project, initiated by the National Institute of Environmental Health Sciences (NIEHS) in 1998, are part of more recent research efforts to understand genetic susceptibility of humans to environmental exposures. This type of research is subject to many of the same regulatory oversight mechanisms as is nongenetic biomedical and epidemiologic research. Publicly funded research and some other forms of research must be in compliance with federal regulations for the protection of human subjects.* The regulations specify that the research undergo prior, independent ethics review by a human subjects review committee, also known as an institutional review board (IRB), before enrolling human subjects. The regulations mandate IRB review and approval of the informed consent process, the minimization of risks and balancing of risks and benefits, including risks to privacy and confidentiality, and considerations related to the selection and recruitment of the subject population, including voluntariness of participation. [See, e.g., Title 45, Part 46 of the *U.S. Code of Federal Regulations* (NIH Dec. 13, 2001)].

Currently, privately funded and conducted research, such as industry-supported research conducted by a private consulting company, does not fall within the purview of federal oversight and regulation, although some states have similar legal requirements that apply to all research conducted within a state. Private companies are, however, legally responsible for notifying the U.S. Environmental Protection Agency of adverse effects of chemical exposures revealed in studies (EPA June 3, 2003). The American Public Health Association has issued a statement in support of institutional review of privately funded research in occupational health and safety because of concerns about the adequacy of worker protections, such as timely disclosure of research results and medical follow-up, and the importance of representation of unions and workers in study design and development (APHA 1948–present). Indeed, some privately funded research efforts have included review by joint labor-management boards assisted by outside scientific reviewers (APHA 1948–present). The beryllium research program supported by National Institute for Occupational Safety and Health (NIOSH) in a public–private partnership with industry is an example of a participatory research model in a private workplace environment that attempts to involve workers in project development, implementation and review, which, if successful, may serve as a model to address concerns about human subjects' protection in the context of genetics research conducted in the private sector (NIOSH June 2003).

The conduct of genetics research in any venue raises ethical issues, such as confidentiality, informed consent, disclosure and reporting of research results that are complicated by predictive uncertainty and clinically unvalidated tests from research laboratories, and the potential for group

* Specifically, the federal regulations apply if the research is conducted by an investigator affiliated with an institution receiving federal research funding (such as an academic medical center), or supported with federal funds (such as funding from the Centers for Disease Control and Prevention), or involves a product that is federally regulated (such as a product regulated by the Food and Drug Administration).

harms related to association of a particular social group with susceptibility to a disease or condition (Schulte et al. 1999). There are, however, no explicit legal protections for human subjects in the federal regulations specifically applicable to genetics research or research conducted in the occupational setting. The IRB, therefore, will assess compliance with regulatory protections and consider ethical issues on a case-by-case basis. IRBs are increasingly sophisticated and sensitive to issues raised in genetics protocols, with some institutions, like the University of Washington, creating IRBs that specialize in review of research proposals involving genetics. Typically an IRB appraisal of an ecogenetics-related research proposal—such as a proposal seeking to understand the level of exposure to an environmental hazard that results in disease among workers with increased genetic susceptibility—would include ensuring that the informed consent form clearly explains the uncertainty and complexity of the research findings, the circumstances under which individual results will be disclosed to the worker, as well as other protections [see, e.g., the University of Washington report on genetic research (UW 1999)].

The IRB will also evaluate measures taken to preserve confidentiality in light of any perceived potential for genetic discrimination by an employer or insurer, such as aggregate reporting of research results and protected storage of data. The potential for research data to have adverse consequences on employment or insurance may provide support for the additional confidentiality protection offered by a National Institutes of Health "certificate of confidentiality," which protects identifiable research data from disclosure in a legal proceeding (NIH March 15, 2002; NIOSH June 2003).

Federal regulations do provide special additional protections for research conducted with certain populations deemed to be vulnerable, including children, prisoners, and pregnant women (because of concerns about the developing fetus). Some commentators have advocated for additional research protections for current and former workers because of the inherent power differential in the relationship between the worker and employer or worker and union (Rose and Pietri 2002; Rothstein 2000a). Regardless of the absence of additional legal protections, investigators involved in research in the occupational setting should be sensitized to the potential for employer or union pressure in study recruitment, and any potential for loss of job or other benefits as a result of participation.

24.3.2. Issues Pertaining to Commercialization

Separate and distinct legal concerns arise at the point of market entry, primarily around the nature and extent of regulatory oversight by the federal

Food and Drug Administration (FDA) during the premarket review and approval process for genetic tests. Government regulation and oversight of genetic tests as they move from the research arena to the marketplace is limited (Gutman 1999; SACGT July 7, 2000). Genetic tests that are manufactured as test kits, to be sold for use in multiple laboratories, must undergo premarket FDA review and in some instances postmarket surveillance. These procedures require documentation of both safety and effectiveness. Although the FDA is authorized to require premarket review and postmarket surveillance for all genetic tests, it has chosen not to enforce this requirement for genetic tests that are manufactured by a laboratory only for in-house use ("home brews") (Gutman 1999; Amos and Patnaik 2002). The bulk of clinically available genetic tests fall into the "home-brew" category (Amos and Patnaik 2002).

The Clinical Laboratory Improvement Act (CLIA) of 1988, which sets quality standards and oversees the proficiency of the laboratories authorized to perform genetic tests, provides another potential source for regulatory oversight. However, it has yet to recognize genetics as a specific laboratory specialty. Thus, CLIA certification of these laboratories is not contingent on a demonstration of proficiencies specific to genetics and genetic testing. Moreover, the laboratory director—while responsible for selecting test methods providing the quality of results required for patient care—is not specifically required to establish diagnostic or clinical effectiveness of the tests (CLIAC 2001). State regulation of laboratories performing genetic tests varies, but, with the exception of the State of New York, does not provide more stringent oversight of genetic tests than does the federal government.

24.3.3. General Issues Pertaining to Genetic Susceptibility Testing

Many of the legal issues that attend genetic testing generally pertain to the application of genetic susceptibility testing, whether through occupational screening programs, through public health programs, or as a result of individual efforts at health promotion and protection. These include the concerns about discrimination with respect to employment; health insurance; and life, disability, and long-term care insurance that have dominated media attention and captured the attention of both state and federal legislative bodies as they rush to pass genetic antidiscrimination legislation limiting access to and use of genetic information by insurers and employers (Collins and Watson 2003; Epps 2003). It remains to be seen whether these potentially precipitous legislative efforts to protect genetic privacy may in fact paradoxically hinder the ability to use genomic advances to improve health outcomes. In the occupational context, employers' motivation to offer testing will depend at least in part on some degree of access to results, combined with an ability to participate in decisions regarding appropriate interventions for those employees who are found to be genetically at risk (Gostin and Hodge 1999).

In addition, the prospect of using genetic susceptibility testing to ameliorate risk for the genetically susceptible raises several issues with somewhat unique legal implications:

1. To what extent should current environmental standards be tightened to protect those member of society whose genetic makeup confers heightened susceptibility to environmental agents?
2. What alternative measures can or should be undertaken to minimize their risk of exposure?
3. Who should bear responsibility (and expense) for protecting the genetically at risk? Should the responsibility to intervene rest with the "polluter," the government, or the individual?
4. Does it make a difference whether the toxic exposure is associated with a recognized social benefit?

For example, society generally benefits from the widespread use of pesticides. However, many of these agents are demonstrated neurotoxins. If a genetic test were available to accurately predict risk for agricultural workers exposed to pesticides, who should be responsible for providing testing? How should the costs associated with minimizing risk for the genetically susceptible worker be allocated among the worker, his or her employer, and the greater community that relies on the crops the pesticide worker is maintaining?

With these more general legal issues in mind, we turn to a focused discussion of the law pertaining to workplace genetic testing.

24.4. GENETIC SUSCEPTIBILITY TESTING IN THE WORKPLACE

Occupational illness resulting from workplace chemicals and other harmful exposures represents an important public health problem and is a significant issue for employers whose production costs include time loss, health insurance, and worker compensation for employees who suffer injuries or illness because of workplace exposures. Employer interest in genetic testing is documented (OTA 1990; AMA 2001). It is seen as a mechanism for protecting workers, improving productivity, reducing liability exposure, and minimizing expenditures for employee healthcare. Workers, on the other hand, voice significant concern about the implications of workplace genetic testing for job dislocation, health insurance, worker's compensation programs, and efforts to reduce workplace hazards for all workers. A variety of federal and state laws attempt to strike a balance between these competing concerns. These laws simultaneously encourage employers to consider genetic screening of employees and attempt to protect workers from the social harms that may attend premature or misinformed use of this burgeoning technology.

24.4.1. Incentives to Apply Genetic Susceptibility Testing in the Workplace

The Occupational Safety and Health Act (OSHA) is a federal mandate to employers to provide a safe and healthy workplace. OSHA regulations, which impose civil penalties on employers for safety and health violations, are particularly relevant in worksites that pose a significant risk of occupational disease or injury—the same worksites where predictive genetic testing might prove beneficial in identifying highly susceptible workers. Indeed, OSHA has included a medical monitoring mandate in some of its occupational exposure standards for certain toxic exposures, although to date none of these require genetic-based testing or monitoring (Marchant 2001). The extent to which OSHA's mandate should oblige employers to use genetic susceptibility screening to promote and protect worker health and safety is debated (Rothstein 2000b; Kaufmann 1999).

State laws governing workers' compensation plans, while not mandating genetic testing, arguably provide a powerful incentive for employers to use genetic screening tests. Workers' compensation is a state-mandated program in which employers are assumed to be responsible for their worker's job-related injuries. These programs require employers to provide compensation and medical coverage to workers who suffer injuries or illness resulting from their job. The claimant must prove that his/her injury or illness is work-related and many states permit the current employer to apportion liability among all the companies (or other causal factors) that could have contributed to the employee's current condition (Rothstein 1983). Thus, an employer could potentially use the results of predictive genetic tests to defend against a worker compensation claim, arguing that the worker's genetic makeup caused or at least contributed to the claimed occupational illness or injury. Alternatively, genetic screening tests could be used to exclude high-risk workers, thus reducing the potential for workers compensation claims or to advise employees of their risk, potentially shifting responsibility for adverse health reactions to employees who knowingly assume the risk and go forward with work.

The law is not clear regarding the impact that an employee's decision to knowingly assume the risk of a hazardous workplace would have on his or her access to worker compensation should the employee become ill following workplace exposure. In some jurisdictions, such as Washington State, an injured employee can elect to waive his or her worker's compensation claim and directly sue their employer. In this instance, a defense based on genetic screening test results and the worker's assumption of the risk could prove beneficial to the employer in offsetting liability for occupationally induced illnesses or injuries. Similarly, an employer might be able to assert assumption of the risk as a defense in instances where an employee who was adequately warned about the potential risks of occupational exposures and the availability of genetic testing to assess individual susceptibility, proceeded to work and subsequently developed an exposure-related health problem—particularly where subsequent testing confirmed the employee's genetic susceptibility. The U.S. Supreme Court's *Johnson Controls* decision arguably supports such an application of the assumption of the risk defense in dicta suggesting that as long as an employer fully informs the employee of the exposure risk and does not thereafter proceed to act negligently, that employer should be protected from liability for workplace injuries related to the exposure (Marchant 2001). However, this application has yet to be directly tested in the courts, and there are several policy reasons why the courts might decline to recognize an assumption of the risk defense in this context. Notable among the arguments against requiring workers to assume the risk of occupational hazards is the fact that those employed at hazardous worksites seldom have the economic freedom that permits a truly independent and informed choice to assume risk. They must assume risk because they need a job. Additionally, the uncertain predictive value of the test results, the likely presence of multiple cofactors in disease etiology (many of which may be unrelated to the occupational exposure), and the potential availability to the employer of less restrictive options for worker protection all militate against placing full responsibility for adverse health outcomes on the worker who "elects" to stay on the job (Rothstein 2000b). At least one state has addressed this issue in the context of pesticide workers who are exposed to the pesticide chlorpryifos. California's mandatory biomonitoring program for agricultural workers expressly provides that those who choose to remain on the job shall receive worker compensation benefits during the periods of work release that occur when biomonitoring results suggest that their exposure has reached toxic levels (as evidenced by cholinesterase suppression to a level below 60% of his or her baseline value) (*California Code, Title 3, Section 6728*).

While the law appears to encourage employers to consider using genetic susceptibility in the workplace, any attempt to do so is subject to federal and state regulation prohibiting employment discrimination, as well as scrutiny under federal and state constitutional provisions protecting individual rights to privacy. We now turn our attention to an overview of the relevant laws in this area, focusing primarily on Title VII of the Civil Rights Act and the Americans with Disabilities Act (ADA).

24.4.2. Limits on the Application of Genetic Susceptibility Testing in the Workplace

Title VII of the Civil Rights Act prohibits discrimination on the basis of race, ethnicity, gender, or religion. In *Norman Bloodsaw v. Lawrence Berkeley Laboratory, 135 F.2d 1260* (*9th Cir. 1998*), the 9th Circuit Court of Appeals concluded that this federal law prohibits workplace genetic testing that has a disparate impact on a protected group, when it is performed without the employee's knowledge or consent. To successfully defend a Title VII claim, an employer must show that its genetic testing program is job-related and a business necessity. This is a difficult defense to make in the context of a hazardous workplace because the alternative of cleaning up the workplace for all workers—which should not have a disparate impact on a protected group—always exists. Moreover, the courts have clearly set a high threshold for business necessity. For example, saving money is not an adequate justification [*Chrapilwy v. Uniroyal, Inc., 458 F. Supp. 252* (*N.D. Ind., 1977*)]. Avoiding fetal injury is also an insufficient business necessity to overcome a claim of disparate impact by a protected class [*United Automobile Workers v. Johnson Controls, Inc., 111 S. Ct. 1196* (*1991*)].

The Americans with Disabilities Act (ADA) regulates the circumstances in which employers can pursue medical screening and monitoring, including genetic testing. The ADA divides the employment process into three stages: (1) application ("preemployment"); (2) conditional offer ("preplacement"); and (3) employed. At the application stage employers cannot make any inquiries regarding a prospective employee's health or medical history and cannot require any medical screening. Once an offer of employment is extended, employers can require medical screening. The scope of the screening process can be as comprehensive as the employer desires—it need not be limited to job-related conditions—and can include genetic testing. However, the information obtained during this process cannot be used to discriminate in job placement or benefits unless the discriminatory action is job-related or dictated by business necessity.

The ADA does not permit genetic testing of existing employees unless the testing is directly related to the employee's qualifications to complete the job, is otherwise necessary for employee safety, is dictated by business necessity, or is voluntary. The Equal Employment Opportunity Commission (EEOC), a federal commission charged with enforcing laws related to job discrimination, settled a lawsuit against Burlington Northern and Santa Fe Railroad (BNSFR) for the alleged use of surreptitious genetic testing of workers who filed compensation claims for carpal tunnel syndrome. Part of the EEOC's concern was that the genetic test lacked sufficient predictive value to provide relevant information regarding the cause of the workers' injuries. BNSFR was testing for variations in the PMP-22 gene that are thought to increase individual susceptibility to peripheral neuropathies, even though the prevalence of the mutations is exceedingly rare (2–5 per 100,000). This case illustrates how the law (in this particular case the "law" is the ADA as enforced by the EEOC) can be used as a deterrent against inappropriate or premature use of susceptibility testing in a potentially hazardous workplace.

The ADA also prohibits employers from discriminating against "qualified individuals with a disability because of the disability" in decisions regarding hiring, termination, compensation, and other terms or conditions of employment. The language of the statute leaves open the question of whether individuals with a genetic predisposition or susceptibility to future illness or injury are "qualified" and therefore protected from employment discrimination on the basis of predictive genetic test results. To be "qualified," a person must have, have had, or be "regarded as" having one or more physical impairments that substantially limit a major life activity. The EEOC believes that the ADA prohibits employers from discriminating against employees on the basis of predictive genetics tests, because employers who do so "regard" a worker's predisposition or susceptibility to illness regard the worker as having a disability. However, its interpretation of the Act has yet to be specifically tested in court. A trio of recent U.S. Supreme Court opinions suggests the court may not apply the EEOC's interpretation.

In *Sutton v. United Airlines Inc., 119 S. Ct. 2139* (*1999*) the court held that to benefit from the ADA's protection the individual must be "presently" substantially limited, as opposed to hypothetically or potentially limited, and that this determination requires consideration of the individual in his or her mitigated state. Thus, severely myopic twins who enjoyed near-20/20 corrected vision were found not to have a disability within the meaning of the ADA. The court's ruling in *Toyota v. Williams, 122 S. Ct. 681* (*2002*) suggests that to qualify for protection, the claimed disability must limit activities that are of central importance to daily life in a considerable way and on a permanent or long-term basis. Finally, in *Chevron v. Echazabal, 122 S. Ct. 2045* (*2002*) the court concluded that the ADA permits employers to exclude workers who pose a direct threat to their own health and safety (as opposed to the health and safety of others). This interpretation of the Act allowed an employer to exclude an individual with hepatitis C from a job involving chemical exposures that could aggravate existing or ongoing liver damage. In each of these cases the claimed disability was not presently limiting the claimant's ability to work, but could potentially do so in the future. This is precisely the situation for workers who are *phenotypically* healthy and capable of functioning on the job but are *genotypically* at increased risk for environmentally induced disease.

In February 2000, President Bill Clinton issued an Executive Order that generally prohibits federal employers from requiring or requesting genetic tests as a condition of a worker's being hired or receiving benefits (DOL Feb. 8, 2000). Moreover, it precludes federal employers from requesting or requiring employees to undergo genetic tests to evaluate their ability to perform a job. However, federal employers may request or require family medical history from employees to the extent necessary to determine whether further medical evaluation is needed to diagnose a current disease, medical condition, or disorder that could prevent the employee from performing an essential function of the job. Whether an employee is qualified to perform essential job functions must be decided on the basis of present ability to safely perform. The scope of the Executive Order has yet to be tested in the courts. However, from the perspective of the EEOC, it is a clear and unequivocal statement that it is inappropriate to base employment decisions on genetic information that has no bearing on the employee's present ability to do the job (Miller 2000).

At least 30 states have laws prohibiting genetic discrimination in employment. These laws impose varying degrees of limits on employers' ability to use genetic testing to screen, monitor, or diagnose workers. Some state laws make exceptions when the discrimination is related to the employee's ability to perform the job. Whether and/or to what extent genetic traits that increase susceptibility to workplace exposures fall into this latter exception is, as in the ADA context, an open question and an ongoing source of debate. To date, only California appears to have addressed this question directly. In 1998 the state legislature amended its Fair Employment and Housing Act (FEHA) to prohibit employment discrimination based on asymptomatic genetic characteristics, in companies with five or more employees. Although California does not specifically prohibit employers from using genetic screening tests for workers, the 1998 FEHA amendments certainly provide greater limitations on the extent to which employers can use test results than the ADA or most other state nondiscrimination statutes.

The ADA, the Rehabilitation Act of 1972, the Executive Order, HIPAA (Health Insurance Portability and Accountability Act) privacy regulations, and various state genetic privacy and nondiscrimination laws contain provisions intended to protect the confidentiality of employees' genetic information. Generally, these provisions require that any genetic information obtained by employers be maintained with the employee's medical information and that release of that information requires written employee consent.

24.5. FORENSIC APPLICATIONS: THE ROLE OF ECOGENETICS IN TOXIC TORT LITIGATION

The preceding sections of this chapter focus on legal issues related to the research and use of genetic susceptibility testing to identify and mitigate individual risk for adverse health outcomes associated with toxic environmental exposures. An alternative application arises in the context of litigation that occurs after an exposure-related injury manifests. Testing that examines genetic susceptibility and genetic biomarkers for exposure has the potential to significantly impact "toxic tort" litigation, in which the plaintiff seeks compensation for injuries he or she alleges were caused by toxic exposures that occurred because of the defendant's negligence. In particular, this type of genetic information could be helpful in proving or disproving both exposure and causation, and in establishing damages, particularly damages for "latent risks."

Causation, meaning the identification of the "more-probable-than-not cause" of the plaintiff's claimed adverse health outcome, is a critical issue in every toxic tort case, and the plaintiff has the burden of proving that the alleged exposure is indeed the causal culprit. This requires evidence establishing each of three propositions: (1) that the alleged exposure did in fact occur ("exposure"), (2) that the exposure is generally capable of causing the type of adverse health outcome involved in the case ("general causation"), and (3) that the exposure caused the plaintiff's adverse health outcome ("specific causation"). The failure to produce evidence establishing any one of these three propositions is fatal to the plaintiff's case. Generally plaintiffs have a difficult time proving both exposure and specific causation, and it is with respect to these two particular elements of a toxic tort case that genetic susceptibility testing and biomarkers could be most relevant.

Claims for latent risk damages seek compensation for an increased risk of future disease, for a present fear of future disease triggered by the increased risk, and for the costs of any periodic medical monitoring necessary to evaluate disease status and progression. Because the law generally precludes recovery for speculative damages, and in particular the threat of a future harm, the courts generally will not permit damage claims for latent risks absent, at a minimum, evidence establishing some degree of present "injury" caused by the exposure and evidence of a quantifiable—and significant—increase in risk for future harm treatment (Marchant 2001). In order to support a claim for medical monitoring, the plaintiff must similarly prove a significantly increased risk of contracting a serious latent disease, that the risk makes periodic medical monitoring reasonably necessary and that the methods used will assist with early detection and treatment (Marchant 2001).

Genetic biomarkers of exposure and effect can provide evidence that exposure to the defendant's product or worksite hazard did in fact occur. For example, the presence of dicentric chromosomes in lymphocytes can be a biomarker of radiation exposure, as can the presence of certain types of chromosomal translocations (Marchant 2001). In some instances, genetic biomarkers of exposure can also demonstrate early effects of exposure that mark the initiation of a disease process. The latter data could potentially be helpful in establishing specific causation. Similarly, evidence that the plaintiff has or does not have a recognized genetic susceptibility to the specific exposure at issue in the case could assist the litigants and the court in resolving issues regarding the cause of the plaintiff's claimed adverse health effects. Finally, these types of data can assist in meeting the threshold requirements for a damages claim predicated on latent risk for disease and the need for medical monitoring.

Although the presence—or absence—of these types of data is potentially relevant to a resolution of toxic tort cases, several complications attend their courtroom applications at the present time, causing many to conclude that testing capabilities for genetic biomarkers and genetic susceptibility are not yet sufficiently mature to support a legal causation analysis (Poulter 2001; Marchant 2002). For example, the absence of biomarkers for exposure can be, in effect, a false-negative response to the question of whether exposure did in fact occur. Current technology does not permit sampling of all potentially relevant cells for signs of exposure effects. Similarly, not all biomarkers of exposure and effect are stable over time. This problem with the timing of testing proved fatal in the Three Mile Island (TMI) litigation involving damages claims associated with the TMI nuclear accident that occurred in 1979. Although the Court of Appeals for the Third Circuit held that the use of genetic biomarkers is a "valid and reliable scientific methodology" for demonstrating exposure, the amount of time that had elapsed between the alleged exposure and the biomarker testing significantly decreased the validity and reliability of the test results, [*In re TMI Litigation, 193 F.3d 613 (3rd Circ., 1999) cert. denied, 120 S. Ct. 2238 (2000)*]. Finally, both the litigants and the courts need to be cautious about the premature application of novel scientific evidence relating to both exposure and effect. Marchant (2001) has described examples of precipitous applications of what turned out to be "junk science" in such legendary cases as the silicone breast implant cases.

In *Daubert v. Merrell Dow Pharms., Inc., 509 U.S. 579 (1973)* the U.S. Supreme Court outlined the test it proposes that courts use when evaluating the admissibility of scientific evidence, particularly in toxic tort cases. At issue was the admissibility of expert testimony linking the pregnant plaintiff's ingestion of the antinausea drug Bendictin with the birth defects of her children. The court designated the judge as a "gatekeeper" who evaluates scientific evidence using a two-prong test: (1) whether the evidence is *reliable* "scientific knowledge," and (2) whether the evidence is *relevant* to the jury's ability to understand and resolve the issues in the case. In assessing the reliability of the evidence the judge is to consider (1) whether the theory or technique can be and has been tested, (2) whether the theory or technique has been subjected to peer review and publication, (3) the known or potential error rate of the methods involved, and (4) whether the theory or technique is generally accepted. The *Daubert* standard for admissibility sets a daunting threshold for plaintiffs, particularly in toxic tort cases where robust epidemiologic evidence supporting causation in human populations is often nascent or altogether lacking. At the very least, *Daubert* significantly limits the admissibility, and therefore forensic applicability, of currently accessible data regarding genetic biomarkers and genetic susceptibility for environmentally induced illness and disease.

24.6. CONCLUSIONS

Ecogenetics has the potential to significantly improve our ability to assess individual and population-based risk for environmentally induced disease. This chapter has identified the types of legal issues that attend the translation of ongoing research in this area into meaningful public health applications, as well as the issues that attend forensic applications in toxic tort litigation. In the discussion of current laws relating to workplace genetic testing, we have attempted to illustrate how the law can help minimize the social harms that may attend susceptibility testing. However, the law could do much more to provide consistent guidance and protection for the interests of workers, employers, and other constituency groups who stand to benefit from our increasing understanding of gene–environment interactions. Ongoing development of a legal framework for workplace genetic testing and other ecogenetic applications will be best accomplished through meaningful multidisciplinary collaboration that involves contributions from the several disciplines contributing to this textbook, including genetics, environmental health, ethics, medicine, law, and epidemiology.

ACKNOWLEDGMENTS

This work was supported by the UW NIEHS sponsored Center for Ecogenetics and Environmental Health, Grant NIEHS P30ES07033. Additional support was provided through NIH Grant HG02263.

25

Risk Assessment and the Impact of Ecogenetics

Elaine M. Faustman
University of Washington, Seattle, WA

Gilbert S. Omenn
University of Michigan, Ann Arbor, MI

> Conceptually, the relationship between genes and the environment is similar to that of a loaded gun and its trigger. A loaded gun by itself causes no harm, it is only when the trigger is pulled that the potential for harm is released or initiated. Likewise, one can inherit a predisposition for a devastating illness, yet never develop the disease unless exposed to the environmental trigger(s).
>
> —*Olden et al. (2001)*

How does the current risk assessment process, designed to predict human risk, address the inherent variability within the human population? How will the growing discipline of ecogenetics impact and potentially improve our evaluation of susceptible populations? How will this ecogenetic information influence the current risk assessment process and the formation of public health policy?

25.1. INTRODUCTION

Fifty years after the discovery of DNA's double-helix structure by Watson and Crick, scientists have systematically decoded the underlying genomic

Gene-Environment Interactions: Fundamentals of Ecogenetics, edited by Lucio G. Costa and David L. Eaton

sequences that are the human biological blueprint. Within this genetic code lie 3 billion basepairs (3×10^9 bp) that serve as the reference for all the proteins and enzymes that sustain human homeostasis and mediate human responses to the environment. Although 99.9% of the human DNA is identical from one person to the next, the other 0.1% [3 million basepairs (3×10^6 bp)] is not identical, thus creating human variability. This difference leads to potential increased individual susceptibility (or resistance) to environmental impacts in risks for diseases. It is critical to evaluate the variability in response and disease occurrence across human populations in order to identify and characterize populations "at risk" for environmental disease.

In addition to the sequencing of the human genome, science has also seen the development and growth of the genomic revolution, allowing for the simultaneous identification and analysis of thousands of expressed genes or proteins in one experiment. Such advanced techniques allow for the evaluation of intra- and interindividual variations in response at the gene, RNA, and protein levels. This powerful technology allows us an increased opportunity to understand human response variability and related variation in susceptibility and allows us to better address the clinical question "Why me, doc?"

Historically, public health scientists and policymakers have used the risk assessment process to understand and decipher potential for adverse health effects by defining risk as "the probability of an adverse health effect." *Environmental health risk assessment* is defined as the "systematic scientific characterization of potential adverse health effects resulting from human exposures to hazardous agents or situations" (Faustman and Omenn 2001; Omenn and Faustman 2002). This evaluation of risk through the risk assessment process has developed over many years. Historically, the Delaney clause, passed in 1958 by the U.S. Congress, instructed the FDA (U.S. Food and Drug Administration) to prohibit the addition of cancer-causing substances to the food supply. Risk assessment methodologies grew from this doctrine in the 1970s and resulted in the 1983 publication of the National Research Council report, *Risk Assessment in the Federal Government: Managing the Process*, widely known as "the red book," which details the key steps of hazard identification, dose–response assessment, exposure assessment, and characterization of risks (NRC 1983). This original framework remains the foundation of the current risk assessment processes (Omenn 2003). We have added genomic information to the list of research inputs for risk assessment (Figure 25.1).

How does the risk assessment framework provide context and best utilize the powerful new data emerging from genomics? To answer this question, one must understand both the foundation of the current risk

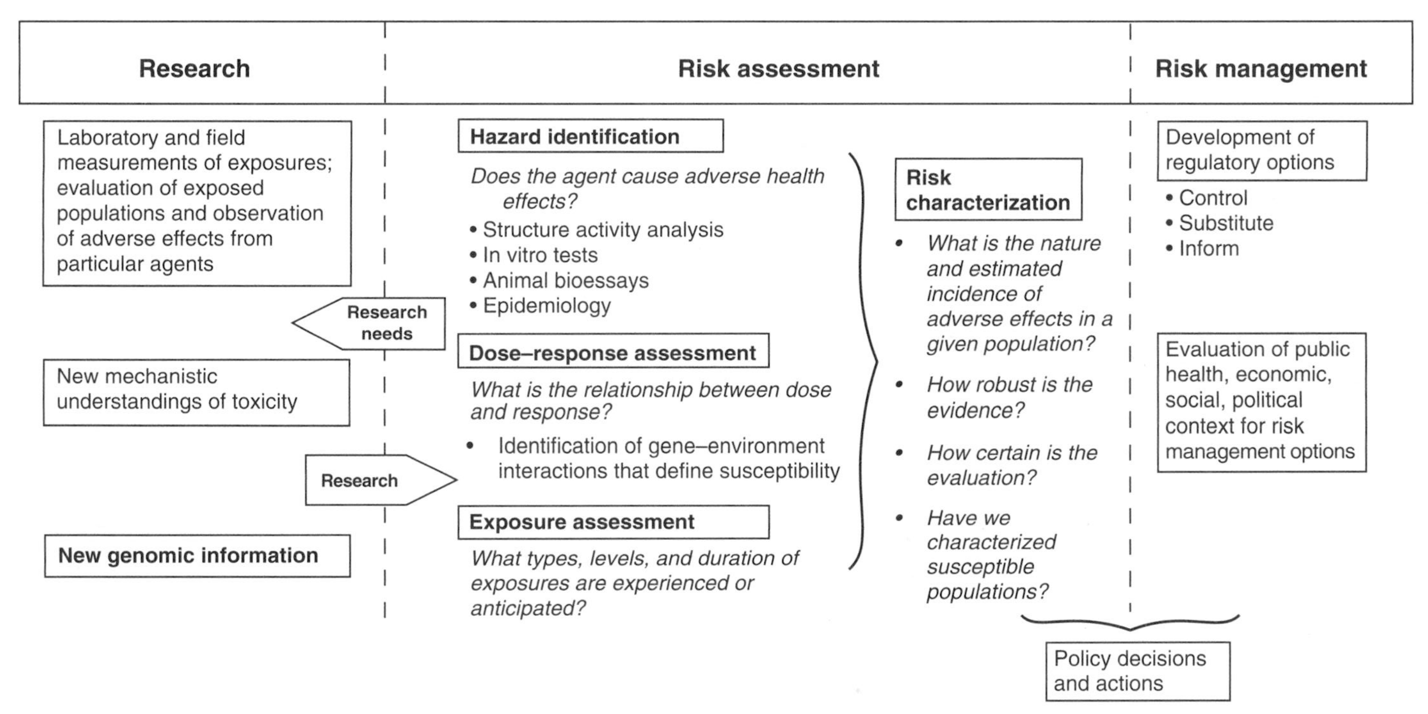

Figure 25.1. Risk assessment/risk management framework. The four key steps of risk assessment as defined by the NAS "red book" report: hazard identification, dose–response assessment, exposure assessment, and risk characterization. This modified figure illustrates an interactive, two-way process where research needs from the risk assessment process drive new research and new research findings modify risk assessment outcomes. Note that genomic information has been added as a new research finding that will help improve hazard identification, dose–response assessment, and exposure assessment. [Adapted from Calkins et al. (1980), Faustman and Omenn (2001), Gargas et al. (1999), NRC (1983, 1994), and Omenn and Faustman (2002).]

assessment process and the type of information that may arise from the use of genomic data. In addition, there must be an understanding of the current regulatory structure with consideration as to where genomic information can best be incorporated. This chapter highlights the integration of this new information into the current risk assessment methods and illustrates how this can improve the overall process of risk assessment and public health policy. We will also identify new directions and information that current risk assessment approaches need to consider but have not been fully incorporated into current methods.

25.2. RISK ASSESSMENT AND RISK MANAGEMENT

The key objectives of risk assessment include (1) identifying and characterizing public health risks, (2) balancing risks and benefits, (3) setting target levels of risk, (4) setting priorities for program activities, and (5) estimating residual risks and/or reduction of risks after steps have been taken to reduce risks. Depending on the regulatory context, each or all of these factors may be considered in order to arrive at an overall risk characterization and public heath policy. With advances in ecogenetics, identification and characterization of potential susceptible populations can be added to the risk assessment objectives.

To achieve these objectives, the risk assessment process is divided into stages representing the key steps used to identify a potential risk (Figure 25.1). Data obtained through laboratory and field studies are used to characterize risk through hazard identification, dose–response evaluations, and exposure assessments. Collectively, this risk characterization contributes to the development of public health policy and risk management considerations. Risk assessments, therefore, provide an approach for translating scientific information into risk characterizations, which in turn provide the basis for *risk management*, defined as the process of developing policies to control hazards that have been identified in the risk assessment and characterization stages. Not only is the scientific evidence for risk considered, but in management, additional factors such as regulations, engineering, economic, social and political factors, and communication processes are important.

25.2.1. Hazard Identification

As shown in Figure 25.1, the key question of hazard identification is "Can the agent or situation cause adverse health effects?" Defining what is an "adverse health effect" is critical in determining risk. Common endpoints considered adverse effects include carcinogenicity, mutations, teratogenicity (birth defects), altered reproductive or immune function, altered respiratory function, and neurobehavioral toxicity. The process of risk assessment uses several different scientific methods to identify an adverse effect. These include the evaluation of chemical structure and activity, in vitro evaluation of chemicals or short-term tests, evaluation of chemicals in animal bioassays, and evaluation of chemical effects in epidemiology studies. Because it can cost millions of dollars and take over 5 years to obtain results in either animal bioassay studies or epidemiologic evaluations, risk assessors frequently have to rely on information from chemical structure or in vitro tests for their initial evaluations. Such information can be used to determine whether subsequent animal or human studies are needed.

25.2.1.1. Structure–Activity Relationships. Structure–activity relationships (SARs) are used to give an early indication of whether a chemical can pose a potential health hazard on the basis of knowledge of chemical structure, solubility, pH sensitivity, electrophilicity, volatility, and chemical reactivity. SAR evaluations determine whether a given agent *may* be toxic, based on comparisons to other chemicals with similar chemical structures and/or properties and their associated toxic effects. Historically, certain key molecular structures provided regulators some readily available information on which to assess hazard potential (Faustman and Omenn 2001; Omenn and Faustman 2002). For example, 8 of the first 14 occupational carcinogens were regulated together by the Occupational Safety and Health Administration (OSHA) as belonging to the aromatic amine chemical class. Chemicals that were identified as structurally related to this particular class of chemicals were evaluated as having potential toxicity. This information guided subsequent calls for additional testing. However, in all cases such information will be predictive only for overall in vivo responses of toxicity.

Environmental mixtures of polychlorinated biphenyls (PCBs) are a classic example of toxicants classified using the SARs method. Compounds are evaluated over a range of relative potency estimates, compared to a reference PCB. These agents were initially classified according to their ability to bind to the arylhydrocarbon (Ah) receptor and to elicit classic gene expression responses in the drug metabolism enzymes, cytochrome P450s.

Early ideas about genetic differences in susceptibility and nonsusceptibility to these compounds came from studies of the Ah-responsive and non-Ah-responsive mouse strains C57BL/6 and DBA, respectively (Nebert 1989). Thus, SAR methods used ecogenetic information to define characteristics of compounds that correlated with susceptibility to toxicity. These methods continue to be used to define relative potency for PCBs (Van den Berg et al. 1998).

The SARs approach has been aided by the growth of genomic technologies. Gene expression microarray technology can be used to identify a chemical exposure through the identification of similar mRNA expression patterns in response to exposure (Waring et al. 2002). Safe et al. (1985) illustrated this approach with structurally and mechanically related chemicals, especially coplanar PCBs and Ah receptor agonists. Rather than relying on chemical structural similarities, investigators have generated common gene expression patterns or "molecular signatures" to identify chemicals that have similar mechanisms of toxicity or functional response (Hamadeh et al. 2002; Thomas et al. 2001; Waring et al. 2002). Another example is toxicants that are believed to cause proxisomal proliferation (Amin et al. 2002).

The hazard identification process (Figure 25.1) can include evaluation of in vitro chemical effects or *short-term tests* performed in systems ranging from simple bacterial or yeast mutation assays in vitro to more complex skin-painting studies in mice in vivo. These tests take from a few hours to more than a year. Results are used to determine whether the agent tested is a *possible* toxic agent in vivo and should be tested further. In addition, these data can provide clues on the mechanisms of toxic effects.

25.2.1.2. Animal Bioassays. Animal bioassays are used to determine whether chemical exposure can cause toxicity such as cancer or birth defects. These in vivo assessments are key components of the hazard identification process. For example, chronic cancer bioassays use both sexes of two rodent species over a near-lifetime exposure as vital aids in the identification of carcinogens (Bucher et al. 1996; EPA 2003). They also require a substantial time and monetary commitment, thus limiting the number of chemicals that can be so thoroughly investigated. Essential considerations in the study design include sex, age, developmental stage, route of exposure, dose, and duration of exposure. All chemicals classified as human carcinogens by the International Agency for Research on Cancer or the U.S. National Toxicology Progam, for example, have demonstrated tumor growth in an animal bioassay; for newly evaluated chemicals, we assume that a carcinogenic response in rodents will predict such an effect in humans [with a few notable exceptions due to species-specific mechanisms (Risk Commission, Vol. 2, 1997)].

Genetic technologies have played a role in the improvement of in vitro and in vivo animal bioassays through development of genetically sensitive cells and transgenic animal models. Specific genes targeted as important in the mechanisms of chemical carcinogenicity have been inserted, deleted, or modified to determine how these genes and their related cellular pathways can influence biological responses to test chemicals, including such critical pathways as cell signaling, apoptosis, and cell cycle progression. $p53^{+/-}$ and Tg-AC models, with modified p53 tumor suppressor and H-ras oncogene, respectively, developed to understand cancer progression, have permitted identification of phenotypes revealing selective responses to carcinogens (Spalding et al. 2000; Tennant 1998). These models have been compared for their ability to identify human carcinogens with the traditional 2-year rodent bioassay (Pritchard et al. 2003). Of 99 chemicals tested, the transgenic lines correctly identified carcinogens and noncarcinogens 74–81% of the time, emphasizing the human relevance of evaluating carcinogenic response in animal models. Thus, transgenic animal bioassays can enhance hazard identification by helping researchers experimentally evaluate the contributions of specific genetic pathways in disease response and susceptibility. Nevertheless, much additional work is needed to validate specific transgenic animal models in order to facilitate their interpretation and use in risk assessment, let alone to replace the lifetime bioassays, which have many limitations (Lave and Omenn 1986).

25.2.1.3. In Vitro Assays. In vitro assays have also been used to evaluate the importance of particular genetic pathways in mediating toxicant response. Such studies use primary cells from transgenic strains or transformed cell lines with targeted gene modification. Such studies have shown dependence and independence of in vitro responses to such key cell response pathways as p21 and p53 (Gribble et al. 2005; Mendoza et al. 2002). Cost-effectiveness criteria have been developed to evaluate use of short-term tests in evaluation of carcinogenic potential of chemicals (Lave and Omenn 1986). There is pressure from certain advocacy groups to move away from toxicity testing in animals to use a wide variety of in vitro models. Similar aims arise because of the high cost of testing as the variety of hazard identification endpoints continues to increase. Validation of such tests, which can incorporate genetic variants, has been difficult, requiring standardization, replication, and comparisons with in vivo results, not to mention concern about missing metabolic conversions of the agent and potential recovery of exposed cells or organs in vivo (Daston and McNamee 2005). Tiered testing strategies that permit use of in vitro tests but require an in vivo test if the in vitro test is negative have been considered a kind of "double jeopardy" by industry. The NIEHS National Toxicology Program administers an Interagency Center for the Evaluation of Alternative Toxicological Methods (NICEATM), including those emerging from the Interagency Coordinating Committee on Validation of Alternative Methods (ICCVAM) and its European counterpart, ECVAM, which faces legislative deadlines to replace most animal testing for substances used in cosmetic products by 2009 and all animal testing by 2013.

25.2.1.4. Epidemiology. A fourth important component of hazard identification is the *epidemiololgic evaluation of exposed human populations.* The major types of epidemiologic studies are cross-sectional, cohort, and case–control studies. These types of studies are described in greater detail in Chapter 4 on epidemiologic approaches. The epidemiologic study types are distinguished from each other by how the human study population is defined and the time period of exposure relative to diagnosis of the disease state. Epidemiology studies are powerful tools for risk assessment in their ability to determine association between exposure and disease in addition to the magnitude and consistency of this association. These studies provide measures of association such as prevalence rate (the number of cases per population per time) and incidence rate (number of new cases per population per time). Since epidemiologic evaluations are within human populations, they are the most convincing form of evidence within the hazard characterization process. However, epidemiologic studies have several inherent limitations. Exposure evidence is often retrospective, requiring information based on past circumstances or events, introducing a considerable degree of uncertainty. In addition, humans are exposed to numerous environmental agents that can confound the data and obscure or enhance an observed association. Delineating the true exposure and assessing whether the outcome is related to that individual exposure can often be difficult (Rebbeck et al. 2004).

Genetic epidemiology studies can help the risk assessor identify and characterize susceptible population groups. Such information can qualitatively identify individuals who are at increased (or decreased) risk of developing a disease or can assist in the identification of individuals who respond at different exposure levels (or dose levels, if exposure to drugs is considered). For ecogenetic considerations, we seek evidence in risk assessment of the extent to which variability in response within a human population arises from genetic variability. For example, 46 specific SNPs have been associated with 39 different cancer-related genes from 166 different molecular epidemiologic studies (Zhu et al. 2004). Genetic variation within the cytochrome P450 gene, CYP19, has been reported as associated with increased risk of developing Alzheimer disease (Iivonen et al. 2004). For additional examples, see Chapters 6–20.

There are hundreds of highly penetrant allelic variants that define "at risk" populations for particular diseases, as documented in *McKusick's Online Mendelian Inheritance in Man* (http://www.ncbi.nlm.nih.gov/entrez/query.fcgi?db=OMIM). Examples include the genetic changes seen in Huntington disease, cystic fibrosis, polyposis of the colon, and muscular dystrophy. Genotyping such genes, as well as breast cancer 1 and 2 genes (BRCA1 and BRCA2) or mutL homolog 1 (MLH-1) genes in relation to hereditary cancer, reveals numerous other variants whose significance must be determined. Some are associated with increased disease risk, while others are inconsequential, producing a variant protein with apparently normal activity, or mutations that do not even change the amino acid sequence. Some common non-disease-causing variants individually or collectively may confer differential response dynamics across the human populations following toxicant exposures. A common example of this type of ecogenetic consideration is seen with the differential sensitivity to ethanol among people who have alcohol and/or aldehyde dehydrogenase variants. (See Chapter 20, on genetic determinants of addition to alcohol, tobacco, and drugs of abuse, for a more complete discussion of this example.) Many of these types of variation have been defined for metabolizing and conjugating enzymes (see Chapter 8, on polymorphisms in xenobiotic conjugation).

It is more complicated for risk assessment when multiple genes contribute to susceptibility. In the case of organophosphate pesticides, polymorphisms in cytochrome P450s contribute to variations in the activation of organophosphate pesticides to oxon metabolites, and polymorphisms in paraoxonase-1 contribute to variation in how humans detoxify the oxon metabolites (Costa et al. 2005). As detailed in Chapter 9 (on paraoxonase, butyrylcholinesterase, and epoxide hydrolase), these polymorphisms are specific to particular organophosphates; hence, any interpretation of "susceptible" populations requires knowledge about the mechanism of action for these pesticides and knowledge about specific exposures (Bradley et al. 2004).

Overall, epidemiologic studies must consider and meet numerous criteria (first proposed by A. B. Hill in the mid-1950s) to establish causality for adverse health effects. These factors include strength of association, consistency of observations (reproducible in time and space), specificity of response, appropriateness of temporal relationship (Did the exposure precede response?), dose responsiveness, biological plausibility and coherence, verification, and analogy (biological extrapolation).

The emerging approaches of *toxicogenomics and toxicoproteomics* can generate molecular signatures of patterns of changes in transcripts (mRNA), proteins, or metabolites that can bridge across in vitro studies, animal bioassays, and epidemiologic studies (Daston and McNamee 2005; Nickerson et al. 2005). Gene-to-gene variation in sequence diversity, linkage disequilibrium, and haplotype blocks is being applied to epidemiologic studies of genotype–phenotype/environment interactions. Common diseases, including asthma, diabetes mellitus, high blood pressure, atherosclerosis, and cancers, are the subjects of such multifactorial investigations. Since 1997 the NIEHS has stimulated research on the relationship between common genetic polymorphisms and environmentally induced diseases through the Environmental Genome Project and the Center for

Toxicogenomics Research (Olden and Wilson 2000; Nickerson et al. 2005). A list of 550 candidate environmental response genes was generated; extensive resequencing of most of these genes has been completed using a panel of 90 specimens from ethnically diverse U.S. subpopulations (http://genome.utah.edu/geneshps). Comparative genomics is being used to discern functions such as transcription factors binding sites in noncoding regions of genes. Indirect association studies based on linkage disequilibrium between genetic markers (SNPs) or blocks of SNPs (haplotypes) will generate "tags" for genes or genomic regions.

25.2.2. Risk Characterization and the Biomarker Paradigm

Risk depends on hazard, exposure, and susceptibility. *Risk characterization*, therefore, integrates all the information and data obtained through hazard identification and dose–response, exposure, and susceptibility assessments to arrive at a narrative description of the adverse effects and a quantitative estimate of risk. Biomarkers can relate risk models to risks for individuals and can link animal and human studies.

The *dose–response* assessment reflects the relationship between the exposure to an agent (dose) and the incidence of an adverse response, or potency. A dose–response curve is used to characterize toxicity using metrics such as NOAELs (no observable effect levels) and LOAELs (lowest observable effect levels) (Figure 25.2) or specific response levels (e.g., one in a million over a lifetime of exposure).

Historically, a *threshold*-based response has been used for noncarcinogens, meaning that doses below the threshold result in no change in biological response (shown as T on Figure 25.2). In contrast, carcinogens have traditionally been regarded as having no "safe" dose; therefore, regulators have used a *nonthreshold* dose response approach where minimization of the risk is set as a goal (de minimus approach). Thus, characterization of carcinogen risk is usually determined by identifying an "acceptable risk" estimate such as "1 in 10,000 to 1 in one million excess incidence of cancer over a lifetime of exposure." More recently, nongenotoxic carcinogens have been examined using a threshold-based risk assessment approach.

The NOAEL reflects the highest dose with no statistically or biologically significant adverse response. The NOAEL commonly serves as the basis for risk assessment calculations for regulatory standards such as reference dose and the acceptable daily intake values. The *reference dose* (RfD), frequently used by EPA, is the estimate of a daily exposure to an agent that is assumed to be without adverse impact on the human population. The *acceptable daily intake* (ADI), commonly used by the FDA for risk assessments, is the daily intake of chemicals estimated over a lifetime that appears to be without appreciable risk:

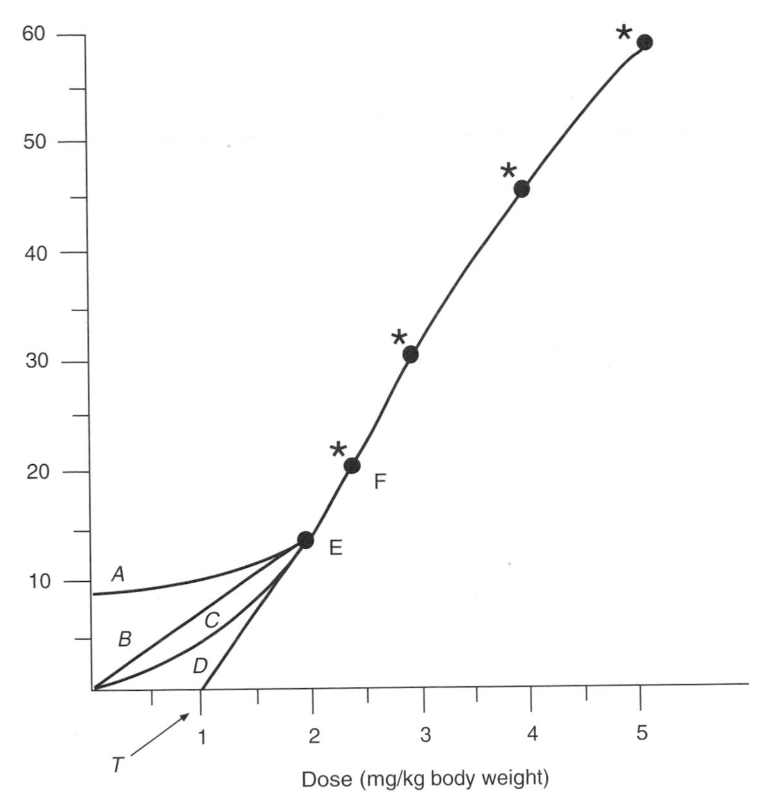

Figure 25.2. Dose–response curve. This figure illustrates a generic dose–response curve with "●" indicating biologically determined responses. The * symbol indicates the statistical significance of these responses. The threshold dose is shown by *T*, a dose below which no change occurs in the biological response. Point *E* represents the highest non–statistically significant response point; hence it is the no observed adverse effect level (NOAEL) in this example. Point *F* is the lowest observed adverse effect level (LOAEL). Curves *A*–*D* show possible options for extrapolating the dose–response relationship below the lowest biologically observed data point *E*. [From Faustman and Omenn (1995, 2001).]

$$\text{RfD} = \frac{\text{NOAEL}}{\text{UF*MF}}$$

$$\text{ADI} = \frac{\text{NOAEL}}{\text{UF*MF}}$$

To calculate an RfD or ADI, the NOAEL is divided by uncertainty factors (UF) and modifying factors (MF) that can be used to address uncertainties

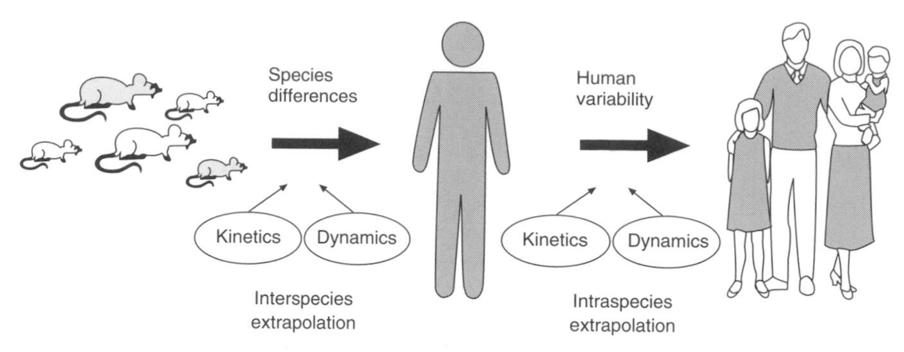

Figure 25.3. Toxicokinetic (TK) and toxicodynamic (TD) considerations inherent in interspecies and interindividual extrapolations. This figure shows how uncertainty in extrapolation both across and within species is due to two key factors: a kinetic component and a dynamic component. *Toxicokinetics* refers to the processes of absorption, distribution, elimination, and metabolism of a toxicant. *Toxicodynamics* refers to the actions and interactions of the toxicant within the organism and describes processes at organ, tissue, cellular, and molecular levels. [Adapted from Renwick (1999).]

in the extrapolation from animal data to human risk, inadequate datasets (i.e., subchronic to chronic risks), and LOAEL to NOAEL values.

Of particular interest for ecogenetics, or susceptibility assessment, are the uncertainty factors that are also used to extrapolate risk estimates from average to sensitive humans. It is in this extrapolation that the impacts of genetic variability have been traditionally considered. Figure 25.3 illustrates sources of variability and factors to consider in extrapolating responses from animal studies to average humans and across diverse human populations. Figure 25.3 epitomizes the following question: "Are humans like big rats or big mice in their response to toxicants?" According to this approach, such a question could be answered by improving our understanding of toxicokinetic and toxicodynamic differences between test species and humans. The variability in these extrapolations should include multiple considerations such as genetically determined differences in toxicokinetics (TK) and toxicodynamics (TD). *Toxicokinetics* describes the processes of absorption, distribution, elimination, and metabolism of toxicants. Many known genetic differences in kinetics between human populations are described in detail in Chapter 7 (on polymorphisms in cytochrome P450 and flavin-containing monooxygenase genes) and Chapter 8 (on polymorphisms in xenobiotic conjugation). *Toxicodynamics* refers to the actions and interactions of toxicants within the organism, reflecting processes and responses at the organ, tissue, cellular, and molecular levels. For example, there are known genetic differences in how humans repair DNA, and such differences can be used to more accurately reflect potential variability in the dynamics of human response (de Boer 2002; Mohrenweiser 2004).

Mechanistic toxicology research has focused on improving the scientific basis of these extrapolations. Ecogenetic research on gene–environment interactions can inform this basis of risk characterization. Traditionally, a tenfold default factor has been proposed for each of the uncertainty factors. Modifying factors (MF) have been used to adjust the UFs when additional information is available.

Figure 25.3 illustrates that each extrapolation for animal to human and across human populations is divided into kinetic and dynamic considerations. Although estimates for the relative contributions of these components to the uncertainty factors have been proposed [see Renwick (1991)], the initial factor of 10 is usually used. The World Health Organization (WHO) recommends the use of 4.0- and 2.5-fold factors for the TK and TD determined from animal data, but changes the TK and TD factors to 3.2 for human data (WHO 1994). With additional genomic information, toxicokinetic models could ideally replace these simple factors with modeled distributions for population variability in response in absorption, distribution, metabolism, and elimination (Meek et al. 2003).

An NRC report (NRC 2000) identified genomically conserved cell-signaling pathways across widely varying organisms from *C. elegans* to humans, suggesting greater similarities in dynamic processes across species development and a potential for modifying interspecies toxicodynamic factors across early life stages.

RfDs can serve as guidelines for public policy and represent an estimate of daily exposure to the human population that is likely to be without appreciable risk of deleterious effects during a lifetime. Although uncertainty factors are included in this estimate, sensitive populations may still need to be identified and characterized. For example, the RfD for methylmercury (MeHg) was derived in 2001 using several uncertainty factors and was considered sufficient protection by EPA for humans. However, the primary human exposure pathway to MeHg occurs through fish consumption; therefore, susceptible populations have been identified through their higher consumption of contaminated fish. In addition, age differences in toxicologic responses to MeHg have been identified, with children and pregnant women at increased risk compared with men. In response, several states use a two-tiered fish advisory with a lower consumption of fish (lower RfD) recommended for women of childbearing age and young children. These considerations are based on the known susceptibility of the developing nervous system to MeHg exposure (Rice 2004). Age-related differences in susceptibility also prompted the tenfold safety factor in the Food Quality and Protection Act (FQPA) that was added to ensure protection of infants and children from pesticides with critical data missing (EPA 1996).

Limitations in the RfD have resulted in the development of benchmark dose (BMD) approaches where the dose–response curve and the lower confidence bound for a dose at a specific response level such as a 10% change in response (e.g., a decrease in body weight) is used in place of a NOAEL or LOAEL value. In most cases, the BMD falls between the NOAEL and the LOAEL levels and has the advantage of considering the whole dose–response curve, the variability, and the use of responses within those doses where measurable effects can be seen (experimental range) (Kavlock et al. 1994; Faustman et al. 1994; Allen et al. 1994a, 1994b). Currently, BMD approaches continue to use uncertainty and modifying factors in calculation of an RfD (with x representing the percent response):

$$\mathrm{RfD} = \frac{\mathrm{BMD}_x}{\mathrm{UF*MF}}$$

Information on mechanism of toxic action has historically been used in risk assessment at many levels including intraspecies, cross-species, cross-compound, and in vitro to in vivo extrapolations. The use of genomic information has the potential to remove some of the inherent uncertainties associated with these extrapolations. For example, the default 10× safety factor used to account for interindividual variation due to susceptibility may be replaced with specific values reflecting mode of action information for a given agent or disease. Improvements in cross-species extrapolations could provide more accurate values to use in place of that default 10× factor. Comparative genomic analysis can measure molecular events across a range of doses that can be compared within species as well as across species. Similarities in gene and protein expression patterns between animal models and human response may provide insights into potential similarities in the biological response at the molecular response level. This information may assist in extrapolating to systems level responses and enhance cross-species and population extrapolations (NRC 2000).

25.2.2.1. Exposure Assessment. Exposure assessment has become recognized as a scientific area complementary to toxicology (Lioy et al. 2005). Studies of exposure in human populations can provide data immediately relevant for decisionmaking about levels that need to be reduced. New technologies and methods for assessing human exposure to chemicals, dietary and lifestyle factors, infectious agents, and other stressors provide an opportunity to extend the range of human health investigations and advance our understanding of the relationship between environmental exposure and disease. An ad hoc Committee on Environmental Exposure Technology Development has identified new technologies and methods for deriving personalized exposure measurements for application to environmental health studies. The committee focused on a "toolbox" of methods for measuring external (environmental) and internal (biologic) exposure and for assessing human behaviors that influence the likelihood of exposure to environmental agents. The methods utilize environmental sensors, geographic information systems, biologic sensors, toxicogenomics and toxicoproteomics, and body burden measurements. Weis et al. (2005) discuss each of the methods in relation to current use in human health research and highlight specific gaps in the development, validation, and application of the methods. They present a conceptual framework for moving these technologies into use and acceptance by the scientific community. The framework uses an integrated approach to exposure assessment to define complex exposure–disease relationships and the interaction of genetic and environmental factors in disease occurrence.

New technologies for personalizing exposure assessment will benefit the scientific and regulatory community by providing range-finding and sensitivity matrices for specific methods, developing baseline data on important environmental factors, and improving the results of exposure–model simulations. Efforts to address genetic or genomic variation alone will have little value in personalizing human risk assessment unless there are effective linkages with information about environmental and behavioral variables that strongly influence the likelihood of exposure and, therefore, risk (Weis et al. 2005). Future studies should require that personal genetic information be linked with estimated or measured personal exposure data, while ensuring that individual privacy is protected (Omenn 2005).

The multifaceted nature of toxic responses can be illustrated within the exposure–disease paradigm (Figure 25.4) developed by the National Research Council (NRC 1983, 1987, 1994). This conceptual framework, embracing the continuum of environmental disease from exposure through early biological response and subsequent disease, is fundamental for understanding how ecogenetics can impact risk assessment. Biomarkers have been defined by the NRC as "xenobiotically-induced variation in cellular or biochemical components or processes, structures, or functions that is measurable in a biological system." In the exposure–disease paradigm, *biomarkers of exposure* provide information linking environmental exposure with internal biological dose. *Biomarkers of effect* relate internal biological dose with early biological response and ultimately with clinical markers of disease. *Biomarkers of susceptibility* identify variations in response across

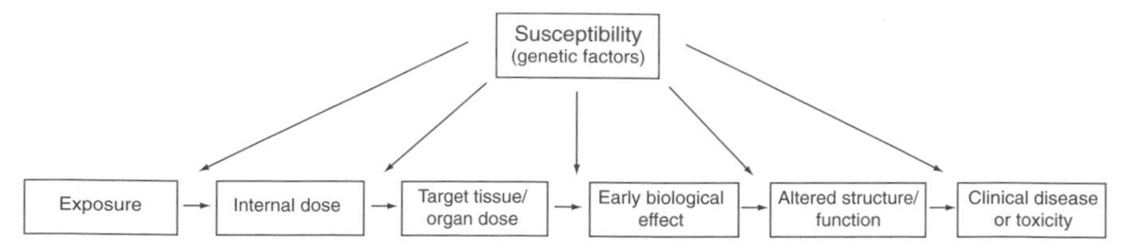

Figure 25.4. Toxicokinetic (TK) and toxicodynamic (TD) considerations inherent in interspecies and interindividual extrapolations. Genetic factors can influence susceptibility at any point along the exposure disease pathway. Polymorphisms within a given gene can perturb the pathway to disease. This information can be used to evaluate risk distribution across populations. [Adapted from NRC Report on Biomarkers (NRC 1983)].

this exposure–disease continuum, from genetic and many other sources of variation (Figure 25.4).

25.2.2.2. Susceptibility Assessment. Each step along the path is influenced by genetic factors. For example, the internal dose can be determined through multiple factors such as exposure, metabolism, or distribution. Thus, target tissue and organ dose are dependent on a variety of kinetic factors, all of which can have genetic variation. This book presents many examples of genetic differences in drug metabolism (see Chapters 7–9).

Once the agent or its metabolite has reached the target organ, dynamic response pathways determine the potential for early biological effect. Many genetic differences in response pathways are known. For example, 115 different human DNA repair genes have been identified as genetic factors for susceptibility, with frequent variants (Ruttan and Glickman 2002). The Environmental Genome Project (see discussion above) is advancing this work rapidly. All individuals have numerous variants in "environmental response genes"; relative susceptibility or resistance to adverse effects from any particular agent will depend on the net influence of the relevant variants.

Diseases arise from both environmental factors and genetic factors and especially from their interactions. This relationship gives rise to the "loaded gun" metaphor for genetic predisposition, awaiting the environmental "trigger" to progress down the disease pathway. Viewing the relationship between exposure and disease highlights the multiple areas in which susceptibility may arise along the continuum.

The "value" within the risk assessment process of genetic knowledge about disease progression can be determined from the sensitivity and specificity of each genetic susceptibility biomarker. The case of occupational exposure to beryllium is a good example. As early as the 1940s, epidemiologic studies associated beryllium exposure with debilitating and incurable chronic beryllium disease (CBD), an immune-mediated granulomatous lung disease. A lymphocyte proliferation test has been used as a biomarker of exposure and early effects for many years. Since the mid-1990s, human studies have revealed that a single-gene variant, substituting glutamate for lysine at position 69 of the human leukocyte antigen antigen HLA-DP beta $1 * 0201$, is highly associated with the presence of CBD, with sensitivity 94% and specificity 70% (Richeldi et al. 1993; Bartell et al. 2000). Using such information, our group has identified the utility of this genetic marker with a comparative evaluation of preplacement screening of workers, voluntary job placement screening, semiannual versus annual screening for sensitization, and exposure reduction risk management options (Bartell et al. 2000; Ponce et al. 2001). A value-of-information (VOI) analysis examined the reduction in total social cost, calculated as the net value of disease reduction and financial expenditures, expected for proposed CBD intervention programs on the basis of the genetic susceptibility test. Despite large parameter uncertainty, probabilistic analysis predicted positive utility for each program when avoidance of a CBD case is valued at $1 million or higher.

The exposure–disease paradigm describes the conceptual relationship between early biological effect, altered structure and function, and clinical disease or toxicity. Along with these considerations there is also an important genetic contribution to disease that must be directly considered. It has long been established that some diseases have an inherited basis. Specific germline mutations are associated with an increased risk of developing particular diseases. For example, the BRCA1 and BRCA2 loci (Narod et al. 1991) are linked to the development of familial breast cancer and account for 5–10% of all breast cancers in the general population (Slattery and Kerber 1993). The population-attributable risk, the proportion of disease incidence associated with a specific risk factor, can be used in both clinical and policy decisions. The breast cancer loci have a low prevalence and, therefore, a low population-attributable risk, which still might be reduced through behavior or environmental exposure changes.

Because of the complex nature of disease and the human biological response, a specific polymorphism may have a variety of associated risks, complicating public policy. The significance of a polymorphism within the metabolizing enzyme glutathione *S*-transferase (GST) and exposure to dichloromethane (DCM) provides such an example. In animal studies, DCM demonstrated carcinogenic properties in populations with a functioning GSTT1 enzyme (Jonsson and Johanson 2001); in contrast, individuals with a null GSTT1 genotype have a nonfunctioning enzyme and demonstrate little risk on exposure. A Monte Carlo simulation generated a distribution of risk based on a physiologically based pharmacokinetic model (PBPK). Approximately 20% of the population has a GSTT1 null phenotype. When the frequency of the polymorphism and its associated protective effect were put into the model, there was a predicted reduced risk of ~24% (El-Masri et al. 1999). However, the same null genotype has been associated with an increased risk of cancer from exposure to ethylene oxide (Fennell et al. 2000).

The ever-expanding wealth of information arising from genomic studies greatly contributes to our understanding of individual and population response, and hence, can contribute to risk assessment and, ultimately, public policy. To date (2005), over 1.8 million SNPs have been identified by the SNP Consortium (www.snp.cshl.org/), a collaboration established to aid the development of genetics-based diagnostics and therapeutics. In addition, as noted above, the Environmental Genome Project is generating SNPs for over 500 genes in specific cellular pathways that were chosen as important for responses to environmental chemical exposures; these are potential clues for identification of susceptible populations. The application of ecogenetic information to risk assessment is not simple, as illustrated above with the GSTT1 polymorphism. Furthermore, data must be in a transparent format suitable for use in policymaking.

25.3. RISK MANAGEMENT

Through hazard identification and risk characterization estimates, the risk assessment process provides a foundation to inform risk management. *Risk management* refers to the process by which policy actions are chosen to control hazards identified in the risk assessment (Faustman and Omenn 2001; Omenn and Faustman 2002). How these data are utilized is mandated by current statutes that vary in their language and regulatory approaches (Table 25.1). For example, the Federal Water and Pollution Control Act (FWPCA) is mandated to protect public health and welfare by eliminating or reducing pollution within water sources and environments. The level of protection is regulated to permit a 1 in 10,000 to 1 in 10 million residual cancer risk. The risk mandates guiding the various government statutes carry a "language of risk." Under FIFRA (Federal Insecticide, Fungicide, and Rodenticide Act), the mandate calls for "risks (to) be less than benefits," requiring a balance between the risks and benefits and the social and economic costs. Other language appears within the mandate under FQPA (Food Quality Protection Act) that requires "reasonable certainty of no harm" and sets a uniform health-based standard for risk. It is within these

Table 25.1 Major Chemical Laws in the United States by Responsible Agency

EPA	Air pollutants	Clean Air Act 1970, 1977, 1990
	Water pollutants	Federal Water Pollution Control Act 1972, 1977
	Drinking water	Safe Drinking Water Act 1974, 1996
	Pesticides	Fungicides, Insecticides, & Rodenticides Act (FIFRA) 1972, Food Quality Protection Act (FQPA) 1996
	Ocean dumping	Marine Protection Research, Sanctuaries Act 1995
	Toxic chemicals	Toxic Substances Control Act (TSCA) 1976
	Hazardous wastes	Resource Conservation and Recovery Act (RCRA) 1976
	Abandoned hazardous wastes	Superfund (CERCLA) 1980, 1986
CEQ	Environmental impacts	National Environmental Policy Act (NEPA) 1969
OSHA	Workplace	Occupational Safety and Health (OSH) Act 1970
FDA	Foods, drugs, and cosmetics	FDS Acts 1906, 1938, 1962, 1977 FDA Modernization Act 1997
CPSC	Dangerous consumer products	Consumer Product Safety Act 1972
DOT	Transport of hazardous materials	THM Act 1975, 1976, 1978, 1979, 1984, 1990 (×2)

[a] *Key:* EPA, Environmental Protection Agency; CEQ, Council for Environmental Quality (now Office of Environmental Policy); OSHA, Occupational Health and Safety Administration; FDA, Food and Drug Administration; CPSC, Consumer Product Safety Commission; DOT, Department of Transportation.

Source: From Faustman and Omenn (2001). Used with permisson.

mandates that genomics and ecogenetics may play a role in improving the overall risk assessment by decreasing uncertainty, identifying susceptible populations, and providing an ample margin of safety for all humans.

The Food and Drug Administration (FDA) has developed a guidance document (www.bio-itworkld.com/archive/files/111203.html) for the use of genomic data in drug development. Although the FDA recognizes the potential ability to identify sources of interindividual variability in drug response, genomic data in isolation are not sufficient to determine regulations. The genomic data used by the FDA must clearly demonstrate well-accepted mechanistic and clinical significance. The U.S. Environmental Protection Agency (EPA) has an interim policy regarding genomics in which genomic data alone are not considered sufficient as the sole basis for a risk assessment decision (www.epa.gov/osp/spc/genomics.html). However, the potential utility of genomics data for risk assessment is recognized, and the current EPA policy allows for its use to be considered in a case-by-case basis with ongoing assessment of its potential as the research advances.

Currently, health-based standards are directed at protecting the most susceptible population, such as the chronically ill, children, or the elderly. Information obtained through genomic data may demand a tightening of standards to include genetically susceptible individuals or populations as well. For example, the 1990 Clean Air Act Amendments require that the EPA set standards for criteria air pollutants at levels to protect public health with an adequate margin of safety. EPA is also required under this act to consider sensitive subpopulations. Several legal cases (*Ober vs. Whitman, 2000*; *ALA vs. EPA, 1998*; *Lead Industries vs. EPA, 1980*) have helped to clarify the need for the EPA to set ambient standard to protect sensitive groups. Figure 25.5 shows population response against two differ-

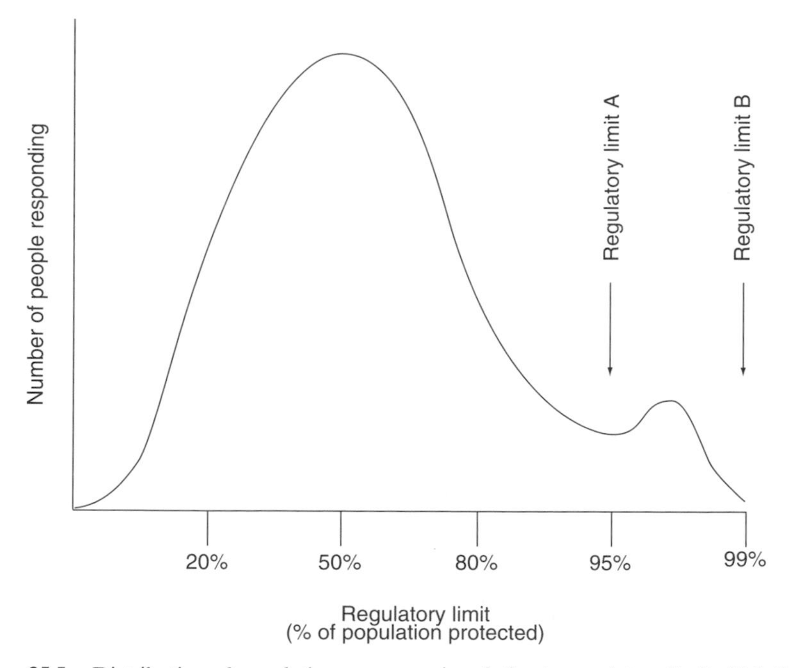

Figure 25.5. Distribution of population responses in relation to regulatory limits. This figure illustrates distribution and population response in relationship to two hypothetical regulatory limits, *A* and *B*. Regulatory limit *A* would be protective for 95% of the population, whereas regulatory limit *B* would be protective for 99% of the population.

ent regulatory limits. At level A, in this hypothetical example, 95% of the human population is protected against an adverse response; at level B, 99% of the human population is protected. The population response distribution in this example is designed to illustrate a distinct population group that exhibits adverse response at exposure between these two possible regulatory levels. Ecogenetic information can help us identify genetically susceptible population groups. Public policy will need to address how this new knowledge will be translated into protective public-health-based risk management decisions (Kramer et al. 2004). In some cases the most susceptible subgroups may be identifiable and may benefit from risk reduction or exposure avoidance not necessary for the rest of the population.

The initial problem formulation step in developing risk assessments is an important component of the process of risk management. Figure 25.6 shows the risk management framework for environmental health from the Presidential/Congressional Commission on Risk Assessment and Risk Management, "Omenn Commission" (Charnley and Omenn 1997; Presidential/Congressional Commission 1997). This framework puts problem formulation at the top of the process. This problem can be stated as simply as a patient asking, "Why me, doc?" in the situation of clinical assessment of occupational exposure hazards. A more complex problem question would be "Are we adequately protecting asthmatics in our population from respiratory disease with the current PM 2.5 air particulate standards?" The problem forms the context in which risks are assessed and management options are developed. Since the overall process of risk management requires data from numerous sources, coordination of information across numerous disciplines, agencies, and organizations is necessary. Stakeholders are the center of the framework, with critical initial input into the problem formulation as well as continued dialog through the subsequent stages of risk assessment, definition of options, decision analysis, management action, and evaluation. Depending on the complexity of the problem, "stakeholders" can consist of patients, community groups, lawyers, local elected officials, judges, businesspeople, labor, government, concerned citizens, and environmental advocates.

The active engagement of stakeholders within the risk management process requires an understanding of the underlying factors that lead to concerns and expectations. One such factor is the perception of risk associated with a given hazard or hazardous situation. This "perceived risk" varies from individual to individual, reflecting the differences in the population's background, education, and experiences (see Figure 25.7). In addition, the "perceived" risk may be quite different from the "true" risk as determined through technical risk estimates and statistical information. Perceived risk strongly reflects psychological factors such as dread, per-

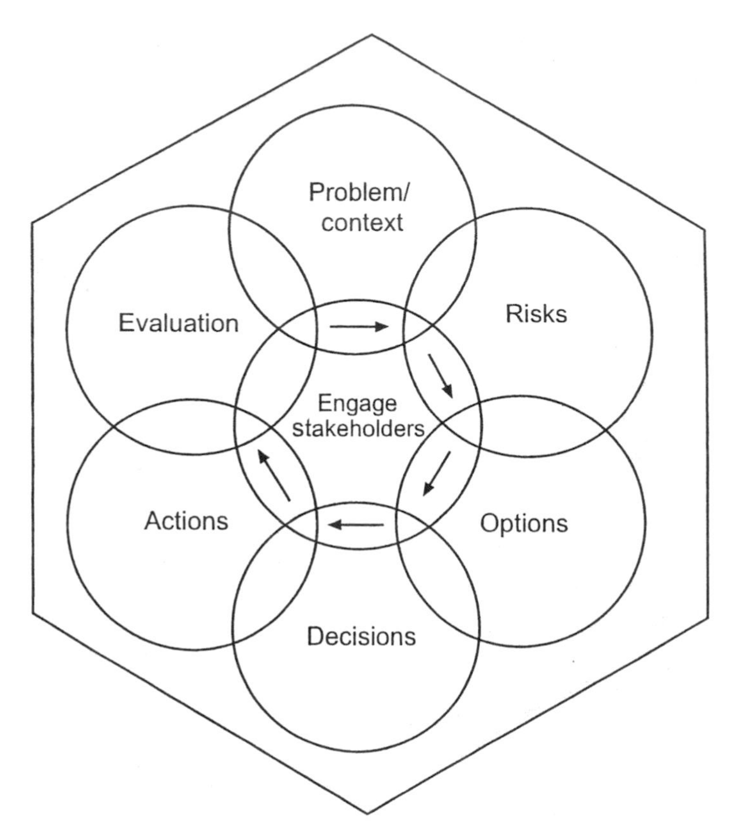

Figure 25.6. Risk management framework for environmental health from the United States Presidential/Congressional Commission on Risk Assessment and Risk Management, "Omenn Commission" (Presidential/Congressional Commission 1997). Six stages constitute this framework: (1) formulating the problem, (2) analyzing the risks, (3) defining the options, (4) making risk reduction decisions, (5) implementing those actions, and (6) evaluating the effectiveness of the actions taken; stakeholder interactions are critical and are at the center of this framework (Charnley and Omenn 1997; Presidential/Congressional Commission 1997; Faustman and Omenn 1995; Faustman and Omenn 2001; Ohanian et al. 1997; Omenn and Faustman 1997).

ceived uncontrollability, and involuntary exposure (Slovic 1987). Understanding this type of behavioral response is critical in constructing public policy and evaluating risk management options. For example, perceptions of risk considerations contributed to the extra safety factor (default value 10) for children in the Food Quality Protection Act of 1996 as a susceptible population group. Risk communication and risk management must consider not only technical risk estimates but also societal concerns and values. Numerous studies have shown that the public perceives greater risks for

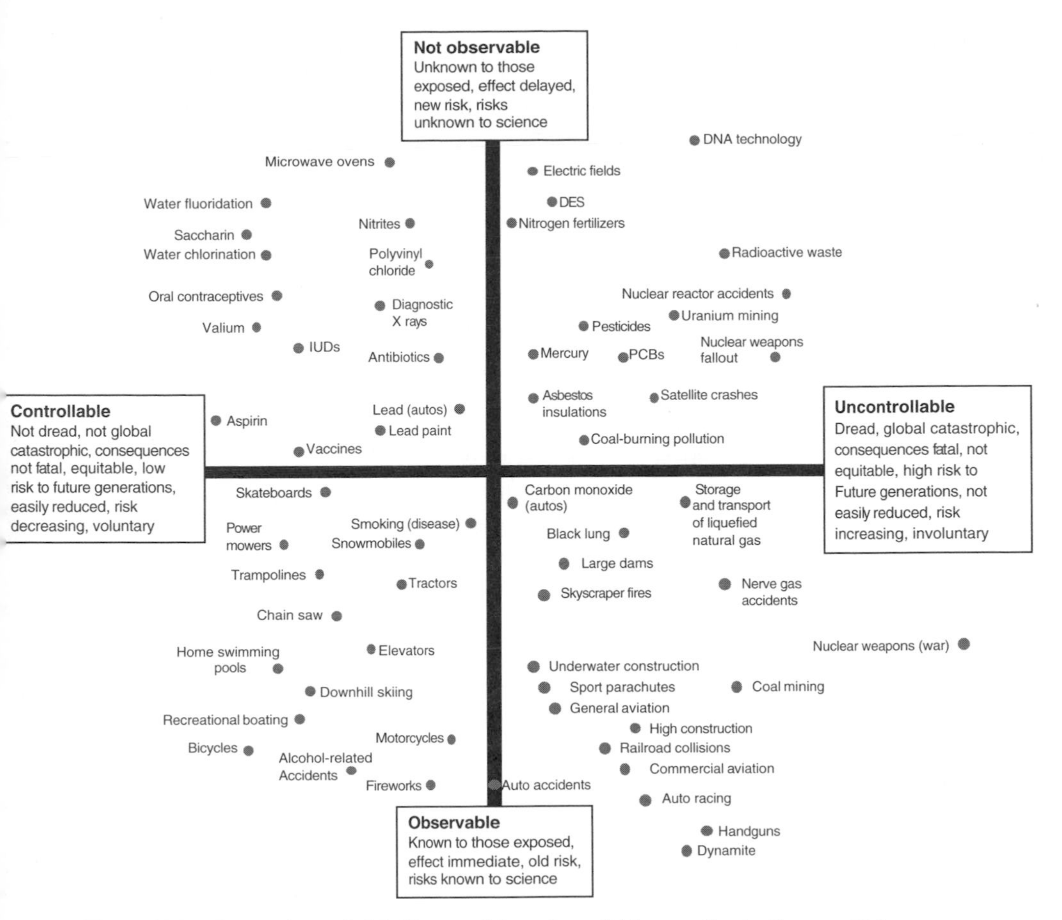

Figure 25.7. Perceptions of risk illustrated by using a "risk space" axis diagram. Risk space has axes that correspond roughly to a hazard's perceived "dreadedness" and to the degree to which it is familiar or observable. Risks in the upper right quadrant of this space, such as DNA technology, are most likely to provoke calls for government regulation (Faustman and Omenn 2001; Morgan 1993; Slovic 1998).

issues that are unfamiliar and nonobservable (Fischhoff et al. 1993; Morgan 1993; Slovic 1987, 1998). Issues of risk perception have been especially important in considerations of DNA testing and DNA technologies. Examples with public controversy include genetically modified crops and foods, and use of nuclear somatic cell transplantation to develop new treatment options ("therapeutic cloning"). Thus, DNA technology falls in the upper right-hand quadrant of the "risk space" diagram shown in Figure 25.7. Without extensive public dialogue and two-way communication, new ecogenetic genomic information could be received in a similarly negative manner. This can be addressed through public forums, which provide an avenue for active involvement and inclusion in the decision process. This process also maximizes transparency in the risk management process leading to increased public confidence.

25.4. CONCLUSIONS AND FUTURE DIRECTIONS

The development and growth of genomics and ecogenetics information will have profound implications for the future of risk assessment and the formation of public policy. With this powerful technology, there will be improved identification of sensitive populations and the setting of health-based standards better directed to protect these susceptible individuals and subgroups. The standards themselves may be enhanced through the refining of uncertainty factors so they are more scientifically based as compared to the standard 10× default value. The promise is that through mechanistic studies of ecogenetics, critical disease response pathways will be identified and lead to a better understanding of variability and range of human response to environmental factors. Identification of polymorphisms that confer individual susceptibility can lead to improved understanding of susceptible populations and how these populations need to be protected by regulatory actions and risk management. In conjunction with the power that genomics brings to risk assessment comes the necessity to engage the public in a dialog regarding genomic information. Education of the public and responsiveness to public concerns will influence perceived risks and perceived benefits and will contribute to the success of applying the knowledge gained from ecogenetics. Finally, genomics gives researchers new tools to understand the interactions of genes and environmental factors that are the key to the multifaceted, complex nature of human predisposition to environmentally triggered diseases.

25.5. SUGGESTED READING

For further information on risk assessment and the impact of ecogenetics, the reader is referred to the contributed chapters by Omenn and Faustman (2002) and Faustman and Omenn (2001) and the journal articles by Ponce et al. (2001), MacGregor (2003), and Faustman et al. (2000).

Bibliography

Abdel-Rahman, S. Z., el-Zein, R. A., Anwar, W. A., and Au, W. W. (1996). A multiplex PCR procedure for polymorphic analysis of GSTM1 and GSTT1 genes in population studies. *Cancer Lett* **107**, 229–233.

Abraham, P. A. and Keane, W. F. (1984). Glomerular and interstitial disease induced by nonsteroidal anti-inflammatory drugs. *Am J Nephrol* **4**, 1–6.

Agarwal, D. P. (2001). Genetic polymorphisms of alcohol metabolizing enzymes. *Pathol Biol* (Paris) **49**, 703–709.

Akerman, B. R., Lemass, H., Chow, L. M., Lambert, D. M., Greenberg, C., Bibeau, C., Mamer, O. A., and Treacy, E. P. (1999). Trimethylaminuria is caused by mutations of the FMO3 gene in a North American cohort. *Mol Genet Metab* **68**, 24–31.

Aklillu, E., Carrillo, J. A., Makonnen, E., Hellman, K., Pitarque, M., Bertilsson, L., and Ingelman-Sundberg, M. (2003). Genetic polymorphism of CYP1A2 in Ethiopians affecting induction and expression: characterization of novel haplotypes with single-nucleotide polymorphisms in intron 1. *Mol Pharmacol* **64**, 659–669.

Albertson, D. G., and Pinkel, D. (2003). Genomic microarrays in human genetic disease and cancer. *Hum Mol Genet* **12**(Spec No 2), R145–R152.

Alexandrov, K., Cascorbi, I., Rojas, M., Bouvier, G., Kriek, E., and Bartsch, H. (2002). CYP1A1 and GSTM1 genotypes affect benzo[a]pyrene DNA adducts in smokers' lung: comparison with aromatic/hydrophobic adduct formation. *Carcinogenesis* **23**, 1969–1977.

Allen, B. C., Kavlock, R. J., Kimmel, C. A., and Faustman, E. M. (1994a). Dose-response assessment for development toxicity. II. Comparison of genetic benchmark dose estimates with no observed adverse effect levels. *Fundam Appl Toxicol* **23**, 487–495.

Allen, B. C., Kavlock, R. J., Kimmel, C. A., and Faustman, E. M. (1994b). Dose-response assessment for developmental toxicity. III. Statistical models. *Fundam Appl Toxical* **23**, 496–509.

Allorge, D., Chevalier, D., Lo-Guidice, J. M., Cauffiez, C., Suard, F., Baumann, P., Eap, C. B., and Broly, F. (2003). Identification of a novel splice-site mutation in the CYP1A2 gene. *Br J Clin Pharmacol* **56**, 341–344.

Gene-Environment Interactions: Fundamentals of Ecogenetics, edited by Lucio G. Costa and David L. Eaton

AMA (2001). Survey Regarding Medical Testing of Employees. In http://www.amanet.org/research/pdfs/mt_2001.pdf (American Management Association).

American Diabetes Association (2002). The prevention or delay of type 2 diabetes. *Diabetes Care* **25**, 742–749.

American Heart Association (2002). *Heart and Stroke Statistics: 2003 Update* (Dallas, TX: American Heart Assoc.).

Ames, S. K., Ellis, K. J., Gunn, S. K., Copeland, K. C., and Abrams, S. A. (1999). Vitamin D receptor gene Fok1 polymorphism predicts calcium absorption and bone mineral density in children. *J Bone Miner Res* **14**, 740–746.

Amin, R., Hamadeh, H., Bushel, P., Bennett, B., Afshari, C., and Paules, R. (2002). Genomic interrogation of mechanism(s) underlying cellular responses of toxicants. *Toxicology* **181–182**, 555–563.

Amos, C. I., Caporaso, N. E., and Weston, A. (1992). Host factors in lung cancer risk: A review of interdisciplinary studies. *Cancer Epidemiol Biomark Prev* **1**, 505–513.

Amos, J., and Patnaik, M. (2002). Commercial molecular diagnostics in the U.S.: The Human Genome Project to the clinical laboratory. *Hum Mutat* **19**, 324–333.

Anderson, R., and Motulsky, A. G. (1966). Adverse effects of raised environment temperature on the expression of hereditary spherocytosis in deer mice. *Blood* **28**, 365–376.

Andjelkovic, D., Taulbee, J., and Symons, M. (1976). Mortality experinece of a cohort of rubber workers, 1964–1973. *J Occup Med* **18**, 387–394.

Annas, G. J., Glantz, L. H., and Roche, P. A. (1995). Drafting the Genetic Privacy Act: Science, policy, and practical considerations. *J Law Med Ethics* **23**, 360–366.

Ando, Y., Tateishi, T., et al. (1999). Re: Modification of clinical presentation of prostate tumors by a novel genetic variant in CYP3A4. *J Natl Cancer Inst* **91**, 1587–1590.

Antequera, F., and Bird, A. (1993). Number of CpG islands and genes in human and mouse. *Proc Natl Acad Sci USA* **90**, 11995–11999.

APHA (1948–present). Position Paper 9503: *Worker Notification and Institutional Review for Privately Funded Research in Occupational Health and Safety* (Washington, DC: American Public Health Assoc.).

Arand, M., Muhlbauer, R., Hengstler, J., Jager, E., Fuchs, J., Winkler, L., and Oesch, F. (1996). A multiplex polymerase chain reaction protocol for the simultaneous analysis of the glutathione S-transferase GSTM1 and GSTT1 polymorphisms. *Anal Biochem* **236**, 184–186.

Arinami, T., Ishiguro, H., and Onaivi, E. S. (2000). Polymorphisms in genes involved in neurotransmission in relation to smoking. *Eur J Pharmacol* **410**, 215–226.

Ariyoshi, N., Miyamoto, M., et al. (2002). Genetic polymorphism of CYP2A6 gene and tobacco-induced lung cancer risk in male smokers. *Cancer Epidemiol Biomarkers Prev* **11**, 890–894.

Armon, C., Kurland, L. T., Daube, J. R., and O'Brien, P. C. (1991). Epidemiologic correlates of sporadic amyotrophic lateral sclerosis. *Neurology* **41**, 1077–1084.

Armstrong, R. N., and Cassidy, C. S. (2000). New structural and chemical insight into the catalytic mechanism of epoxide hydrolases. *Drug Metab Rev* **32**, 327–338.

Armstrong, S. C., and Cozza, K. L. (2003). Pharmacokinetic drug interactions of morphine, codeine, and their derivatives: Theory and clinical reality, part I. *Psychosomatics* **44**, 167–171.

Arruda, V. R., Lima, C. S., Grignoli, C. R., de Melo, M. B., Lorand-Metze, I., Alberto, F. L., Saad, S. T., and Costa, F. F. (2001). Increased risk for acute myeloid leukaemia in individuals with glutathione S-transferase mu 1 (GSTM1) and theta 1 (GSTT1) gene defects. *Eur J Haematol* **66**, 383–388.

Ashfield-Watt, P. A., Pullin, C. H., Whiting, J. M., Clark, Z. E., Moat, S. J., Newcombe, R. G., Burr, M. L., Lewis, M. J., Powers, H. J., and McDowell, I. F. (2002). Methylenetetrahydrofolate reductase 677C→T genotype modulates homocysteine responses to a folate-rich diet or a low-dose folic acid supplement: A randomized controlled trial. *Am J Clin Nutr* **76**, 180–186.

Assicot, M., and Bohuon, C. (1971). Presence of two distinct catechol-O-methyltransferase activities in red blood cells. *Biochimie* **53**, 871–874.

Austin, M. A., Hutter, C., Zimmern, R. L., and Humphries, S. E. (2004). Familial hypercholesterolemia and coronary heart disease: A Human Genome Epidemiology Association review. *Am J Epidemiol* **160**, 421–429.

Austin, M. A., King, M. C., Bawol, R. D., Hulley, S. B., and Friedman, G. D. (1987). Risk factors for coronary heart disease in adult female twins. Genetic heritability and shared environmental influences. *Am J Epidemiol* **125**, 308–318.

Ayesh, R., Mitchell, S. C., Zhang, A., and Smith, R. L. (1993). The fish odour syndrome: biochemical, familial, and clinical aspects. *BMJ* **307**, 655–657.

Azmitia, E. C., and Whitaker-Azmitia, P. M. (1991). Awakening the sleeping giant: Anatomy and plasticity of the brain serotonergic system. *J Clin Psychiatry* **52**, 4–16.

Backstrom, A., Pirttila-Backman, A. M., and Tuorila, H. (2003). Dimensions of novelty: A social representation approach to new foods. *Appetite* **40**, 299–307.

Bader, A., Hansen, T., Kirchner, G., Allmeling, C., Haverich, A., and Borlak, J. T. (2000). Primary porcine enterocyte and hepatocyte cultures to study drug oxidation reactions. *Br J Pharmacol* **129**, 331–342.

Baier, L. J., Permana, P. A., Yang, X., Pratley, R. E., Hanson, R. L., Shen, G. Q., Mott, D., Knowler, W. C., Cox, N. J., Horikawa, Y., et al. (2000). A calpain-10 gene polymorphism is associated with reduced muscle mRNA levels and insulin resistance. *J Clin Invest* **106**, R69–R73.

Bailey, M. J., and Dickinson, R. G. (2003). Acyl glucuronide reactivity in perspective: Biological consequences. *Chem Biol Interact* **145**, 117–137.

Baquet, C. R., Hammond, C., Commiskey, P., Brooks, S., and Mullins, C. D. (2002). Health disparities research—a model for conducting research on cancer disparities: Characterization and reduction. *J Assoc Acad Minor Physicians* **13**, 33–40.

Barker, D. J. P. (1998). *Mothers, Babies and Health in Later Life*, 2nd ed. (London: Churchill Livingston).

Bartell, S. M., Ponce, R. A., Takaro, T. K., Zerbe, R. O., Omenn, G. S., and Faustman, E. M. (2000). Risk estimation and value-of-information analysis for three proposed genetic screening programs for chronic beryllium disease prevention. *Risk Anal* **20**, 87–99.

Bartoli, D., Battista, G., De Santis, M., Iaia, T. E., Orsi, D., Tarchi, M., Pirastu, R., and Valiani, M. (1998). Cohort study of art glass workers in Tuscany, Italy: Mortality from non-malignant diseases. *Occup Med* (Lond) **48**, 441–445.

Bartsch, H., Nair, U., Risch, A., Rojas, M., Wikman, H., and Alexandrov, K. (2000). Genetic polymorphism of CYP genes, alone or in combination, as a risk modifier of tobacco-related cancers. *Cancer Epidemiol Biomarkers Prev* **9**, 3–28.

Batra, V., Patkar, A. A., Berrettini, W. H., Weinstein, S. P., and Leone, F. T. (2003). The genetic determinants of smoking [comment]. *Chest* **123**, 1730–1739.

Baum, A., Gatchel, R. J., and Schaeffer, M. A. (1983). Emotional, behavioral, and physiological effects of chronic stress at Three Mile Island. *J Consult Clin Psychol* **51**, 565–572.

Beachy, R., Bennetzen, J. L., Chassy, B. M., Chrispeels, M., Chory, J., Ecker, J. R., Noel, J. P., Kay, S. A., Dean, C., Lamb, C., et al. (2002). Divergent perspectives on GM food. *Nature Biotechnol* **20**, 1195–1196.

Beauchamp, T. L., and Childress, J. F. (2001). *Principles of Biomedical Ethics*, 5th ed. (New York: Oxford Univ. Press).

Beck-Nielsen, H., Vaag, A., Poulsen, P., and Gaster, M. (2003). Metabolic and genetic influence on glucose metabolism in type 2 diabetic subjects—experiences from relatives and twin studies. *Best Pract Res Clin Endocrinol Metab* **17**, 445–467.

Beier, V., Bauer, A., Baum, M., and Hoheisel, J. D. (2004). Fluorescent sample labeling for DNA microarray analyses. *Meth Mol Biol* **283**, 127–135.

Bejjani, B. A., Lewis, R. A., et al. (1998). Mutations in CYP1B1, the gene for cytochrome P4501B1, are the predominant cause of primary congenital glaucoma in Saudi Arabia. *Am J Hum Genet* **62**, 325–333.

Benhamou, S., Lee, W. J., Alexandrie, A. K., Boffetta, P., Bouchardy, C., Butkiewicz, D., Brockmoller, J., Clapper, M. L., Daly, A., Dolzan, V., et al. (2002). Meta- and pooled analyses of the effects of glutathione S-transferase M1 polymorphisms and smoking on lung cancer risk. *Carcinogenesis* **23**, 1343–1350.

Benjamini, Y., and Hochberg, Y. (1995). Controlling the false discovery rate: A practical and powerful approach to multiple testing. *J Roy Stat Soc* **57**, 289–300.

Beresford, S. A. (1994). How do we get enough folic acid to prevent some neural tube defects? *Am J Publ Health* **84**, 348–350.

Bergen, A. W., Korczak, J. F., Weissbecker, K. A., and Goldstein, A. M. (1999). A genome-wide search for loci contributing to smoking and alcoholism. *Genet Epidemiol* **17**(Suppl 1), S55–S60.

Bertilsson, L., Dahl, M. L., Dalen, P., and Al-Shurbaji, A. (2002). Molecular genetics of CYP2D6: clinical relevance with focus on psychotropic drugs. *Br J Clin Pharmacol* **53**, 111–122.

Best, S. M., and Kerr, P. J. (2000). Coevolution of host and virus: The pathogenesis of virulent and attenuated strains of myxoma virus in resistant and susceptible European rabbits. *Virology* **267**, 36–48.

Betarbet, R., Sherer, T. B., MacKenzie, G., Garcia-Osuna, M., Panov, A. V., and Greenamyre, J. T. (2000). Chronic systemic pesticide exposure reproduces features of Parkinson's disease. *Nat Neurosci* **3**, 1301–1306.

Beutler, E. (1957). The glutathione instability of drug-sensitive red cells; a new method for the in vitro detection of drug sensitivity. *J Lab Clin Med* **49**, 84–95.

Beutler, E. (1996). G6PD: population genetics and clinical manifestations. *Blood Rev* **10**, 45–52.

Beutler, E., Felitti, V., Gelbart, T., and Ho, N. (2001). Genetics of iron storage and hemochromatosis. *Drug Metab Dispos* **29**, 495–499.

Bierut, L. J., Rice, J. P., Edenberg, H. J., Goate, A., Foroud, T., Cloninger, C. R., Begleiter, H., Conneally, P. M., Crowe, R. R., Hesselbrock, V., et al. (2000). Family-based study of the association of the dopamine D2 receptor gene (DRD2) with habitual smoking. *Am J Med Genet* **90**, 299–302.

Bis, J. C., Smith, N. L., Psaty, B. M., Heckbert, S. R., Edwards, K. L., Lemaitre, R. N., Lumley, T., and Rosendaal, F. R. (2003). Angiotensinogen Met235Thr polymorphism, angiotensin-converting enzyme inhibitor therapy, and the risk of nonfatal stroke or myocardial infarction in hypertensive patients. *Am J Hypertens* **16**, 1011–1017.

Bjertness, E., Candy, J. M., Torvik, A., Ince, P., McArthur, F., Taylor, G. A., Johansen, S. W., Alexander, J., Gronnesby, J. K., Bakketeig, L. S., and Edwardson, J. A. (1996). Content of brain aluminum is not elevated in Alzheimer disease. *Alzheimer Dis Assoc Disord* **10**, 171–174.

Blackwelder, W. C., and Elston, R. C. (1985). A comparison of sib-pair linkage tests for disease susceptibility loci. *Genet Epidemiol* **2**, 85–97.

Blair, A., Decoufle, P., and Grauman, D. (1979). Causes of death among laundry and dry cleaning workers. *Am J Publ Health* **69**, 508–511.

Blot, W. J. and Fraumeni, J. F. (1996). *Cancers of the lung and pleura*. Cancer Epidemiology and Prevention. D. Schottenfeld and J. F. Fraumeni. New York, Oxford University Press, 637–665.

Bock, K. W. (2003). Vertebrate UDP-glucuronosyltransferases: Functional and evolutionary aspects. *Biochem Pharmacol* **66**, 691–696.

Bockenhauer, D. (2001). Ion channels in disease. *Curr Opin Pediatr* **13**, 142–149.

Bodnar, R. J., and Hadjimarkou, M. M. (2002). Endogenous opiates and behavior: 2001. *Peptides* **23**, 2307–2365.

Boehm, D., Herold, S., Kuechler, A., Liehr, T., and Laccone, F. (2004). Rapid detection of subtelomeric deletion/duplication by novel real-time quantitative PCR using SYBR-green dye. *Hum Mutat* **23**, 368–378.

Boersma, M. G., van der Woude, H., Bogaards, J., Boeren, S., Vervoort, J., Cnubben, N. H. P., van Iersel, M. L. P. S., van Bladeren, P. J., and Rietjens, I. M. C. M. (2002). Regioselectivity of phase II metabolism of luteolin and quercetin by UDP-glucuronosyl transferases. *Chem Res Toxicol* **15**, 662–670.

Boerwinkle, E., and Utermann, G. (1988). Simultaneous effects of the apolipoprotein E polymorphism on apolipoprotein E, apolipoprotein B, and cholesterol metabolism. *Am J Hum Genet* **42**, 104–112.

Bollschweiler, E., Wolfgarten, E., Gutschow, C., and Holscher, A. H. (2001). Demographic variations in the rising incidence of esophageal adenocarcinoma in white males. *Cancer* **92**, 549–555.

Bond, C., LaForge, K. S., Tian, M., Melia, D., Zhang, S., Borg, L., Gong, J., Schluger, J., Strong, J. A., Leal, S. M., et al. (1998). Single-nucleotide polymorphism in the human mu opioid receptor gene alters beta-endorphin binding and activity: Possible implications for opiate addiction. *Proc Natl Acad Sci USA* **95**, 9608–9613.

Bonham, V. L., and Nathan, V. R. (2002). Environmental public health research: Engaging communities. *Int J Hygiene Environ Health* **205**, 11–18.

Bosma, P. J. (2003). Inherited disorders of bilirubin metabolism. *J Hepatol* **38**, 107–117.

Botto, L. D., and Khoury, M. J. (2004). Facing the challenge of complex genotypes and gene-environment interaction: The basic epidemiologic units in case-control and case-only designs. In *Human Genome Epidemiology: A Scientific Foundation for Using Genetic Information to Improve Health and Prevent Disease*, Khoury, M. J., Little, J., and Burke, W., eds. (Oxford: Oxford University Press) pp. 111–126.

Bouchardy, C., Mitrunen, K., Wikman, H., Husgafvel-Pursiainen, K., Dayer, P., Benhamou, S., and Hirvonen, A. (1998). N-acetyltransferase NAT1 and NAT2 genotypes and lung cancer risk. *Pharmacogenetics* **8**, 291–298.

Boulware, L. E., Cooper, L. A., Ratner, L. E., LaVeist, T. A., and Powe, N. R. (2003). Race and trust in the health care system. *Publ Health Rep* **118**, 358–365.

Boustead, C., Taber, H., Idle, J. R., and Cholerton, S. (1997). CYP2D6 genotype and smoking behaviour in cigarette smokers. *Pharmacogenetics* **7**, 411–414.

Bowen, D., Moinpour, C., Thompson, B., Andersen, M. R., Meischke, H., and Cochrane, B. (in press). Creating a framework for public health intervention design. In *Handbook of Behavioral Science and Cancer*, Miller, S. M., Bowen, D., Croyle, R. T., and Rowland, J. H., eds. (Washington, DC: American Psychological Associ.).

Bowen, W. (2002). An analytical review of environmental justice research: What do we really know? *Environ Manage* **29**, 3–15.

Bradley, A. B., Cullen, A. C., Burke, W., and Faustman, E. M. (2004). Impact and policy implications of genetic information in regulation: A case study of organophosphate pesticides [abstract]. Society For Risk Analysis Annual Meeting 2004, Palms Springs, CA, Society For Risk Analysis.

Brandi, M. L., Becherini, L., Gennari, L., Racchi, M., Bianchetti, A., Nacmias, B., Sorbi, S., Mecocci, P., Senin, U., and Govoni, S. (1999). Association of the estrogen receptor alpha gene polymorphisms with sporadic Alzheimer's disease. *Biochem Biophys Res Commun* **265**, 335–338.

Bray, M. S., Krushkal, J., Li, L., Ferrell, R., Kardia, S., Sing, C. F., Turner, S. T., and Boerwinkle, E. (2000). Positional genomic analysis identifies the beta(2)-adrenergic receptor gene as a susceptibility locus for human hypertension. *Circulation* **101**, 2877–2882.

Bray, N. J., Buckland, P. R., Williams, N. M., Williams, H. J., Norton, N., Owen, M. J., and O'Donovan, M. C. (2003). A haplotype implicated in schizophrenia susceptibility is associated with reduced COMT expression in human brain. *Am J Hum Genet* **73**, 152–161.

Brennan, P., Lewis, S., Hashibe, M., Bell, D. A., Boffetta, P., Bouchardy, C., Caporaso, N., Chen, C., Coutelle, C., Diehl, S. R., et al. (2004). Pooled analysis of alcohol dehydrogenase genotypes and head and neck cancer: A HuGE review. *Am J Epidemiol* **159**, 1–16.

Brewer, G. (1971). Human ecology, an expanding role for the human geneticist. *Am J Hum Genet* **23**, 92–94.

Brewer, G. J. (1971). Annotation: Human ecology, an expanding role for the human geneticist. *Am J Hum Genet* **23**, 92–94.

Brockmoller, J., Cascorbi, I., Henning, S., Meisel, C., and Roots, I. (2000). Molecular genetics of cancer susceptibility. *Pharmacology* **61**, 212–227.

Brockmoller, J., Cascorbi, I., Kerb, R., Sachse, C., and Roots, I. (1998). Polymorphisms in xenobiotic conjugation and disease predisposition. *Toxicol Lett* **102–103**, 173–183.

Brodde, O. E., and Stein, C. M. (2003). The Gly389Arg beta1-adrenergic receptor polymorphism: A predictor of response to beta-blocker treatment? *Clin Pharmacol Ther* **74**, 299–302.

Brown, A. J., Dusso, A., and Slatopolsky, E. (1999). Vitamin D. *Am J Physiol* **277**, F157–F175.

Brown, J. L., and Ping, Y. (2003). Consumer perception of risk associated with eating genetically engineered soybeans is less in the presence of a perceived consumer benefit. *J Am Dietetic Assoc* **103**, 208–214.

Brown, P., Mayer, B., Zavestoski, S., Luebke, T., Mandelbaum, J., and McCormick, S. (2003). The health politics of asthma: Environmental justice and collective illness experience in the United States. *Soc Sci Med* **57**, 453–464.

Brownson, R. C., Chang, J. C., and Davis, J. R. (1992). Gender and histologic type variations in smoking-related risk of lung cancer. *Epidemiology* **3**, 61–64.

Brzezinski, M. R., Spink, B. J., Dean, R. A., Berkman, C. E., Cashman, J. R., and Bosron, W. F. (1997). Human liver carboxylesterase hCE-1: Binding specificity for cocaine, heroin, and their metabolites and analogs. *Drug Metab Dispos* **25**, 1089–1096.

Bucher, J. R., Portier, C. J., Goodman, J. I., Faustman, E. M., and Lucier, G. W. (1996). Workshop overview. National Toxicology Program Studies: Principles of dose selection and applications to mechanistic based risk assessment. *Fundam Appl Toxicol* **31**, 1–8.

Buller, D. B., Callister, M. A., and Reichert, T. (1995). Skin cancer prevention by parents of young children: Health information sources, skin cancer knowledge, and sun-protection practices. *Oncol Nurs Forum* **22**, 1559–1566.

Bunning, R. D. and Barth, W. F. (1982). Sulindac. A potentially renal-sparing nonsteroidal anti-inflammatory drug. *Jama* **248**, 2864–2867.

Burrill, G. S., ed. (2003). *Biotech 2003: 17th Annual Report on the Industry, Life Sciences: Revaluation and Restructuring* (San Francisco: Burrill & Co).

Burk, O., Tegude, H., et al. (2002). Molecular mechanisms of polymorphic CYP3A7 expression in adult human liver and intestine. *J Biol Chem* **277**, 24280–24288.

Busch, C. P., and Hegele, R. A. (2001). Genetic determinants of type 2 diabetes mellitus. *Clin Genet* **60**, 243–254.

Butkiewicz, D., Rusin, M., Enewold, L., Shields, P. G., Chorazy, M., and Harris, C. C. (2001). Genetic polymorphisms in DNA repair genes and risk of lung cancer. *Carcinogenesis* **22**, 593–597.

Butterworth, C. E., Jr., and Tamura, T. (1989). Folic acid safety and toxicity: A brief review. *Am J Clin Nutr* **50**, 353–358.

Cadoret, R. J., Troughton, E., O'Gorman, T. W., and Heywood, E. (1986). An adoption study of genetic and environmental factors in drug abuse. *Arch Gen Psychiatry* **43**, 1131–1136.

Cai, Q., Gao, Y. T., Wen, W., Shu, X. O., Jin, F., Smith, J. R., and Zheng, W. (2003a). Association of breast cancer risk with a GT dinucleotide repeat polymorphism upstream of the estrogen receptor-alpha gene. *Cancer Res* **63**, 5727–5730.

Cai, Q., Shu, X. O., Jin, F., Dai, Q., Wen, W., Cheng, J. R., Gao, Y. T., and Zheng, W. (2003b). Genetic polymorphisms in the estrogen receptor alpha gene and risk of breast cancer: Results from the Shanghai Breast Cancer Study. Cancer *Epidemiol Biomark Prev* **12**, 853–859.

Calabrese, E. J. (1984). Ecogenetics: Genetics Variation in Susceptibility to Environmental Agents. New York, John Wiley and Sons.

Calafell, F. (2003). Classifying humans. *Nat Genet* **33**, 435–436.

Calkins, D., Dixon, R., Gerber, C., Zarin, D., and Omenn, G. (1980). Identification, characterization, and control of potential human carcinogens: A framework for federal decision-making. *J Natl Cancer Inst* **61**, 169–175.

Calvert, G. M., Sweeney, M. H., Deddens, J., and Wall, D. K. (1999). Evaluation of diabetes mellitus, serum glucose, and thyroid function among United States workers exposed to 2,3,7,8-tetrachlorodibenzo-p-dioxin. *Occup Environ Med* **56**, 270–276.

Cami, J., and Farre, M. (2003). Drug addiction. *New Engl J Med* **349**, 975–986.

Cannon, S. C. (1996). Sodium channel defects in myotonia and periodic paralysis. *Annu Rev Neurosci* **19**, 141–164.

Caporaso, N. E., Lerman, C., Main, D., Audrain, J., Boyd, N. R., Bowman, E., Lockshin, B., and Shields, P. (1997). The genetics of smoking: the dopamine receptor (DRD2) and transporter (DAT) polymorphisms in a smoking cessation study. *Proc Am Assoc Cancer Res* **38**, 168–169.

Cardon, L. R., and Palmer, L. J. (2003). Population stratification and spurious allelic association. *Lancet* **361**, 598–604.

Carlborg, O., and Haley, C. S. (2004). Epistasis: Too often neglected in complex trait studies? *Nat Rev Genet* **5**, 618–625.

Carlson, C. S., Eberle, M. A., Rieder, M. J., Yi, Q., Kruglyak, L., and Nickerson, D. A. (2004). Selecting a maximally informative set of single-nucleotide polymorphisms for association analyses using linkage disequilibrium. *Am J Hum Genet* **74**, 106–120.

Caro, A. A. and Cederbaum, A. I. (2004). Oxidative stress, toxicology, and pharmacology of CYP2E1. *Annu Rev Pharmacol Toxicol* **44**, 27–42.

Carriere, V., Berthou, F., Baird, S., Belloc, C., Beaune, P., and de Waziers, I. (1996). Human cytochrome P450 2E1 (CYP2E1): from genotype to phenotype. *Pharmacogenetics* **6**, 203–211.

Carrington, M., and O'Brien, S. J. (2003). The influence of HLA genotype on AIDS. *Annu Rev Med* **54**, 535–551.

Carson, P. E., Flanagan, C. L., Ickes, C. E., and Alving, A. S. (1956). Enzymatic deficiency in primaquine-sensitive erythrocytes. *Science* **124**, 484–485.

Cartwright, M., Wardle, J., Steggles, N., Simon, A. E., Croker, H., and Jarvis, M. J. (2003). Stress and dietary practices in adolescents. *Health Psychology* **22**, 362–369.

Carvalho, B., Ouwerkerk, E., Meijer, G. A., and Ylstra, B. (2004). High resolution microarray comparative genomic hybridisation analysis using spotted oligonucleotides. *J Clin Pathol* **57**, 644–646.

Casanova, J. L., and Abel, L. (2002). Genetic dissection of immunity to mycobacteria: the human model. *Annu Rev Immunol* **20**, 581–620.

Casanova, J. L., and Abel, L. (2004). The human model: A genetic dissection of immunity to infection in natural conditions. *Nat Rev Immunol* **4**, 55–66.

Cascorbi, I., Brockmoller, J., Mrozikiewicz, P. M., Bauer, S., Loddenkemper, R., and Roots, I. (1996). Homozygous rapid arylamine N-acetyltransferase (NAT2) genotype as a susceptibility factor for lung cancer. *Cancer Res* **56**, 3961–3966.

Cashman, J. R. (1998). Stereoselectivity in S- and N-oxygenation by the mammalian flavin-containing and cytochrome P-450 monooxygenases. *Drug Metab Rev* **30**, 675–707.

Cashman, J. R., Camp, K., et al. (2003). Biochemical and clinical aspects of the human flavin-containing monooxygenase form 3 (FMO3) related to trimethylaminuria. *Curr Drug Metab* **4**, 151–170.

Cashman, J. R., Celestial, J. R., and Leach, A. R. (1992). Enantioselective N-oxygenation of chlorpheniramine by the flavin-containing monooxygenase from hog liver. *Xenobiotica* **22**, 459–469.

Catterall, W. A. (2000). From ionic currents to molecular mechanisms: the structure and function of voltage-gated sodium channels. *Neuron* **26**, 13–25.

Cauffiez, C., Lo-Guidice, J. M., et al. (2004). Genetic polymorphism of the human cytochrome CYP2A13 in a French population: implication in lung cancer susceptibility. *Biochem Biophys Res Commun* **317**, 662–669.

Centers for Disease Control and Prevention (1992). Recommendations for the use of folic acid to reduce the number of cases of spina bifida and other neural tube defects. *Morb Mort Wkly Rep Recommend Rep* **41**, 1–7.

Centers for Disease Control and Prevention (1997). Trends in the prevalence and incidence of self-reported diabetes mellitus-United States: 1980–1994. *Morb Mort Wkly Rep* **46**, 1014–1018.

Centers for Disease Control and Prevention (2004). *National Diabetes Fact Sheet: General Information and National Estimates on Diabetes in the United States, 2003* (revised edition) (Atlanta, GA: U.S. Dept. Health and Human Services, Centers for Disease Control and Prevention).

Centers for Disease Control and Prevention (2005). Epi Info. Atlanta, GA, U.S. Department of Health and Human Services, Centers for Disease Control and Prevention, Epidemiology Program Office, Division of Public Health Surveillance and Informatics: http://www.cdc.gov/epiinfo/.

Cestele, S., and Catterall, W. A. (2000). Molecular mechanisms of neurotoxin action on voltage-gated sodium channels. *Biochimie* **82**, 883–892.

Chamberlain, J. S., Gibbs, R. A., Ranier, J. E., Nguyen, P. N., and Caskey, C. T. (1988). Deletion screening of the Duchenne muscular dystrophy locus via multiplex DNA amplification. *Nucleic Acids Res* **16**, 11141–11156.

Chan, A. O., Luk, J. M., Hui, W. M., and Lam, S. K. (1999). Molecular biology of gastric carcinoma: From laboratory to bedside. *J Gastroenterol Hepatol* **14**, 1150–1160.

Charnley, G., and Omenn, G. (1997). A summary of the findings and recommendations of the Commission of Risk Assessment and Risk Management (and accompanying papers prepared for the Commission). *Hum Ecol Risk Assess* **3**, 701–711.

Checkoway, H., Farin, F. M., Costa-Mallen, P., Kirchner, S. C., and Costa, L. G. (1998a). Genetic polymorphisms in Parkinson's disease. *Neurotoxicology* **19**, 635–643.

Checkoway, H., Franklin, G. M., Costa-Mallen, P., Smith-Weller, T., Dilley, J., Swanson, P. D., and Costa, L. G. (1998b). A genetic polymorphism of MAO-B modifies the association of cigarette smoking and Parkinson's disease. *Neurology* **50**, 1458–1461.

Chen, A. C., LaForge, K. S., Ho, A., McHugh, P. F., Kellogg, S., Bell, K., Schluger, R. P., Leal, S. M., and Kreek, M. J. (2002). Potentially functional polymorphism in the promoter region of prodynorphin gene may be associated with protection against cocaine dependence or abuse. *Am J Med Genet* **114**, 429–435.

Chen, Q., Kirsch, G. E., Zhang, D., Brugada, R., Brugada, J., Brugada, P., Potenza, D., Moya, A., Borggrefe, M., Breithardt, G., et al. (1998). Genetic basis and molecular mechanism for idiopathic ventricular fibrillation. *Nature* **392**, 293–296.

Chen, S., Wu, K., and Knox, R. (2000). Structure-function studies of DT-diaphorase (NQO1) and NRH: Quinone oxidoreductase (NQO2). *Free Radic Biol Med* **29**, 276–284.

Cheng, S., Fockler, C., Barnes, W. M., and Higuchi, R. (1994). Effective amplification of long targets from cloned inserts and human genomic DNA. *Proc Natl Acad Sci USA* **91**, 5695–5699.

Chettle, D. R., Scott, M. C., and Somervaille, L. J. (1991). Lead in bone: Sampling and quantitation using K X-rays excited by 109Cd. *Environ Health Perspect* **91**, 49–55.

Chiba-Falek, O., and Nussbaum, R. L. (2001). Effect of allelic variation at the NACP-Rep1 repeat upstream of the alpha-synuclein gene (SNCA) on transcription in a cell culture luciferase reporter system. *Hum Mol Genet* **10**, 3101–3109.

Chien, J. Y., Thummel, K. E., and Slattery, J. T. (1997). Pharmacokinetic consequences of induction of CYP2E1 by ligand stabilization. *Drug Metab Dispos* **25**, 1165–1175.

Cholerton, S., Boustead, C., Taber, H., Arpanahi, A., and Idle, J. R. (1996). CYP2D6 genotypes in cigarette smokers and non-tobacco users. *Pharmacogenetics* **6**, 261–263.

Chow, W. H., Blot, W. J., Vaughan, T. L., Risch, H. A., Gammon, M. D., Stanford, J. L., Dubrow, R., Schoenberg, J. B., Mayne, S. T., Farrow, D. C., et al. (1998). Body mass index and risk of adenocarcinomas of the esophagus and gastric cardia. *J Natl Cancer Inst* **90**, 150–155.

Christensen, P. M., Gotzsche, P. C., and Brosen, K. (1997). The sparteine/debrisoquine (CYP2D6) oxidation polymorphism and the risk of lung cancer: a meta-analysis. *Eur J Clin Pharmacol* **51**, 389–393.

Christensen, P. M., Gotzsche, P. C., and Brosen, K. (1998). The sparteine/debrisoquine (CYP2D6) oxidation polymorphism and the risk of Parkinson's disease: A meta-analysis. *Pharmacogenetics* **8**, 473–479.

Churchill, G. A. (2002). Fundamentals of experimental design for cDNA microarrays. *Nat Genet* **32**(Suppl), 490–495.

Cigler, T., LaForge, K. S., McHugh, P. F., Kapadia, S. U., Leal, S. M., and Kreek, M. J. (2001). Novel and previously reported single-nucleotide polymorphisms in the human 5-HT(1B) receptor gene: No association with cocaine or alcohol abuse or dependence. *Am J Med Genet* **105**, 489–497.

Clayton, D., and McKeigue, P. M. (2001). Epidemiological methods for studying genes and environmental factors in complex diseases. *Lancet* **358**, 1356–1360.

Clayton, E. W. (2003). Ethical, legal, and social implications of genomic medicine. *New Engl J Med* **349**, 562–569.

Cleaver, J. E., and Kraemer, K. H. (1989). Xeroderma pigmentosum. In *Metabolic Basis of Inherited Disease*, Scriver, C. R., Beudet, A. L., Sktm, W. S., and Valle, D., eds. (New York: McGraw-Hill), pp. 2949–2971.

Clegg, J. B., and Weatherall, D. J. (1999). Thalassemia and malaria: New insights into an old problem. *Proc Assoc Am Physicians* **111**, 278–282.

CLIAC (Clinical Laboratory Improvement Advisory Committee) (2001). *Summary Report.*

Coffman, B. L., Rios, G. R., King, C. D., and Tephly, T. R. (1997). Human UGT2B7 catalyzes morphine glucuronidation. *Drug Metab Dispos* **25**, 1–4.

Colditz, G. A., Manson, J. E., and Hankinson, S. E. (1997). The Nurses' Health Study: 20-year contribution to the understanding of health among women. *J Womens Health* **6**, 49–62.

Coles, B., Yang, M., Lang, N. P., and Kadlubar, F. F. (2000). Expression of hGSTP1 alleles in human lung and catalytic activity of the native protein variants towards 1-chloro-2,4-dinitrobenzene, 4-vinylpyridine and (+)-anti benzo[a]pyrene-7,8-diol-9,10-oxide. *Cancer Lett* **156**, 167–175.

Coles, B. F., Morel, F., Rauch, C., Huber, W. W., Yang, M., Teitel, C. H., Green, B., Lang, N. P., and Kadlubar, F. F. (2001). Effect of polymorphism in the human glutathione S-transferase A1 promoter on hepatic GSTA1 and GSTA2 expression. *Pharmacogenetics* **11**, 663–669.

Colhoun, H. M., McKeigue, P. M., and Davey Smith, G. (2003). Problems of reporting genetic associations with complex outcomes. *Lancet* **361**, 865–872.

Collins, F. S. (1999). Shattuck lecture - medical and societal consequences of the human genome project. *New England Journal of Medicine* **341**, 28–37.

Collins, F. S., Green, E. D., Guttmacher, A. E., and Guyer, M. S. (2003). A vision for the future of genomics research. *Nature* **422**, 835–847.

Collins, F. S., and Mansoura, M. K. (2001). The Human Genome Project. Revealing the shared inheritance of all humankind. *Cancer* **91**, 221–225.

Collins, F. S., and Watson, J. D. (2003). Genetic discrimination: Time to act. *Science* **302**, 745.

Comings, D. E., Blake, H., Dietz, G., Gade-Andavolu, R., Legro, R. S., Saucier, G., Johnson, P., Verde, R., and MacMurray, J. P. (1999a). The proenkephalin gene (PENK) and opioid dependence. *Neuroreport* **10**, 1133–1135.

Comings, D. E., Ferry, L., Bradshaw-Robinson, S., Burchette, R., Chiu, C., and Muhleman, D. (1996). The dopamine D2 receptor (DRD2) gene: A genetic risk factor in smoking. *Pharmacogenetics* **6**, 73–79.

Comings, D. E., Gade, R., Wu, S., Chiu, C., Dietz, G., Muhleman, D., Saucier, G., Ferry, L., Rosenthal, R. J., Lesieur, H. R., et al. (1997a). Studies of the potential role of the dopamine D1 receptor gene in addictive behaviors. *Mol Psychiatry* **2**, 44–56.

Comings, D. E., Gonzalez, N., Wu, S., Saucier, G., Johnson, P., Verde, R., and MacMurray, J. P. (1999b). Homozygosity at the dopamine DRD3 receptor gene in cocaine dependence. *Mol Psychiatry* **4**, 484–487.

Comings, D. E., Muhleman, D., Gade, R., Johnson, P., Verde, R., Saucier, G., and MacMurray, J. (1997b). Cannabinoid receptor gene (CNR1): Association with i.v. drug use. *Mol Psychiatry* **2**, 161–168.

Committee on Genetics (1996). Newborn screening fact sheets. *Pediatrics* **98**, 473–501.

Conaway, C. C., Jiao, D., Kohri, T., Liebes, L., and Chung, F. (1999). Disposition and pharmacokinetics of phenethyl isothiocyanate and 6-phenylhexyl isothiocyanate in F344 rats. *Drug Metab Dispos* **27**, 13–20.

Connor, J. P., Young, R. M., Lawford, B. R., Ritchie, T. L., and Noble, E. P. (2002). D(2) dopamine receptor (DRD2) polymorphism is associated with severity of alcohol dependence. *Eur Psychiatry* **17**, 17–23.

Connor, S. (2003). Glaxo Chief: Our drugs do not work on most patients. In *Independent/UK* (London).

Cooper, G. S., and Umbach, D. M. (1996). Are vitamin D receptor polymorphisms associated with bone mineral density? A meta-analysis. *J Bone Miner Res* **11**, 1841–1849.

Corburn, J. (2002). Environmental justice, local knowledge, and risk: The discourse of a community-based cumulative exposure assessment. *Environ Manage* **29**, 451–466.

Correa, P. (2003). Chemoprevention of gastric cancer: Has the time come? *J Clin Oncol* **21**, 270s-271s.

Cosma, C. L., Sherman, D. R., and Ramakrishnan, L. (2003). The secret lives of pathogenic mycobacteria. *Annu Rev Microbiol* **57**, 641–676.

Costa, L. G. (2000). The emerging field of ecogenetics. *Neurotoxicology* **21**, 85–89.

Costa, L. G., Cole, T. B., and Furlong, C. E. (2003a). Polymorphisms of paraoxonase (PON1) and their significance in clinical toxicology of organophosphates. *J Toxicol Clin Toxicol* **41**, 37–45.

Costa, L. G., Cole, T. B., Jarvik, G. P., and Furlong, C. E. (2003b). Functional genomic of the paraoxonase (PON1) polymorphisms: Effects on pesticide sensitivity, cardiovascular disease, and drug metabolism. *Annu Rev Med* **54**, 371–392.

Costa, L. G., Cole, T. B., Vitalone, A., and Furlong, C. E. (2005). Measurement of paraoxonase (PON1) status as a potential biomarker of susceptibility to organophosphate toxicity. *Clin Chim Acta* **352**, 37–47.

Costa, L. G., and Furlong, C. E., eds. (2002). *Paraoxonase (PON1) in Health and Disease: Basic and Clinical Aspects* (Boston: Kluwer Academic Press).

Costa, L. G., Omiecinski, C. J., Faustman, E. M., and Omenn, G. S. (1993). Ecogenetics: determining susceptibility to chemical-induced diseases. *Wash Publ Health* **22**, 8–11.

Costa, L. G., Richter, R. J., Li, W. F., Cole, T., Guizzetti, M., and Furlong, C. E. (2003c). Paraoxonase (PON 1) as a biomarker of susceptibility for organophosphate toxicity. *Biomarkers* **8**, 1–12.

Costa, P., Checkoway, H., Levy, D., Smith-Weller, T., Franklin, G. M., Swanson, P. D., and Costa, L. G. (1997). Association of a polymorphism in intron 13 of the monoamine oxidase B gene with Parkinson disease. *Am J Med Genet* **74**, 154–156.

Costa-Mallen, P., Afsharinejad, Z., Kelada, S. N., Costa, L. G., Franklin, G. M., Swanson, P. D., Longstreth, W. T., Jr., Viernes, H. M., Farin, F. M., Smith-Weller, T., and Checkoway, H. (2004). DNA sequence analysis of monoamine oxidase B gene coding and promoter regions in Parkinson's disease cases and unrelated controls. *Mov Disord* **19**, 76–83.

Coulston, A. M., Rock, C. L., and Monsen, E. R., eds. (2001). *Nutrition in the Prevention and Treatment of Disease* (San Diego, CA: Academic Press).

Covault, J., Gelernter, J., and Kranzler, H. (2001). Association study of cannabinoid receptor gene (CNR1) alleles and drug dependence. *Mol Psychiatry* **6**, 501–502.

Cranmer, M., Louie, S., Kennedy, R. H., Kern, P. A., and Fonseca, V. A. (2000). Exposure to 2,3,7,8-tetrachlorodibenzo-p-dioxin (TCDD) is associated with hyperinsulinemia and insulin resistance. *Toxicol Sci* **56**, 431–436.

Crook, M. (1981). Migraine: a biochemical headache? *Biochem Soc Trans* **9**, 351–357.

Csaszar, A., and Abel, T. (2001). Receptor polymorphisms and diseases. *Eur J Pharmacol* **414**, 9–22.

Cubells, J. F., Kranzler, H. R., McCance-Katz, E., Anderson, G. M., Malison, R. T., Price, L. H., and Gelernter, J. (2000). A haplotype at the DBH locus, associated with low plasma dopamine beta-hydroxylase activity, also associates with cocaine-induced paranoia. *Mol Psychiatry* **5**, 56–63.

Curl, C. L., Fenske, R. A., Kissel, J. C., Shirai, J. H., Moate, T. F., Griffith, W., Coronado, G., and Thompson, B. (2002). Evaluation of take-home organophosphorus pesticide exposure among agricultural workers and their children. *Environ Health Perspects* **110**, A787–A792.

Cuzick, J., Sasieni, P., and Evans, S. (1992). Ingested arsenic, keratoses, and bladder cancer. *Am J Epidemiol* **136**, 417–421.

Dai, D., Tang, J., Rose, R., Hodgson, E., Bienstock, R. J., Mohrenweiser, H. W., and Goldstein, J. A. (2001). Identification of variants of CYP3A4 and characterization of their abilities to metabolize testosterone and chlorpyrifos. *J Pharmacol Exp Ther* **299**, 825–831.

Daly, A. K. (2003). Pharmacogenetics of the major polymorphic metabolizing enzymes. *Fundam Clin Pharmacol* **17**, 27–41.

Darvesh, S., Hopkins, D. A., and Geula, C. (2003). Neurobiology of butyrylcholinesterase. *Nat Rev Neurosci* **4**, 131–138.

Daston, G. P. and McNamee, P. (2005). *Alternatives to toxicity testing in animals: Challenges and opportunities*. Essays on the Future of Environmental Health Research: A Tribute to Dr. Kenneth Olden. T. J. Goehl. Research Triangle Park, NC, Environmental Health Perspectives/National Institute of Environmental Health Sciences, 6–15.

David-Beabes, G. L., and London, S. J. (2001). Genetic polymorphism of XRCC1 and lung cancer risk among African-Americans and Caucasians. *Lung Cancer* **34**, 333–339.

Davies, M. (2003). The role of GABAA receptors in mediating the effects of alcohol in the central nervous system. *J Psychiatry Neurosci* **28**, 263–274.

Daw, E. W., Heath, S. C., and Wijsman, E. M. (1999). Multipoint oligogenic analysis of age-at-onset data with applications to Alzheimer disease pedigrees. *Am J Hum Genet* **64**, 839–851.

Dawling, S., Roodi, N., Mernaugh, R. L., Wang, X., and Parl, F. F. (2001). Catechol-O-methyltransferase (COMT)-mediated metabolism of catechol estrogens: Comparison of wild-type and variant COMT isoforms. *Cancer Res* **61**, 6716–6722.

Dawson-Hughes, B., Harris, S. S., and Finneran, S. (1995). Calcium absorption on high and low caclium intakes in relation to vitamin D receptor genotype. *J Clin Endocrinol Metab* **80**, 3657–3661.

de Boer, J. G. (2002). Polymorphisms in DNA repair and environmental interactions. *Mutat Res* **509**, 201–210.

de Wildt, S. N., Kearns, G. L., Leeder, J. S., and van den Anker, J. N. (1999). Glucuronidation in humans. Pharmacogenetic and developmental aspects. *Clin Pharmacokinet* **36**, 439–452.

Deitz, A. C., Rothman, N., Rebbeck, T. R., Hayes, R. B., Chow, W. H., Zheng, W., Hein, D. W., and Garcia-Closas, M. (2004). Impact of misclassification in genotype-exposure interaction studies: example of N-acetyltransferase 2 (NAT2), smoking, and bladder cancer. *Cancer Epidemiol Biomarkers Prev* **13**, 1543–1546.

DeMille, M. M., Kidd, J. R., Ruggeri, V., Palmatier, M. A., Goldman, D., Odunsi, A., Okonofua, F., Grigorenko, E., Schulz, L. O., Bonne-Tamir, B., Lu, R., Parnas, J., Pakstis, A., and Kidd, K. (2002). Population variation in linkage disequilibrium across the COMT gene considering promoter region and coding region variation. *Hum Genet* **111**, 521–537.

Dhanasekaran, S. M., Barrette, T. R., Ghosh, D., Shah, R., Varambally, S., Kurachi, K., Pienta, K. J., Rubin, M. A., and Chinnaiyan, A. M. (2001). Delineation of prognostic biomarkers in prostate cancer. *Nature* **412**, 822–826.

Dick, D. M., and Foroud, T. (2003). Candidate genes for alcohol dependence: A review of genetic evidence from human studies. *Alcohol Clin Exp Res* **27**, 868–879.

Dieterich, W., Esslinger, B., and Schuppan, D. (2003). Pathomechanisms in celiac disease. *Int Arch Allergy Immunol* **132**, 98–108.

Diffey, B. L., Gibson, C. J., Haylock, R., and McKinlay, A. F. (1996). Outdoor ultraviolet exposure of children and adolescents. *Br J Dermatol* **134**, 1030–1034.

Dinkova-Kostova, A. T., and Talalay, P. (2000). Persuasive evidence that quinone reductase type 1 (DT diaphorase) protects cells against the toxicity of electrophiles and reactive forms of oxygen. *Free Radic Biol Med* **29**, 231–240.

Divine, K. K., Gilliland, F. D., Crowell, R. E., Stidley, C. A., Bocklage, T. J., Cook, D. L., and Belinsky, S. A. (2001). The XRCC1 399 glutamine allele is a risk factor for adenocarcinoma of the lung. *Mutat Res* **461**, 273–278.

Dobashi, I., Inada, T., and Hadano, K. (1997). Alcoholism and gene polymorphisms related to central dopaminergic transmission in the Japanese population. *Psychiatr Genet* **7**, 87–91.

DOL (Feb. 8, 2000). Federal Executive Order 13145 to Prohibit Discrimination in Federal Employment Based on Genetic Information. *Federal Register* **65**.

Doll, R., Peto, R., Wheatley, K., Gray, R., and Sutherland, I. (1994). Mortality in relation to smoking: 40 years' observations on male British doctors. *Br Med J* **309**, 901–911.

Dolphin, C. T., Beckett, D. J., Janmohamed, A., Cullingford, T. E., Smith, R. L., Shephard, E. A., and Phillips, I. R. (1998). The flavin-containing monooxygenase 2 gene (FMO2) of humans, but not of other primates, encodes a truncated, nonfunctional protein. *J Biol Chem* **273**, 30599–30607.

Dolphin, C. T., Cullingford, T. E., Shephard, E. A., Smith, R. L., and Phillips, I. R. (1996). Differential developmental and tissue-specific regulation of expression of the genes encoding three members of the flavin-containing monooxygenase family of man, FMO1, FMO3 and FM04. *Eur J Biochem* **235**, 683–689.

Dolphin, C. T., Janmohamed, A., Smith, R. L., Shephard, E. A., and Phillips, I. R. (1997). Missense mutation in flavin-containing mono-oxygenase 3 gene, FMO3, underlies fish-odour syndrome. *Nat Genet* **17**, 491–494.

Dolphin, C. T., Janmohamed, A., Smith, R. L., Shephard, E. A. and Phillips, I. R. (2000). Compound heterozygosity for missense mutations in the flavin-containing monooxygenase 3 (FM03) gene in patients with fish-odour syndrome. *Pharmacogenetics* **10**, 799–807.

Doniger, S. W., Salomonis, N., Dahlquist, K. D., Vranizan, K., Lawlor, S. C., and Conklin, B. R. (2003). MAPPFinder: Using gene ontology and GenMAPP to create a global gene-expression profile from microarray data. *Genome Biol* **4**, R7.

Dosemeci, M., McLaughlin, J. K., Chen, J. Q., Hearl, F., McCawley, M., Wu, Z., Chen, R. G., Peng, K. L., Chen, A. L., Rexing, S. H., et al. (1994). Indirect validation of a retrospective method of exposure assessment used in a nested case-control study of lung cancer and silica exposure. *Occup Environ Med* **51**, 136–138.

Draper, E. (1991). *Risky Busines: Genetic Testing and Exclusionary Practices in the Hazardous Workplace* (Cambridge, UK: Cambridge Univ. Press).

Drewnowski, A., and Gomez-Carneros, C. (2000). Bitter taste, phyotnutrients, and the consumer: A review. *Am J Clin Nutr* **72**, 1424–1435.

Drewnowski, A., Henderson, S. A., and Barratt-Fornell, A. (2001). Genetic taste markers and food preferences. *Drug Metab Dispos* **29**, 535–538.

Druilhe, P., Hagan, P., and Rook, G. A. (2002). The importance of models of infection in the study of disease resistance. *Trends Microbiol* **10**, S38–S46.

Dryer, M., and Rüdiger, H. W. (1988). Genetic defects of human receptor function. *Trends Pharmacol Sci* **9**, 98–102.

Duaux, E., Gorwood, P., Griffon, N., Bourdel, M. C., Sautel, F., Sokoloff, P., Schwartz, J. C., Ades, J., Loo, H., and Poirier, M. F. (1998). Homozygosity at the dopamine D3 receptor gene is associated with opiate dependence. *Mol Psychiatry* **3**, 333–336.

Duggirala, R., Blangero, J., Almasy, L., Dyer, T. D., Williams, K. L., Leach, R. J., O'Connell, P., and Stern, M. P. (1999). Linkage of type 2 diabetes mellitus and of age at onset to a genetic location on chromosome 10q in Mexican Americans. *Am J Hum Genet* **64**, 1127–1140.

Dumenco, L. L., Allay, E., Norton, K., and Gerson, S. L. (1993). The prevention of thymic lymphomas in transgenic mice by human O6-alkylguanine-DNA alkyltransferase. *Science* **259**, 219–222.

Duncan, R. L., Grogan, W. M., Kramer, L. B., and Watlington, C. O. (1988). Corticosterone's metabolite is an agonist for Na+ transport stimulation in A6 cells. *Am J Physiol* **255**, F736–748.

Durfy, S. J., Bowen, D. J., McTiernan, A., Sporleder, J., and Burke, W. (1999). Attitudes and interest in genetic testing for breast and ovarian cancer susceptibility in diverse groups of women in western Washington. *Cancer Epidemiol Biomark Prev* **8**, 369–375.

Eaton, D. L. (2000). Biotransformation enzyme polymorphism and pesticide susceptibility. *Neurotoxicology* **21**, 101–111.

Eaton, D. L., and Bammler, T. K. (1999). Concise review of the glutathione S-transferases and their significance to toxicology. *Toxicol Sci* **49**, 156–164.

Eaton, D. L., Farin, F., Omiecinski, C. J., and Omenn, G. S. (1998). Genetic susceptibility. In *Environmental and Occupational Medicine*, 3rd ed., Rom, W. N., ed. (Philadelphia: Lippincott-Raven), pp. 209–221.

Eaton, D. L., Gallagher, E. P., Bammler, T. K., and Kunze, K. L. (1995). Role of cytochrome P4501A2 in chemical carcinogenesis: implications for human variability in expression and enzyme activity. *Pharmacogenetics* **5**, 259–274.

Efron, B., and Tibshirani, R. (2002). Empirical bayes methods and false discovery rates for microarrays. *Genet Epidemiol* **23**, 70–86.

Eichner, J. E., Dunn, S. T., Perveen, G., Thompson, D. M., Stewart, K. E., and Stroehla, B. C. (2002). Apolipoprotein E polymorphism and cardiovascular disease: A HuGE review. *Am J Epidemiol* **155**, 487–495.

Eigen, M. (2002). Error catastrophe and antiviral strategy. *Proc Natl Acad Sci USA* **99**, 13374–13376.

Eisen, M. B., Spellman, P. T., Brown, P. O., and Botstein, D. (1998). Cluster analysis and display of genome-wide expression patterns. *Proc Natl Acad Sci USA* **95**, 14863–14868.

Eisenbarth, G. S., and Rewers, M. (1995). Refining genetic analysis of type I diabetes. *J Clin Endocrinol Metab* **80**, 2564–2566.

Eisman, J. A. (2001). Pharmacogenetics of the vitamin D receptor and osteoporosis. *Drug Metab Dispos* **29**, 505–512.

El-Masri, H., Bell, D., and Portier, C. (1999). Replication potential of cells via the protein kinase C-MAPK pathway: Application of a mathematical model. *Bull Math Biol* **61**, 379–398.

El-Omar, E. M., Rabkin, C. S., Gammon, M. D., Vaughan, T. L., Risch, H. A., Schoenberg, J. B., Stanford, J. L., Mayne, S. T., Goedert, J., Blot, W. J., et al. (2003). Increased risk of noncardia gastric cancer associated with proinflammatory cytokine gene polymorphisms. *Gastroenterology* **124**, 1193–1201.

Elbein, S. C., Hoffman, M. D., Teng, K., Leppert, M. F., and Hasstedt, S. J. (1999). A genome-wide search for type 2 diabetes susceptibility genes in Utah Caucasians. *Diabetes* **48**, 1175–1182.

Elsby, R., Kitteringham, N. R., Goldring, C. E., Lovatt, C. A., Chamberlain, M., Henderson, C. J., Wolf, C. R., and Park, B. K. (2003). Increased constitutive c-Jun N-terminal kinase signaling in mice lacking glutathione S-transferase Pi. *J Biol Chem* **278**, 22243–22249.

Emilien, G., Maloteaux, J. M., Geurts, M., Hoogenberg, K., and Cragg, S. (1999). Dopamine receptors—physiological understanding to therapeutic intervention potential. *Pharmacol Ther* **84**, 133–156.

Enan, E., Lasley, B., Stewart, D., Overstreet, J., and Vandevoort, C. A. (1996). 2,3,7,8-Tetracgkiridubebzo-r-dioxin (TCDD) modulates function of human luteinizing granulosa cells via cAMP signaling and early reduction of glucose transporting activity. *Reprod Toxicol* **10**, 191–198.

EPA (1996). Guidelines for reproductive toxicity risk assessment. In EPA/630/R-96/009. National Center for Environment Assessment, Office of Research and Development. (Washington, DC).

EPA (June 3, 2003). *Toxic Substances Control Act, Section 8(3): Notice to Administrator of Substantial Risks* (U.S. Environmental Protection Agency), pp. 33129–33140.

Epps, P. G. (2003). Policy before practice: Genetic discrimination reviewed. *Am J Pharmacogenom* **3**, 405–418.

Epstein, M. P., and Satten, G. A. (2003). Inference on haplotype effects in case-control studies using unphased genotype data. *Am J Hum Genet* **73**, 1316–1329.

Erlich, H. A., Gelfand, D., and Sninsky, J. J. (1991). Recent advances in the polymerase chain reaction. *Science* **252**, 1643–1651.

Ewald, P. W. (1993). The evolution of virulence. *Sci Am* **268**, 86–93.

Excoffier, L., and Slatkin, M. (1995). Maximum-likelihood estimation of molecular haplotype frequencies in a diploid population. *Mol Biol Evol* **12**, 921–927.

Faber, D. R., and Krieg, E. J. (2002). Unequal exposure to ecological hazards: Environmental injustices in the Commonwealth of Massachusetts. *Environ Health Perspect* **110**(Suppl 2), 277–288.

Fairbrother, K. S., Grove, J., de Waziers, I., Steimel, D. T., Day, C. P., Crespi, C. L. and Daly, A. K. (1998). Detection and characterization of novel polymorphisms in the CYP2E1 gene. *Pharmacogenetics* **8**, 543–552.

Fallin, D., Cohen, A., Essioux, L., Chumakov, I., Blumenfeld, M., Cohen, D., and Schork, N. J. (2001). Genetic analysis of case/control data using estimated haplotype frequencies: Application to APOE locus variation and Alzheimer's disease. *Genome Res* **11**, 143–151.

Farrell, J. (1998). *Invisible Enemies: Stories of Infectious Disease* (New York: Farrar, Straus, Giroux).

Farrer, L. A., Cupples, L. A., Haines, J. L., Hyman, B., Kukull, W. A., Mayeux, R., Myers, R. H., Pericak-Vance, M. A., Risch, N., and van Duijn, C. M. (1997). Effects of age, sex, and ethnicity on the association between apolipoprotein E genotype and Alzheimer disease. A meta-analysis. APOE and Alzheimer Disease Meta Analysis Consortium. *JAMA* **278**, 1349–1356.

Farrer, M., Maraganore, D. M., Lockhart, P., Singleton, A., Lesnick, T. G., de Andrade, M., West, A., de Silva, R., Hardy, J., and Hernandez, D. (2001). alpha-Synuclein gene haplotypes are associated with Parkinson's disease. *Hum Mol Genet* **10**, 1847–1851.

Farrer, M., Skipper, L., Berg, M., Bisceglio, G., Hanson, M., Hardy, J., Adam, A., Gwinn-Hardy, K., and Aasly, J. (2002). The tau H1 haplotype is associated with Parkinson's disease in the Norwegian population. *Neurosci Lett* **322**, 83–86.

Farrow, D. C., Vaughan, T. L., Sweeney, C., Gammon, M. D., Chow, W. H., Risch, H. A., Stanford, J. L., Hansten, P. D., Mayne, S. T., Schoenberg, J. B., et al. (2000). Gastroesophageal reflux disease, use of H2 receptor antagonists, and risk of esophageal and gastric cancer. *Cancer Causes Control* **11**, 231–238.

Faustman, E., and Omenn, G. (1995). Risk Assessment. In *Casarett and Doull's Toxicology: The Basic Science of Poisons*, 5th edn. Klaassen, C., ed. (New York: McGraw-Hill), pp. 75–88.

Faustman, E. M., Allen, B. C., Kavlock, R. J., and Kimmel, C. A. (1994). Dose-response assessment for developmental toxicity. I. Characterization of database and determination of no observed adverse effect levels. *Fundam Appl Toxicol* **23**, 478–486.

Faustman, E. M., and Omenn, G. S. (2001). Risk assessment. In *Casarett and Doull's Toxicology: The Basic Science of Poisons*, 6th ed., Klaassen, C. D., ed. (New York: McGraw-Hill), pp. 83–104.

Faustman, E. M., Silbernagel, S. M., Fenske, R. A., Burbacher, T. M., and Ponce, R. A. (2000). Mechanisms underlying children's susceptibility to environmental toxicants. *Environ Health Perspect* **108**(Suppl 1), 13–21.

Feinleib, M., Beresford, S. A., Bowman, B. A., Mills, J. L., Rader, J. I., Selhub, J., and Yetley, E. A. (2001). Folate fortification for the prevention of birth defects: Case study. *Am J Epidemiol* **154**, S60–S69.

Fenech, A., and Hall, I. P. (2002). Pharmacogenetics of asthma. *Br J Clin Pharmacol* **53**, 3–15.

Fennell, T., MacNeela, J., Morris, R., Watson, M., Thompson, C., and Bell, D. (2000). Hemoglobin adducts from acrylonitrile and ethylene oxide in cigarette smokers: Effects of glutathione S-transferase T1-null and M1-null genotypes. *Cancer Epidemiol Biomark Prev* **9**, 705–712.

Feychting, M., Jonsson, F., Pedersen, N. L., and Ahlbom, A. (2003). Occupational magnetic field exposure and neurodegenerative disease. *Epidemiology* **14**, 413–419; discussion 427–418.

Feyler, A., Voho, A., Bouchardy, C., Kuokkanen, K., Dayer, P., Hirvonen, A., and Benhamou, S. (2002). Point: Myeloperoxidase –463G → a polymorphism and lung cancer risk. *Cancer Epidemiol Biomark Prev* **11**, 1550–1554.

Finefrock, A. E., Bush, A. I., and Doraiswamy, P. M. (2003). Current status of metals as therapeutic targets in Alzheimer's disease. *J Am Geriatr Soc* **51**, 1143–1148.

Finucane, M. L. (2002). Mad cows, mad corn and mad communities: The role of socio-cultural factors in the perceived risk of genetically-modified food. *Proc Nutr Soc* **61**, 31–37.

Fischhoff, B., Bostrom, A., and Quadrel, M. J. (1993). Risk perception and communication. *Annu Rev Publ Health* **14**, 183–203.

Fontana, R. J., Lown, K. S., Paine, M. F., Fortlage, L., Santella, R. M., Felton, J. S., Knize, M. G., Greenberg, A., and Watkins, P. B. (1999). Effects of a chargrilled meat diet on expression of CYP3A, CYP1A, and P-glycoprotein levels in healthy volunteers. *Gastroenterology* **117**, 89–98.

Food and Drug Administration (2005). Guidance for Industry: Pharmacogenomic Data Submissions. Washington, D.C., Food and Drug Administration, Center for Biologics Evaluation and Research. http://www.fda.gov/cber/gdlns/pharmdtasub.htm.

Fortin, A., Stevenson, M. M., and Gros, P. (2002). Susceptibility to malaria as a complex trait: Big pressure from a tiny creature. *Hum Mol Genet* **11**, 2469–2478.

Foster, C. E., Bianchet, M. A., Talalay, P., Faig, M., and Amzel, L. M. (2000). Structures of mammalian cytosolic quinone reductases. *Free Radic Biol Med* **29**, 241–245.

Fowler, J. S., Volkow, N. D., Wang, G. J., Pappas, N., Logan, J., MacGregor, R., Alexoff, D., Shea, C., Schlyer, D., Wolf, A. P., et al. (1996). Inhibition of monoamine oxidase B in the brains of smokers. *Nature* **379**, 733–736.

Franke, P., Nothen, M. M., Wang, T., Neidt, H., Knapp, M., Lichtermann, D., Weiffenbach, O., Mayer, P., Hollt, V., Propping, P., and Maier, W. (1999a). Human delta-opioid receptor gene and susceptibility to heroin and alcohol dependence. *Am J Med Genet* **88**, 462–464.

Franke, P., Schwab, S. G., Knapp, M., Gansicke, M., Delmo, C., Zill, P., Trixler, M., Lichtermann, D., Hallmayer, J., Wildenauer, D. B., and Maier, W. (1999b). DAT1 gene polymorphism in alcoholism: A family-based association study. *Biol Psychiatry* **45**, 652–654.

Franke, P., Wang, T., Nothen, M. M., Knapp, M., Neidt, H., Albrecht, S., Jahnes, E., Propping, P., and Maier, W. (2001). Nonreplication of association between mu-opioid-receptor gene (OPRM1) A118G polymorphism and substance dependence. *Am J Med Genet* **105**, 114–119.

Freimer, M., Kranzler, H., Satel, S., Lacobelle, J., Skipsey, K., Charney, D., and Gelernter, J. (1996). No association between D3 dopamine receptor (DRD3) alleles and cocaine dependence. *Addict Biol* **1**, 281–287.

Fretland, A. J., and Omiecinski, C. J. (2000). Epoxide hydrolases: Biochemistry and molecular biology. *Chem Biol Interact* **129**, 41–59.

Frewer, L. J., Miles, S., and Marsh, R. (2002). The media and genetically modified foods: Evidence in support of social amplification of risk. *Risk Anal* **22**, 701–711.

Fride, E. (2002). Endocannabinoids in the central nervous system—an overview. *Prostagland Leukotrienes Essent Fatty Acids* **66**, 221–233.

Friedberg, E. C., Wagner, R., and Radman, M. (2002). Specialized DNA polymerases, cellular survival, and the genesis of mutations. *Science* **296**, 1627–1630.

Friedlander, Y., Leitersdorf, E., Vecsler, R., Funke, H., and Kark, J. (2000). The contribution of candidate genes to the response of plasma lipids and lipoproteins to dietary challenge. *Atherosclerosis* **152**, 239–248.

Friedman, M. J., and Trager, W. (1981). The biochemistry of resistance to malaria. *Sci Am* **244**, 154–155, 158–164.

Friso, S., and Choi, S.-W. (2002). Gene-nutrient interactions and DNA methylation. *J Nutr* **132**, 2382S–2387S.

Fu, Y. P., Yu, J. C., Cheng, T. C., Lou, M. A., Hsu, G. C., Wu, C. Y., Chen, S. T., Wu, H. S., Wu, P. E., and Shen, C. Y. (2003). Breast cancer risk associated with genotypic polymorphism of the nonhomologous end-joining genes: A multigenic study on cancer susceptibility. *Cancer Res* **63**, 2440–2446.

Furnes, B., Feng, J., Sommer, S. S., and Schlenk, D. (2003). Identification of novel variants of the flavin-containing monooxygenase gene family in African Americans. *Drug Metab Dispos* **31**, 187–193.

Furnes, B. and Schlenk, D. (2004). Evaluation of xenobiotic N- and S-oxidation by variant flavin-containing monooxygenase 1 (FMO1) enzymes. *Toxicol Sci* **78**, 196–203.

Gabriel, S. B., Schaffner, S. F., Nguyen, H., Moore, J. M., Roy, J., Blumenstiel, B., Higgins, J., DeFelice, M., Lochner, A., Faggart, M., et al. (2002). The structure of haplotype blocks in the human genome. *Science* **296**, 2225–2229.

Gadea, A., and Lopez-Colome, A. M. (2001). Glial transporters for glutamate, glycine, and GABA: II. GABA transporters. *J Neurosci Res* **63**, 461–468.

Galeeva, A. R., Gareeva, A. E., Iur'ev, E. B., and Khusnutdinova, E. K. (2002). [VNTR polymorphisms of the serotonin transporter and dopamine transporter genes in male opiate addicts]. *Molekuliarnaia Biologiia* **36**, 593–598.

Gallagher, E. P., and Gardner, J. L. (2002). Comparative expression of two alpha class glutathione S-transferases in human adult and prenatal liver tissues. *Biochem Pharmacol* **63**, 2025–2036.

Gammon, M. D., Schoenberg, J. B., Ahsan, H., Risch, H. A., Vaughan, T. L., Chow, W. H., Rotterdam, H., West, A. B., Dubrow, R., Stanford, J. L., et al. (1997). Tobacco, alcohol, and socioeconomic status and adenocarcinomas of the esophagus and gastric cardia [see comments]. *J Natl Cancer Inst* **89**, 1277–1284.

Garcia-Closas, M., Kelsey, K. T., Wiencke, J. K., Xu, X., Wain, J. C., and Christiani, D. C. (1997). A case-control study of cytochrome P450 1A1, glutathione S-transferase M1, cigarette smoking and lung cancer susceptibility (Massachusetts, USA). *Cancer Causes Control* **8**, 544–553.

Garcia-Closas, M., and Lubin, J. H. (1999). Power and sample size calculations in case-control studies of gene-environment interactions: Comments on different approaches. *Am J Epidemiol* **149**, 689–692.

Gargas, M., Finley, B., Paustenback, D., and Long, T. (1999). Environmental health risk assessment: Theory and practice. In *General and Applied Toxicology*, Ballantyne, B., Marrs, T., and Syversen, T., eds. (New York: Grove's Dictionaries), pp. 1749–1809.

Garrod, A. E. (1902). Alkaptonuria. *Lancet* 653–656.

Garrod, A. E. (1909). *Inborn Errors of Metabolism* (London: H. Frowde and Hodder & Stoughton).

Garrod, A. E. (1931). *The Inborn Factors in Disease: An Essay* (Oxford: Clarendon Press).

Gelernter, J., Kranzler, H., and Cubells, J. (1999a). Genetics of two mu opioid receptor gene (OPRM1) exon I polymorphisms: Population studies, and allele frequencies in alcohol- and drug-dependent subjects. *Mol Psychiatry* **4**, 476–483.

Gelernter, J., Kranzler, H., and Satel, S. L. (1999b). No association between D2 dopamine receptor (DRD2) alleles or haplotypes and cocaine dependence or severity of cocaine dependence in European- and African-Americans. *Biol Psychiatry* **45**, 340–345.

George, J., Murray, M., Byth, K., and Farrell, G. C. (1995). Differential alterations of cytochrome P450 proteins in livers from patients with severe chronic liver disease. *Hepatology* **21**, 120–128.

Georgiadis, P., Demopoulos, N. A., et al. (2004). Impact of phase I or phase II enzyme polymorphisms on lymphocyte DNA adducts in subjects exposed to urban air pollution and environmental tobacco smoke. *Toxicol Lett* **149**, 269–280.

Gerbal-Chaloin, S., Daujat, M., Pascussi, J. M., Pichard-Garcia, L., Vilarem, M. J., and Maurel, P., (2002). Transcriptional regulation of CYP2C9 gene. Role of glucocorticoid receptor and constitutive androstane receptor. *J Biol Chem* **277**, 209–217.

Ghosh, S. S., Basu, A. K., Ghosh, S., Hagley, R., Kramer, L., Schuetz, J., Grogan, W. M., Guzelian, P. and Watlington, C. O. (1995). Renal and hepatic family 3A cytochromes P450 (CYP3A) in spontaneously hypertensive rats. *Biochem Pharmacol* **50**, 49–54.

Ghosh, S., Watanabe, R. M., Valle, T. T., Hauser, E. R., Magnuson, V. L., Langefeld, C. D., Ally, D. S., Mohlke, K. L., Silander, K., Kohtamaki, K., et al. (2000). The Finland-United States investigation of non-insulin-dependent diabetes mellitus genetics (FUSION) study. I. An autosomal genome scan for genes that predispose to type 2 diabetes. *Am J Hum Genet* **67**, 1174–1185.

Gianoulakis, C. (2001). Influence of the endogenous opioid system on high alcohol consumption and genetic predisposition to alcoholism. Journal of Psychiatry & *Neuroscience* **26**, 304–318.

Gillam, E. M., Guo, Z., Ueng, Y. F., Yamazaki, H., Cock, I., Reilly, P. E., Hooper, W. D., and Guengerich, F. P. (1995). Expression of cytochrome P450 3A5 in Escherichia coli: effects of 5′ modification, purification, spectral characterization, reconstitution conditions, and catalytic activities. *Arch Biochem Biophys* **317**, 374–384.

Ginsberg, H. N., Kris-Etherton, P., Dennis, B., Elmer, P. J., Ershow, A., Lefevre, M., Pearson, T., Roheim, P., Ramakrishnan, R., Reed, R., et al. (1998). Effects of reducing dietary saturated fatty acids on plasma lipids and lipoproteins in healthy subjects: The DELTA Study, protocol 1. *Arterioscler Thromb Vasc Biol* **18**, 441–449.

Gitan, R. S., Shi, H., Chen, C. M., Yan, P. S., and Huang, T. H. (2002). Methylation-specific oligonucleotide microarray: A new potential for high-throughput methylation analysis. *Genome Res* **12**, 158–164.

Givens, R. C., Lin, Y. S., Dowling, A. L., Thummel, K. E., Lamba, J. K., Schuetz, E. G., Stewart, P. W., and Watkins, P. B. (2003). CYP3A5 genotype predicts renal CYP3A activity and blood pressure in healthy adults. *J Appl Physiol* **95**, 1297–1300.

Glatt, H., Boeing, H., Engelke, C. E., Ma, L., Kuhlow, A., Pabel, U., Pomplun, D., Teubner, W., and Meinl, W. (2001). Human cytosolic sulphotransferases: Genetics, characteristics, toxicological aspects. *Mutat Res* **482**, 27–40.

Glatt, H., Engelke, C. E., Pabel, U., Teubner, W., Jones, A. L., Coughtrie, M. W., Andrae, U., Falany, C. N., and Meinl, W. (2000). Sulfotransferases: Genetics and role in toxicology. *Toxicol Lett* **112–113**, 341–348.

Glatt, H., and Meinl, W. (2004). Pharmacogenetics of soluble sulfotransferases (SULTs). *Naunyn-Schmiedebergs Arch Pharmacol* **369**, 55–68 (first published online in 2003 at http://www.springerlink.com).

Glatt, S. J., Faraone, S. V., and Tsuang, M. T. (2003). Association between a functional catechol O-methyltransferase gene polymorphism and schizophrenia: Meta-analysis of case-control and family-based studies. *Am J Psychiatry* **160**, 469–476.

Goedert, M. (2001). Alpha-synuclein and neurodegenerative diseases. *Nat Rev Neurosci* **2**, 492–501.

Goff, D. C., and Coyle, J. T. (2001). The emerging role of glutamate in the pathophysiology and treatment of schizophrenia. *Am J Psychiatry* **158**, 1367–1377.

Goldin, A. L., Barchi, R. L., Caldwell, J. H., Hofmann, F., Howe, J. R., Hunter, J. C., Kallen, R. G., Mandel, G., Meisler, M. H., Netter, Y. B., et al. (2000). Nomenclature of voltage-gated sodium channels. *Neuron* **28**, 365–368.

Goldstein, A. M., Falk, R. T., Korczak, J. F., and Lubin, J. H. (1997). Detecting gene-environment interactions using a case-control design. *Genet Epidemiol* **14**, 1085–1089.

Goldstein, D. B. (2001). Islands of linkage disequilibrium. *Nat Genet* **29**, 109–111.

Golub, T. R., Slonim, D. K., Tamayo, P., Huard, C., Gaasenbeek, M., Mesirov, J. P., Coller, H., Loh, M. L., Downing, J. R., Caligiuri, M. A., et al. (1999). Molecular classification of cancer: Class discovery and class prediction by gene expression monitoring. *Science* **286**, 531–537.

Goode, E. L., Ulrich, C. M., and Potter, J. D. (2002). Polymorphisms in DNA repair genes and associations with cancer risk. *Cancer Epidemiol Biomark Prev* **11**, 1513–1530.

Gooderham, N. J., Murray, S., Lynch, A. M., Yadllahi-Farsani, M., Zhao, K., Boobis, A. R., and Davies, D. S. (2001). Food-derived heterocyclic amine mutagens: Variable metabolism and signficance to humans. *Drug Metab Dispos* **29**, 529–534.

Goodwin, B., Redinbo, M. R., and Kliewer, S. A. (2002). Regulation of cyp3a gene transcription by the pregnane × receptor. *Annu Rev Pharmacol Toxicol* **42**, 1–23.

Gorell, J. M., Johnson, C. C., Rybicki, B. A., Peterson, E. L., Kortsha, G. X., Brown, G. G., and Richardson, R. J. (1997). Occupational exposures to metals as risk factors for Parkinson's disease. *Neurology* **48**, 650–658.

Gorwood, P., Aissi, F., Batel, P., Ades, J., Cohen-Salmon, C., Hamon, M., Boni, C., and Lanfumey, L. (2002). Reappraisal of the serotonin 5-HT(1B) receptor gene in alcoholism: Of mice and men. *Brain Res Bull* **57**, 103–107.

Gorwood, P., Limosin, F., Batel, P., Duaux, E., Gouya, L., and Ades, J. (2001). The genetics of addiction: Alcohol-dependence and D3 dopamine receptor gene. *Pathol Biol* (Paris) **49**, 710–717.

Gostin, L. O., and Hodge, J. G. (1999). Genetic privacy and the law: An end to genetics exceptionalism. *Jurimetrics* **21**, 58.

Grandjean, P. (1991). Ecogenetics: Genetic Predisposition to the Toxic Effects of Chemicals. London: Chapman and Hall.

Greenland, P., Knoll, M. D., Stamler, J., Neaton, J. D., Dyer, A. R., Garside, D. B., and Wilson, P. W. (2003). Major risk factors as antecedents of fatal and nonfatal coronary heart disease events. *JAMA* **290**, 891–897.

Gribble, E. J., Hong, S. W., and Faustman, E. M. (2005). The magnitude of methylmercury-induced cytotoxicity and cell cycle arrest is p53-dependent. *Birth Defects Res A Clin Mol Teratol* **73**, 29–38.

Gropman, A. L., and Batshaw, M. L. (2004). Cognitive outcome in urea cycle disorders. *Mol Genet Metab* **81**, S58–S62.

Grundberg, E., Brandstrom, H., Ribom, E. L., Ljunggren, O., Mallmin, H., and Kindmark, A. (2004). Genetic variation in the human vitamin D receptor is associated with muscle strength, fat mass and body weight in Swedish women. *Eur J Endocrinol* **150**, 323–328.

Guengerich, F. P. (2003a). Cytochrome P450 oxidations in the generation of reactive electrophiles: epoxidation and related reactions. *Arch Biochem Biophys* **409**, 59–71.

Guengerich, F. P. (2003b). Cytochromes P450, drugs, and diseases. *Mol Interv* **3**, 194–204.

Guengerich, F. P., Chun, Y. J., Kim, D., Gillam, E. M., and Shimada, T. (2003). Cytochrome P450 1B1: a target for inhibition in anticarcinogenesis strategies. *Mutat Res* **523–524**, 173–182.

Guengerich, F. P., and Shimada, T. (1998). Activation of procarcinogens by human cytochrome P450 enzymes. *Mutat Res* **400**, 201–213.

Guo, S.-W. (2000). Gene-environment interaction and the mapping of complex traits: Some statistical models and their implications. *Hum Hered* **50**, 286–303.

Gutierrez, P. L. (2000). The role of NAD(P)H oxidoreductase (DT-Diaphorase) in the bioactivation of quinone-containing antitumor agents: A review. *Free Radic Biol Med* **29**, 263–275.

Gutman, S. (1999). The role of Food and Drug Administration regulation of in vitro diagnostic devices—applications to genetics testing. *Clin Chem* **45**, 746–749.

Habermas, J. (1987). *Theory of Communicative Competence*, Vol. II: *Lifeworld and System* (Boston: Beacon Books).

Hadi, M. Z., Coleman, M. A., Fidelis, K., Mohrenweiser, H. W., and Wilson, D. M., 3rd (2000). Functional characterization of Ape1 variants identified in the human population. *Nucleic Acids Res* **28**, 3871–3879.

Hadidi, H. F., Cholerton, S., Atkinson, S., Irshaid, Y. M., Rawashdeh, N. M., and Idle, J. R. (1995). The N-oxidation of trimethylamine in a Jordanian population. *Br J Clin Pharmacol* **39**, 179–181.

Haehner, B. D., Gorski, J. C., Vandenbranden, M., Wrighton, S. A., Janardan, S. K., Watkins, P. B., and Hall, S. D. (1996). Bimodal distribution of renal cytochrome P450 3A activity in humans. *Mol Pharmacol* **50**, 52–59.

Haldane, J. B. S. (1938). *Heredity and Politics* (New York: Norton).

Haldane, J. B. S. (1949). The rate of mutation of human genes. *Proc VIII Int Congr Genet Hereditas* **35**, 267–273.

Haldane, J. B. S. (1954). *Biochemistry of Genetics* (London: Allyn & Unwin).

Haliassos, A., Chomel, J. C., Grandjouan, S., Kruh, J., Kaplan, J. C., and Kitzis, A. (1989). Detection of minority point mutations by modified PCR technique: A new approach for a sensitive diagnosis of tumor-progression markers. *Nucleic Acids Res* **17**, 8093–8099.

Hamadeh, H. K., Bushel, P. R., Jayadev, S., Martin, K., DiSorbo, O., Sieber, S., Bennett, L., Tennant, R., Stoll, R., Barrett, J. C., et al. (2002). Gene expression analysis reveals chemical-specific profiles. *Toxicol Sci* **67**, 219–231.

Hamajima, N., Ito, H., Matsuo, K., Saito, T., Tajima, K., Ando, M., Yoshida, K., and Takahashi, T. (2002). Association between smoking habits and dopamine receptor D2 taqI A A2 allele in Japanese males: A confirmatory study. *J Epidemiol* **12**, 297–304.

Hamman, M. A., Haehner-Daniels, B. D., Wrighton, S. A., Rettie, A. E. and Hall, S. D. (2000). Stereoselective sulfoxidation of sulindac sulfide by flavin-containing monooxygenases. Comparison of human liver and kidney microsomes and mammalian enzymes. *Biochem Pharmacol* **60**, 7–17.

Hanawalt, P. C., and Spivak, G. (1999). Transcription-coupled DNA repair. In *Advances in DNA Damage and Repair*, Dizdaroglu, M., and Karakaya, E., eds. (New York: Plenum).

Hanna, I. H., Dawling, S., Roodi, N., Guengerich, F. P. and Parl, F. F. (2000). Cytochrome P450 1B1 (CYP1B1) pharmacogenetics: association of polymorphisms with functional differences in estrogen hydroxylation activity. *Cancer Res* **60**, 3440–3444.

Hanis, C. L., Boerwinkle, E., Chakraborty, R., Ellsworth, D. L., Concannon, P., Stirling, B., Morrison, V. A., Wapelhorst, B., Spielman, R. S., Gogolin-Ewens, K. J., et al. (1996). A genome-wide search for human non-insulin-dependent (type 2) diabetes genes reveals a major susceptibility locus on chromosome 2. *Nat Genet* **13**, 161–166.

Hanson, R. L., Ehm, M. G., Pettitt, D. J., Prochazka, M., Thompson, D. B., Timberlake, D., Foroud, T., Kobes, S., Baier, L., Burns, D. K., et al. (1998). An autosomal genomic scan for loci linked to type II diabetes mellitus and body-mass index in Pima Indians. *Am J Hum Genet* **63**, 1130–1138.

Harding, J. E. (2001). The nutritional basis of the fetal origins of adult disease. *Int J Epidemiol* **30**, 15–23.

Harper, P. A., Wong, J. Y., Lam, M. S., and Okey, A. B. (2002). Polymorphisms in the human AH receptor. *Chem Biol Interact* **141**, 161–187.

Harries, L. W., Stubbins, M. J., Forman, D., Howard, G. C., and Wolf, C. R. (1997). Identification of genetic polymorphisms at the glutathione S-transferase Pi locus and association with susceptibility to bladder, testicular and prostate cancer. *Carcinogenesis* **18**, 641–644.

Harris, H. (1963). *Garrod's Inborn Errors of Metabolism* (London: Oxford Univ. Press).

Harris, M. J., Coggan, M., Langton, L., Wilson, S. R., and Board, P. G. (1998). Polymorphism of the Pi class glutathione S-transferase in normal populations and cancer patients. *Pharmacogenetics* **8**, 27–31.

Harris, S. B., Gittelsohn, J., Hanley, A., Barnie, A., Wolever, T. M., Gao, J., Logan, A., and Zinman, B. (1997). The prevalence of NIDDM and associated risk factors in native Canadians. *Diabetes Care* **20**, 185–187.

Harrison, T. A., Hindorff, L. A., Kim, H., Wines, R. C., Bowen, D. J., McGrath, B. B., and Edwards, K. L. (2003). Family history of diabetes as a potential public health tool. *Am J Prev Med* **24**, 152–159.

Haseman, J. K., and Elston, R. C. (1972). The investigation of linkage between a quantitative trait and a marker locus. *Behav Genet* **2**, 3–19.

Hashimoto, H., Yanagawa, Y., Sawada, M., Itoh, S., Deguchi, T., and Kamataki, T. (1995). Simultaneous expression of human CYP3A7 and N-acetyltransferase in Chinese hamster CHL cells results in high cytotoxicity for carcinogenic heterocyclic amines. *Arch Biochem Biophys* **320**, 323–329.

Haskell, W. L. (1986). The influence of exercise training on plasma lipids and lipoproteins in health and disease. *Acta Med Scand Suppl* **711**, 25–37.

Hastbacka, J., de la Chapelle, A., Kaitila, I., Sistonen, P., Weaver, A., and Lander, E. (1992). Linkage disequilibrium mapping in isolated founder populations: Diastrophic dysplasia in Finland. *Nat Genet* **2**, 204–211.

Haufroid, V., Mourad, M., et al. (2004). The effect of CYP3A5 and MDR1 (ABCB1) polymorphisms on cyclosporine and tacrolimus dose requirements and trough blood levels in stable renal transplant patients. *Pharmacogenetics* **14**, 147–154.

Haugen, A., Ryberg, D., Mollerup, S., Zienolddiny, S., Skaug, V., and Svendsrud, D. H. (2000). Gene-environment interactions in human lung cancer. *Toxicol Lett* **112–113**, 233–237.

Hayashi, S., Watanabe, J., and Kawajiri, K. (1992). High susceptibility to lung cancer analyzed in terms of combined genotypes of P450IA1 and Mu-class glutathione S-transferase genes. *Jpn J Cancer Res* **83**, 866–870.

Hayes, R. A., and Richardson, B. J. (2001). Biological control of the rabbit in Australia: Lessons not learned? *Trends Microbiol* **9**, 459–460.

Haynes, E. N., Kalkwarf, H. J., Hornung, R., Wenstrup, R., Dietrich, K., and Lanphear, B. P. (2003). Vitamin D receptor Fok1 polymorphism and blood lead concentration in children. *Environ Health Perspect* **111**, 1665–1669.

Heath, A.-L. M., and Fairweather-Tait, S. J. (2003). Health implications of iron overload: The role of diet and genotype. *Nutr Rev* **61**, 45–62.

Heath, S. C. (1997). Markov chain Monte Carlo segregation and linkage analysis for oligogenic models. *Am J Hum Genet* **61**, 748–760.

Health Insurance Portability and Accountability Act (1996). Public Law 104–191. *United States Statutes at Large* **110**, 1936.

Hegele, R. A., Harris, S. B., Zinman, B., Hanley, A. J., and Cao, H. (2001). Absence of association of type 2 diabetes with CAPN10 and PC-1 polymorphisms in Oji-Cree. *Diabetes Care* **24**, 1498–1499.

Hegele, R. A., Sun, F., Harris, S. B., Anderson, C., Hanley, A. J., and Zinman, B. (1999). Genome-wide scanning for type 2 diabetes susceptibility in Canadian Oji-Cree, using 190 microsatellite markers. *J Hum Genet* **44**, 10–14.

Hein, D. W. (2002). Molecular genetics and function of NAT1 and NAT2: Role in aromatic amine metabolism and carcinogenesis. *Mutat Res* **506–507**, 65–77.

Hein, D. W., Doll, M. A., Fretland, A. J., Leff, M. A., Webb, S. J., Xiao, G. H., Devanaboyina, U. S., Nangju, N. A., and Feng, Y. (2000). Molecular genetics and epidemiology of the NAT1 and NAT2 acetylation polymorphisms. *Cancer Epidemiol Biomarkers Prev* **9**, 29–42.

Helbock, H. J., Beckman, K. B., Shigenaga, M. K., Walter, P. B., Woodall, A. A., Yeo, H. C., and Ames, B. N. (1998). DNA oxidation matters: the HPLC-electrochemical detection assay of 8-oxo-deoxyguanosine and 8-oxo-guanine. *Proc Natl Acad Sci USA* **95**, 288–293.

Heller, D. A., de Faire, U., Pedersen, N. L., Dahlen, G., and McClearn, G. E. (1993). Genetic and environmental influences on serum lipid levels in twins. *New Engl J Med* **328**, 1150–1156.

Henegariu, O., Heerema, N. A., Dlouhy, S. R., Vance, G. H., and Vogt, P. H. (1997). Multiplex PCR: Critical parameters and step-by-step protocol. *BioTechniques* **23**, 504–511.

Hengstler, J. G., Arand, M., Herrero, M. E., and Oesch, F. (1998). Polymorphisms of N-acetyltransferases, glutathione S-transferases, microsomal epoxide hydrolase and sulfotransferases: influence on cancer susceptibility. *Recent Results Cancer Res* **154**, 47–85.

Henriksen, G. L., Ketchum, N. S., Michalek, J. E., and Swaby, J. A. (1997). Serum dioxin and diabetes mellitus in veterans of Operation Ranch Hand. *Epidemiology* **8**, 252–258.

Hernan, M. A., Checkoway, H., O'Brien, R., Costa-Mallen, P., De Vivo, I., Colditz, G. A., Hunter, D. J., Kelsey, K. T., and Ascherio, A. (2002a). MAOB intron 13 and COMT codon 158 polymorphisms, cigarette smoking, and the risk of PD. *Neurology* **58**, 1381–1387.

Hernan, M. A., Takkouche, B., Caamano-Isorna, F., and Gestal-Otero, J. J. (2002b). A meta-analysis of coffee drinking, cigarette smoking, and the risk of Parkinson's disease. *Ann Neurol* **52**, 276–284.

Herrington, D. M., and Howard, T. D. (2003). ER-alpha variants and the cardiovascular effects of hormone replacement therapy. *Pharmacogenomics* **4**, 269–277.

Hesselink, D. A., van Schaik, R. H., van der Heiden, I. P., van der Werf, M., Gregoor, P. J., Lindemans, J., Weimar, W., and van Gelder, T. (2003). Genetic polymorphisms of the CYP3A4, CYP3A5, and MDR-1 genes and pharmacokinetics of the calcineurin inhibitors cyclosporine and tacrolimus. *Clin Pharmacol Ther* **74**, 245–254.

Higashi, M. K., Veenstra, D. L., Kondo, L. M., Wittkowsky, A. K., Srinouanprachanh, S. L., Farin, F. M., and Rettie, A. E. (2002). Association between CYP2C9 genetic variants and anticoagulation-related outcomes during warfarin therapy. *Jama* **287**, 1690–1698.

Hildebrandt, M. A., Salavaggione, O. E., Martin, Y. N., Flynn, H. C., Jalal, S., Wieben, E. D., and Weinshilboum, R. M. (2004). Human SULT1A3 pharmacogenetics: Gene duplication and functional genomic studies. *Biochem Biophys Res Commun* **321**, 870–878.

Hill, A. V. (2001). The genomics and genetics of human infectious disease susceptibility. *Annu Rev Genom Hum Genet* **2**, 373–400.

Hines, R. N., Hopp, K. A., Franco, J., Saeian, K., and Begun, F. P. (2002). Alternative processing of the human FMO6 gene renders transcripts incapable of encoding a functional flavin-containing monooxygenase. *Mol Pharmacol* **62**, 320–325.

Hines, R. N., Luo, Z., Hopp, K. A., Cabacungan, E. T., Koukouritaki, S. B., and McCarver, D. G. (2003). Genetic variability at the human FMO1 locus: significance of a basal promoter yin yang 1 element polymorphism (FMO1*6). *J Pharmacol Exp Ther* **306**, 1210–1218.

Hoehe, M. R., Kopke, K., Wendel, B., Rohde, K., Flachmeier, C., Kidd, K. K., Berrettini, W. H., and Church, G. M. (2000a). Sequence variability and candidate gene analysis in complex disease: Association of mu opioid receptor gene variation with substance dependence. *Hum Mol Gen* **9**, 2895–2908.

Hoehe, M. R., Rinn, T., Flachmeier, C., Heere, P., Kunert, H. J., Timmermann, B., Kopke, K., and Ehrenreich, H. (2000b). Comparative sequencing of the human CB1 cannabinoid receptor gene coding exon: no structural mutations in individuals exhibiting extreme responses to cannabis. *Psychiatr Genet* **10**, 173–177.

Holmes, H. C., Burns, S. P., Michelakakis, H., Kordoni, V., Bain, M. D., Chalmers, R. A., Rafter, J. E., and Iles, R. A. (1997). Choline and L-carnitine as precursors of trimethylamine. *Biochem Soc Trans* **25**, 96S.

Holtzman, N. A., and Watson, M. S. (1997). *Promoting Safe and Effective Genetic Testing in the United States: Final Report of the Task Force on Genetic Testing*, PIC ID No. 6090 (Bethesda, MD: National Institutes of Health).

Hopenhayn-Rich, C., Biggs, M. L., and Smith, A. H. (1998). Lung and kidney cancer mortality associated with arsenic in drinking water in Cordoba, Argentina. *Int J Epidemiol* **27**, 561–569.

Horikawa, Y., Oda, N., Cox, N. J., Li, X., Orho-Melander, M., Hara, M., Hinokio, Y., Lindner, T. H., Mashima, H., Schwarz, P. E., et al. (2000). Genetic variation in the gene encoding calpain-10 is associated with type 2 diabetes mellitus. *Nat Genet* **26**, 163–175.

Hou, S. M., Lambert, B., and Hemminki, K. (1995). Relationship between hprt mutant frequency, aromatic DNA adducts and genotypes for GSTM1 and NAT2 in bus maintenance workers. *Carcinogenesis* **16**, 1913–1917.

Houghton, P. (1991). Selective influences and morphological variation amongst Pacific Homo sapiens. *J Hum Evol* **20**, 41–51.

Houlston, R. S. (1999). Glutathione S-transferase M1 status and lung cancer risk: A meta-analysis. *Cancer Epidemiol Biomark Prev* **8**, 675–682.

Houlston, R. S. (2000). CYP1A1 polymorphisms and lung cancer risk: a meta-analysis. *Pharmacogenetics* **10**, 105–114.

Howard, L. A., Ahluwalia, J. S., Lin, S. K., Sellers, E. M., and Tyndale, R. F. (2003). CYP2E1*1D regulatory polymorphism: Association with alcohol and nicotine dependence. *Pharmacogenetics* **13**, 321–328.

Howard, L. A., Sellers, E. M., and Tyndale, R. F. (2002). The role of pharmacogenetically-variable cytochrome P450 enzymes in drug abuse and dependence. *Pharmacogenomics* **3**, 185–199.

Huang, W., Lin, Y. S., McConn, D. J., 2nd, Calamia, J. C., Totah, R. A., Isoherranen, N., Glodowski, M., and Thummel, K. E. (2004). Evidence of significant contribution from cyp3a5 to hepatic drug metabolism. *Drug Metab Dispos* **32**, 1434–1445.

Huang, Z., Guengerich, F. P.,and Kaminsky, L. S. (1998). 16Alpha-hydroxylation of estrone by human cytochrome P4503A4/5. *Carcinogenesis* **19**, 867–872.

Human Cytochrome P450 (CYP) Allele Nomenclature Committee (2004). Nomenclature Files for Human Cytochrome P450 Alleles. http://www.imm.ki.se/CYPalleles.htm.

Humpel, N., Owen, N., and Leslie, E. (2002). Environmental factors associated with adults' participation in physical activity. A review. *Am J Prev Med* **22**, 188–199.

Humphries, S. E., Talmud, P. J., Hawe, E., Bolla, M., Day, I. N., and Miller, G. J. (2001). Apolipoprotein E4 and coronary heart disease in middle-aged men who smoke: A prospective study. *Lancet* **358**, 115–119.

Hung, R. J., Boffetta, P., Brennan, P., Malaveille, C., Hautefeuille, A., Donato, F., Gelatti, U., Spaliviero, M., Placidi, D., Carta, A., et al. (2004). GST, NAT, SULT1A1, CYP1B1 genetic polymorphisms, interactions with environmental exposures and bladder cancer risk in a high-risk population. *Int J Cancer* **110**, 598–604.

Hunt, S. C., Cook, N. R., Oberman, A., Cutler, J. A., Hennekens, C. H., Allender, P. S., Walker, W. G., Whelton, P. K., and Williams, R. R. (1998). Angiotensinogen genotype, sodium reduction, weight loss, and prevention of hypertension: Trials of hypertension prevention, phase II. *Hypertension* **32**, 393–401.

Hutchison, K. E., LaChance, H., Niaura, R., Bryan, A., and Smolen, A. (2002). The DRD4 VNTR polymorphism influences reactivity to smoking cues. *J Abnorm Psychol* **111**, 134–143.

Iivonen, S., Corder, E., Lehtovirta, M., Helisalmi, S., Mannermaa, A., Vepsalainen, S., Hanninen, T., Soininen, H., and Hiltunen, M. (2004). Polymorphisms in the

CYP19 gene confer increased risk for Alzheimer disease. *Neurology* **62**, 1170–1176.

Infante-Rivard, C., Amre, D., and Sinnett, D. (2002). GSTT1 and CYP2E1 polymorphisms and trihalomethanes in drinking water: Effect on childhood leukemia. *Environ Health Perspect* **110**, 591–593.

Infante-Rivard, C., Olson, E., Jacques, L., and Ayotte, P. (2001). Drinking water contaminants and childhood leukemia. *Epidemiology* **12**, 13–19.

Innocenti, F., and Ratain, M. J. (2002). Update on pharmacogenetics in cancer chemotherapy. *Eur J Cancer* **38**, 639–644.

Inoue, T., Kusumi, I., and Yoshioka, M. (2002). Serotonin transporters. *Curr Drug Targets CNS Neurol Disord* **1**, 519–529.

Insel, T. R., and Collins, F. S. (2003). Psychiatry in the genomics era. *Am J Psychiatry* **160**, 616–620.

Institute of Medicine (1999). *Toward Environmental Justice: Research, Education, and Health Policy Needs* (Washington, DC: National Academy Press).

Institute of Medicine (2003). *Unequal Treatment: Confronting Racial and Ethnic Disparities in Health Care* (Washington, DC: National Academy Press).

Inturrisi, C. E. (2002). Clinical pharmacology of opioids for pain. *Clin J Pain* **18**, S3–S13.

Ioannidis, J. P., Ntzani, E. E., Trikalinos, T. A., and Contopoulos-Ioannidis, D. G. (2001). Replication validity of genetic association studies. *Nat Genet* **29**, 306–309.

Ishikawa, H., Ohtsuki, T., Ishiguro, H., Yamakawa-Kobayashi, K., Endo, K., Lin, Y. L., Yanagi, H., Tsuchiya, S., Kawata, K., Hamaguchi, H., and Arinami, T. (1999). Association between serotonin transporter gene polymorphism and smoking among Japanese males. *Cancer Epidemiol Biomark Prev* **8**, 831–833.

Itagaki, K., Carver, G. T., and Philpot, R. M. (1996). Expression and characterization of a modified flavin-containing monooxygenase 4 from humans. *J Biol Chem* **271**, 20102–20107.

Ito, H., Hamajima, N., Matsuo, K., Okuma, K., Sato, S., Ueda, R., and Tajima, K. (2003). Monoamine oxidase polymorphisms and smoking behaviour in Japanese. *Pharmacogenetics* **13**, 73–79.

Jaiswal, A. K. (2000). Regulation of genes encoding NAD(P)H:quinone oxidoreductases. *Free Radic Biol Med* **29**, 254–262.

Janssens, S., and Beyaert, R. (2003). Role of toll-like receptors in pathogen recognition. *Clin Microbiol Rev* **16**, 637–646.

Jarvik, G. P. (1997). Genetic predictors of common disease: Apolipoprotein E genotype as a paradigm. *Ann Epidemiol* **7**, 357–362.

Jarvik, G. P., Jampsa, R., Richter, R. J., Carlson, C. S., Rieder, M. J., Nickerson, D. A., and Furlong, C. E. (2003). Novel paraoxonase (PON1) nonsense and missense mutations predicted by functional genomic assay of PON1 status. *Pharmacogenetics* **13**, 291–295.

Jarvik, G. P., Rozek, L. S., Brophy, V. H., Hatsukami, T. S., Richter, R. J., Schellenberg, G. D., and Furlong, C. E. (2000). Paraoxonase (PON1) phenotype is a better predictor of vascular disease than is PON1(192) or PON1(55) genotype. *Arterioscler Thromb Vasc Biol* **20**, 2441–2447.

Jarvik, G. P., Wijsman, E. M., Kukull, W. A., Schellenberg, G. D., Yu, C., and Larson, E. B. (1995). Interactions of apolipoprotein E genotype, total cholesterol level, age, and sex in prediction of Alzheimer's disease: A case-control study. *Neurology* **45**, 1092–1096.

Jennings, I. G., Cotton, R. G. H., and Kobe, B. (2000). Structural interpretation of mutations in phenylalanine hydroxylase protein aids in identifying genotype-phenotype correlations in phenylketonuria. *Eur J Hum Genet* **8**, 683–696.

Johns, L. E., and Houlston, R. S. (2000). Glutathione S-transferase mu1 (GSTM1) status and bladder cancer risk: A meta-analysis. *Mutagenesis* **15**, 399–404.

Johnson, B. A., and Ait-Daoud, N. (2000). Neuropharmacological treatments for alcoholism: Scientific basis and clinical findings. *Psychopharmacology* **149**, 327–344.

Johnson, G. C., Esposito, L., Barratt, B. J., Smith, A. N., Heward, J., Di Genova, G., Ueda, H., Cordell, H. J., Eaves, I. A., Dudbridge, F., et al. (2001). Haplotype tagging for the identification of common disease genes. *Nat Genet* **29**, 233–237.

Johnson, J. A., Zineh, I., Puckett, B. J., McGorray, S. P., Yarandi, H. N., and Pauly, D. F. (2003). Beta 1-adrenergic receptor polymorphisms and antihypertensive response to metoprolol. *Clin Pharmacol Ther* **74**, 44–52.

Johnson, S. W., and North, R. A. (1992). Opioids excite dopamine neurons by hyperpolarization of local interneurons. *J Neurosci* **12**, 483–488.

Johnstone, E. C., Clark, T. G., Griffiths, S. E., Murphy, M. F., and Walton, R. T. (2002a). Polymorphisms in dopamine metabolic enzymes and tobacco consumption in smokers: Seeking confirmation of the association in a follow-up study. *Pharmacogenetics* **12**, 585–587.

Johnstone, E. C., York, E. E., and Walton, R. T. (2002b). Genetic testing: The future of smoking cessation therapy? *Expert Rev Mol Diagn* **2**, 60–68.

Jonsson, F., and Johanson, G. (2001). A Bayesian analysis of the influence of GSTT1 polymorphism on the cancer risk estimate for dichloromethane. *Toxicol Appl Pharmacol* **174**, 99–112.

Jorm, A. F., Henderson, A. S., Jacomb, P. A., Christensen, H., Korten, A. E., Rodgers, B., Tan, X., and Easteal, S. (2000). Association of smoking and personality with a polymorphism of the dopamine transporter gene: Results from a community survey. *Am J Med Genet* **96**, 331–334.

Jugessur, A., Wilcox, A. J., Lie, R. T., Murray, J. C., Taylor, J. A., Ulvik, A., Drevon, C. A., Vindenes, H. A., and Abyholm, F. E. (2003). Exploring the effects of methylenetetrahydrofolate reductase gene variants C677T and A1298C on the risk of orofacial clefts in 261 Norwegian case-parent triads. *Am J Epidemiol* **157**, 1083–1091.

Kakourou, T., Bakoula, C., Kavadias, G., Gatos, A., Bilalis, L., Krikos, X., and Matsaniotis, N. (1995). Mothers' knowledge and practices related to sun protection in Greece. *Pediatr Dermatol* **12**, 207–210.

Kallioniemi, A., Kallioniemi, O. P., Sudar, D., Rutovitz, D., Gray, J. W., Waldman, F., and Pinkel, D. (1992). Comparative genomic hybridization for molecular cytogenetic analysis of solid tumors. *Science* **258**, 818–821.

Kalow, W. (1962). *Pharmacogenetics: Heredity and Response to Drugs* (Philadelphia: Saunders).

Kalow, W. (1990). Pharmacogenetics: Past and future. *Life Sci* **47**, 1385–1397.

Kalow, W., and Staron, N. (1957). On distribution and inheritance of atypical forms of human serum cholinesterase, as indicated by dibucaine numbers. *Can J Med Sci* **35**, 1305–1320.

Kamboh, M. I., Aston, C. E., and Hamman, R. F. (2000). DNA sequence variation in human apolipoprotein C4 gene and its effect on plasma lipid profile. *Atherosclerosis* **152**, 193–201.

Kamel, F., Umbach, D. M., Lehman, T. A., Park, L. P., Munsat, T. L., Shefner, J. M., Sandler, D. P., Hu, H., and Taylor, J. A. (2003). Amyotrophic lateral sclerosis, lead, and genetic susceptibility: polymorphisms in the delta-aminolevulinic acid dehydratase and vitamin D receptor genes. *Environ Health Perspect* **111**, 1335–1339.

Kamel, F., Umbach, D. M., Munsat, T. L., Shefner, J. M., Hu, H., and Sandler, D. P. (2002). Lead exposure and amyotrophic lateral sclerosis. *Epidemiology* **13**, 311–319.

Kamel, F., Umbach, D. M., Munsat, T. L., Shefner, J. M., and Sandler, D. P. (1999). Association of cigarette smoking with amyotrophic lateral sclerosis. *Neuroepidemiology* **18**, 194–202.

Kang, J. H., Chung, W. G., et al. (2000). Phenotypes of flavin-containing monooxygenase activity determined by ranitidine N-oxidation are positively correlated with genotypes of linked FM03 gene mutations in a Korean population. *Pharmacogenetics* **10**, 67–78.

Kaprio, J., Ferrell, R. E., Kottke, B. A., Kamboh, M. I., and Sing, C. F. (1991). Effects of polymorphisms in apolipoproteins E, A-IV, and H on quantitative traits related to risk for cardiovascular disease. *Arterioscler Thromb* **11**, 1330–1348.

Katz, R. M., and Jowett, D. (1981). Female laundry and dry cleaning workers in Wisconsin: A mortality analysis. *Am J Publ Health* **71**, 305–307.

Kaufmann, M. B. (1999). Genetic discrimination in the workplace: An overview of existing protections. *Loyola Univ Chicago Law J* **30**, 393–438.

Kavlock, R. J., Allen, B. C., Faustman, E. M., and Kimmel, C. A. (1995). Dose-response assessments for developmental toxicity. IV. Benchmark doses for fetal weight changes. *Fundam Appl Toxicol* **26**, 211–222.

Kelada, S. N., Costa-Mallen, P., Costa, L. G., Smith-Weller, T., Franklin, G. M., Swanson, P. D., Longstreth, W. T., Jr., and Checkoway, H. (2002). Gender difference in the interaction of smoking and monoamine oxidase B intron 13 genotype in Parkinson's disease. *Neurotoxicology* **23**, 515–519.

Kelada, S. N., Eaton, D. L., Wang, S. S., Rothman, N., and Khoury, M. J. (2004). Applications of human genome epidemiology to environmental health. In *Human Genome Epidemiology: A Scientific Foundation for Using Genetic Information to*

Improve Health and Prevent Disease (Oxford, UK: Oxford Univ. Press), pp. 145–167.

Kelley, J. R., and Duggan, J. M. (2003). Gastric cancer epidemiology and risk factors. *J Clin Epidemiol* **56**, 1–9.

Kendler, K. S., Jacobson, K. C., Prescott, C. A., and Neale, M. C. (2003). Specificity of genetic and environmental risk factors for use and abuse/dependence of cannabis, cocaine, hallucinogens, sedatives, stimulants, and opiates in male twins. *Am J Psychiatry* **160**, 687–695.

Kerr, M. K. (2003). Design considerations for efficient and effective microarray studies. *Biometrics* **59**, 822–828.

Kerr, M. K., and Churchill, G. A. (2001). Bootstrapping cluster analysis: Assessing the reliability of conclusions from microarray experiments. *Proc Natl Acad Sci USA* **98**, 8961–8965.

Kerr, M. K., Martin, M., and Churchill, G. A. (2000). Analysis of variance for gene expression microarray data. *J Comput Biol* **7**, 819–837.

Keshava, C., McCanlies, E. C., and Weston, A. (2004). CYP3A4 polymorphisms–potential risk factors for breast and prostate cancer: a HuGE review. *Am J Epidemiol* **160**, 825–841.

Khalil, A. A., Steyn, S., and Castagnoli, N., Jr. (2000). Isolation and characterization of a monoamine oxidase inhibitor from tobacco leaves. *Chem Res Toxicol* **13**, 31–35.

Khanna, K. K., and Jackson, S. P. (2001). DNA double-strand breaks: signaling, repair and the cancer connection. *Nat Genet* **27**, 247–254.

Khot, U. N., Khot, M. B., Bajzer, C. T., Sapp, S. K., Ohman, E. M., Brener, S. J., Ellis, S. G., Lincoff, A. M., and Topol, E. J. (2003). Prevalence of conventional risk factors in patients with coronary heart disease. *JAMA* **290**, 898–904.

Khoury, M. J., Beaty, T. H., and Cohen, B. H. (1993). *Fundamentals of Genetic Epidemiology* (New York: Oxford Univ. Press).

Khoury, M. J., Burke, W., and Thomson, E., Eds. (2000). Genetics and Public Health in the 21st Century. New York, Oxford University Press.

Khoury, M. J., and Flanders, W. D. (1996). Nontraditional epidemiologic approaches in the analysis of gene-environment interaction: Case-control studies with no controls! *Am J Epidemiol* **144**, 207–213.

Khoury, M. J., Little, J., and Burke, W., eds. (2004). *Human Genome Epidemiology: A Scientific Foundation for Using Genetic Information to Improve Health and Prevent Disease* (New York: Oxford Univ. Press).

Kim, D. and Guengerich, F. P. (2005). Cytochrome P450 activation of arylamines and heterocyclic amines. *Annu Rev Pharmacol Toxicol* **45**, 27–49.

Kimber, I., and Dearman, R. J. (2002). Factors affecting the development of food allergy. *Proc Nutr Soc* **61**, 435–439.

King, C. D., Rios, G. R., Assouline, J. A., and Tephly, T. R. (1999a). Expression of UDP-glucuronosyltransferases (UGTs) 2B7 and 1A6 in the human brain and identification of 5-hydroxytryptamine as a substrate. *Arch Biochem Biophys* **365**, 156–162.

King, H., Aubert, R. E., and Herman, W. H. (1998). Global burden of diabetes, 1995–2025: Prevalence, numerical estimates, and projections. *Diabetes Care* **21**, 1414–1431.

Kim, S. Y., Choi, J. K., Cho, Y. H., Chung, E. J., Paek, D. and Chung, H. W. (2004). Chromosomal aberrations in workers exposed to low levels of benzene: association with genetic polymorphisms. *Pharmacogenetics* **14**, 453–463.

Kirk, S. F., Greenwood, D., Cade, J. E., and Pearman, A. D. (2002). Public perception of a range of potential food risks in the United Kingdom. *Appetite* **38**, 189–197.

Kirke, P. N., Molloy, A. M., Daly, L. E., Burke, H., Weir, D. G., and Scott, J. M. (1993). Maternal plasma folate and vitamin B12 are independent risk factors for neural tube defects. *Quart J Med* **86**, 703–708.

Kiss, I., Sándor, J., and Ember, I. (2000). Allelic polymorphism of GSTM1 and NAT2 genes modifies dietary-induced DNA damage in colorectal mucosa. *Eur J Cancer Prev* **9**, 429–432.

Kiyohara, C., Otsu, A., Shirakawa, T., Fukuda, S., and Hopkin, J. M. (2002). Genetic polymorphisms and lung cancer susceptibility: A review. *Lung Cancer* **37**, 241–256.

Klausner, R. D. (1999). Foreword. *J Natl Cancer Inst Monogr* **25**, 1.

Kobayashi, K., Yamamoto, T., Chiba, K., Tani, M., Shimada, N., Ishizaki, T., and Kuroiwa, Y. (1998). Human buprenorphine N-dealkylation is catalyzed by cytochrome P450 3A4. *Drug Metab Dispos* **26**, 818–821.

Koepsell, T. D., and Weiss, N. S. (2003). *Epidemiologic Methods: Studying the Occurrence of Illness* (New York: Oxford Univ. Press).

Kolonel, L. N., Henderson, B. E., Hankin, J. H., Nomura, A. M., Wilkens, L. R., Pike, M. C., Stram, D. O., Monroe, K. R., Earle, M. E., and Nagamine, F. S. (2000). A multiethnic cohort in Hawaii and Los Angeles: Baseline characteristics. *Am J Epidemiol* **151**, 346–357.

Koob, G. F., and Le Moal, M. (1997). Drug abuse: Hedonic homeostatic dysregulation. *Science* **278**, 52–58.

Koob, G. F., and Le Moal, M. (2001). Drug addiction, dysregulation of reward, and allostasis. *Neuropsychopharmacology* **24**, 97–129.

Koskela, S., Hakkola, J., et al. (1999). Expression of CYP2A genes in human liver and extrahepatic tissues. *Biochem Pharmacol* **57**, 1407–1413.

Kostraba, J. N., Gay, E. C., Rewers, M., and Hamman, R. F. (1992). Nitrate levels in community drinking waters and risk of IDDM. An ecological analysis. *Diabetes Care* **15**, 1505–1508.

Kotler, M., Cohen, H., Segman, R., Gritsenko, I., Nemanov, L., Lerer, B., Kramer, I., Zer-Zion, M., Kletz, I., and Ebstein, R. P. (1997). Excess dopamine D4 receptor (D4DR) exon III seven repeat allele in opioid-dependent subjects. *Mol Psychiatry* **2**, 251–254.

Koukouritaki, S. B., Simpson, P., Yeung, C. K., Rettie, A. E., and Hines, R. N. (2002). Human hepatic flavin-containing monooxygenases 1 (FMO1) and 3 (FMO3) developmental expression. *Pediatr Res* **51**, 236–243.

Kramer, C. B., Cullen, A. C., and Faustman, E. M. (2004). Policy implications of genetic information in regulation under the Clean Air Act: The case of particulate matter and asthmatics [abstract]. Society For Risk Analysis Annual Meeting 2004, Palms Springs, CA, Society For Risk Analysis.

Kranzler, H., Lappalainen, J., Nellissery, M., and Gelernter, J. (2002a). Association study of alcoholism subtypes with a functional promoter polymorphism in the serotonin transporter protein gene. *Alcoholism Clin Exp Res* **26**, 1330–1335.

Kranzler, H. R., Hernandez-Avila, C. A., and Gelernter, J. (2002b). Polymorphism of the 5-HT1B receptor gene (HTR1B): Strong within-locus linkage disequilibrium without association to antisocial substance dependence. *Neuropsychopharmacology* **26**, 115–122.

Krause, R. M. (2002). Evolving microbes and re-emerging streptococcal disease. *Clin Lab Med* **22**, 835–848.

Krause, R. J., Lash, L. H., and Elfarra, A. A. (2003). Human kidney flavin-containing monooxygenases and their potential roles in cysteine s-conjugate metabolism and nephrotoxicity. *J Pharmacol Exp Ther* **304**, 185–191.

Kroetz, D. L. and Xu, F. (2004). Regulation and inhibition of arachidonic acid-hydroxylases and 20-HETE formation. *Annu Rev Pharmacol Toxicol* **45**, 413–438.

Kroeze, W. K., Kristiansen, K., and Roth, B. L. (2002). Molecular biology of serotonin receptors structure and function at the molecular level. *Curr Topics Med Chem* **2**, 507–528.

Kruger, R., Vieira-Saecker, A. M., Kuhn, W., Berg, D., Muller, T., Kuhnl, N., Fuchs, G. A., Storch, A., Hungs, M., Woitalla, D., et al. (1999). Increased susceptibility to sporadic Parkinson's disease by a certain combined alpha-synuclein/apolipoprotein E genotype. *Ann Neurol* **45**, 611–617.

Krueger, S. K., Martin, S. R., Yueh, M. F., Pereira, C. B., and Williams, D. E. (2002). Identification of active flavin-containing monooxygenase isoform 2 in human lung and characterization of expressed protein. *Drug Metab Dispos* **30**, 34–41.

Kuehl, P., Zhang, J., et al. (2001). Sequence diversity in CYP3A promoters and characterization of the genetic basis of polymorphic CYP3A5 expression. *Nat Genet* **27**, 383–391.

Kuiper, G. G., Lemmen, J. G., Carlsson, B., Corton, J. C., Safe, S. H., van der Saag, P. T., van der Burg, B., and Gustafsson, J. A. (1998). Interaction of estrogenic chemicals and phytoestrogens with estrogen receptor beta. *Endocrinology* **139**, 4252–4263.

Kukull, W. A. (2001). The association between smoking and Alzheimer's disease: Effects of study design and bias. *Biol Psychiatry* **49**, 194–199.

Kukull, W. A., Larson, E. B., Bowen, J. D., McCormick, W. C., Teri, L., Pfanschmidt, M. L., Thompson, J. D., O'Meara, E. S., Brenner, D. E., and van Belle, G. (1995).

Solvent exposure as a risk factor for Alzheimer's disease: A case-control study. *Am J Epidemiol* **141**, 1059–1071; discussion 1072–1059.

Kwiatkowski, D. (2000). Genetic susceptibility to malaria getting complex. *Curr Opin Genet Devel* **10**, 320–324.

Laden, F., Ishibe, N., Hankinson, S. E., Wolff, M. S., Gertig, D. M., Hunter, D. J., and Kelsey, K. T. (2002). Polychlorinated biphenyls, cytochrome P450 1A1, and breast cancer risk in the Nurses' Health Study. *Cancer Epidemiol Biomark Prev* **11**, 1560–1565.

Ladona, M. G., Gonzalez, M. L., Rane, A., Peter, R. M., and de la Torre, R. (2000). Cocaine metabolism in human fetal and adult liver microsomes is related to cytochrome P450 3A expression. *Life Sci* **68**, 431–443.

Lamartiniere, C. A., Cotroneo, M. S., Fritz, W. A., Wang, J., Mentor-Marcel, R., and Elgavish, A. (2002). Genistein chemoprevention: Timing and mechanisms of action in murine mammary and prostate. *J Nutr* **132**, 552S–558S.

Lamba, J. K., Lin, Y. S., Schuetz, E. G., and Thummel, K. E. (2002). Genetic contribution to variable human CYP3A-mediated metabolism. *Adv Drug Deliv Rev* **54**, 1271–1294.

Lamba, J. K., Lin, Y. S., Thummel, K., Daly, A., Watkins, P. B., Strom, S., Zhang, J., and Schuetz, E. G. (2002). Common allelic variants of cytochrome P4503A4 and their prevalence in different populations. *Pharmacogenetics* **12**, 121–132.

Lampe, J. W. and Peterson, S. (2002). Brassica, biotransformation and cancer risk: genetic polymorphisms alter the preventive effects of cruciferous vegetables. *J Nutr* **132**, 2991–2994.

Lan, Q., Rothman, N., Chow, W. H., Lissowska, J., Doll, M. A., Xiao, G. H., Zatonski, W., and Hein, D. W. (2003). No apparent association between NAT1 and NAT2 genotypes and risk of stomach cancer. *Cancer Epidemiol Biomark Prev* **12**, 384–386.

Lan, Q., Zhang, L., et al. (2004). Hematotoxicity in workers exposed to low levels of benzene. *Science* **306**, 1774–1776.

Lander, E., and Kruglyak, L. (1995). Genetic dissection of complex traits: Guidelines for interpreting and reporting linkage results. *Nat Genet* **11**, 241–247.

Lander, E. S., Linton, L. M., Birren, B., Nusbaum, C., Zody, M. C., Baldwin, J., Devon, K., Dewar, K., Doyle, M., FitzHugh, W., et al. (2001). Initial sequencing and analysis of the human genome. *Nature* **409**, 860–921.

Lang, N. P., Butler, M. A., Massengill, J., Lawson, M., Stotts, R. C., Hauer-Jensen, M., and Kadlubar, F. F. (1994). Rapid metabolic phenotypes for acetyltransferase and cytochrome P4501A2 and putative exposure to food-borne heterocyclic amines increase the risk for colorectal cancer or polyps. *Cancer Epidemiol Biomark Prev* **3**, 675–682.

Lang, D. H. and Rettie, A. E. (1998). Isoform selective in vitro probes for human FMO1 and FMO4. 5th International ISSX Meeting, Cairns, Australia.

Langsenlehner, U., Krippl, P., Renner, W., Yazdani-Biuki, B., Eder, T., Wolf, G., Wascher, T. C., Paulweber, B., Weitzer, W., and Samonigg, H. (2004). Genetic variants of the sulfotransferase 1A1 and breast cancer risk. *Breast Cancer Res Treat* **87**, 19–22.

Langston, J. W., Ballard, P., Tetrud, J. W., and Irwin, I. (1983). Chronic Parkinsonism in humans due to a product of meperidine-analog synthesis. *Science* **219**, 979–980.

Lash, T. L., and Fink, A. K. (2003). Semi-automated sensitivity analysis to assess systematic errors in observational data. *Epidemiology* **14**, 451–458.

Launoy, G., Milan, C., Day, N. E., Faivre, J., Pienkowski, P., and Gignoux, M. (1997). Oesophageal cancer in France: Potential importance of hot alcoholic drinks. *Int J Cancer* **71**, 917–923.

Lave, L. B., and Omenn, G. S. (1981). *Clearing the Air: Reforming the Clean Air Act: A Staff Paper* (Washington, DC: Brookings Inst.).

Lave, L. B., and Omenn, G. S. (1986). Cost-effectiveness of short-term tests for carcinogenicity. *Nature* **324**, 29–34.

Lawford, B. R., Young, R. M., Noble, E. P., Sargent, J., Rowell, J., Shadforth, S., Zhang, X., and Ritchie, T. (2000). The D(2) dopamine receptor A(1) allele and opioid dependence: Association with heroin use and response to methadone treatment. *Am J Med Genet* **96**, 592–598.

Lawton, M. P., Cashman, J. R., et al. (1994). A nomenclature for the mammalian flavin-containing monooxygenase gene family based on amino acid sequence identities. *Arch Biochem Biophys* **308**, 254–257.

Le Marchand, L., Donlon, T., Lum-Jones, A., Seifried, A., and Wilkens, L. R. (2002). Association of the hOGG1 Ser326Cys polymorphism with lung cancer risk. *Cancer Epidemiol Biomarkers Prev* **11**, 409–412.

Le Marchand, L., Seifried, A., Lum, A. and Wilkens, L. R. (2000). Association of the myeloperoxidase-463G→a polymorphism with lung cancer risk. *Cancer Epidemiol Biomarkers Prev* **9**, 181–184.

Le Marchand, L., Sivaraman, L., Pierce, L., Seifried, A., Lum, A., Wilkens, L. R., and Lau, A. F. (1998). Associations of CYP1A1, GSTM1, and CYP2E1 polymorphisms with lung cancer suggest cell type specificities to tobacco carcinogens. *Cancer Res* **58**, 4858–4863.

Le Marchand, L., Wilkinson, G. R. and Wilkens, L. R. (1999). Genetic and dietary predictors of CYP2E1 activity: a phenotyping study in Hawaii Japanese using chlorzoxazone. *Cancer Epidemiol Biomarkers Prev* **8**, 495–500.

Le Novere, N., Corringer, P. J., and Changeux, J. P. (2002). The diversity of subunit composition in nAChRs: Evolutionary origins, physiologic and pharmacologic consequences. *J Neurobiol* **53**, 447–456.

Leclercq, G. (2002). Molecular forms of the estrogen receptor in breast cancer. *J Steroid Biochem Mol Biol* **80**, 259–272.

Lee, A. J., Kosh, J. W., Conney, A. H., and Zhu, B. T. (2001). Characterization of the NADPH-dependent metabolism of 17beta-estradiol to multiple metabolites by human liver microsomes and selectively expressed human cytochrome P450 3A4 and 3A5. *J Pharmacol Exp Ther* **298**, 420–432.

Lee, H. S. (2003). Gender-specific molecular heterosis and association studies: Dopamine D2 receptor gene and smoking. *Am J Med Genet* **118B**, 55–59.

Lee, M. L., Kuo, F. C., Whitmore, G. A., and Sklar, J. (2000). Importance of replication in microarray gene expression studies: Statistical methods and evidence from repetitive cDNA hybridizations. *Proc Natl Acad Sci USA* **97**, 9834–9839.

Lee, W. J., Brennan, P., Boffetta, P., London, S. J., Benhamou, S., Rannug, A., To-Figueras, J., Ingelman-Sundberg, M., Shields, P., Gaspari, L., and Taioli, E. (2002). Microsomal epoxide hydrolase polymorphisms and lung cancer risk: A quantitative review. *Biomarkers* **7**, 230–241.

Lehmann, H., and Ryan, E. (1956). The familial incidence of low pseudocholinesterase level. *Lancet* **271**, 124.

Leineweber, K., and Brodde, O. E. (2004). Beta2-adrenoceptor polymorphisms: Relation between in vitro and in vivo phenotypes. *Life Sci* **74**, 2803–2814.

LeMarquand, D., Pihl, R. O., and Benkelfat, C. (1994). Serotonin and alcohol intake, abuse, and dependence: Clinical evidence. *Biol Psychiatry* **36**, 326–337.

Leonard, S., and Bertrand, D. (2001). Neuronal nicotinic receptors: From structure to function. *Nicotine Tob Res* **3**, 203–223.

Lesch, K. P., Bengel, D., Heils, A., Sabol, S. Z., Greenberg, B. D., Petri, S., Benjamin, J., Muller, C. R., Hamer, D. H., and Murphy, D. L. (1996). Association of anxiety-related traits with a polymorphism in the serotonin transporter gene regulatory region. *Science* **274**, 1527–1531.

Levy-Lahad, E., Lahad, A., Eisenberg, S., Dagan, E., Paperna, T., Kasinetz, L., Catane, R., Kaufman, B., Beller, U., Renbaum, P., and Gershoni-Baruch, R. (2001). A single nucleotide polymorphism in the RAD51 gene modifies cancer risk in BRCA2 but not BRCA1 carriers. *Proc Natl Acad Sci USA* **98**, 3232–3236.

Lewis, F. M., and Hammond, M. A. (1996). The father's, mother's, and adolescent's functioning with breast cancer. *Family Relations: J Appl Family Child Stud* **45**, 456–465.

Li, M. D., Cheng, R., Ma, J. Z., and Swan, G. E. (2003a). A meta-analysis of estimated genetic and environmental effects on smoking behavior in male and female adult twins. *Addiction* **98**, 23–31.

Li, T., Liu, X., Zhu, Z. H., Zhao, J., Hu, X., Ball, D. M., Sham, P. C., and Collier, D. A. (2000a). No association between (AAT)n repeats in the cannabinoid receptor gene (CNR1) and heroin abuse in a Chinese population. *Mol Psychiatry* **5**, 128–130.

Li, T., Xu, K., Deng, H., Cai, G., Liu, J., Liu, X., Wang, R., Xiang, X., Zhao, J., Murray, R. M., et al. (1997). Association analysis of the dopamine D4 gene exon III VNTR and heroin abuse in Chinese subjects. *Mol Psychiatry* **2**, 413–416.

Li, W. F., Costa, L. G., Richter, R. J., Hagen, T., Shih, D. M., Tward, A., Lusis, A. J., and Furlong, C. E. (2000b). Catalytic efficiency determines the in-vivo efficacy of PON1 for detoxifying organophosphorus compounds. *Pharmacogenetics* **10**, 767–779.

Liddle, C., Goodwin, B. J., George, J., Tapner, M., and Farrell, G. C. (1998). Separate and interactive regulation of cytochrome P450 3A4 by triiodothyronine, dexamethasone, and growth hormone in cultured hepatocytes. *J Clin Endocrinol Metab* **83**, 2411–2416.

Lieber, C. S. (1994). Hepatic and metabolic effects of ethanol: Pathogenesis and prevention. *Ann Med* **26**, 325–330.

Lieber, C. S. (1997). Cytochrome P-4502E1: its physiological and pathological role. *Physiol Rev* **77**, 517–544.

Lieber, C. S. (2004). The discovery of the microsomal ethanol oxidizing system and its physiologic and pathologic role. *Drug Metab Rev* **36**, 511–529.

Liggett, S. B. (2000). Pharmacogenetics of beta-1- and beta-2-adrenergic receptors. *Pharmacology* **61**, 167–173.

Liggett, S. B., Wagoner, L. E., Craft, L. L., Hornung, R. W., Hoit, B. D., McIntosh, T. C., and Walsh, R. A. (1998). The Ile164 beta2-adrenergic receptor polymorphism adversely affects the outcome of congestive heart failure. *J Clin Invest* **102**, 1534–1539.

Limosin, F., Gorwood, P., Loze, J. Y., Dubertret, C., Gouya, L., Deybach, J. C., and Ades, J. (2002). Male limited association of the dopamine receptor D2 gene TaqI a polymorphism and alcohol dependence. *Am J Med Genet* **112**, 343–346.

Lin, J. and Cashman, J. R. (1997a). Detoxication of tyramine by the flavin-containing monooxygenase: stereoselective formation of the trans oxime. *Chem Res Toxicol* **10**, 842–852.

Lin, J. and Cashman, J. R. (1997b). N-oxygenation of phenethylamine to the trans-oxime by adult human liver flavin-containing monooxygenase and retroreduction of phenethylamine hydroxylamine by human liver microsomes. *J Pharmacol Exp Ther* **282**, 1269–1279.

Lin, H. J., Probst-Hensch, N. M., Louie, A. D., Kau, I. H., Witte, J. S., Ingles, S. A., Frankl, H. D., Lee, E. R., and Haile, R. W. (1998). Glutathione transferase null genotype, broccoli, and lower prevalence of colorectal adenomas. *Cancer Epidemiol Biomark Prev* **7**, 647–652.

Lin, Y. S., Dowling, A. L., Quigley, S. D., Farin, F. M., Zhang, J., Lamba, J., Schuetz, E. G., and Thummel, K. E. (2002). Co-regulation of CYP3A4 and CYP3A5 and contribution to hepatic and intestinal midazolam metabolism. *Mol Pharmacol* **62**, 162–172.

Lindahl, T., and Nyberg, B. (1972). Rate of depurination of native deoxyribonucleic acid. *Biochemistry* **11**, 3610–3618.

Ling, L. L., Keohavong, P., Dias, C., and Thilly, W. G. (1991). Optimization of the polymerase chain reaction with regard to fidelity: Modified T7, Taq, and vent DNA polymerases. *PCR Meth Appl* **1**, 63–69.

Lioy, P., Leaderer, B., Graham, J., Lebret, E., Sheldon, L., Needham, L., Pellizzari, E., and Lebowitz, M. (2005). The Major Themes from the Plenary Panel Session of the International Society of Exposure Analysis—2004 Annual Meeting on: The Application of Exposure Assessment to Environmental Health Science and

Public Policy—What has been Accomplished and What Needs to Happen before Our 25th Anniversary in 2014. *J Expo Anal Environ Epidemiol* **15**, 121–122.

Listgarten, J., Damaraju, S., et al. (2004). Predictive models for breast cancer susceptibility from multiple single nucleotide polymorphisms. *Clin Cancer Res* **10**, 2725–2737.

Livak, K. J., Flood, S. J., Marmaro, J., Giusti, W., and Deetz, K. (1995). Oligonucleotides with fluorescent dyes at opposite ends provide a quenched probe system useful for detecting PCR product and nucleic acid hybridization. *PCR Meth Appl* **4**, 357–362.

Liu, G., Miller, D. P., et al. (2001). Differential association of the codon 72 p53 and GSTM1 polymorphisms on histological subtype of non-small cell lung carcinoma. *Cancer Res* **61**, 8718–8722.

Lochner, K., Kawachi, I., and Kennedy, B. P. (1999). Social capital: A guide to its measurement. *Health Place* **5**, 259–270.

Lockridge, O., and Masson, P. (2000). Pesticides and susceptible populations: People with butyrylcholinesterase genetic variants may be at risk. *Neurotoxicology* **21**, 113–126.

Loeb, L. A. (2001). A mutator phenotype in cancer. *Cancer Res* **61**, 3230–3239.

Loeb, L. A., Loeb, K. R., and Anderson, J. P. (2003). Multiple mutations and cancer. *Proc Natl Acad Sci USA* **100**, 776–781.

Loeb, L. A., Preston, B. D., Snow, E. T., and Schaaper, R. M. (1986). Apurinic sites as common intermediates in mutagenesis. *Basic Life Sci* **38**, 341–347.

London, S. J., Daly, A. K., Leathart, J. B., Navidi, W. C., Carpenter, C. C., and Idle, J. R. (1997). Genetic polymorphism of CYP2D6 and lung cancer risk in African-Americans and Caucasians in Los Angeles County. *Carcinogenesis* **18**, 1203–1214.

London, S. J., Idle, J. R., Daly, A. K., and Coetzee, G. A. (1999). Genetic variation of CYP2A6, smoking, and risk of cancer. *Lancet* **353**, 898–899.

Long, J. C., Knowler, W. C., Hanson, R. L., Robin, R. W., Urbanek, M., Moore, E., Bennett, P. H., and Goldman, D. (1998). Evidence for genetic linkage to alcohol dependence on chromosomes 4 and 11 from an autosome-wide scan in an American Indian population. *Am J Med Genet* **81**, 216–221.

Longnecker, M. P., and Daniels, J. L. (2001). Environmental contaminants as etiologic factors for diabetes. *Environ Health Perspect* **109** (Suppl 6), 871–876.

Longnecker, M. P., and Michalek, J. E. (2000). Serum dioxin level in relation to diabetes mellitus among Air Force veterans with background levels of exposure. *Epidemiology* **11**, 44–48.

Loriot, M. A., Rebuissou, S., et al. (2001). Genetic polymorphisms of cytochrome P450 2A6 in a case-control study on lung cancer in a French population. *Pharmacogenetics* **11**, 39–44.

Lossin, C., Wang, D. W., Rhodes, T. H., Vanoye, C. G., and George, A. L., Jr. (2002). Molecular basis of an inherited epilepsy. *Neuron* **34**, 877–884.

Lotta, T., Vidgren, J., Tilgmann, C., Ulmanen, I., Melen, K., Julkunen, I., and Taskinen, J. (1995). Kinetics of human soluble and membrane-bound catechol

O-methyltransferase: A revised mechanism and description of the thermolabile variant of the enzyme. *Biochemistry* **34**, 4202–4210.

Love-Gregory, L. D., Wasson, J., Ma, J., Jin, C. H., Glaser, B., Suarez, B. K., and Permutt, M. A. (2004). A common polymorphism in the upstream promoter region of the hepatocyte nuclear factor-4 alpha gene on chromosome 20q is associated with type 2 diabetes and appears to contribute to the evidence for linkage in an Ashkenazi Jewish population. *Diabetes* **53**, 1134–1140.

Lucas, D., Menez, C., Girre, C., Berthou, F., Bodenez, P., Joannet, I., Hispard, E., Bardou, L. G., and Menez, J. F. (1995). Cytochrome P450 2E1 genotype and chlorzoxazone metabolism in healthy and alcoholic Caucasian subjects. *Pharmacogenetics* **5**, 298–304.

Lueders, K. K., Hu, S., McHugh, L., Myakishev, M. V., Sirota, L. A., and Hamer, D. H. (2002). Genetic and functional analysis of single nucleotide polymorphisms in the beta2-neuronal nicotinic acetylcholine receptor gene (CHRNB2). *Nicotine Tob Res* **4**, 115–125.

Lundstrom, K., Tenhunen, J., Tilgmann, C., Karhunen, T., Panula, P., and Ulmanen, I. (1995). Cloning, expression and structure of catechol-O-methyltransferase. *Biochim Biophys Acta* **1251**, 1–10.

Luo, T. H., Zhao, Y., Li, G., Yuan, W. T., Zhao, J. J., Chen, J. L., Huang, W., and Luo, M. (2001). A genome-wide search for type II diabetes susceptibility genes in Chinese Hans. *Diabetologia* **44**, 501–506.

Luo, X., Kranzler, H. R., Zhao, H., and Gelernter, J. (2003). Haplotypes at the OPRM1 locus are associated with susceptibility to substance dependence in European-Americans. *Am J Med Genet* **120B**, 97–108.

Lussier-Cacan, S., Bolduc, A., Xhignesse, M., Niyonsenga, T., and Sing, C. F. (2002). Impact of alcohol intake on measures of lipid metabolism depends on context defined by gender, body mass index, cigarette smoking, and apolipoprotein E genotype. *Arterioscler Thromb Vasc Biol* **22**, 824–831.

Lutz, B. (2002). Molecular biology of cannabinoid receptors. *Prostagland Leukotrienes Essen Fatty Acids* **66**, 123–142.

Lynch, M. and Walsh, B. (1998). Genetics and Analysis of Quantitative Traits. Sunderland, MA, Sinauer Associates.

Ma, Q., and Lu, A. Y. (2003). Origins of individual variability in P4501A induction. *Chem Res Toxicol* **16**, 249–260.

Ma, Q. W., Lin, G. F., Chen, J. G., Xiang, C. Q., Guo, W. C., Golka, K., and Shen, J. H. (2004). Polymorphism of N-acetyltransferase 2 (NAT2) gene polymorphism in Shanghai population: Occupational and non-occupational bladder cancer patient groups. *Biomed Environ Sci* **17**, 291–298.

Macarthur, M., Hold, G. L., and El-Omar, E. M. (2004). Inflammation and Cancer II. Role of chronic inflammation and cytokine gene polymorphisms in the pathogenesis of gastrointestinal malignancy. *Am J Physiol Gastrointest Liver Physiol* **286**, G515–G520.

MacGregor, J. T. (2003). The future of regulatory toxicology: Impact of the biotechnology revolution. *Toxicol Sci* **75**, 236–248.

Mackenzie, P. I., Gregory, P. A., Gardner-Stephen, D. A., Lewinsky, R. H., Jorgensen, B. R., Nishiyama, T., Xie, W., and Radominska-Pandya, A. (2003). Regulation of UDP glucuronosyltransferase genes. *Curr Drug Metab* **4**, 249–257.

Mackness, B., Durrington, P., Povey, A., Thomson, S., Dippnall, M., Mackness, M., Smith, T., and Cherry, N. (2003). Paraoxonase and susceptibility to organophosphorus poisoning in farmers dipping sheep. *Pharmacogenetics* **13**, 81–88.

Magnusson, M. K., and Koivisto Hursti, U. K. (2002). Consumer attitudes towards genetically modified foods. *Appetite* **39**, 9–24.

Mahtani, M. M., Widen, E., Lehto, M., Thomas, J., McCarthy, M., Brayer, J., Bryant, B., Chan, G., Daly, M., Forsblom, C., et al. (1996). Mapping of a gene for type 2 diabetes associated with an insulin secretion defect by a genome scan in Finnish families. *Nat Genet* **14**, 90–94.

Maimone, D., Dominici, R., and Grimaldi, L. M. (2001). Pharmacogenomics of neurodegenerative diseases. *Eur J Pharmacol* **413**, 11–29.

Makishima, M., Lu, T. T., Xie, W., Whitfield, G. K., Domoto, H., Evans, R. M., Haussler, M. R., and Mangelsdorf, D. J. (2002). Vitamin D receptor as an intestinal bile acid sensor. *Science* **296**, 1313–1316.

Mantripragada, K. K., Buckley, P. G., de Stahl, T. D., and Dumanski, J. P. (2004). Genomic microarrays in the spotlight. *Trends Genet* **20**, 87–94.

Mapp, C. E., Beghe, B., Balboni, A., Zamorani, G., Padoan, M., Jovine, L., Baricordi, O. R., and Fabbri, L. M. (2000). Association between HLA genes and susceptibility to toluene diisocyanate-induced asthma. *Clin Exp Allergy* **30**, 651–656.

Mapp, C. E., Fryer, A. A., De Marzo, N., Pozzato, V., Padoan, M., Boschetto, P., Strange, R. C., Hemmingsen, A., and Spiteri, M. A. (2002). Glutathione S-transferase GSTP1 is a susceptibility gene for occupational asthma induced by isocyanates. *J Allergy Clin Immunol* **109**, 867–872.

Maraganore, D. M., de Andrade, M., Lesnick, T. G., Farrer, M. J., Bower, J. H., Hardy, J. A., and Rocca, W. A. (2003). Complex interactions in Parkinson's disease: A two-phased approach. *Mov Disord* **18**, 631–636.

Marchant, G. E. (2001). Genetics and toxic torts. *Seton Hall Law Rev* **31**, 949–982.

Marchant, G. E. (2002). Toxicogenomics and toxic torts. *Trends Biotechnol* **20**, 329–332.

Marcy, T. W., Stefanek, M., and Thompson, K. M. (2002). Genetic testing for lung cancer risk: If physicians can do it, should they? *J General Internal Med* **17**, 946–951.

Marenberg, M. E., Risch, N., Berkman, L. F., Floderus, B., and de Faire, U. (1994). Genetic susceptibility to death from coronary heart disease in a study of twins. *New Engl J Med* **330**, 1041–1046.

Markoulatos, P., Siafakas, N., and Moncany, M. (2002). Multiplex polymerase chain reaction: A practical approach. *J Clin Lab Anal* **16**, 47–51.

Marks, R., Dudley, F., and Wan, A. (1978). Trimethylamine metabolism in liver disease. *Lancet* **1**, 1106–1107.

Marks, R., Greaves, M. W., Danks, D., and Plummer, V. (1976). Proceedings: Trimethylaminuria or fish odour syndrome in a child. *Br J Dermatol* **95 Suppl 14**, 11–12.

Marsh, G. M., Enterline, P. E., and McCraw, D. (1991). Mortality patterns among petroleum refinery and chemical plant workers. *Am J Ind Med* **19**, 29–42.

Marteau, T. M., and Lerman, C. (2001). Genetic risk and behavioural change. *Br Med J* **322**, 1056–1059.

Martin, E. R., Scott, W. K., Nance, M. A., Watts, R. L., Hubble, J. P., Koller, W. C., Lyons, K., Pahwa, R., Stern, M. B., Colcher, A., et al. (2001). Association of single-nucleotide polymorphisms of the tau gene with late-onset Parkinson disease. *JAMA* **286**, 2245–2250.

Masson, L. F., McNeill, G., and Avenell, A. (2003). Genetic variation and the lipid response to dietary intervention: a systematic review. *Am J Clin Nutr* **77**, 1098–1111.

Matsumura, K., Saito, T., Takahashi, Y., Ozeki, T., Kiyotani, K., Fujieda, M., Yamazaki, H., Kunitoh, H., and Kamataki, T. (2004). Identification of a novel polymorphic enhancer of the human CYP3A4 gene. *Mol Pharmacol* **65**, 326–334.

Matsuo, K., Hamajima, N., Shinoda, M., Hatooka, S., Inoue, M., Takezaki, T., and Tajima, K. (2001). Gene-environment interaction between an aldehyde dehydrogenase-2 (ALDH2) polymorphism and alcohol consumption for the risk of esophageal cancer. *Carcinogenesis* **22**, 913–916.

Mayatepek, E. and Kohlmuller, D. (1998). Transient trimethylaminuria in childhood. *Acta Paediatr* **87**, 1205–1207.

Mayer, P., and Hollt, V. (2001). Allelic and somatic variations in the endogenous opioid system of humans. *Pharmacol Ther* **91**, 167–177.

Mayer, P., Rochlitz, H., Rauch, E., Rommelspacher, H., Hasse, H. E., Schmidt, S., and Hollt, V. (1997). Association between a delta opioid receptor gene polymorphism and heroin dependence in man. *Neuroreport* **8**, 2547–2550.

Mayeux, R., Ottman, R., Maestre, G., Ngai, C., Tang, M. X., Ginsberg, H., Chun, M., Tycko, B., and Shelanski, M. (1995). Synergistic effects of traumatic head injury and apolipoprotein-epsilon 4 in patients with Alzheimer's disease. *Neurology* **45**, 555–557.

McBride, C. M., Bepler, G., Lipkus, I. M., Lyna, P., Samsa, G., Albright, J., Datta, S., and Rimer, B. K. (2002). Incorporating genetic susceptibility feedback into a smoking cessation program for African-American smokers with low income. *Cancer Epidemiol Biomark Prev* **11**, 521–528.

McCarver, D. G., Byun, R., Hines, R. N., Hichme, M., and Wegenek, W. (1998). A genetic polymorphism in the regulatory sequences of human CYP2E1: association with increased chlorzoxazone hydroxylation in the presence of obesity and ethanol intake. *Toxicol Appl Pharmacol* **152**, 276–281.

McClure, J. B. (2001). Are biomarkers a useful aid in smoking cessation? A review and analysis of the literature. *Behav Med* **27**, 37–47.

McCombie, R. R., Dolphin, C. T., Povey, S., Phillips, I. R., and Shephard, E. A. (1996). Localization of human flavin-containing monooxygenase genes FMO2 and FMO5 to chromosome 1q. *Genomics* **34**, 426–429.

McConnell, H. W., Mitchell, S. C., Smith, R. L., and Brewster, M. (1997). Trimethylaminuria associated with seizures and behavioural disturbance: a case report. *Seizure* **6**, 317–321.

McGlynn, K. A., Hunter, K., LeVoyer, T., Roush, J., Wise, P., Michielli, R. A., Shen, F.-M., Evans, A. A., London, W. T., and Buetow, K. H. (2003). Susceptibility to aflatoxin B_1-related primary hepatocellular carcinoma in mice and humans. *Cancer Res* **63**, 4594–4601.

McGuire, V., Longstreth, W. T., Jr., Nelson, L. M., Koepsell, T. D., Checkoway, H., Morgan, M. S., and van Belle, G. (1997). Occupational exposures and amyotrophic lateral sclerosis. A population-based case-control study. *Am J Epidemiol* **145**, 1076–1088.

McKinney, E. F., Walton, R. T., Yudkin, P., Fuller, A., Haldar, N. A., Mant, D., Murphy, M., Welsh, K. I., and Marshall, S. E. (2000). Association between polymorphisms in dopamine metabolic enzymes and tobacco consumption in smokers. *Pharmacogenetics* **10**, 483–491.

McKusick, V. A. et al., eds. *Online Mendelian Inheritance in Man* (OMIM) (Baltimore, MD: National Center for Biotechnology Information, Johns Hopkins University) [developed for the World Wide Web by the National Center for Biotechnology Information (NCBI); available online at http://www.ncbi.nlm.nih.gov/entrez/query.fcgi?db=OMIM].

McLeroy, K. R., Bibeau, D., Steckler, A., and Glanz, K. (1988). An ecological perspective on health promotion programs. *Health Educ Quart* **15**, 351–377.

McMichael, A. J., Spirtas, R., and Kupper, L. L. (1974). An epidemiologic study of mortality within a cohort of rubber workers, 1964–72. *J Occup Med* **16**, 458–464.

McNicholl, J. M., Downer, M. V., Udhayakumar, V., Alper, C. A., and Swerdlow, D. L. (2000). Host-pathogen interactions in emerging and re-emerging infectious diseases: A genomic perspective of tuberculosis, malaria, human immunodeficiency virus infection, hepatitis B, and cholera. *Annu Rev Publ Health* **21**, 15–46.

McWilliams, J. E., Sanderson, B. J., Harris, E. L., Richert-Boe, K. E., and Henner, W. D. (1995). Glutathione S-transferase M1 (GSTM1) deficiency and lung cancer risk. *Cancer Epidemiol Biomark Prev* **4**, 589–594.

Mechoulam, R., and Parker, L. (2003). Cannabis and alcohol—a close friendship. *Trends Pharmacol Sci* **24**, 266–268.

Medeiros, R., Vasconcelos, A., Costa, S., Pinto, D., Ferreira, P., Lobo, F., Morais, A., Oliveira, J., and Lopes, C. (2004). Metabolic susceptibility genes and prostate cancer risk in a southern European population: The role of glutathione S-transferases GSTM1, GSTM3, and GSTT1 genetic polymorphisms. *Prostate* **58**, 414–420.

Medical Research Council Human Genetics Unit (2003). The Allele Variant Database Web Site. December 7, 2004. http://human-fmo3.biochem.ucl.ac.uk/Human_FMO3.

Meek, B., Renwick, A., and Sonich-Mullin, C. (2003). Practical application of kinetic data in risk assessment—an IPCS initiative. *Toxicol Lett* **138**, 151–160.

Meier, H. (1963). *Experimental Pharmacogenetics: Physiopathology of Heredity and Pharmacologic Responses* (New York: Academic Press).

Meischke, H., and Johnson, J. D. (1995). Women's selection of sources for information on breast cancer detection. *Health Values* **19**, 30–39.

Meldrum, B. S., Akbar, M. T., and Chapman, A. G. (1999). Glutamate receptors and transporters in genetic and acquired models of epilepsy. *Epilepsy Res* **36**, 189–204.

Mellick, G. D., McCann, S. J., and Le Couter, D. G. (1999). Parkinson's disease, MAOB, and smoking. *Neurology* **53**, 658.

Mendoza, M. A., Ponce, R. A., Ou, Y. C., and Faustman, E. M. (2002). p21(WAF1/CIP1) inhibits cell cycle progression but not G2/M-phase transition following methylmercury exposure. *Toxicol Appl Pharmacol* **178**, 117–125.

Merikangas, K. R., and Avenevoli, S. (2000). Implications of genetic epidemiology for the prevention of substance use disorders. *Addict Behav* **25**, 807–820.

Merikangas, K. R., and Risch, N. (2003). Will the genomics revolution revolutionize psychiatry? *Am J Psychiatry* **160**, 625–635.

Messina, E. S., Tyndale, R. F., and Sellers, E. M. (1997). A major role for CYP2A6 in nicotine C-oxidation by human liver microsomes. *J Pharmacol Exp Ther* **282**, 1608–1614.

Mettimano, M., Specchia, M. L., Ianni, A., Arzani, D., Ricciardi, G., Savi, L., and Romano-Spica, V. (2003). CCR5 and CCR2 gene polymorphisms in hypertensive patients. *Br J Biomed Sci* **60**, 19–21.

Meyer, U. A. (2004). Pharmacogenetics: Five decades of therapeutic lessons from genetic diversity. *Nat Rev Genet* **5**, 669–676.

Mihara, K., Kondo, T., Suzuki, A., Yasui-Furukori, N., Ono, S., Otani, K., and Kaneko, S. (2001). Effects of genetic polymorphism of CYP1A2 inducibility on the steady-state plasma concentrations of trazodone and its active metabolite m-chlorophenylpiperazine in depressed Japanese patients. *Pharmacol Toxicol* **88**, 267–270.

Miller, D. P., Neuberg, D., de Vivo, I., Wain, J. C., Lynch, T. J., Su, L., and Christiani, D. C. (2003). Smoking and the risk of lung cancer: Susceptibility with GSTP1 polymorphisms. *Epidemiology* **14**, 545–551.

Miller, G. M., and Madras, B. K. (2002). Polymorphisms in the 3′-untranslated region of human and monkey dopamine transporter genes affect reporter gene expression. *Mol Psychiatry* **7**, 44–55.

Miller, L. H., Good, M. F., and Milon, G. (1994). Malaria pathogenesis. *Science* **264**, 1878–1883.

Miller, M. C., 3rd, Mohrenweiser, H. W., and Bell, D. A. (2001). Genetic variability in susceptibility and response to toxicants. *Toxicol Lett* **120**, 269–280.

Miller, M. M., James, R. A., Richer, J. K., Gordon, D. F., Wood, W. M., and Horwitz, K. B. (1997). Progesterone regulated expression of flavin-containing monooxygenase 5 by the B-isoform of progesterone receptors: implications for tamoxifen carcinogenicity. *J Clin Endocrinol Metab* **82**, 2956–2961.

Miller, P. S. (2000). Is there a pink slip in my gene? Genetic discrimination in the workplace. *J Health Care Law Policy* **3**, 225–265.

Miller, S. M., Bowen, D. J., Campbell, M. K., Diefenbach, M. A., Gritz, E. R., Jacobsen, P. B., Stefanek, M., Fang, C. Y., Lazovich, D., Sherman, K. A., and Wang, C. (2004). Current research promises and challenges in behavioral oncology: Report from the American Society of Preventive Oncology Annual Meeting, 2002. *Cancer Epidemiol Biomark Prev* **13**, 171–180.

Mimura, J., and Fujii-Kuriyama, Y. (2003). Functional role of AhR in the expression of toxic effects by TCDD. *Biochim Biophys Acta* **1619**, 263–268.

Minematsu, N., Nakamura, H., Iwata, M., Tateno, H., Nakajima, T., Takahashi, S., Fujishima, S., and Yamaguchi, K. (2003). Association of CYP2A6 deletion polymorphism with smoking habit and development of pulmonary emphysema. *Thorax* **58**, 623–628.

Miners, J. O., McKinnon, R. A., and Mackenzie, P. I. (2002). Genetic polymorphisms of UDP-glucuronosyltransferases and their functional significance. *Toxicology* **181–182**, 453–456.

Mitchell, J. D. (2000). Amyotrophic lateral sclerosis: Toxins and environment. *Amyotroph Lateral Scler Other Motor Neuron Disord* **1**, 235–250.

Mitchell, S., Ayesh, R., Barrett, T., and Smith, R. (1999). Trimethylamine and foetor hepaticus. *Scand J Gastroenterol* **34**, 524–528.

Mitchell, S. C., Zhang, A. Q., Barrett, T., Ayesh, R., and Smith, R. L. (1997). Studies on the discontinuous N-oxidation of trimethylamine among Jordanian, Ecuadorian and New Guinean populations. *Pharmacogenetics* **7**, 45–50.

Miyoshi, Y. and Noguchi, S. (2003). Polymorphisms of estrogen synthesizing and metabolizing genes and breast cancer risk in Japanese women. *Biomed Pharmacother* **57**, 471–481.

Modrich, P., and Lahue, R. (1996). Mismatch repair in replication fidelity, genetic recombination and cancer biology. *Annu Rev Biochem* **65**, 101–133.

Mohrenweiser, H. W. (2004). Genetic variation and exposure related risk estimation: Will toxicology enter a new era? DNA repair and cancer as a paradigm. *Toxicol Pathol* **32** (Suppl 1), 136–145.

Mokdad, A. H., Ford, E. S., Bowman, B. A., Nelson, D. E., Engelgau, M. M., Vinicor, F., and Marks, J. S. (2000). Diabetes trends in the U.S.: 1990–1998. *Diabetes Care* **23**, 1278–1283.

Mol, C. D., Arval, A. S., Slupphaug, G., Kavli, B., Alseth, I., Krokan, H. E., and Tainer, J. A. (1995). Crystal structure and mutational analysis of human uracil-DNA glycosylase: Structural basis for specificity and catalysis. *Cell* **80**, 869–878.

Mollerup, S., Ryberg, D., Hewer, A., Phillips, D. H., and Haugen, A. (1999). Sex differences in lung CYP1A1 expression and DNA adduct levels among lung cancer patients. *Cancer Res* **59**, 3317–3320.

Molloy, A. M., Kirke, P., Hillary, I., Weir, D. G., and Scott, J. M. (1985). Maternal serum folate and vitamin B12 concentrations in pregnancies associated with neural tube defects. *Arch Dis Childhood* **60**, 660–665.

Moonen, H. J., Moonen, E. J., Maas, L., Dallinga, J. W., Kleinjans, J. C., and de Kok, T. M. (2004). CYP1A2 and NAT2 genotype/phenotype relations and urinary excretion of 2-amino-1-methyl-6-phenylimidazo[4,5-b]pyridine (PhIP) in a human dietary intervention study. *Food Chem Toxicol* **42**, 869–878.

Moreno, V., Glatt, H., Guino, E., Fisher, E., Meinl, W., Navarro, M., Badosa, J. M., and Boeing, H. (2005). Polymorphisms in sulfotransferases SULT1A1 and SULT1A2 are not related to colorectal cancer. *Int J Cancer* **113**, 683–686.

Morgan, D. P., Lin, L. I., and Saikaly, H. H. (1980). Morbidity and mortality in workers occupationally exposed to pesticides. *Arch Environ Contam Toxicol* **9**, 349–382.

Morgan, M. G. (1993). Risk analysis and management. *Sci Am* **269**, 32–35, 38–41.

Morishima-Kawashima, M., Oshima, N., Ogata, H., Yamaguchi, H., Yoshimura, M., Sugihara, S., and Ihara, Y. (2000). Effect of apolipoprotein E allele epsilon4 on the initial phase of amyloid beta-protein accumulation in the human brain. *Am J Pathol* **157**, 2093–2099.

Morrow, A. C., and Motulsky, A. G. (1968). Rapid screening method for the common atypical pseudocholinesterase variant. *J Lab Clin Med* **71**, 350–356.

Morton, N. E. (1955). Sequential tests for the detection of linkage. *Am J Hum Genet* **7**, 277–318.

Motulsky, A. G. (1957). Drug reactions enzymes, and biochemical genetics. *JAMA Assoc* **165**, 835–837.

Motulsky, A. G. (1968). Genetics and environmental health. *Arch Environ Health* **16**, 75–76.

Motulsky, A. G. (1971). The William Allan Memorial Award Lecture. Human and medical genetics: A scientific discipline and an expanding horizon. *Am J Hum Genet* **23**, 107–123.

Motulsky, A. G. (1972). History and current status of pharmacogenetics. In *Human Genetics, Proc. 4th Int. Congr. Human Genetics,* Grouchy, J. de, Ebling, F. J. G., and Henderson, I. W. eds. (Amsterdam: Excerpta Medica), pp. 381–390.

Motulsky, A. G. (1991). Pharmacogenetics and ecogenetics in 1991. *Pharmacogenetics* **1**, 2–3.

Motulsky, A. G. (2002). From pharmacogenetics and ecogenetics to pharmacogenomics. *Medicina nei Secoli Arte e Scienza* (*Journal of History of Medicine*) **14**, 683–705.

MRC (1991). *Prevention of Neural Tube Defects: Results of the Medical Research Council Vitamin Study.* MRC Vitamin Study Research Group.

MRC (1992). *Recommendations for the Use of Folic Acid to Reduce the Number of Cases of Spina Bifida and Other Neural Tube Defects.* MRC Vitamin Study Research Group.

Mueller, R. F., Hornung, S., Furlong, C. E., Anderson, J., Giblett, E. R., and Motulsky, A. G. (1983). Plasma paraoxonase polymorphism: A new enzyme assay, population, family, biochemical, and linkage studies. *Am J Hum Genet* **35**, 393–408.

Mukherjee, D., and Topol, E. J. (2002). Pharmacogenomics in cardiovascular diseases. *Prog Cardiovasc Dis* **44**, 479–498.

Mullis, K., Faloona, F., Scharf, S., Saiki, R., Horn, G., and Erlich, H. (1986). Specific enzymatic amplification of DNA in vitro: The polymerase chain reaction. *Cold Spring Harb Symp Quant Biol* **51**, 263–273.

Munoz, D. G. (1998). Is exposure to aluminum a risk factor for the development of Alzheimer disease?—No. *Arch Neurol* **55**, 737–739.

Murray, T. H. (1997). Genetic exceptionalism and "future diaries": Is genetic information different from other medical information? In *Genetic Secrets: Protecting Privacy and Confidentiality in the Genetic Era*, Rothstein, M. A., ed. (New Haven, CT: Yale Univ. Press).

Mushiroda, T., Douya, R., Takahara, E., and Nagata, O. (2000). The involvement of flavin-containing monooxygenase but not CYP3A4 in metabolism of itopride hydrochloride, a gastroprokinetic agent: comparison with cisapride and mosapride citrate. *Drug Metab Dispos* **28**, 1231–1237.

Muskhelishvili, L., Thompson, P. A., Kusewitt, D. F., Wang, C., and Kadlubar, F. F. (2001). In situ hybridization and immunohistochemical analysis of cytochrome P450 1B1 expression in human normal tissues. *J Histochem Cytochem* **49**, 229–236.

Mutter, G. L., and Boyton, K. A. (1995). PCR bias in amplification of androgen receptor alleles, a trinucleotide repeat marker used in clonality studies. *Nucleic Acids Res* **23**, 1411–1418.

Nabel, E. G. (2003). Genomic medicine: Cardiovascular disease. *New Engl J Med* **349**, 60–72.

Nakachi, K., Imai, K., Hayashi, S., and Kawajiri, K. (1993). Polymorphisms of the CYP1A1 and glutathione S-transferase genes associated with susceptibility to lung cancer in relation to cigarette dose in a Japanese population. *Cancer Res* **53**, 2994–2999.

Nakajima, M., Yamamoto, T., et al. (1996). Role of human cytochrome P4502A6 in C-oxidation of nicotine. *Drug Metab Dispos* **24**, 1212–1217.

Nakajima, M., Yokoi, T., Mizutani, M., Kinoshita, M., Funayama, M. and Kamataki, T. (1999). Genetic polymorphism in the 5′-flanking region of human CYP1A2 gene: effect on the CYP1A2 inducibility in humans. *J Biochem (Tokyo)* **125**, 803–808.

Nakayama, M., Bennett, C. J., Hicks, J. L., Epstein, J. I., Platz, E. A., Nelson, W. G., and De Marzo, A. M. (2003). Hypermethylation of the human glutathione S-transferase-pi gene (GSTP1) CpG island is present in a subset of proliferative inflammatory atrophy lesions but not in normal or hyperplastic epithelium of the prostate: A detailed study using laser-capture microdissection. *Am J Pathol* **163**, 923–933.

Napolitano, C., Rivolta, I., and Priori, S. G. (2003). Cardiac sodium channel diseases. *Clin Chem Lab Med* **41**, 439–444.

Narod, S., Feunteun, J., Lynch, H., Watson, P., Conway, T., and Lenoir, G. (1991). Familial breast-ovarian cancer locus on chromosome 17q12-q23. *Lancet* **388**, 82–83.

National Commission for the Protection of Human Subjects of Biomedical and Behavioral Research (1979). *The Belmont Report: Ethical Principles and Guidelines for the Protection of Human Subjects of Research* (Washington, DC: Dept. Health, Education, and Welfare, available from U.S. Government Printing Office).

National Heart Lung and Blood Institute (2002). *NHLBI Morbidity and Mortality Chartbook, 2002* (Bethesda, MD: National Heart, Lung, and Blood Inst.).

National Research Council (NRC) (1989). *Biomarkers of Exposure, Effect, and Susceptibility* (Washington, DC: National Academy Press).

Nazar-Stewart, V., Vaughan, T. L., Stapleton, P., Van Loo, J., Nicol-Blades, B., and Eaton, D. L. (2003). A population-based study of glutathione S-transferase M1, T1 and P1 genotypes and risk for lung cancer. *Lung Cancer* **40**, 247–258.

Nebert, D. W. (1989). The Ah locus: Genetic differences in toxicity, cancer, mutation, and birth defects. *Crit Rev Toxicol* **20**, 153–174.

Nebert, D. W., Dalton, T. P., Okey, A. B., and Gonzalez, F. J. (2004). Role of aryl hydrocarbon receptor-mediated induction of the CYP1 enzymes in environmental toxicity and cancer. *J Biol Chem* **279**, 23847–23850.

Nebert, D. W., Roe, A. L., Dieter, M. Z., Solis, W. A., Yang, Y., and Dalton, T. P. (2000). Role of the aromatic hydrocarbon receptor and [Ah] gene battery in the oxidative stress response, cell cycle control, and apoptosis. *Biochem Pharmacol* **59**, 65–85.

Nebert, D. W., Roe, A. L., Vandale, S. E., Bingham, E., and Oakley, G. G. (2002). NAD(P)H:quinone oxidoreductase (NQO1) polymorphism, exposure to benzene, and predisposition to disease: A HuGE review. *Genet Med* **4**, 62–70.

Nebert, D. W., and Vasiliou, V. (2004). Analysis of the glutathione S-transferase (GST) gene family. *Hum Genom* **1**, 460–464.

Need, A. C., Motulsky, A. G,. and Goldstein, D. B. (2005). Priorities and standards in pharmacogenetic research. *Nat Genet* **37**, 671–681.

Neel, J. V. (1962). Diabetes mellitus: A thrifty genotype rendered detrimental by "progress"? *Am J Hum Genet* **14**, 353–362.

Negishi, M., Pedersen, L. G., Petrotchenko, E., Shevtsov, S., Gorokhov, A., Kakuta, Y., and Pedersen, L. C. (2001). Minireview: Structure and function of sulfotransferases. *Arch Biochem Biophys* **390**, 149–157.

Neibergs, H. L., Hein, D. W., and Spratt, J. S. (2002). Genetic profiling of colon cancer. *J Surg Oncol* **80**, 204–213.

Nelson, G., Chandrashekar, J., Hoon, M. A., Feng, L., Zhao, G., Ryba, N. J. P., and Zuker, C. S. (2002a). An amino-acid taste receptor. *Nature* **416**, 199–202.

Nelson, H. H., Kelsey, K. T., Mott, L. A., and Karagas, M. R. (2002b). The XRCC1 Arg399Gln polymorphism, sunburn, and non-melanoma skin cancer: Evidence of gene-environment interaction. *Cancer Res* **62**, 152–155.

Nelson, L. M. (1995). Epidemiology of ALS. *Clin Neurosci* **3**, 327–331.

Nelson, L. M., Longstreth, W. T., Jr., Koepsell, T. D., and van Belle, G. (1990). Proxy respondents in epidemiologic research. *Epidemiol Rev* **12**, 71–86.

Nelson, L. M., McGuire, V., Longstreth, W. T., Jr., and Matkin, C. (2000). Population-based case-control study of amyotrophic lateral sclerosis in western Washington State. I. Cigarette smoking and alcohol consumption. *Am J Epidemiol* **151**, 156–163.

Newill, C. A., Khoury, M. J., and Chase, G. A. (1986). Epidemiological approach to the evaluation of genetic screening in the workplace. *J Occup Med* **28**, 1108–1111.

Nickerson, D. A., Rieder, M. J., Crawford, D. C., Carlson, C. S., and Livingston, R. J. (2005). *An overview of the Environmental Genome Project.* Essays on the Future of Environmental Health Research: A Tribute to Dr. Kenneth Olden. Research Triangle Park, NC, Environmental Health Perspectives/National Institute of Environmental Health Sciences, 42–53.

Nickerson, D. A., Taylor, S. L., Fullerton, S. M., Weiss, K. M., Clark, A. G., Stengard, J. H., Salomaa, V., Boerwinkle, E., and Sing, C. F. (2000). Sequence diversity and large-scale typing of SNPs in the human apolipoprotein E gene. *Genome Res* **10**, 1532–1545.

NIH (Dec. 13, 2001). Title 45 Public Welfare, Part 46 Protection of Human Subjects. *U.S. Code of Federal Regulations* (http://ohrp.osophs.dhhs.gov/humansubjects/guidance/45cfr46.htm).

NIH (March 15, 2002). Office of Extramural Research, Certificates of Confidentiality Kiosk (National Institutes of Health) (http://grants1.hih.gov/grants/policy/coc).

NIOSH (June 2003). *Beryllium Research Newsletter*, issue 1 (National Institute for Occupational Safety and Health) (http://www.cdc.gov/niosh/topics/beryllium/pdfs/news-issue1.pdf).

Noble, E. P., Zhang, X., Ritchie, T., Lawford, B. R., Grosser, S. C., Young, R. M., and Sparkes, R. S. (1998). D2 dopamine receptor and GABA(A) receptor beta3 subunit genes and alcoholism. *Psychiatry Res* **81**, 133–147.

NRC (1983). *Risk Assessment in the Federal Government: Managing the Process* (Washington, DC: National Academy Press).

NRC (1987). *Committee of Biological Markers* (Washington, DC: National Academy Press).

NRC (1994). *Science and Judgment in Risk Assessment* (Washington, DC: National Academy Press).

NRC (2000). *Scientific Frontiers in Developmental Toxicology and Risk Assessment* (Washington, DC: National Academy Press).

Nuffield Council on Bioethics (2003). *Pharmacogenetics: Ethical Issues* (Oxford, Nuffield Trust) (www.nuffieldbioethics.org/pharmacogenetics).

Nyberg, F., Hou, S. M., Hemminki, K., Lambert, B., and Pershagen, G. (1998). Glutathione S-transferase mu1 and N-acetyltransferase 2 genetic polymorphisms and exposure to tobacco smoke in nonsmoking and smoking lung cancer patients and population controls. *Cancer Epidemiol Biomark Prev* **7**, 875–883.

O'Brien, S. J., and Dean, M. (1997). In search of AIDS-resistance genes. *Sci Am* **277**, 44–51.

O'Brien, S. J., and Moore, J. P. (2000). The effect of genetic variation in chemokines and their receptors on HIV transmission and progression to AIDS. *Immunol Rev* **177**, 99–111.

O'Byrne, K. J., and Dalgleish, A. G. (2001). Chronic immune activation and inflammation as the cause of malignancy. *Br J Cancer* **85**, 473–483.

Occupational Safety and Health Act (December 29, 1970). Public Law 91–596. *United States Statues at Large* **84**, 1590–1619.

Ohanian, E., Moore, J., Fowle, J., Omenn, G., Lewis, S., Gray, G., and North, D. (1997). Risk characterization: A bridge in informed decision-making. *Fundam Appl Toxicol* **39**, 81–88.

Olden, K., Guthrie, J., and Newton, S. (2001). A bold new direction for environmental health research. *Am J Publ Health* **91**, 1964–1967.

Olden, K., and Wilson, S. (2000). Environmental health and genomics: Visions and implications. *Nat Rev Genet* **1**, 149–153.

Oller, A. R., Rastogi, P., Morgenthaler, S., and Thilly, W. G. (1989). A statistical model to estimate variance in long term-low dose mutation assays: Testing of the model in a human lymphoblastoid mutation assay. *Mutat Res* **216**, 149–161.

Olson, A. L., Dietrick, A. J., Sox, C. H., Stevens, M. M., Winchell, W., and Ahles, T. A. (1997). Solar protection of chldren at the beach. *Pediatrics* **99**, 1–5.

Omenn, G. S. (1982). Predictive identification of hypersusceptible individuals. *J Occup Med* **24**, 369–374.

Omenn, G. S. (2000a). The genomic era: A crucial role for the public health sciences. *Environ Health Perspect* **108**, A204–A205.

Omenn, G. S. (2000b). Public health genetics: An emerging interdisciplinary field for the post-genomic era. *Annu Rev Publ Health* **21**, 1–13.

Omenn, G. S. (2001). Prospects for pharmacogenetics and ecogenetics in the new millennium. *Drug Metab Dispos* **29**, 611–614.

Omenn, G. S. (2002). The crucial role of the public health sciences in the postgenomic era. *Genet Med* **4**, 21S-26S.

Omenn, G., and Faustman, E. (1997). Risk assessment and risk management. In *Oxford Textbook of Public Health*, Detels, R., Holland, W., McEwen, J., and Omenn, G., eds. (New York: Oxford Univ. Press), pp. 969–986.

Omenn, G. S. (2003). On the significance of "the red book" in the evolution of risk assessment and risk management. *Hum Ecol Risk Assess* **9**, 1155–1167.

Omenn, G. S. (2005). Genomics and public health. *Issues Sci Technol* **21**, 42–48.

Omenn, G. S., and Faustman, E. M. (2002). Risk assessment and risk management. In *Oxford Textbook of Public Health*, 4th ed., Detels, R., McEwen, J., Beaglehole, R., and Tanaka, H., eds. (New York: Oxford Univ. Press), pp. 1083–1103.

Omenn, G. S., and Motulsky, A. G. (1978). Eco-genetics: Genetic variation in susceptibility to environmental agents. In *Genetics, Environment and Behavior: Implications for Educational Policy*, Ehrman, L., Omenn, G. S., and Caspari, E., eds. (New York: Academic Press), pp. 129–179.

Omenn, G. S., and Motulsky, A. G. (2003). Integration of pharmacogenomics into medical practice. In *Pharmacogenomics: Social, Ethical, and Clinical Dimensions*, Rothstein, M. A., ed. (New York: Wiley), pp. 137–162.

Omiecinski, C. J., Hassett, C., and Hosagrahara, V. (2000). Epoxide hydrolase—polymorphism and role in toxicology. *Toxicol Lett* **112–113**, 365–370.

Onaivi, E. S., Leonard, C. M., Ishiguro, H., Zhang, P. W., Lin, Z., Akinshola, B. E., and Uhl, G. R. (2002). Endocannabinoids and cannabinoid receptor genetics. *Prog Neurobiol* **66**, 307–344.

Ongphiphadhanakul, B., Rajatanavin, R., Chanprasertyothin, S., Chailurkit, L., Piaseu, N., Teerarungsikul, K., Sirisriro, R., Komindr, S., and Puavilai, G. (1997). Vitamin D receptor gene polymorphism is associated with urinary calcium excretion but not with bone mineral density in postmenopausal women. *J Endocrinol Invest* **20**, 592–596.

Online Mendelian Inheritance in Man (OMIM) (2004). *Online Mendelian Inheritance in Man Statistics* (Baltimore, MD: National Center for Biotechnology Information, Johns Hopkins Univ.).

Ordovas, J. M. (2001). Genetic influences on blood lipids and cardiovascular disease risk. In *Nutrition in the Prevention and Treatment of Disease*, Coulston, A. M., Rock, C. L., and Monsen, E. R., eds. (San Diego, CA: Academic Press), pp. 157–182.

Ostrerova-Golts, N., Petrucelli, L., Hardy, J., Lee, J. M., Farer, M., and Wolozin, B. (2000). The A53T alpha-synuclein mutation increases iron-dependent aggregation and toxicity. *J Neurosci* **20**, 6048–6054.

Ostroff, J. S., Hay, J. L., Primavera, L. H., Bivona, P., Cruz, G. D., and LeGeros, R. (1999). Motivating smoking cessation among dental patients: Smokers' interest in biomarker testing for susceptibility to tobacco-related cancers. *Nicotine Tob Res* **1**, 347–355.

OTA (1983). *The Role of Genetic Testing in the Prevention of Occupational Disease* (Washington, DC: U.S. Government Printing Office, U.S. Congress Office of Technology Assessment).

OTA (1990). *Genetic Monitoring and Screening in the Workplace* (Washington, DC: U.S. Government Printing Office, U.S. Congress Office of Technology Assessment).

Ott, A., Slooter, A. J., Hofman, A., van Harskamp, F., Witteman, J. C., Van Broeckhoven, C., van Duijn, C. M., and Breteler, M. M. (1998). Smoking and risk of dementia and Alzheimer's disease in a population-based cohort study: The Rotterdam Study. *Lancet* **351**, 1840–1843.

Ott, J. (1999). *Analysis of Human Genetic Linkage*, 3rd ed. (Baltimore, MD: John Hopkins Univ. Press).

Ott, M. G., Zober, A., and Germann, C. (1994). Laboratory results for selected target organs in 138 individuals occupationally exposed to TCDD. *Chemosphere* **29**, 2423–2437.

Ottman, R. (1996). Gene-environment interaction: Definitions and study designs. *Prev Med* **25**, 764–770.

Overby, L. H., Buckpitt, A. R., Lawton, M. P., Atta-Asafo-Adjei, E., Schulze, J., and Philpot, R. M. (1995). Characterization of flavin-containing monooxygenase 5 (FMO5) cloned from human and guinea pig: evidence that the unique catalytic properties of FMO5 are not confined to the rabbit ortholog. *Arch Biochem Biophys* **317**, 275–284.

Owen, N., Leslie, E., Salmon, J., and Fotheringham, M. J. (2000). Environmental determinants of physical activity and sedentary behavior. *Exercise Sport Sci Rev* **28**, 153–158.

Oyama, T., Matsumoto, A., Isse, T., Kim, Y. D., Ozaki, S., Osaki, T., Sugio, K., Yasumoto, K., and Kawamoto, T. (2003). Evidence-based prevention (EBP): approach to lung cancer prevention based on cytochrome 1A1 and cytochrome 2E1 polymorphism. *Anticancer Res* **23**, 1731–1737.

Ozdemir, V., Basile, V. S., Masellis, M., and Kennedy, J. L. (2001). Pharmacogenetic assessment of antipsychotic-induced movement disorders: contribution of the dopamine D3 receptor and cytochrome P450 1A2 genes. *J Biochem Biophys Methods* **47**, 151–157.

Ozdemir, V., Kalowa, W., Tang, B. K., Paterson, A. D., Walker, S. E., Endrenyi, L., and Kashuba, A. D. (2000). Evaluation of the genetic component of variability in CYP3A4 activity: a repeated drug administration method. *Pharmacogenetics* **10**, 373–388.

Pagani, F., and Baralle, F. E. (2004). Genomic variants in exons and introns: Identifying the splicing spoilers. *Nat Rev Genet* **5**, 389–396.

Page, G. P., George, V., Go, R. C., Page, P. Z., and Allison, D. B. (2003). "Are We There Yet?": Deciding when one has demonstrated specific genetic causation in complex diseases and quantitative traits. *Am J Hum Genet* **73**, 711–719.

Paine, M. F. and Thummel, K. E. (2003). *Role of intestinal cytochromes P450 in drug disposition*. Drug Metabolizing Enzymes. Cytochrome P450 and Other Enzymes in Drug Discovery and Development. J. S. Lee, R. S. Obach and M. B. Fisher. New York, Marcel Dekker, 421–451.

Palmer, L. J., Silverman, E. S., Weiss, S. T., and Drazen, J. M. (2002). Pharmacogenetics of asthma. *Am J Respir Crit Care Med* **165**, 861–866.

Pan, S. S., Han, Y., Farabaugh, P., and Xia, H. (2002). Implication of alternative splicing for expression of a variant NAD(P)H:quinone oxidoreductase-1 with a single nucleotide polymorphism at 465C>T. *Pharmacogenetics* **12**, 479–488.

Pardo, R., Midden, C., and Miller, J. D. (2002). Attitudes toward biotechnology in the European Union. *J Biotechnol* **98**, 9–24.

Park, R. M., and Mirer, F. E. (1996). A survey of mortality at two automotive engine manufacturing plants. *Am J Ind Med* **30**, 664–673.

Parker, A., Meyer, J., Lewitzky, S., Rennich, J. S., Chan, G., Thomas, J. D., Orho-Melander, M., Lehtovirta, M., Forsblom, C., Hyrkko, A., et al. (2001). A gene conferring susceptibility to type 2 diabetes in conjunction with obesity is located on chromosome 18p11. *Diabetes* **50**, 675–680.

Parkin, D. M., Pisani, P., and Ferlay, J. (1999). Estimates of the worldwide incidence of 25 major cancers in 1990. *Int J Cancer* **80**, 827–841.

Parkinson, A. (1996). Biotransformation of Xenobiotics. In *Casarett and Doull's Toxicology. The Basic Science of Poisons*, Klaassen, C. D., ed. (New York: McGraw-Hill).

Parsian, A., and Cloninger, C. R. (2001). Serotonergic pathway genes and subtypes of alcoholism: Association studies. *Psychiatr Genet* **11**, 89–94.

Parsian, A., and Zhang, Z. H. (1999). Human chromosomes 11p15 and 4p12 and alcohol dependence: Possible association with the GABRB1 gene. *Am J Med Genet* **88**, 533–538.

Pasanen, M. (1999). The expression and regulation of drug metabolism in human placenta. *Adv Drug Deliv Rev* **38**, 81–97.

Pascussi, J. M., Gerbal-Chaloin, S., Pichard-Garcia, L., Daujat, M., Fabre, J. M., Maurel, P., and Vilarem, M. J. (2000). Interleukin-6 negatively regulates the expression of pregnane X receptor and constitutively activated receptor in primary human hepatocytes. *Biochem Biophys Res Commun* **274**, 707–713.

Patil, N., Berno, A. J., Hinds, D. A., Barrett, W. A., Doshi, J. M., Hacker, C. R., Kautzer, C. R., Lee, D. H., Marjoribanks, C., McDonough, D. P., et al. (2001). Blocks of limited haplotype diversity revealed by high-resolution scanning of human chromosome 21. *Science* **294**, 1719–1723.

Patkar, A. A., Berrettini, W. H., Hoehe, M., Hill, K. P., Gottheil, E., Thornton, C. C., and Weinstein, S. P. (2002). No association between polymorphisms in the serotonin transporter gene and susceptibility to cocaine dependence among African-American individuals. *Psychiatr Genet* **12**, 161–164.

Patrick, D. L., Cheadle, A., Thompson, D. C., Diehr, P., Koepsell, T., and Kinne, S. (1994). The validity of self-reported smoking: a review and meta-analysis. *Am J Public Health* **84**, 1086–1093.

Patrick, D. L., Danis, M., Southerland, L. I., and Hong, G. (1988). Quality of life following intensive care. *J General Internal Med* **3**, 218–223.

Pazderova-Vejlupkova, J., Lukas, E., Nemcova, M., Pickova, J., and Jirasek, L. (1981). The development and prognosis of chronic intoxication by tetrachlordibenzo-p-dioxin in men. *Arch Environ Health* **36**, 5–11.

Pearson, E. R., Liddell, W. G., Shepherd, M., Corrall, R. J., and Hattersley, A. T. (2000). Sensitivity to sulphonylureas in patients with hepatocyte nuclear factor-1alpha gene mutations: Evidence for pharmacogenetics in diabetes. *Diabet Med* **17**, 543–545.

Pearson, E. R., Starkey, B. J., Powell, R. J., Gribble, F. M., Clark, P. M., and Hattersley, A. T. (2003). Genetic cause of hyperglycaemia and response to treatment in diabetes. *Lancet* **362**, 1275–1281.

Permutt, M. A., Wasson, J. C., Suarez, B. K., Lin, J., Thomas, J., Meyer, J., Lewitzky, S., Rennich, J. S., Parker, A., DuPrat, L., et al. (2001). A genome scan for type 2 diabetes susceptibility loci in a genetically isolated population. *Diabetes* **50**, 681–685.

Persico, A. M., Bird, G., Gabbay, F. H., and Uhl, G. R. (1996). D2 dopamine receptor gene TaqI A1 and B1 restriction fragment length polymorphisms: Enhanced frequencies in psychostimulant-preferring polysubstance abusers. *Biol Psychiatry* **40**, 776–784.

Persson, I., Johansson, I., Bergling, H., Dahl, M. L., Seidegard, J., Rylander, R., Rannug, A., Hogberg, J., and Sundberg, M. I. (1993). Genetic polymorphism of cytochrome P4502E1 in a Swedish population. Relationship to incidence of lung cancer. *FEBS Lett* **319**, 207–211.

Perucho, M. (1996). Cancer of the microsatellite mutator phenotype. *Biol Chem* **377**, 675–684.

Petersen, G. M., Larkin, E., Codori, A.-M., Wang, C.-Y., Booker, S. V., Bacon, J., Giardiello, F. M., and Boyd, P. A. (1999). Attitudes toward colon cancer gene testing: Survey of relatives of colon cancer patients. *Cancer Epidemiol Biomark Prev* **8**, 337–344.

Pfeiffer, R. M., and Gail, M. H. (2003). Sample size calculations for population- and family-based case-control association studies on marker genotypes. *Genet Epidemiol* **25**, 136–148.

Phillips, I. R., Dolphin, C. T., Clair, P., Hadley, M. R., Hutt, A. J., McCombie, R. R., Smith, R. L. and Shephard, E. A. (1995). The molecular biology of the flavin-containing monooxygenases of man. *Chem Biol Interact* **96**, 17–32.

Phillips, J., and Eberwine, J. H. (1996). Antisense RNA amplification: A linear amplification method for analyzing the mRNA population from single living cells. *Methods* **10**, 283–288.

Picciotto, M. R. (2003). Nicotine as a modulator of behavior: Beyond the inverted U. *Trends Pharmacol Sci* **24**, 493–499.

Picciotto, M. R., and Corrigall, W. A. (2002). Neuronal systems underlying behaviors related to nicotine addiction: Neural circuits and molecular genetics. *J Neurosci* **22**, 3338–3341.

Piegorsch, W. W., Weinberg, C. R., and Taylor, J. A. (1994). Non-hierarchical logistic models and case-only designs for assessing susceptibility in population-based case-control studies. *Stat Med* **13**, 153–162.

Piirila, P., Wikman, H., Luukkonen, R., Kaaria, K., Rosenberg, C., Nordman, H., Norppa, H., Vainio, H., and Hirvonen, A. (2001). Glutathione S-transferase genotypes and allergic responses to diisocyanate exposure. *Pharmacogenetics* **11**, 437–445.

Pinkel, D., Segraves, R., Sudar, D., Clark, S., Poole, I., Kowbel, D., Collins, C., Kuo, W. L., Chen, C., Zhai, Y., et al. (1998). High resolution analysis of DNA copy number variation using comparative genomic hybridization to microarrays. *Nat Genet* **20**, 207–211.

Pirmohamed, M. and Park, B. K. (2003). Cytochrome P450 enzyme polymorphisms and adverse drug reactions. *Toxicology* **192**, 23–32.

Pisani, P., Parkin, D. M., Bray, F., and Ferlay, J. (1999). Estimates of the worldwide mortality from 25 cancers in 1990. *Int J Cancer* **83**, 18–29.

Plummer, S. J., Conti, D. V., Paris, P. L., Curran, A. P., Casey, G., and Witte, J. S. (2003). CYP3A4 and CYP3A5 genotypes, haplotypes, and risk of prostate cancer. *Cancer Epidemiol Biomarkers Prev* **12**, 928–932.

Poirier, J., Delisle, M. C., Quirion, R., Aubert, I., Farlow, M., Lahiri, D., Hui, S., Bertrand, P., Nalbantoglu, J., Gilfix, B. M., et al. (1995). Apolipoprotein E4 allele as a predictor of cholinergic deficits and treatment outcome in Alzheimer disease. *Proc Natl Acad Sci USA* **92**, 12260–12264.

Pollack, J. R., Perou, C. M., Alizadeh, A. A., Eisen, M. B., Pergamenschikov, A., Williams, C. F., Jeffrey, S. S., Botstein, D., and Brown, P. O. (1999). Genome-wide analysis of DNA copy-number changes using cDNA microarrays. *Nat Genet* **23**, 41–46.

Pollack, J. R., Sorlie, T., Perou, C. M., Rees, C. A., Jeffrey, S. S., Lonning, P. E., Tibshirani, R., Botstein, D., Borresen-Dale, A. L., and Brown, P. O. (2002). Microarray analysis reveals a major direct role of DNA copy number alteration in the transcriptional program of human breast tumors. *Proc Natl Acad Sci USA* **99**, 12963–12968.

Pompeo, F., Brooke, E., Kawamura, A., Mushtaq, A., and Sim, E. (2002). The pharmacogenetics of NAT: Structural aspects. *Pharmacogenomics* **3**, 19–30.

Ponce, R. A., Wang, E. Y., and Faustman, E. M. (2001). Quality adjusted life years (QALYs) and dose-response models in environmental health policy analysis—methodological considerations. *Sci Total Environ* **274**, 79–91.

Poulter, S. (2001). Genetic testing in toxic injury litigation: The path to scientific certainty or blind alley? *Jurimetrics* **41**, 211.

Powers, K. M., Smith-Weller, T., Franklin, G. M., Longstreth, W. T., Jr., Swanson, P. D., and Checkoway, H. (2003). Parkinson's disease risks associated with dietary iron, manganese, and other nutrient intakes. *Neurology* **60**, 1761–1766.

Presidential/Congressional Commission on Risk Assessment and Risk Management (1997). A framework for environmental health risk management (Vol. 1); Risk assessment and risk management in regulatory decision-making (Vol. 2). (Washington, DC: Government Printing Office).

Price, E. A., Bourne, S. L., Radbourne, R., Lawton, P. A., Lamerdin, J., Thompson, L. H., and Arrand, J. E. (1997). Rare microsatellite polymorphisms in the DNA repair genes XRCC1, XRCC3 and XRCC5 associated with cancer in patients of varying radiosensitivity. *Somat Cell Mol Genet* **23**, 237–247.

Pritchard, J., French, J., Davis, B., and Haseman, J. (2003). The role of transgenic mouse models in carcinogen identification. *Environ Health Perspect* **111**, 444–454.

Priyadarshi, A., Khuder, S. A., Schaub, E. A., and Shrivastava, S. (2000). A meta-analysis of Parkinson's disease and exposure to pesticides. *Neurotoxicology* **21**, 435–440.

Public Health Seattle & King County Healthy Homes: Asthma Project. Seattle, Wash., Public Health Seattle & King County. http://www.metrokc.gov/health/asthma/healthyhomes/. Accessed August 2005.

Qin, L. X., and Kerr, K. F. (2004). Empirical evaluation of data transformations and ranking statistics for microarray analysis. *Nucleic Acids Res* **32**, 5471–5479.

Qiu, L. O., Linder, M. W., Antonino-Green, D. M. and Valdes, R., Jr. (2004). Suppression of cytochrome P450 2E1 promoter activity by interferon-gamma and loss of response due to the -71G>T nucleotide polymorphism of the CYP2E1*7B allele. *J Pharmacol Exp Ther* **308**, 284–288.

Quinn, R. W. (1989). Comprehensive review of morbidity and mortality trends for rheumatic fever, streptococcal disease, and scarlet fever: The decline of rheumatic fever. *Rev Infect Dis* **11**, 928–953.

Qureshi, S. T., Skamene, E., and Malo, D. (1999). Comparative genomics and host resistance against infectious diseases. *Emerg Infect Dis* **5**, 36–47.

Racky, J., Schmitz, H. J., Kauffmann, H. M., and Schrenk, D. (2004). Single nucleotide polymorphism analysis and functional characterization of the human Ah receptor (AhR) gene promoter. *Arch Biochem Biophys* **421**, 91–98.

Rahman, M., and Axelson, O. (1995). Diabetes mellitus and arsenic exposure: A second look at case-control data from a Swedish copper smelter. *Occup Environ Med* **52**, 773–774.

Rahman, M., Tondel, M., Ahmad, S. A., and Axelson, O. (1998). Diabetes mellitus associated with arsenic exposure in Bangladesh. *Am J Epidemiol* **148**, 198–203.

Rahman, M., Wingren, G., and Axelson, O. (1996). Diabetes mellitus among Swedish art glass workers—an effect of arsenic exposure? *Scand J Work Environ Health* **22**, 146–149.

Ratnasinghe, D., Yao, S. X., Tangrea, J. A., Qiao, Y. L., Andersen, M. R., Barrett, M. J., Giffen, C. A., Erozan, Y., Tockman, M. S., and Taylor, P. R. (2001). Polymorphisms of the DNA repair gene XRCC1 and lung cancer risk. *Cancer Epidemiol Biomark Prev* **10**, 119–123.

Rawls, J. (1971). *A Theory of Justice* (Cambridge, MA: Harvard Univ. Press).

Rebbeck, T. R., Martinez, M. E., Sellers, T. A., Shields, P. G., Wild, C. P., and Potter, J. D. (2004). Genetic variation and cancer: improving the environment for publication of association studies. *Cancer Epidemiol Biomarkers Prev* **13**, 1985–1986.

Reich, D. E., Gabriel, S. B., and Altshuler, D. (2003). Quality and completeness of SNP databases. *Nat Genet* **33**, 457–458.

Reich, T., Hinrichs, A., Culverhouse, R., and Bierut, L. (1999). Genetic studies of alcoholism and substance dependence. *Am J Hum Genet* **65**, 599–605.

Relling, M. V., and Dervieux, T. (2001). Pharmacogenetics and cancer therapy. *Nat Rev Cancer* **1**, 99–108.

Renan, M. J. (1993). How many mutations are required for tumorigenesis? Implications from human cancer data. *Mol Carcinog* **7**, 139–146.

Rene, A. A., Daniels, D. E., and Martin, S. A., Jr. (2000). Impact of environmental inequity on health outcome: Where is the epidemiological evidence? *J Natl Med Assoc* **92**, 275–280.

Renwick, A. (1999). Toxicokinetics. In *General and Applied Toxicology*, Ballantyne, B., Marrs, T., and Syversen, T., eds. (New York: Grove's Dictionary), pp. 67–95.

Renwick, A. G. (1991). Safety factors and establishment of acceptable daily intakes. *Food Addit Contain* **8**, 135–149.

Rettie, A. E. and Fisher, M. B. (1999). *Transformation enzymes: oxidative; non-P450*. Handbook of Drug Metabolism. T. F. Woolf. New York, Marcel-Dekker, 131–145.

Rewers, M., and Hamman, R. F. (1995). *Risk Factors for Non-Insulin-Dependent Diabetes*, 2nd ed. [Bethesda, MD: National Institutes of Health, National Institute of Diabetes and Digestive and Kidney Diseases (NIH Publication 95-1468)].

Riboli, E., Hunt, K. J., Slimani, N., Ferrari, P., Norat, T., Fahey, M., Charrondiere, U. R., Hemon, B., Casagrande, C., Vignat, J., et al. (2002). European Prospective Investigation into Cancer and Nutrition (EPIC): Study populations and data collection. *Public Health Nutr* **5**, 1113–1124.

Rice, D. (2004). The US EPA reference dose for methylmercury: Sources of uncertainty. *Environ Res* **95**, 406–413.

Richeldi, L., Kreiss, K., Mroz, M. M., Zhen, B., Tartoni, P., and Saltini, C. (1997). Interaction of genetic and exposure factors in the prevalence of berylliosis. *Am J Ind Med* **32**, 337–340.

Richeldi, L., Sorrentino, R., and Saltini, C. (1993). HLA-DPB1 glutamate 69: A genetic marker of beryllium disease. *Science* **262**, 242–244.

Richter, R. J., and Furlong, C. E. (1999). Determination of paraoxonase (PON1) status requires more than genotyping. *Pharmacogenetics* **9**, 745–753.

Riedl, A. G., Watts, P. M., Jenner, P,. and Marsden, C. D. (1998). P450 enzymes and Parkinson's disease: the story so far. *Mov Disord* **13**, 212–220.

Ries, L., Eisner, M. P., Kosary, C. L., Hankey, B. F., Miller, B. A., Clegg, L., Mariotto, A., Feuer, E. J., and Edwards, B. K. (2004). SEER Cancer Statistics Review, 1975–2001 (Bethesda, MD: National Cancer Inst.).

Risch, H. A., Howe, G. R., Jain, M., Burch, J. D., Holowaty, E. J. and Miller, A. B. (1993). Are female smokers at higher risk for lung cancer than male smokers? A case-control analysis by histologic type. *Am J Epidemiol* **138**, 281–293.

Risch, N. (1990). Linkage strategies for genetically complex traits. I. Multilocus models. *Am J Hum Genet* **46**, 222–228.

Roberts, M. C., and McElreath, L. H. (1992). The role of families in the prevention of physical and mental health problems. In *Family Health Psychology*, Akamatsu, T. J., Stephens, M. A. P., Hobfoll, S. E., and Crowther, J. H., eds. (Washington, DC: Hemisphere).

Roddam, P. L., Rollinson, S., O'Driscoll, M., Jeggo, P. A., Jack, A., and Morgan, G. J. (2002). Genetic variants of NHEJ DNA ligase IV can affect the risk of developing multiple myeloma, a tumour characterised by aberrant class switch recombination. *J Med Genet* **39**, 900–905.

Rollerova, E., and Urbancikova, M. (2000). Intracellular estrogen receptors, their characterization and function (review). *Endocr Regul* **34**, 203–218.

Rose, S. L., and Pietri, C. E. (2002). Workers as research subjects: A vulnerable population. *J Occup Environ Med* **44**, 801–805.

Rosendaal, F. R., Siscovick, D. S., Schwartz, S. M., Beverly, R. K., Psaty, B. M., Longstreth, W. T. J., Raghunathan, T. E., Koepsell, T. D., and Reitsma, P. H. (1997). Factor V Leiden (resistance to activated protein C) increases the risk of myocardial infarction in young women. *Blood* **89**, 2817–2821.

Roses, A. (2002). Genome-based pharmacogenetics and the pharmaceutical industry. *Nature Rev Drug Disc* **1**, 541–549.

Roses, A. D. (2004). Pharmacogenetics and drug development: The path to safer and more effective drugs. *Nat Rev Genet* **5**, 645–656.

Rostami-Hodjegan, A., Lennard, M. S., Woods, H. F., and Tucker, G. T. (1998). Meta-analysis of studies of the CYP2D6 polymorphism in relation to lung cancer and Parkinson's disease. *Pharmacogenetics* **8**, 227–238.

Rothman, K. J. (2002). *Epidemiology: An Introduction* (New York: Oxford Univ. Press).

Rothman, K. J., and Greenland, S. (1998). *Modern Epidemiology*, 2nd ed. (Philadelphia: Lippincott-Williams & Wilkins).

Rothman, N., Garcia-Closas, M., Stewart, W. T., and Lubin, J. (1999). The impact of misclassification in case-control studies of gene-environment interactions. In *Metabolic Polymorphisms and Susceptibility to Cancer*, Vineis, P., Malats, N., Lang, M., d'Errico, A., Caporaso, N., Cuzick, J., and Boffetta, P., eds. (Lyon: IARC Scientific Publications), pp. 89–96.

Rothman, N., Wacholder, S., Caporaso, N. E., Garcia-Closas, M., Buetow, K., and Fraumeni, J. F., Jr. (2001). The use of common genetic polymorphisms to enhance the epidemiologic study of environmental carcinogens. *Biochim Biophys Acta* **1471**, C1–C10.

Rothschild, J. G. and Hansen, R. C. (1985). Fish odor syndrome: trimethylaminuria with milk as chief dietary factor. *Pediatr Dermatol* **3**, 38–39.

Rothstein, M. A. (1983). Employee selection based on susceptibility to occupational illness. *Mich Law Rev* **81**, 1379–1496.

Rothstein, M. A. (2000a). Ethical guidelines for medical research on workers. *J Occup Environ Med* **42**, 1166–1171.

Rothstein, M. A. (2000b). Genetics and the workforce of the next hundred years. *Columbia Bus Law Rev* **2000**, 371.

Rothstein, M. A., and Epps, P. G. (2001). Ethical and legal implications of pharmacogenomics. *Nat Rev Genet* **2**, 228–231.

Ruocco, V., Florio, M., Filioli, F. G., Guerrera, V., and Prota, G. (1989). An unusual case of trimethylaminuria. *Br J Dermatol* **120**, 459–461.

Ruttan, C. C., and Glickman, B. W. (2002). Coding variants in human double-strand break DNA repair genes. *Mutat Res* **509**, 175–200.

Ryberg, D., Skaug, V., et al. (1997). Genotypes of glutathione transferase M1 and P1 and their significance for lung DNA adduct levels and cancer risk. *Carcinogenesis* **18**, 1285–1289.

Saarikoski, S. T., Sata, F., Husgafvel-Pursiainen, K., Rautalahti, M., Haukka, J., Impivaara, O., Jarvisalo, J., Vainio, H., and Hirvonen, A. (2000). CYP2D6 ultrarapid metabolizer genotype as a potential modifier of smoking behaviour. *Pharmacogenetics* **10**, 5–10.

SACGT (July 7, 2000). *Enhancing the Oversight of Genetic Tests: Recommendations of the SACGT* (Secretary's Advisory Committee on Genetic Testing).

Sachidanandam, R., Weissman, D., Schmidt, S. C., Kakol, J. M., Stein, L. D., Marth, G., Sherry, S., Mullikin, J. C., Mortimore, B. J., Willey, D. L., et al. (2001). A map of human genome sequence variation containing 1.42 million single nucleotide polymorphisms. *Nature* **409**, 928–933.

Sachse, C., Brockmoller, J., Bauer, S., and Roots, I. (1999). Functional significance of a C→A polymorphism in intron 1 of the cytochrome P450 CYP1A2 gene tested with caffeine. *Br J Clin Pharmacol* **47**, 445–449.

Safe, S., Bandiera, S., Sawyer, T., Robertson, L., Safe, L., Parkinson, A., Thomas, P. E., Ryan, D. E., Reik, L. M., Levin, W., et al. (1985). PCBs: Structure-function relationships and mechanism of action. *Environ Health Perspect* **60**, 47–56.

Saha, S., Arbelaez, J. J., and Cooper, L. A. (2003). Patient-physician relationships and racial disparities in the quality of health care. *Am J Publ Health* **93**, 1713–1719.

Saiki, R. K., Scharf, S., Faloona, F., Mullis, K. B., Horn, G. T., Erlich, H. A., and Arnheim, N. (1985). Enzymatic amplification of beta-globin genomic sequences and restriction site analysis for diagnosis of sickle cell anemia. *Science* **230**, 1350–1354.

Salama, S. A., Sierra-Torres, C. H., Oh, H. Y., Hamada, F. A., and Au, W. W. (1999). A multiplex-PCR/RFLP procedure for simultaneous CYP2E1, mEH and GSTM1 genotyping. *Cancer Lett* **143**, 51–56.

Salmon, A. Y., Goren, Z., Avissar, Y., and Soreq, H. (1999). Human erythrocyte but not brain acetylcholinesterase hydrolyses heroin to morphine. *Clin Exp Pharmacol Physiol* **26**, 596–600.

Salvan, A., Thomaseth, K., Bortot, P., and Sartori, N. (2001). Use of a toxicokinetic model in the analysis of cancer mortality in relation to the estimated absorbed dose of dioxin (2,3,7,8-tetrachlorodibenzo-p-dioxin, TCDD). *Sci Total Environ* **274**, 21–35.

Sander, T., Harms, H., Podschus, J., Finckh, U., Nickel, B., Rolfs, A., Rommelspacher, H., and Schmidt, L. G. (1995). Dopamine D1, D2 and D3 receptor genes in alcohol dependence. *Psychiatr Genet* **5**, 171–176.

Sander, T., Harms, H., Podschus, J., Finckh, U., Nickel, B., Rolfs, A., Rommelspacher, H., and Schmidt, L. G. (1997). Allelic association of a dopamine transporter gene polymorphism in alcohol dependence with withdrawal seizures or delirium. *Biol Psychiatry* **41**, 299–304.

Sanger, F., Nicklen, S., and Coulson, A. R. (1977). DNA sequencing with chain-terminating inhibitors. *Proc Natl Acad Sci USA* **74**, 5463–5467.

Sasaki, M., Tanaka, Y., Okino, S. T., Nomoto, M., Yonezawa, S., Nakagawa, M., Fujimoto, S., Sakuragi, N., and Dahiya, R. (2004). Polymorphisms of the CYP1B1 gene as risk factors for human renal cell cancer. *Clin Cancer Res* **10**, 2015–2019.

Savitz, D. A. (2003). *Interpreting Epidemiologic Evidence: Strategies for Study Design and Analysis* (New York: Oxford Univ. Press).

Savolainen, M. J., and Kesaniemi, Y. A. (1995). Effects of alcohol on lipoproteins in relation to coronary heart disease. *Curr Opin Lipidol* **6**, 243–250.

Sawyer, M. B., Innocenti, F., Das, S., Cheng, C., Ramirez, J., Pantle-Fisher, F. H., Wright, C., Badner, J., Pei, D., Boyett, J. M., et al. (2003). A pharmacogenetic study of uridine diphosphate-glucuronosyltransferase 2B7 in patients receiving morphine. *Clin Pharmacol Ther* **73**, 566–574.

Scanlon, P. D., Raymond, F. A., and Weinshilboum, R. M. (1979). Catechol-O-methyltransferase: Thermolabile enzyme in erythrocytes of subjects homozygous for allele for low activity. *Science* **203**, 63–65.

Scariano, J. K., Simplicio, S. G., Montoya, G. D., Garry, P. J., and Baumgartner, R. N. (2004). Estrogen receptor beta dinucleotide (CA) repeat polymorphism is significantly associated with bone mineral density in postmenopausal women. *Calcif Tissue Int.* **74**, 501–508.

Schaid, D. J. (1999). Case-parents design for gene-environment interaction. *Genet Epidemiol* **16**, 261–273.

Schaid, D. J., Rowland, C. M., Tines, D. E., Jacobson, R. M., and Poland, G. A. (2002). Score tests for association between traits and haplotypes when linkage phase is ambiguous. *Am J Hum Genet* **70**, 425–434.

Scheel, J., Hussong, R., Schrenk, D., and Schmitz, H. J. (2002). Variability of the human aryl hydrocarbon receptor nuclear translocator (ARNT) gene. *J Hum Genet* **47**, 217–224.

Schena, M., Shalon, D., Davis, R. W., and Brown, P. O. (1995). Quantitative monitoring of gene expression patterns with a complementary DNA microarray. *Science* **270**, 467–470.

Schena, M., Shalon, D., Heller, R., Chai, A., Brown, P. O., and Davis, R. W. (1996). Parallel human genome analysis: microarray-based expression monitoring of 1000 genes. *Proc Natl Acad Sci USA* **93**, 10614–10619.

Scheuner, M. T. (2003). Genetic evaluation for coronary artery disease. *Genet Med* **5**, 269–285.

Schill, A. L. (2000). Genetic information in the workplace. Implications for occupational health surveillance. *AAOHN J* **48**, 80–91.

Schinka, J. A., Town, T., Abdullah, L., Crawford, F. C., Ordorica, P. I., Francis, E., Hughes, P., Graves, A. B., Mortimer, J. A., and Mullan, M. (2002). A functional polymorphism within the mu-opioid receptor gene and risk for abuse of alcohol and other substances. *Mol Psychiatry* **7**, 224–228.

Schlenk, D., Cashman, J. R., Yeung, C. K., Zhang, X., and Rettie, A. E. (2002). Role of human flavin-containing monooxygenases in the sulfoxidation of [14C] aldicarb. *Pesticide Biochemistry and Physiology* **73**, 67–73.

Schlesselman, J. (1982). Case Control Studies: Design, Conduct, Analysis (Monographs in Epidemiology and Biostatistics). New York, Oxford University Press.

Schmidt, S., and Schaid, D. J. (1999). Potential misinterpretation of the case-only study to assess gene-environment interaction. *Am J Epidemiol* **150**, 878–885.

Schneider, J., Bernges, U., Philipp, M., and Woitowitz, H. J. (2004). GSTM1, GSTT1, and GSTP1 polymorphism and lung cancer risk in relation to tobacco smoking. *Cancer Lett* **208**, 65–74.

Schuckit, M. A., and Smith, T. L. (1996). An 8-year follow-up of 450 sons of alcoholic and control subjects. *Arch Gen Psychiatry* **53**, 202–210.

Schuckit, M. A., Tsuang, J. W., Anthenelli, R. M., Tipp, J. E., and Nurnberger, J. I., Jr. (1996). Alcohol challenges in young men from alcoholic pedigrees and control families: a report from the COGA project. *J Stud Alcohol* **57**, 368–377.

Schuetz, E. G., Schuetz, J. D., Grogan, W. M., Naray-Fejes-Toth, A., Fejes-Toth, G., Raucy, J., Guzelian, P., Gionela, K., and Watlington, C. O. (1992). Expression of cytochrome P450 3A in amphibian, rat, and human kidney. *Arch Biochem Biophys* **294**, 206–214.

Schulte, P. A., Lomax, G. P., Ward, E. M., and Colligan, M. J. (1999). Ethical issues in the use of genetic markers in occupational epidemiologic research. *J Occup Environ Med* **41**, 639–646.

Schulze, T. G., Schumacher, J., et al. (2001). Lack of association between a functional polymorphism of the cytochrome P450 1A2 (CYP1A2) gene and tardive dyskinesia in schizophrenia. *Am J Med Genet* **105**, 498–501.

Schumann, G., Rujescu, D., Szegedi, A., Singer, P., Wiemann, S., Wellek, S., Giegling, I., Klawe, C., Anghelescu, I., Heinz, A., et al. (2003). No association of alcohol dependence with a NMDA-receptor 2B gene variant. *Mol Psychiatry* **8**, 11–12.

Schwartz, B. S., Lee, B. K., Lee, G. S., Stewart, W. F., Simon, D., Kelsey, K., and Todd, A. C. (2000). Associations of blood lead, dimercaptosuccinic acid-chelatable lead, and tibia lead with polymorphisms in the vitamin D receptor and [delta]-aminolevulinic acid dehydratase genes. *Environ Health Perspect* **108**, 949–954.

Schwartz, E. (1988). A proportionate mortality ratio analysis of pulp and paper mill workers in New Hampshire. *Br J Ind Med* **45**, 234–238.

Schwarz, M., Glick, D., Loewenstein, Y., and Soreq, H. (1995). Engineering of human cholinesterases explains and predicts diverse consequences of administration of various drugs and poisons. *Pharmacol Ther* **67**, 283–322.

Scriver, C. R., and Childs, B. (1989). *Garrod's Inborn Factors in Disease* (London: Oxford Univ. Press).

Sealey, J., and Laragh, J. (1990). The renin-angiotensin-aldosterone system for normal regulation of blood pressure and sodium and potassium homeostasis. In *Hypertension: Pathophysiology, Diagnosis and Management*, Laragh, J., and Brenner, B., eds. (New York: Raven Press), pp. 1287–1317.

Secretary's Advisory Committee on Genetic Testing (2000). *Enhancing the Oversight of Genetic Tests: Recommendations of the SACGT* (Bethesda, MD: National Institutes of Health).

Sellers, E. M., Tyndale, R. F., and Fernandes, L. C. (2003). Decreasing smoking behaviour and risk through CYP2A6 inhibition. *Drug Discov Today* **8**, 487–493.

Seow, A., Shi, C.-Y., Chung, F.-L., Jiao, D., Hankin, J. H., Lee, H.-P., Coetzee, G. A., and Yu, M. C. (1998). Urinary total isothiocyanate (ITC) in a population-based sample of middle-aged and older Chinese in Singapore: Relationship with dietary total ITC and glutathione *S*-transferase *M1/T1/P1* genotypes. *Cancer Epidemiol Biomark Prev* **7**, 775–781.

Shannon, W., Culverhouse, R., and Duncan, J. (2003). Analyzing microarray data using cluster analysis. *Pharmacogenomics* **4**, 41–52.

Sharp, R. R. (2003). Ethical issues in environmental health research. *Environ Health Perspect* **111**, 1786–1788.

Sheehan, D., Meade, G., Foley, V. M., and Dowd, C. A. (2001). Structure, function and evolution of glutathione transferases: Implications for classification of non-mammalian members of an ancient enzyme superfamily. *Biochem J* **360**, 1–16.

Shelby, J. P., White, J., Ganesan, K., Rathod, P. K., and Chiu, D. T. (2003). A microfluidic model for single-cell capillary obstruction by *Plasmodium falciparum*-infected erythrocytes. *Proc Natl Acad Sci USA* **100**, 14618–14622.

Shi, H., Maier, S., Nimmrich, I., Yan, P. S., Caldwell, C. W., Olek, A., and Huang, T. H. (2003). Oligonucleotide-based microarray for DNA methylation analysis: Principles and applications. *J Cell Biochem* **88**, 138–143.

Shield, A. J., Thomae, B. A., Eckloff, B. W., Wieben, E. D., and Weinshilboum, R. M. (2004). Human catechol O-methyltransferase genetic variation: Gene resequencing and functional characterization of variant allozymes. *Mol Psychiatry* **9**, 151–160.

Shields, P. G., Lerman, C., Audrain, J., Bowman, E. D., Main, D., Boyd, N. R., and Caporaso, N. E. (1998). Dopamine D4 receptors and the risk of cigarette smoking in African-Americans and Caucasians. *Cancer Epidemiol Biomark Prev* **7**, 453–458.

Shin, A., Kang, D., Nishio, H., Lee, M. J., Park, S. K., Kim, S. U., Noh, D. Y., Choe, K. J., Ahn, S. H., Hirvonen, A., et al. (2003). Estrogen receptor alpha gene polymorphisms and breast cancer risk. *Breast Cancer Res Treat* **80**, 127–131.

Shin, R. W., Lee, V. M., and Trojanowski, J. Q. (1995). Neurofibrillary pathology and aluminum in Alzheimer's disease. *Histol Histopathol* **10**, 969–978.

Shoaib, M., Gommans, J., Morley, A., Stolerman, I. P., Grailhe, R., and Changeux, J. P. (2002). The role of nicotinic receptor beta-2 subunits in nicotine discrimination and conditioned taste aversion. *Neuropharmacology* **42**, 530–539.

Shojania, A. M. (1980). Problems in the diagnosis and investigation of megaloblastic anemia. *Can Med Assoc J* **122**, 999–1004.

Shu, X. O., Jin, F., Dai, Q., Wen, W., Potter, J. D., Kushi, L. H., Ruan, Z., Gao, Y.-T., and Zheng, W. (2001). Soyfood intake during adolescence and subsequent risk of breast cancer among Chinese women. *Cancer Epidemiol Biomark Prev* **10**, 483–488.

Shuber, A. P., Skoletsky, J., Stern, R., and Handelin, B. L. (1993). Efficient 12-mutation testing in the CFTR gene: A general model for complex mutation analysis. *Hum Mol Genet* **2**, 153–158.

Sigvardsson, S., Bohman, M., and Cloninger, C. R. (1996). Replication of the Stockholm Adoption Study of alcoholism. Confirmatory cross-fostering analysis. *Arch Genet Psychiatry* **53**, 681–687.

Silander, K., Mohlke, K. L., Scott, L. J., Peck, E. C., Hollstein, P., Skol, A. D., Jackson, A. U., Deloukas, P., Hunt, S., Stavrides, G., et al. (2004a). Genetic variation near the hepatocyte nuclear factor-4alpha gene predicts susceptibility to type 2 diabetes. *Diabetes* **53**, 1141–1149.

Silander, K., Scott, L. J., Valle, T. T., Mohlke, K. L., Stringham, H. M., Wiles, K. R., Duren, W. L., Doheny, K. F., Pugh, E. W., Chines, P., et al. (2004b). A large set of Finnish affected sibling pair families with type 2 diabetes suggests susceptibility loci on chromosomes 6, 11, and 14. *Diabetes* **53**, 821–829.

Silverman, M. A., Neale, M. C., Sullivan, P. F., Harris-Kerr, C., Wormley, B., Sadek, H., Ma, Y., Kendler, K. S., and Straub, R. E. (2000). Haplotypes of four novel single nucleotide polymorphisms in the nicotinic acetylcholine receptor beta2-subunit (CHRNB2) gene show no association with smoking initiation or nicotine dependence. *Am J Med Genet* **96**, 646–653.

Sim, E., Payton, M., Noble, M., and Minchin, R. (2000). An update on genetic, structural and functional studies of arylamine N-acetyltransferases in eucaryotes and procaryotes. *Hum Mol Genet* **9**, 2435–2441.

Simeone, T. A., Donevan, S. D., and Rho, J. M. (2003). Molecular biology and ontogeny of gamma-aminobutyric acid (GABA) receptors in the mammalian central nervous system. *J Child Neurol* **18**, 39–48; discussion 49.

Sing, C. F. and Davignon, J. (1985). Role of the apolipoprotein E polymorphism in determining normal plasma lipid and lipoprotein variation. *Am J Hum Genet* **37**, 268–285.

Sing, C. F., Stengard, J. H., and Kardia, S. L. R. (2003). Genes, environment, and cardiovascular disease. *Arterioscler Thromb Vasc Biol* **23**, 1190–1196.

Singer, B., and Grunberger, D. (1983). *The Molecular Biology of Mutagens and Carcinogens* (New York; Plenum Press).

Singhal, A., Cole, T. J., Fewtrell, M., and Lucas, A. (2004). Breastmilk feeding and lipoprotein profile in adolescents born preterm: Follow-up of a prospective randomised study. *Lancet* **363**, 1571–1578.

Singhal, R. K., Prasad, R., and Wilson, S. H. (1995). DNA polymerase b conducts the gap-filling step in uracil-initiated base excision repair in bovine testis nuclear extract. *J Biol Chem* **270**, 949–957.

Singleton, A. B., Thomson, J. H., Morris, C. M., Court, J. A., Lloyd, S., and Cholerton, S. (1998). Lack of association between the dopamine D2 receptor gene allele DRD2*A1 and cigarette smoking in a United Kingdom population. Pharmacogenetics **8**, 125–128.

Sinha, R., Cloninger, C. R., and Parsian, A. (2003). Linkage disequilibrium and haplotype analysis between serotonin receptor 1B gene variations and subtypes of alcoholism. *Am J Med Genet* **121B,** 83–88.

Sinha, R., Rothman, N., Mark, S. D., Murray, S., Brown, E. D., Levander, O. A., Davies, D. S., Lang, N. P., Kadlubar, F. F., and Hoover, R. N. (1995). Lower levels of urinary 2-amino-3,8-dimethylimidazo[4,5-f]-quinoxaline (MeIQx) in humans with higher CYP1A2 activity. *Carcinogenesis* **16**, 2859–2861.

Slattery, M., and Kerber, R. (1993). A comprehensive evaluation of family history and breast cancer risk. The Utah Population Database. *JAMA* **270**, 1563–1568.

Slattery, M. L., Kampman, E., Samowitz, W., Caan, B. J., and Potter, J. D. (2000). Interplay between dietary inducers of GST and the *GSTM-1* genotype in colon cancer. *Int J Cancer* **87**, 728–733.

Slonim, D. K. (2002). From patterns to pathways: gene expression data analysis comes of age. *Nat Genet* **32** (Suppl), 502–508.

Slovic, P. (1987). Perception of risk. *Science* **236**, 280–285.

Slovic, P. (1998). If hormesis exists: Implications for risk perception and communication. *Hum Exp Toxicol* **17**, 439–440.

Small, K. M., and Liggett, S. B. (2001). Identification and functional characterization of alpha(2)-adrenoceptor polymorphisms. *Trends Pharmacol Sci* **22**, 471–477.

Small, K. M., McGraw, D. W., and Liggett, S. B. (2003). Pharmacology and physiology of human adrenergic receptor polymorphisms. *Annu Rev Pharmacol Toxicol* **43**, 381–411.

Small, K. M., Wagoner, L. E., Levin, A. M., Kardia, S. L., and Liggett, S. B. (2002). Synergistic polymorphisms of beta1- and alpha2C-adrenergic receptors and the risk of congestive heart failure. *New Engl J Med* **347**, 1135–1142.

Smedley, B. D., Syme, L. S., et al., (eds.) (2000). *Promoting Health: Intervention Strategies from Social and Behavioral Research* (Washington, DC: National Academy Press).

Smith, A. H., Goycolea, M., Haque, R., and Biggs, M. L. (1998). Marked increase in bladder and lung cancer mortality in a region of Northern Chile due to arsenic in drinking water. *Am J Epidemiol* **147**, 660–669.

Smith, E., Bianco-Miotto, T., Drew, P., and Watson, D. (2003). Method for optimizing methylation-specific PCR. *Biotechniques* **35**, 32–33.

Smith, L. M., Sanders, J. Z., Kaiser, R. J., Hughes, P., Dodd, C., Connell, C. R., Heiner, C., Kent, S. B., and Hood, L. E. (1986). Fluorescence detection in automated DNA sequence analysis. *Nature* **321**, 674–679.

Snijders, A. M., Pinkel, D., and Albertson, D. G. (2003). Current status and future prospects of array-based comparative genomic hybridisation. *Brief Funct Genom Proteom* **2**, 37–45.

Sobel, E., Davanipour, Z., Sulkava, R., Erkinjuntti, T., Wikstrom, J., Henderson, V. W., Buckwalter, G., Bowman, J. D., and Lee, P. J. (1995). Occupations with exposure to electromagnetic fields: A possible risk factor for Alzheimer's disease. *Am J Epidemiol* **142**, 515–524.

Social Issues Subcommittee of the American Society of Human Genetics (ASHG) on Familial Disclosure (1998). Professional disclosure of familial genetic information. *Am J Hum Genet* **62**, 474–483.

Sofowora, G. G., Dishy, V., Muszkat, M., Xie, H. G., Kim, R. B., Harris, P. A., Prasad, H. C., Byrne, D. W., Nair, U. B., Wood, A. J., and Stein, C. M. (2003). A common beta1-adrenergic receptor polymorphism (Arg389Gly) affects blood pressure response to beta-blockade. *Clin Pharmacol Ther* **73**, 366–371.

Sogawa, K. and Fujii-Kuriyama, Y. (1997). Ah receptor, a novel ligand-activated transcription factor. *J Biochem (Tokyo)* **122**, 1075–1079.

Sohda, T. (1999). Allele-specific polymerase chain reaction for genotyping human cytochrome P450 2E1. *J Clin Lab Anal* **13**, 205–208.

Solinas, M., Panlilio, L. V., Antoniou, K., Pappas, L. A., and Goldberg, S. R. (2003). The cannabinoid CB1 antagonist N-piperidinyl-5-(4-chlorophenyl)-1-(2,4-dichlorophenyl)-4-methylpyrazole-3-carboxamide (SR-141716A) differentially alters the reinforcing effects of heroin under continuous reinforcement, fixed ratio, and progressive ratio schedules of drug self-administration in rats. *J Pharmacol Exp Ther* **306**, 93–102.

Solinas-Toldo, S., Lampel, S., Stilgenbauer, S., Nickolenko, J., Benner, A., Dohner, H., Cremer, T., and P., L. (1997). Matrix-based comparative genomic hybridization: biochips to screen for genomic imbalances. *Genes Chromosomes Cancer* **20**, 399–407.

Song, J., Koller, D. L., Foroud, T., Carr, K., Zhao, J., Rice, J., Nurnberger, J. I., Jr., Begleiter, H., Porjesz, B., Smith, T. L., et al. (2003). Association of GABAA receptors and alcohol dependence and the effects of genetic imprinting. *Am J Med Genet* **117B**, 39–45.

Soreq, H., and Zakut, H. (1993). *Human Cholinesterases and Anticholinesterases* (San Diego, CA: Academic Press).

Spalding, J. W., French, J. E., Stasiewicz, S., Furedi-Machacek, M., Conner, F., Tice, R. R., and Tennant, R. W. (2000). Responses of transgenic mouse lines p53(+/–) and Tg.AC to agents tested in conventional carcinogenicity bioassays. *Toxicol Sci* **53**, 213–223.

Sparks, D. L., and Schreurs, B. G. (2003). Trace amounts of copper in water induce beta-amyloid plaques and learning deficits in a rabbit model of Alzheimer's disease. *Proc Natl Acad Sci USA* **100**, 11065–11069.

Spielman, R. S., McGinnis, R. E., and Ewens, W. J. (1993). Transmission test for linkage disequilibrium: The insulin gene region and insulin-dependent diabetes mellitus (IDDM). *Am J Hum Genet* **52**, 506–516.

Spitz, M. R., Duphorne, C. M., Detry, M. A., Pillow, P. C., Amos, C. I., Lei, L., de Andrade, M., Gu, X., Hong, W. K., and Wu, X. (2000). Dietary intake of isothiocyanates: Evidence of a joint effect with glutathione S-transferase polymorphisms in lung cancer risk. *Cancer Epidemiol Biomark Prev* **9**, 1017–1020.

Splawski, I., Timothy, K. W., Tateyama, M., Clancy, C. E., Malhotra, A., Beggs, A. H., Cappuccio, F. P., Sagnella, G. A., Kass, R. S., and Keating, M. T. (2002). Variant of SCN5A sodium channel implicated in risk of cardiac arrhythmia. *Science* **297**, 1333–1336.

Spurr, N. K., Gough, A. C., Stevenson, K., and Wolf, C. R. (1987). Msp-1 polymorphism detected with a cDNA probe for the P-450 I family on chromosome 15. *Nucleic Acids Res* **15**, 5901.

Sreenan, S. K., Zhou, Y. P., Otani, K., Hansen, P. A., Currie, K. P., Pan, C. Y., Lee, J. P., Ostrega, D. M., Pugh, W., Horikawa, Y., et al. (2001). Calpains play a role in insulin secretion and action. *Diabetes* **50**, 2013–2020.

Steinmetz, K. A., and Potter, J. D. (1996). Vegetables, fruit, and cancer prevention: A review. *J Am Diet Assoc* **96**, 1027–1039.

Stevens, J. C., Hines, R. N., Gu, C., Koukouritaki, S. B., Manro, J. R., Tandler, P. J., and Zaya, M. J. (2003). Developmental expression of the major human hepatic CYP3A enzymes. *J Pharmacol Exp Ther* **307**, 573–582.

Stewart, R. D. (1999). On the complexity of the DNA damages created by endogenous processes. *Radiat Res* **152**, 101–104.

Stoilov, I., Akarsu, A. N., and Sarfarazi, M. (1997). Identification of three different truncating mutations in cytochrome P4501B1 (CYP1B1) as the principal cause of primary congenital glaucoma (Buphthalmos) in families linked to the GLC3A locus on chromosome 2p21. *Hum Mol Genet* **6**, 641–647.

Stoler, D. L., Chen, N., Basik, M., Kahlenberg, M. S., Rodriguez-Bigas, M. A., Petrelli, N. J., and Anderson, G. R. (1999). The onset and extent of genomic instability in sporadic colorectal tumor progression. *Proc Natl Acad Sci USA* **96**, 15121–15126.

Stone, A. N., Mackenzie, P. I., Galetin, A., Houston, J. B., and Miners, J. O. (2003). Isoform selectivity and kinetics of morphine 3- and 6-glucuronidation by human udp-glucuronosyltransferases: Evidence for atypical glucuronidation kinetics by UGT2B7. *Drug Metab Dispos* **31**, 1086–1089.

Storey, J. D., and Tibshirani, R. (2003). Statistical significance for genomewide studies. *Proc Natl Acad Sci USA* **100**, 9440–9445.

Stormer, E., Roots, I., and Brockmoller, J. (2000). Benzydamine N-oxidation as an index reaction reflecting FMO activity in human liver microsomes and impact of FMO3 polymorphisms on enzyme activity. *Br J Clin Pharmacol* **50**, 553–561.

Stram, D. O., Haiman, C. A., Hirschhorn, J. N., Altshuler, D., Kolonel, L. N., Henderson, B. E., and Pike, M. C. (2003a). Choosing haplotype-tagging SNPS based on unphased genotype data using a preliminary sample of unrelated subjects with an example from the Multiethnic Cohort Study. *Hum Hered* **55**, 27–36.

Stram, D. O., Leigh Pearce, C., Bretsky, P., Freedman, M., Hirschhorn, J. N., Altshuler, D., Kolonel, L. N., Henderson, B. E., and Thomas, D. C. (2003b). Modeling and

E-M estimation of haplotype-specific relative risks from genotype data for a case-control study of unrelated individuals. *Hum Hered* **55**, 179–190.

Strange, R. C., Spiteri, M. A., Ramachandran, S., and Fryer, A. A. (2001). Glutathione-S-transferase family of enzymes. *Mutat Res* **482**, 21–26.

Strott, C. A. (2002). Sulfonation and molecular action. *Endocr Rev* **23**, 703–732.

Su, T., Bao, Z., Zhang, Q. Y., Smith, T. J., Hong, J. Y., and Ding, X. (2000). Human cytochrome P450 CYP2A13: predominant expression in the respiratory tract and its high efficiency metabolic activation of a tobacco-specific carcinogen, 4-(methylnitrosamino)-1-(3-pyridyl)-1-butanone. *Cancer Res* **60**, 5074–5079.

Sugimura, H., Hamada, G. S., Suzuki, I., Iwase, T., Kiyokawa, E., Kino, I., and Tsugane, S. (1995). CYP1A1 and CYP2E1 polymorphism and lung cancer, case-control study in Rio de Janeiro, Brazil. *Pharmacogenetics* **5**, Spec No 5, S145–S148.

Sull, J. W., Ohrr, H., Kang, D. R., and Nam, C. M. (2004). Glutathione S-transferase M1 status and breast cancer risk: A meta-analysis. *Yonsei Med* J **45**, 683–689.

Surveillance, Epidemiology, and End Results Program (2004). *SEER*Stat Database: Incidence—SEER 9 Regs Public-Use* (Bethesda, MD: Surveillance, Epidemiology, and End Results Program, National Cancer Inst.). www.seer.cancer.gov.

Suskind, R. R., and Hertzberg, V. S. (1984). Human health effects of 2,4,5-T and its toxic contaminants. *JAMA* **251**, 2372–2380.

Swallow, D. M., Poulter, M., and Hollox, E. J. (2001). Intolerance to lactose and other dietary sugars. *Drug Metab Dispos* **29**, 513–516.

Sweeney, C., Nazar-Stewart, V., Stapleton, P. L., Eaton, D. L., and Vaughan, T. L. (2003). Glutathione S-transferase M1, T1, and P1 polymorphisms and survival among lung cancer patients. *Cancer Epidemiol Biomarkers Prev* **12**, 527–533.

Swinburn, B. A. (1996). The thrifty genotype hypothesis: How does it look after 30 years? *Diabet Med* **13**, 698–699.

Talmud, P. J., and Humphries, S. E. (2002). Gene:environment interaction in lipid metabolism and effect on coronary heart disease risk. *Curr Opin Lipidol* **13**, 149–154.

Talmud, P. J., and Stephens, J. W. (2004). Lipoprotein lipase gene variants and the effect of environmental factors on cardiovascular disease risk. *Diabetes Obes Metab* **6**, 1–7.

Tan, E. K., Khajavi, M., Thornby, J. I., Nagamitsu, S., Jankovic, J., and Ashizawa, T. (2000). Variability and validity of polymorphism association studies in Parkinson's disease. *Neurology* **55**, 533–538.

Tanaka, Y., Sasaki, M., Kaneuchi, M., Fujimoto, S., and Dahiya, R. (2003). Estrogen receptor alpha polymorphisms and renal cell carcinoma—a possible risk. *Mol Cell Endocrinol* **202**, 109–116.

Tanis, B. C., Bloemenkamp, D. G., van den Bosch, M. A., Kemmeren, J. M., Algra, A., van de Graaf, Y., and Rosendaal, F. R. (2003). Prothrombotic coagulation defects and cardiovascular risk factors in young women with acute myocardial infarction. *Br J Haematol* **122**, 471–478.

Tanner, C. M., Ottman, R., Goldman, S. M., Ellenberg, J., Chan, P., Mayeux, R., and Langston, J. W. (1999). Parkinson disease in twins: An etiologic study. *JAMA* **281**, 341–346.

Taylor, S. L. (1986). Histamine food poisoning: toxicology and clinical aspects. *Crit Rev Toxicol* **17**, 91–128.

Tenhunen, J., Salminen, M., Lundstrom, K., Kiviluoto, T., Savolainen, R., and Ulmanen, I. (1994). Genomic organization of the human catechol O-methyltransferase gene and its expression from two distinct promoters. *Eur J Biochem* **223**, 1049–1059.

Tennant, R. W. (1998). Evaluation and validation issues in the development of transgenic mouse carcinogenicity bioassays. *Environ Health Perspect* **106** (Suppl 2), 473–476.

Teschke, K., Smith, J. C., and Olshan, A. F. (2000). Evidence of recall bias in volunteered vs. prompted responses about occupational exposures. *Am J Ind Med* **38**, 385–388.

The International HapMap Consortium (2003). The International HapMap Project. *Nature* **426**, 789–796.

Thier, R., Bruning, T., Roos, P. H., Rihs, H. P., Golka, K., Ko, Y., and Bolt, H. M. (2003). Markers of genetic susceptibility in human environmental hygiene and toxicology: The role of selected CYP, NAT and GST genes. *Int J Hyg Environ Health* **206**, 149–171.

Thiruchelvam, M., McCormack, A., Richfield, E. K., Baggs, R. B., Tank, A. W., Di Monte, D. A., and Cory-Slechta, D. A. (2003). Age-related irreversible progressive nigrostriatal dopaminergic neurotoxicity in the paraquat and maneb model of the Parkinson's disease phenotype. *Eur J Neurosci* **18**, 589–600.

Thithapandha, A. (1997). A pharmacogenetic study of trimethylaminuria in Orientals. *Pharmacogenetics* **7**, 497–501.

Thomae, B. A., Eckloff, B. W., Freimuth, R. R., Wieben, E. D., and Weinshilboum, R. M. (2002). Human sulfotransferase SULT2A1 pharmacogenetics: Genotype-to-phenotype studies. *Pharmacogenom J* **2**, 48–56.

Thomae, B. A., Rifki, O. F., Theobald, M. A., Eckloff, B. W., Wieben, E. D., and Weinshilboum, R. M. (2003). Human catecholamine sulfotransferase (SULT1A3) pharmacogenetics: Functional genetic polymorphism. *J Neurochem* **87**, 809–819.

Thomas, D. B. (1995). Alcohol as a cause of cancer. *Environ Health Perspect* **103** (Suppl 8), 153–160.

Thomas, D. C. (2004). Statistical issues in the design and analysis of gene-disease association studies. In *Human Genome Epidemiology: A Scientific Foundation for Using Genetic Information to Improve Health and Prevent Disease*, Khoury, M. J., Little, J., and Burke, W., eds. (Oxford: Oxford University Press) pp. 92–110.

Thomas, D. C., and Witte, J. S. (2002). Population stratification: A problem for case-control studies of candidate gene associations? *Cancer Epidemiol Biomark Prev* **11**, 505–512.

Thomas, P. S. (1980). Hybridization of denatured RNA and small DNA fragments transferred to nitrocellulose. *Proc Natl Acad Sci USA* **77**, 5201–5205.

Thomas, R. S., Rank, D. R., Penn, S. G., Zastrow, G. M., Hayes, K. R., Pande, K., Glover, E., Silander, T., Craven, M. W., Reddy, J. K., et al. (2001). Identification of toxicologically predictive gene sets using cDNA microarrays. *Mol Pharmacol* **60**, 1189–1194.

Thompson, E. E., Kuttab-Boulos, H., Witonsky, D., Yang, L., Roe, B. A., and Di Rienzo, A. (2004). CYP3A Variation and the Evolution of Salt-Sensitivity Variants. *Am J Hum Genet* **75**, 1059–1069.

Thornquist, M. D., Edelstein, C., Goodman, G. E., and Omenn, G. S. (2002). Streamlining IRB review in multisite trials through single-study IRB cooperative agreements: Experience of the Beta-Carotene and Retinol Efficacy Trial (CARET). *Control Clin Trials* **23**, 80–86.

Thummel, K. E., Brimer, C., et al. (2001). Transcriptional control of intestinal cytochrome P-4503A by 1alpha,25-dihydroxy vitamin D3. *Mol Pharmacol* **60**, 1399–1406.

Tiwari, H. K., Zhu, X., Elston, R. C., Shu, Y., and George, V. (1999). Association and linkage analysis of ICD-10 diagnosis for alcoholism. *Genet Epidemiol* **17** (Suppl 1), S343–S347.

Tomlin, C. (1994). The Pesticide Manual. (Surrey, UK: Crop Protection Publications, British Crop Protection Council).

Town, T., Abdullah, L., Crawford, F., Schinka, J., Ordorica, P. I., Francis, E., Hughes, P., Duara, R., and Mullan, M. (1999). Association of a functional mu-opioid receptor allele (+118A) with alcohol dependency. *Am J Med Genet* **88**, 458–461.

Townsend, D., and Tew, K. (2003). Cancer drugs, genetic variation and the glutathione-S-transferase gene family. *Am J pharmacogenom* **3**, 157–172.

Treacy, E., Johnson, D., Pitt, J. J., and Danks, D. M. (1995). Trimethylaminuria, fish odour syndrome: a new method of detection and response to treatment with metronidazole. *J Inherit Metab Dis* **18**, 306–312.

Treacy, E. P., Akerman, B. R., et al. (1998). Mutations of the flavin-containing monooxygenase gene (FMO3) cause trimethylaminuria, a defect in detoxication. *Hum Mol Genet* **7**, 839–845.

Tricker, A. R. (2003). Nicotine metabolism, human drug metabolism polymorphisms, and smoking behaviour. *Toxicology* **183**, 151–173.

Troelsen, J. T., Olsen, J., Møller, J., and Sjöström, H. (2003). An upstream polymorphism associated with lactase persistence has increased enhancer activity. *Gastroenterology* **125**, 1686–1694.

Trojanowski, J. Q., and Lee, V. M. (2000). "Fatal attractions" of proteins. A comprehensive hypothetical mechanism underlying Alzheimer's disease and other neurodegenerative disorders. *Ann NY Acad Sci* **924**, 62–67.

True, W. R., Xian, H., Scherrer, J. F., Madden, P. A., Bucholz, K. K., Heath, A. C., Eisen, S. A., Lyons, M. J., Goldberg, J., and Tsuang, M. (1999). Common genetic vulnerability for nicotine and alcohol dependence in men. *Arch Gen Psychiatry* **56**, 655–661.

Tsai, S. M., Wang, T. N., and Ko, Y. C. (1999). Mortality for certain diseases in areas with high levels of arsenic in drinking water. *Arch Eviron Health* **54**, 186–193.

Tseng, C. H., Tai, T. Y., Chong, C. K., Tseng, C. P., Lai, M. S., Lin, B. J., Chiou, H. Y., Hsueh, Y. M., Hsu, K. H., and Chen, C. J. (2000). Long-term arsenic exposure and incidence of non-insulin-dependent diabetes mellitus: A cohort study in arseniasis-hyperendemic villages in Taiwan. *Environ Health Perspect* **108**, 847–851.

Tsuda, T., Nagira, T., Yamamoto, M., and Kume, Y. (1990). An epidemiological study on cancer in certified arsenic poisoning patients in Toroku. *Ind Health* **28**, 53–62.

Tsujino, A., Maertens, C., Ohno, K., Shen, X. M., Fukuda, T., Harper, C. M., Cannon, S. C., and Engel, A. G. (2003). Myasthenic syndrome caused by mutation of the SCN4A sodium channel. *Proc Natl Acad Sci USA* **100**, 7377–7382.

Tugnait, M., Hawes, E. M., McKay, G., Rettie, A. E., Haining, R. L., and Midha, K. K. (1997). N-oxygenation of clozapine by flavin-containing monooxygenase. *Drug Metab Dispos* **25**, 524–527.

Tyfield, L., Reichardt, J. F.-K., J., Croke, D. T., Elsas, L. J. I., Strobl, W., Kozak, L., Coskun, T., Novelli, G., Okano, Y., Zekanowski, C., et al. (1999). Classical galactosemia and mutations at the galactose-1-phosphate uridyl transferase (*GALT*) gene. *Hum Mutat* **13**, 417–430.

Tyndale, R. F., Droll, K. P., and Sellers, E. M. (1997). Genetically deficient CYP2D6 metabolism provides protection against oral opiate dependence. *Pharmacogenetics* **7**, 375–379.

Tyndale, R. F., and Sellers, E. M. (2001). Variable CYP2A6-mediated nicotine metabolism alters smoking behavior and risk. *Drug Metab Dispos* **29**, 548–552.

Ubeaud, G., Schiller, C. D., Hurbin, F., Jaeck, D., and Coassolo, P. (1999). Estimation of flavin-containing monooxygenase activity in intact hepatocyte monolayers of rat, hamster, rabbit, dog and human by using N-oxidation of benzydamine. *Eur J Pharm Sci* **8**, 255–260.

Uematsu, F., Kikuchi, H., Motomiya, M., Abe, T., Sagami, I., Ohmachi, T., Wakui, A., Kanamaru, R., and Watanabe, M. (1991). Association between restriction fragment length polymorphism of the human cytochrome P450IIE1 gene and susceptibility to lung cancer. *Jpn J Cancer Res* **82**, 254–256.

Ueno, S., Nakamura, M., Mikami, M., Kondoh, K., Ishiguro, H., Arinami, T., Komiyama, T., Mitsushio, H., Sano, A., and Tanabe, H. (1999). Identification of a novel polymorphism of the human dopamine transporter (DAT1) gene and the significant association with alcoholism. *Mol Psychiatry* **4**, 552–557.

Uitterlinden, A. G., Fang, Y., Bergink, A. P., van Meurs, J. B., van Leeuwen, H. P., and Pols, H. A. (2002). The role of vitamin D receptor gene polymorphisms in bone biology. *Mol Cell Endocrinol* **197**, 15–21.

Ulrich, C. M., Kampman, E., Bigler, J., Schwartz, S. M., Chen, C., Bostick, R., Fosdick, L., Beresford, S. A., Yasui, Y., and Potter, J. D. (1999). Colorectal adenomas and the C677T MTHFR polymorphism: Evidence for gene-environment interaction? *Cancer Epidemiol Biomark Prev* **8**, 659–668.

Umbach, D. M., and Weinberg, C. R. (2000). The use of case-parent triads to study joint effects of genotype and exposure. *Am J Hum Genet* **66**, 251–261.

University of Washington Center for Genomics and Public Health (2004). *Asthma Genomics: Implications for Public Health.*

Uversky, V. N., Li, J., and Fink, A. L. (2001a). Metal-triggered structural transformations, aggregation, and fibrillation of human alpha-synuclein. A possible molecular NK between Parkinson's disease and heavy metal exposure. *J Biol Chem* **276**, 44284–44296.

Uversky, V. N., Li, J., and Fink, A. L. (2001b). Pesticides directly accelerate the rate of alpha-synuclein fibril formation: A possible factor in Parkinson's disease. *FEBS Lett* **500**, 105–108.

UW (1999). *Genetic Research* (Seattle: Univ. Washington Human Subjects Manual VII.N, http://depts.washington.edu/hsd/INFO/MANUAL/99-VII.htm#VIIn).

Vadigepalli, R., Chakravarthula, P., Zak, D. E., Schwaber, J. S., and Gonye, G. E. (2003). PAINT: A promoter analysis and interaction network generation tool for gene regulatory network identification. *Omics* **7**, 235–252.

Vale T., Tuomilehto, J., Bergman, R. N., Ghosh, S., Hauser, E. R., Eriksson, J., Nylund, S. J., Kohtamaki, K., Toivanen, L., Vidgren, G., et al. (1998). Mapping genes for DIDDM: Design of the Finland-United States Investigation of NIDDM Genetics (FUSION) study. *Diabetes Care* **21**, 949–958.

Valerie, K., and Povirk, L. F. (2003). Regulation and mechanisms of mammalian double-strand break repair. *Oncogene* **22**, 5792–5812.

Van Damme, K., Casteleyn, L., Heseltine, E., Huici, A., Sorsa, M., van Larebeke, N., and Vineis, P. (1995). Individual susceptibility and prevention of occupational diseases: scientific and ethical issues. *J Occup Environ Med* **37**, 91–99.

Van den Berg, M., Birnbaum, L., Bosveld, A. T., Brunstrom, B., Cook, P., Feeley, M., Giesy, J. P., Hanberg, A., Hasegawa, R., Kennedy, S. W., et al. (1998). Toxic equivalency factors (TEFs) for PCBs, PCDDs, PCDFs for humans and wildlife. *Environ Health Perspect* **106**, 775–792.

van der Put, N. M., Eskes, T. K., and Blom, H. J. (1997). Is the common 677C→T mutation in the methylenetetrahydrofolate reductase gene a risk factor for neural tube defects? A meta-analysis. *Quart J Med* **90**, 111–115.

Vandenbergh, D. J., Bennett, C. J., Grant, M. D., Strasser, A. A., O'Connor, R., Stauffer, R. L., Vogler, G. P., and Kozlowski, L. T. (2002). Smoking status and the human dopamine transporter variable number of tandem repeats (VNTR) polymorphism: Failure to replicate and finding that never-smokers may be different. *Nicotine Tob Res* **4**, 333–340.

Vandenbergh, D. J., Thompson, M. D., Cook, E. H., Bendahhou, E., Nguyen, T., Krasowski, M. D., Zarrabian, D., Comings, D., Sellers, E. M., Tyndale, R. F., et al.

(2000). Human dopamine transporter gene: Coding region conservation among normal, Tourette's disorder, alcohol dependence and attention-deficit hyperactivity disorder populations. *Mol Psychiatry* **5**, 283–292.

Vandenbroucke, J. P., Koster, T., Briet, E., Reitsma, P. H., Bertina, R. M., and Rosendaal, F. R. (1994). Increased risk of venous thrombosis in oral-contraceptive users who are carriers of factor V Leiden mutation. *Lancet* **344**, 1453–1457.

Varadaraj, K., and Skinner, D. H. (1994). Denaturation or cosolvents improve the specificity of PCR amplification of a G + C rich DNA using genetically engineered DNA polymerases. *Gene* **140**, 1–5.

Vaughan, T. L. (2002). Esophagus. In *Cancer Precursors*, Franco, E. L., and Rohan, T. E., eds. (New York: Springer-Verlag), pp. 96–116.

Venter, J. C., Adams, M. D., Myers, E. W., Li, P. W., Mural, R. J., Sutton, G. G., Smith, H. O., Yandell, M., Evans, C. A., Holt, R. A., et al. (2001). The sequence of the human genome. *Science* **291**, 1304–1351.

Vesell, E. S. (1973). Advances in pharmacogenetics. *Prog Med Genet* **9**, 291–367.

Vesell, E. S. (1981). The influence of host factors on drug response. V. Endocrinological, gastrointestinal, and pulmonary diseases. *Ration Drug Ther* **15**, 1–6.

Vesell, E. S., and Page, J. G. (1968). Genetic control of drug levels in man: Phenylbutazone. *Science* **159**, 1479–1480.

Veverka, K. A., Johnson, K. L., Mays, D. C., Lipsky, J. J., and Naylor, S. (1997). Inhibition of aldehyde dehydrogenase by disulfiram and its metabolite methyl diethylthiocarbamoyl-sulfoxide. *Biochem Pharmacol* **53**, 511–518.

Viken, R. J., Rose, R. J., Morzorati, S. L., Christian, J. C., and Li, T. K. (2003). Subjective intoxication in response to alcohol challenge: Heritability and covariation with personality, breath alcohol level, and drinking history. *Alcohol Clin Exp Res* **27**, 795–803.

Vineis, P., Veglia, F., et al. (2003). CYP1A1 T3801 C polymorphism and lung cancer: a pooled analysis of 2451 cases and 3358 controls. *Int J Cancer* **104**, 650–657.

Vogel, F. (1959). Moderne problem der humangenetik. *Ergeb Inn Med U Kinderheilk* **12**, 52–125.

Volkert, M. R. (1988). Adaptive response of Escherichia coli to alkylation damage. *Environ Mol Mutagen* **11**, 241–255.

Volles, M. J., and Lansbury, P. T., Jr. (2003). Zeroing in on the pathogenic form of alpha-synuclein and its mechanism of neurotoxicity in Parkinson's disease. *Biochemistry* **42**, 7871–7878.

Wacholder, S., Chatterjee, N., and Hartge, P. (2002). Joint effect of genes and environment distorted by selection biases: Implications for hospital-based case-control studies. *Cancer Epidemiol Biomark Prev* **11**, 885–889.

Wacholder, S., Rothman, N., and Caporaso, N. (2000). Population stratification in epidemiologic studies of common genetic variants and cancer: Quantification of bias. *J Natl Cancer Inst* **92**, 1151–1158.

Wall, J. D., and Pritchard, J. K. (2003). Haplotype blocks and linkage disequilibrium in the human genome. *Nat Rev Genet* **4**, 587–597.

Wall, T. L., Carr, L. G., and Ehlers, C. L. (2003). Protective association of genetic variation in alcohol dehydrogenase with alcohol dependence in Native American Mission Indians. *Am J Psychiatry* **160**, 41–46.

Wallace, R. H., Wang, D. W., Singh, R., Scheffer, I. E., George, A. L., Jr., Phillips, H. A., Saar, K., Reis, A., Johnson, E. W., Sutherland, G. R., et al. (1998). Febrile seizures and generalized epilepsy associated with a mutation in the Na+-channel beta1 subunit gene SCN1B. *Nat Genet* **19**, 366–370.

Wan, J., Shi, J., Hui, L., Wu, D., Jin, X., Zhao, N., Huang, W., Xia, Z., and Hu, G. (2002). Association of genetic polymorphisms in CYP2E1, MPO, NQO1, GSTM1, and GSTT1 genes with benzene poisoning. *Environ Health Perspect* **110**, 1213–1218.

Wang, H., Tan, W., Hao, B., Miao, X., Zhou, G., He, F., and Lin, D. (2003a). Substantial reduction in risk of lung adenocarcinoma associated with genetic polymorphism in CYP2A13, the most active cytochrome P450 for the metabolic activation of tobacco-specific carcinogen NNK. *Cancer Res* **63**, 8057–8061.

Wang, J., Deng, Y., et al. (2003b). Association of GSTM1, CYP1A1 and CYP2E1 genetic polymorphisms with susceptibility to lung adenocarcinoma: a case-control study in Chinese population. *Cancer Sci* **94**, 448–452.

Wani, A. A., Wani, G., and D'Ambrosio, S. M., eds. (1990). *Repair of O^4-Alkylthymine Damage in Human Cells* (New York: Plenum Press).

Wardle, J., Steptoe, A., Oliver, G., and Lipsey, Z. (2000). Stress, dietary restraint and food intake. *J Psychosom Res* **48**, 195–202.

Waring, J. F., Ciurlionis, R., Jolly, R. A., Heindel, M., and Ulrich, R. G. (2001a). Microarray analysis of hepatotoxins in vitro reveals a correlation between gene expression profiles and mechanisms of toxicity. *Toxicol Lett* **120**, 359–368.

Waring, J. F., Jolly, R. A., Ciurlionis, R., Lum, P. Y., Praestgaard, J. T., Morfitt, D. C., Buratto, B., Roberts, C., Schadt, E., and Ulrich, R. G. (2001b). Clustering of hepatotoxins based on mechanism of toxicity using gene expression profiles. *Toxicol Appl Pharmacol* **175**, 28–42.

Waring, J. F., Gum, R., Morfitt, D., Jolly, R. A., Ciurlionis, R., Heindel, M., Gallenberg, L., Buratto, B., and Ulrich, R. G. (2002). Identifying toxic mechanisms using DNA microarrays: Evidence that an experimental inhibitor of cell adhesion molecule expression signals through the aryl hydrocarbon nuclear receptor. *Toxicology* **181–182**, 537–550.

Warner, A., and Norman, A. B. (2000). Mechanisms of cocaine hydrolysis and metabolism in vitro and in vivo: A clarification. *Ther Drug Monit* **22**, 266–270.

Watanabe, J., Hayashi, S., and Kawajiri, K. (1994). Different regulation and expression of the human CYP2E1 gene due to the RsaI polymorphism in the 5′-flanking region. *J Biochem (Tokyo)* **116**, 321–326.

Waterland, R. A., and Jirtle, R. L. (2003). Transposable elements: Targets for early nutritional effects on epigenetic gene regulation. *Mol Cell Biol* **23**, 5293–5300.

Waters, M. D., Olden, K., and Tennant, R. W. (2003). Toxicogenomic approach for assessing toxicant-related disease. *Mutat Res* **544**, 415–424.

Watlington, C. O., Kramer, L. B., Schuetz, E. G., Zilai, J., Grogan, W. M., Guzelian, P., Gizek, F., and Schoolwerth, A. C. (1992). Corticosterone 6 beta-hydroxylation correlates with blood pressure in spontaneously hypertensive rats. *Am J Physiol* **262**, F927–931.

Wei, Q., Eicher, S. A., Guan, Y., Cheng, L., Xu, J., Young, L. N., Saunders, K. C., Jiang, H., Hong, W. K., Spitz, M. R., and Strom, S. S. (1998). Reduced expression of hMLH1 and hGTBP/hMSH6: A risk factor for head and neck cancer. *Cancer Epidemiol Biomark Prev* **7**, 309–314.

Wei, Q. and Spitz, M. R. (1997). The role of DNA repair capacity in susceptibility to lung cancer: a review. *Cancer Metastasis Rev* **16**, 295–307.

Weiland, S. K., Mundt, K. A., Keil, U., Kraemer, B., Birk, T., Person, M., Bucher, A. M., Straif, K., Schumann, J., and Chambless, L. (1996). Cancer mortality among workers in the German rubber industry: 1981–91. *Occup Environ Med* **53**, 289–298.

Weir, B. S. (1996). *Genetic Data Analysis II: Methods for Discrete Population Genetic Data* (Sunderland, MA: Sinauer Assoc.).

Weis, B. K., Balshaw, D., Barr, J., Brown, D., Ellisman, M., Lioy, P., Omenn, G. S., Potter, J., Smith, M., Sohn, L., et al. (2005). Personalized exposure assessment: Promising approaches for human environmental health research. *Environ Health Perspect*, (in press; online March 3, 2005 at http://ehp.niehs.nih.gov/docs/2005/7651/abstract.html).

Weiss, J. F., and Landauer, M. R. (2003). Protection against ionizing radiation by antioxidant nutrients and phytochemicals. *Toxicology* **189**, 1–20.

Weiss, K. M., and Clark, A. G. (2002). Linkage disequilibrium and the mapping of complex human traits. *Trends Genet* **18**, 19–24.

Welton, W. E., Kantner, T. A., and Katz, S. M. (1997). Developing tomorrow's integrated community health systems: A leadership challenge for public health and primary care. *Milbank Quart* **75**, 261–288.

Wen, L. (2001). Two-step cycle sequencing improves base ambiguities and signal dropouts in DNA sequencing reactions using energy-transfer-based fluorescent dye terminators. *Mol Biotechnol* **17**, 135–142.

Wendo, C. (2003). Uganda tries to learn from Zambia's GM food controversy. *Lancet* **361**, 500.

Wendorf, M. (1989). Diabetes, the ice free corridor, and the Paleoindian settlement of North America. *Am J Phys Anthropol* **79**, 503–520.

Westlind, A., Lofberg, L., Tindberg, N., Andersson, T. B., and Ingelman-Sundberg, M. (1999). Interindividual differences in hepatic expression of CYP3A4: relationship to genetic polymorphism in the 5′-upstream regulatory region. *Biochem Biophys Res Commun* **259**, 201–205.

Westlind, A., Malmebo, S., Johansson, I., Otter, C., Andersson, T. B., Ingelman-Sundberg, M., and Oscarson, M. (2001). Cloning and tissue distribution of a novel human cytochrome p450 of the CYP3A subfamily, CYP3A43. *Biochem Biophys Res Commun* **281**, 1349–1355.

Weston, A. D., and Hood, L. (2004). Systems biology, proteomics, and the future of health care: Toward predictive, preventative, and personalized medicine. *J Proteome Res* **3**, 179–196.

Whitfield, J. B. (1997). Meta-analysis of the effects of alcohol dehydrogenase genotype on alcohol dependence and alcoholic liver disease. *Alcohol Alcohol* **32**, 613–619.

Whittaker, M. (1986). *Cholinesterase*, Monographs in Human Genetics, Vol. 11 (Basel: Karger).

WHO (1994). Environmental Health Criteria, 170. In *Environmental Health Criteria for Guidance Values for Human Exposure Limits* (Geneva: WHO).

Wikman, H., Piirila, P., Rosenberg, C., Luukkonen, R., Kaaria, K., Nordman, H., Norppa, H., Vainio, H., and Hirvonen, A. (2002). N-Acetyltransferase genotypes as modifiers of diisocyanate exposure-associated asthma risk. *Pharmacogenetics* **12**, 227–233.

Wilke, K., Duman, B., and Horst, J. (2000). Diagnosis of haploidy and triploidy based on measurement of gene copy number by real-time PCR. *Hum Mutat* **16**, 431–436.

Williams, R. J. (1956). *Biochemical Individuality: The Basis for the Genetotrophic Concept* (New York: J Wiley).

Wilson, P. W., Schaefer, E. J., Larson, M. G., and Ordovas, J. M. (1997). Apolipoprotein E alleles and risk of coronary disease. A meta-analysis. *Arterioscler Thromb Vasc Biol* **16**, 1250–1255.

Wingren, G., Persson, B., Thoren, K., and Axelson, O. (1991). Mortality pattern among pulp and paper mill workers in Sweden: A case-referent study. *Am J Ind Med* **20**, 769–774.

Witte, J. S., Gauderman, W. J., and Thomas, D. C. (1999). Asymptotic bias and efficiency in case-control studies of candidate genes and gene-environment interactions: Basic family designs. *Am J Epidemiol* **149**, 693–705.

Wojnowski, L., Hustert, E., et al. (2002). Re: modification of clinical presentation of prostate tumors by a novel genetic variant in CYP3A4. *J Natl Cancer Inst* **94**, 630–631; author reply 631–632.

Wojnowski, L., Turner, P. C., Pedersen, B., Hustert, E., Brockmoller, J., Mendy, M., Whittle, H. C., Kirk, G., and Wild, C. P. (2004). Increased levels of aflatoxin-albumin adducts are associated with CYP3A5 polymorphisms in The Gambia, West Africa. *Pharmacogenetics* **14**, 691–700.

Women's Health Initiative (WHI) Study Group (1998). Design of the Women's Health Initiative clinical trial and observational study. The Women's Health Initiative Study Group. *Control Clin Trials* **19**, 61–109.

Wong, J. M., Okey, A. B., and Harper, P. A. (2001). Human aryl hydrocarbon receptor polymorphisms that result in loss of CYP1A1 induction. *Biochem Biophys Res Commun* **288**, 990–996.

Wong, O., Brocker, W., Davis, H. V., and Nagle, G. S. (1984). Mortality of workers potentially exposed to organic and inorganic brominated chemicals, DBCP, TRIS, PBB, and DDT. *Br J Ind Med* **41**, 15–24.

Wong, O., Ragland, D. R., and Marcero, D. H. (1996). An epidemiologic study of employees at seven pulp and paper mills. *Int Arch Occup Environ Health* **68**, 498–507.

Wood, M. L., Esteve, A., Morningstar, M. L., Kuziemko, G. M., and Essigmann, J. M. (1992). Genetic effects of oxidative DNA damage: Comparative mutagenesis of 7,8,-dihydro-8-oxoguanine and 7,8-dihydro-8-oxoadenine in *Escherichia coli. Nucleic Acids Res* **20**, 6023–6032.

Wood, R. D., Mitchell, M., Sgouros, J., and Lindahl, T. (2001). Human DNA repair genes. *Science* **291**, 1284–1289.

Wu, M. T., Huang, S. L., Ho, C. K., Yeh, Y. F., and Christiani, D. C. (1998). Cytochrome P450 1A1 MspI polymorphism and urinary 1-hydroxypyrene concentrations in coke-oven workers. *Cancer Epidemiol Biomarkers Prev* **7**, 823–829.

Wu, R. M., Cheng, C. W., Chen, K. H., Lu, S. L., Shan, D. E., Ho, Y. F., and Chern, H. D. (2001). The COMT L allele modifies the association between MAOB polymorphism and PD in Taiwanese. *Neurology* **56**, 375–382.

Wu, X., Hudmon, K. S., Detry, M. A., Chamberlain, R. M., and Spitz, M. R. (2000). D2 dopamine receptor gene polymorphisms among African-Americans and Mexican-Americans: A lung cancer case-control study. *Cancer Epidemiol Biomark Prev* **9**, 1021–1026.

Wu, X., Shi, H., Jiang, H., Kemp, B., Hong, W. K., Delclos, G. L., and Spitz, M. R. (1997). Associations between cytochrome P4502E1 genotype, mutagen sensitivity, cigarette smoking and susceptibility to lung cancer. *Carcinogenesis* **18**, 967–973.

Wu, Z., and Irizarry, R. A. (2004). Preprocessing of oligonucleotide array data. *Nat Biotechnol* **22**, 656–658; author reply 658.

Xu, C., Goodz, S., Sellers, E. M. and Tyndale, R. F. (2002). CYP2A6 genetic variation and potential consequences. *Adv Drug Deliv Rev* **54**, 1245–1256.

Xu, J., Zheng, S. L., Turner, A., Isaacs, S. D., Wiley, K. E., Hawkins, G. A., Chang, B. L., Bleecker, E. R., Walsh, P. C., Meyers, D. A., and Isaacs, W. B. (2002). Associations between hOGG1 sequence variants and prostate cancer susceptibility. *Cancer Res* **62**, 2253–2257.

Yamada, Y., Izawa, H., Ichihara, S., Takatsu, F., Ishihara, H., Hirayama, H., Sone, T., Tanaka, M., and Yokota, M. (2002). Prediction of the risk of myocardial infarction from polymorphisms in candidate genes. *New Engl J Med* **347**, 1916–1923.

Yamaori, S., Yamazaki, H., Iwano, S., Kiyotani, K., Matsumura, K., Honda, G., Nakagawa, K., Ishizaki, T., and Kamataki, T. (2004). CYP3A5 Contributes significantly to CYP3A-mediated drug oxidations in liver microsomes from Japanese subjects. *Drug Metab Pharmacokinet* **19**, 120–129.

Yang, Y. H., and Speed, T. (2002). Design issues for cDNA microarray experiments. *Nat Rev Genet* **3**, 579–588.

Yeung, C. K., Lang, D. H., Thummel, K. E., and Rettie, A. E. (2000). Immunoquantitation of FMO1 in human liver, kidney, and intestine. *Drug Metab Dispos* **28**, 1107–1111.

Yokoyama, A., Kato, H., Yokoyama, T., Tsujinaka, T., Muto, M., Omori, T., Haneda, T., Kumagai, Y., Igaki, H., Yokoyama, M., et al. (2002). Genetic polymorphisms of alcohol and aldehyde dehydrogenases and glutathione S-transferase M1 and drinking, smoking, and diet in Japanese men with esophageal squamous cell carcinoma. *Carcinogenesis* **23**, 1851–1859.

Yoshida, K., Hamajima, N., Kozaki, K., Saito, H., Maeno, K., Sugiura, T., Ookuma, K., and Takahashi, T. (2001). Association between the dopamine D2 receptor A2/A2 genotype and smoking behavior in the Japanese. *Cancer Epidemiol Biomark Prev* **10**, 403–405.

Young, T. K., Szathmary, E. J., Evers, S., and Wheatley, B. (1990). Geographical distribution of diabetes among the native population of Canada: A national survey. *Soc Sci Med* **31**, 129–139.

Yu, F. H., and Catterall, W. A. (2003). Overview of the voltage-gated sodium channel family. *Genome Biol* **4**, 207.

Yu, Y., Okayasu, R., Weil, M. M., Silver, A., McCarthy, M., Zabriskie, R., Long, S., Cox, R., and Ullrich, R. L. (2001). Elevated breast cancer risk in irradiated BALB/c mice associates with unique functional polymorphism of the Prkdc (DNA-dependent protein kinase catalytic subunit) gene. *Cancer Res* **61**, 1820–1824.

Yuzaki, M. (2003). New insights into the structure and function of glutamate receptors: The orphan receptor delta2 reveals its family's secrets. *Keio J Med* **52**, 92–99.

Zang, E. A. and Wynder, E. L. (1996). Differences in lung cancer risk between men and women: examination of the evidence. *J Natl Cancer Inst* **88**, 183–192.

Zeeberg, B. R., Feng, W., Wang, G., Wang, M. D., Fojo, A. T., Sunshine, M., Narasimhan, S., Kane, D. W., Reinhold, W. C., Lababidi, S., et al. (2003). GoMiner: A resource for biological interpretation of genomic and proteomic data. *Genome Biol* **4**, R28.

Zhang, A. Q., Mitchell, S., and Smith, R. (1995). Fish odour syndrome: verification of carrier detection test. *J Inherit Metab Dis* **18**, 669–674.

Zhang, A. Q., Mitchell, S. C., and Smith, R. L. (1996). Exacerbation of symptoms of fish-odour syndrome during menstruation. *Lancet* **348**, 1740–1741.

Zhang, J., Kuehl, P., et al. (2001). The human pregnane X receptor: genomic structure and identification and functional characterization of natural allelic variants. *Pharmacogenetics* **11**, 555–572.

Zhang, M. and Robertus, J. D. (2002). Molecular cloning and characterization of a full-length flavin-dependent monooxygenase from yeast. *Arch Biochem Biophys* **403**, 277–283.

Zhang, X., Chen, L. and Hardwick, J. P. (2000). Promoter activity and regulation of the CYP4F2 leukotriene B(4) omega-hydroxylase gene by peroxisomal proliferators and retinoic acid in HepG2 cells. *Arch Biochem Biophys* **378**, 364–376.

Zhang, Y., Kolm, R. H., Mannervik, B., and Talalay, P. (1995). Reversible conjugation of isothiocyanates with glutathione catalyzed by human glutathione transferases. *Biochem Biophys Res Commun* **206**, 748–755.

Zhao, J. H., Curtis, D., and Sham, P. C. (2000). Model-free analysis and permutation tests for allelic associations. *Hum Hered* **50**, 133–139.

Zhao, X., Li, C., Paez, J. G., Chin, K., Janne, P. A., Chen, T. H., Girard, L., Minna, J., Christiani, D., Leo, C., et al. (2004). An integrated view of copy number and allelic alterations in the cancer genome using single nucleotide polymorphism arrays. *Cancer Res* **64**, 3060–3071.

Zhu, Y., Spitz, M., Amos, C., Lin, J., Schabath, M., and Wu, X. (2004). An evolutionary perspective on single-nucleotide polymorphism screening in molecular cancer epidemiology. *Cancer Res* **64**, 2251–2257.

Ziegler, D. M. (1990). Flavin-containing monooxygenases: enzymes adapted for multisubstrate specificity. *Trends Pharmacol Sci* **11**, 321–324.

Ziegler, D. M. (1993). Recent studies on the structure and function of multisubstrate flavin-containing monooxygenases. *Annu Rev Pharmacol Toxicol* **33**, 179–199.

Zimmerman, P. A., Patel, S. S., Maier, A. G., Bockarie, M. J., and Kazura, J. W. (2003). Erythrocyte polymorphisms and malaria parasite invasion in Papua New Guinea. *Trends Parasitol* **19**, 250–252.

Zinman, R., Schwartz, S., Gordon, K., Fitzpatrick, E., and Camfield, C. (1995). Predictors of sunscreen use in childhood. *Arch Pediatr Adol Med* **149**, 804–807.

Zmuda, J. M., Cauley, J. A., and Ferrell, R. E. (2000). Molecular epidemiology of vitamin D receptor gene variants. *Epidemiol Rev* **22**, 203–217.

Zober, A., Ott, M. G., et al. (1994). Morbidity follow up study study of BASF employees exposed to 2,3,7,8-tetrachlorodibenzo-p-dioxin (TCDD) after a 1953 chemical reactor incident. *Occup Environ Med* **51**, 479–486.

Zochbauer-Muller, S., Lam, S., Toyooka, S., Virmani, A. K., Toyooka, K. O., Seidl, S., Minna, J. D., and Gazdar, A. F. (2003). Aberrant methylation of multiple genes in the upper aerodigestive tract epithelium of heavy smokers. *Int J Cancer* **107**, 612–616.

Index

Page numbers followed by f denote figures; page numbers followed by t denote tables.

Gene-Environment Interactions: Fundamentals of Ecogenetics, edited by Lucio G. Costa and David L. Eaton